Soil and Groundwater Remediation

Soil and Groundwater Remediation

Fundamentals, Practices,
and Sustainability

Chunlong (Carl) Zhang, Ph.D., PE

College of Science and Engineering,
University of Houston-Clear Lake,
Houston, TX, USA

This edition first published 2020

Registered Office
John Wiley & Sons, Inc., 111 River Street, Hoboken, NJ 07030, USA

Editorial Office
111 River Street, Hoboken, NJ 07030, USA

For details of our global editorial offices, customer services, and more information about Wiley products visit us at www.wiley.com.

Wiley also publishes its books in a variety of electronic formats and by print-on-demand. Some content that appears in standard print versions of this book may not be available in other formats.

Library of Congress Cataloging-in-Publication Data

Names: Zhang, Chunlong, 1964– author.
Title: Soil and groundwater remediation : fundamentals, practices and sustainability / Chunlong (Carl) Zhang, University of Houston-Clear Lake, College of Science and Engineering.
Description: First edition. | Hoboken, NJ : John Wiley & Sons, Inc., [2020] | Includes bibliographical references and index. |
Identifiers: LCCN 2019014818 (print) | LCCN 2019017605 (ebook) | ISBN 9781119393160 (Adobe PDF) | ISBN 9781119393177 (ePub) | ISBN 9781119393153 (hardcover)
Subjects: LCSH: Soil remediation. | Groundwater–Purification.
Classification: LCC TD878 (ebook) | LCC TD878 .Z42 2019 (print) | DDC 628.5/5–dc23
LC record available at https://lccn.loc.gov/2019014818

Cover Design: Wiley
Cover Image: © Archy Grey/Shutterstock

Set in 10/12pt Warnock by SPi Global, Pondicherry, India

10 9 8 7 6 5 4 3 2 1

To my mother Yunfeng Pan (1926–2018)
To my wife Sue and sons Richard and Arnold

Contents

About the Author

Dr. Zhang is a professor of environmental science at the College of Science and Engineering, University of Houston-Clear Lake. He has a combined three decades of experience in academia, industries, and consulting in the environmental field. He is the author and co-author of more than 150 papers, proceedings, and technical reports in diverse areas including contaminant fate and transport, environmental remediation, sampling and analysis, and environmental assessment. In his current position at the University of Houston, he lectures extensively in the area of environmental chemistry, environmental sampling and analysis, soil and groundwater remediation, and environmental engineering at both the undergraduate and graduate levels. He is the author of the popular textbook "Fundamentals of Environmental Sampling and Analysis" published by Wiley in 2007. His expertise includes a variety of practical experience in both field and lab on contaminant behavior in soil and groundwater, emerging chemical analysis, and remediation feasibility studies. Dr. Zhang is a registered professional engineer in environmental engineering. He also serves as an adjunct professor in the College of Environmental and Resource Sciences at Zhejiang University.

Preface

This book is written primarily to equip our undergraduate and graduate students with essential knowledge in the field of soil and groundwater remediation (SGWR). Current environmental science and engineering curriculums in most universities have well-established course work in the traditional fields of air pollution control, water/wastewater and solid/hazardous waste treatment and disposal. That is obviously not the case for this equally important field of soil and groundwater remediation. Clean air, surface water, soil, and groundwater are all essential to the quality of our life.

SGWR has emerged and it has become an increasingly important and popular topic since the 1980s. SGWR or its equivalent courses are becoming required or elective courses for students at both the undergraduate and graduate levels in many universities, especially in the North America and European countries where sufficient experience has been accumulated following the decades-long research and development of remediation techniques at various contaminated sites. These sites have received public scrutiny since the 1950s and the remediation technologies started to be developed and implemented since then. The issue of contaminated soil and groundwater has also become increasingly important in many developing countries in the recent decade because of contaminated sites associated with the rapid economic development in Asia and Latin America.

Unfortunately, there is no book that can be considered as a suitable "textbook" for this increasingly important subject. The development of this textbook is to fill this gap. As a textbook, it introduces the underlying principles and essential components of soil and groundwater remediation in a comprehensible but not overwhelming manner. Ample numbers of worked and practice problems are included toward the understanding of fundamental chemistry, microbiology and hydrogeology, practical applications, easy-to-understand design equations, and calculations to be useful in preparing our students for the workplace in the environmental remediation industries. In addition to the principles and practices, sustainable remediation is also the theme of this textbook. It is expected that this book will serve as a single source for an instructor to introduce this interdisciplinary subject, including chemical fate and transport, hydrogeology, regulations, cost analysis, risk assessment, site characterization, modeling, and various conventional and innovative remediation technologies.

Chapter 1 serves as an introductory chapter in describing the importance of groundwater resource, groundwater quality, contaminant sources and types from both the US and the global perspectives, and the scope of soil and groundwater remediation. Chapter 2 introduces frequent soil and groundwater contaminants and their fate and transport processes in the subsurface environment, including abiotic and biotic chemical processes, interphase and intraphase chemical movement. Chapter 3 provides some essential background information for readers to understand basic soil and aquifer properties, groundwater wells, and equations in describing groundwater flow under various conditions. Chapter 4 introduces three separate but all essential components of SGWR, i.e. the regulatory framework as a driving force to site cleanup, the cost analysis for the comparison of remedial scenarios, and the risk assessment to help define the cleanup goal. Chapter 5 introduces the methodology to characterize contaminated sites, field techniques to survey soil/geological/hydrogeological parameters, soil and groundwater sampling, and analysis of subsurface contaminants. Chapter 6 provides a general framework for the development of site-specific remediation technologies, a screening matrix for the selection of remediation technologies, and an overview of various SGWR technologies.

The next six chapters will detail soil and groundwater remediation technologies that are in the common practice in remediation industries. Each chapter will accompany with case studies and example design calculations whenever appropriate. Chapter 7 introduces pump-and-treat systems with a focus on design equations on the belowground capture zone and aboveground treatment using air stripping and activated carbon adsorption. The conditions of why conventional pump-and-treat systems work/do not work, and how to improve pump-and-treat are discussed. Chapter 8 describes soil vapor behavior and gas flow in the subsurface, and how to use design equations to determine well number, flow rate, and well locations. Chapter 9 is devoted to bioremediation and environmental biotechnology in general. The fundamentals of bacterial growth, stoichiometry, kinetics, pathways, and optimal conditions are delineated first, followed by various bioremediation/biotechnology applications such as *in situ* and *ex situ* biological treatment, landfills, and phytoremediation. Design calculations such as nutrient/oxygen delivery for bioremediation and landfill design basics will be discussed. Chapter 10 introduces thermally enhanced remediation (i.e. use of hot air, steam, hot water, and electro-heating) and thermal destruction (vitrification and incineration). Practical design calculations that focus primarily on incinerators are included. Chapter 11 discusses the principles, applications, design, and cost-effectiveness of soil washing and *in situ* soil flushing using water, surfactants, and cosolvents. Chapter 12 describes the chemical reaction mechanisms, hydraulics, configuration, design, and construction of permeable reactive barriers.

The last chapter (Chapter 13) is dedicated to the basics about modeling of groundwater flow and contaminant transport. Our focus will be the Darcy's law and mass balance approach in developing the governing equations for groundwater flow in saturated and unsaturated zones (e.g. the Laplace equation and the Richards equation). This is followed by the same approach in the development of governing equations for contaminant transport in saturated and unsaturated zones, with the consideration of advection, dispersion, adsorption, and reaction, as well as more complicated multiphase flow and transport. Finally, the analytical solutions to several simplified flow and transport processes are provided, followed by the mathematical framework in reaching the numerical solutions.

This book is unique in several aspects to serve as a textbook for senior undergraduate and graduate students, as well as a valuable reference book for general audiences. Several approaches are used to ensure the book for a wide usage of readers. (i) Each chapter will have a set of learning objectives, and the discussion of key theories/principles followed by example problems will be provided to help readers' understanding of subjects for classroom use. (ii) When remediation techniques are introduced, case studies will be provided so readers can relate the principles to the applications of relevant remediation techniques. (iii) End-of-Chapter Questions and Problems are included to further help understand the materials. (iv) Supplemental materials are provided in the format of Box in each chapter, such as Superfund sites versus Brown fields, emerging contaminants, hydraulic head, and Bernoulli's equation, and terms relevant to environmental legislature. In Chapters 7 through 12, a particular emphasis is placed on the best management practices and green/sustainable remediation in Box format. (v) A bibliography is given at the end of each chapter for those who need specific details from guidelines (e.g. EPA) or recent development from peer-reviewed journal articles. This list of references is intended to provide an up-to-date or in-depth discussion of remediation topics, such that researchers and experienced remediation practitioners will also find it to be useful.

This book should be an appropriate textbook for a Soil and Groundwater Remediation course perhaps with various focuses and variations depending on the specific discipline, such as Soil and Groundwater Restoration, Groundwater Engineering, Environmental Remediation/Restoration, Remediation Technologies, Remediation/Environmental Geotechnics, Reclamation of Contaminated Land, Site Remediation Technologies, and Site Assessment and Remediation. This book is also appropriate for use as reference or supplemental material for courses such as Environmental Geology, Applied Hydrogeology, Subsurface Fate and Transport, Environmental Engineering, and Groundwater Contamination. Besides a textbook, this book should also be appropriate as a single source reference for environmental professionals to quickly grasp the fundamental principles, practices, and sustainable concepts of soil and groundwater remediation. As a single source, the readers can comprehend the basic science and engineering principles related to site remediation without going into numerous detailed standard methods, handbooks, and technical reports currently available from various sources.

The author will be happy to receive comments and suggestions about this book at his e-mail address: zhang@uhcl.edu.

January 2019

Chunlong (Carl) Zhang
Houston, TX, USA

Acknowledgments

Materials of this book have been used in several courses at the University of Houston. I first would like to thank my students for their comments, suggestions, and encouragement. These feedbacks are typically not technically detailed, but help me immeasurably to improve its readability. Certainly I would like to thank technical reviewers for the review of this book from the beginning to the final draft, including Dr. K.J. Reimer of the Royal Military College of Canada, Dr. Ming Zhang, Geological Survey of Japan, Dr. W. Andrew Jackson, Texas Tech University, Dr. Jianying Zhang, Zhejiang University, and a dozen of anonymous reviewers at the U.S. EPA, Argonne National Laboratory, environmental consulting firms, and academia.

My special thanks to Wiley's Executive Editors Mr. Bob Esposito and Mr. Michael Leventhal for their vision and guidance of this project. Ms. Beryl Mesiadas, Project Editor has been very helpful in insuring me the right format of this writing even from the beginning of this project. Thanks to the Production Editor Ms. Gayathree Sekar who exhibited dedicated assistance throughout the production process of the book. It has been a pleasant experience in working with this editorial team of high professional standards and experience.

This book would not be accomplished without the support and love of my wife Sue and the joys and emotions I have shared with my two sons, Richard and Arnold. Even during many hours of my absence in the past years for this project, I felt the drive and inspiration. This book is written as a return for their love and encouragement. With that, I felt at some points the obligation of fulfilling and delivering what is beyond my capability.

Whom This Book Is Written For

College students and graduate students are the primary targeted readers of this book. Suitable readers who are routinely involved in various aspects of soil and groundwater remediation may include project managers, field personnel, hydrogeologists, monitoring personnel, remediation investigators/engineers, environmental consultants, regulatory personnel, environmental attorneys, expert witness, industrial compliance officers, industrial hygienists, occupational health professionals, and managers/supervisors who will interact with remediation personnel on a daily basis. Other interested readers may also include allied disciplines such as applied chemists, microbiologists, toxicologists, hydrologists, soil scientists, statisticians, universities researchers, and site owners and professionals in various industries for regulatory compliance.

To the Instructor

This book is designed to have more materials than needed for a one-semester course. It can be taught as a one-semester course with various focuses and selected chapters depending on the specific need. For example, not all the remediation technologies in Chapters 7 through 12 should be taught, and Chapter 13 regarding the modeling of flow and transport can be generally disregarded at the undergraduate level. In properly using this text, instructors are provided with a solution manual and lecture slides through Wiley. Students and readers of other interesting groups may also find the answers to selected problems at the end of this book. For visual learners, additional audiovisual links can be found through the book Web site from Wiley, located at www.wiley.com/go/Zhang/Remediation_1e. A word of caution is the unit system used in this book. The International System of Units (SI) is used with US units in parenthesis where appropriate. This makes the book useful to professionals outside the United States and to those within the United States.

List of Symbols

Letters Symbols

$1/n$	Freundlich isotherm constant
A	Area (m^2)
ABS	Absorption rate for dust (%)
AT	Averaging time of exposure (years)
b	Thickness of a confined aquifer or saturated thickness of an unconfined aquifer (m)
b	Reactive cell thickness (m)
BCF	Bioconcentration factor (L/kg)
BR	Breathing rate for dust (%)
BW	Body weight (kg)
C	Concentration in water (mg/L), air (mg/m^3), or soil/dust (mg/kg)
C	Weight percentage of carbon (%)
C	Cost
C	A constant in equation
$\hat{C}$	Concentration in adsorbed phase (mg/kg)
CDI	Chronic daily intake (mg/kg-day)
CE	Combustion efficiency (%)
CMC	Critical micellar concentration (mol/L)
C_p	Specific heat capacity (J/kg-K; Btu/lb-F)
C_s	Surfactant concentration (mol/L)
CSF	Cancer slope factor $(mg/kg\text{-}d)^{-1}$
d	Diameter; Infinitesimally small change (m)
D	Molecular diffusion coefficient (m^2/s)
D	Distance (m)
dC/dx	Derivative of concentration with respect to x (concentration gradient)
D_e	Effective diffusion coefficient (m^2/s)
D_h	Hydrodynamic dispersion (m^2/s)
dh	The infinitesimally small change in hydraulic head (m)
dh/dl	Derivative of head with respect to distance (hydraulic gradient) (unitless)
dl	The infinitesimally small change in distance (m)
dq/dA	Heat flux (W/m^2)
DRE	Destruction and removal efficiencies (%)
dx	Infinitesimally small change in x coordinate (m)
dz/dx	Derivative of head with respect to x (potential head change) (unitless)
E	Activation energy (kJ/mol)
E	Electrode potential (volt)

e	2.71828
EC	Exposure concentration (water: μg/L; air mg/m^3)
ED	Exposure duration (years)
EF	Exposure frequency (days/yr)
ET	Evapotranspiration rate (L/day, in-acre/yr)
erf	The error function
$\exp(x)$	Exponential of x, $\exp(x) = e^x$
f	Fraction (soil component, cosolvent, and heat loss) (unitless)
f_{oc}	Fraction of organic carbon in soil
FV	Future value
f_w	Fraction of contaminant remaining in soil water
g	Gravitational constant (9.81 m/s^2)
G	Air flow rate (m^3/m^2-hr) in a stripping tower
G	Gibbs free energy (J)
h	Potential head/water level/soil depth (m)
H	Henry's law constant: atm/(mg/L), atm/M, atm/(mol/m^3), or dimensionless
H	Total head/water level (m)
ΔH	Enthalpy change (J)
ΔH_v	Heat of vaporization (J/kg; Btu/lb)
H	Weight percentage of hydrogen (%)
h_c	Heat transfer coefficient (W/m^2-K)
HI	Hazard index (dimensionless)
HQ	Hazard quotient (dimensionless)
i	Interest rate
I	Cost index value
IR	Intake (ingestion) rate (water: L/d; air: m^3/d; soil and dust: kg/d)
J	Mass (mole) flux per unit area and time (mg/m^2-s; mol/m^2-h)
k	Intrinsic permeability (m^2)
k	First-order rate constant (s^{-1})
k_b	Biodegradation rate constant (s^{-1})
K	Hydraulic conductivity (m/s)
K	Equilibrium constant
K	Freundlich isotherm partitioning coefficient
K	Thermal conductivity (W/m-K)
$K(\theta)$	Moisture-dependent unsaturated hydraulic conductivity
K_d	Soil–water partition coefficient, adsorption coefficient (L/kg)
K_L	Mass transfer coefficient (m/h), concentration driving force
K_{La}	Overall mass transfer coefficient (T^{-1}) in a stripping tower
K_m	Micelle–water partition coefficient (mol/mol)
K_{ow}	Octanol–water partitioning coefficient (unitless)
K_{sp}	Solubility product constant
l	Distance (m)
L	Liquid loading (m^3/m^2-hr) in a stripping tower
L	Length of a flow path (m)
L	Reactive cell thickness (m)
L_e	Effective length of well screen (m)
m	Mass rate (kg/s)

M	Mass or mass per unit area (kg, kg/m^2)
MW	Molecular weight (g/mol)
n	Soil porosity (%)
n	Number of moles/electrons/years/wells
N	Weight percentage of nitrogen (%)
N_c	Capillary number (dimensionless)
n_e	Effective soil porosity (%)
NOAEL	No-observed-adverse-effect level (mg/kg-d)
N_{wells}	Well number
O	Weight percentage of oxygen (%)
P	Pressure (atm)/Other properties
P_{atm}	Absolute ambient pressure (1 atm or 1.01 × 10^6 g/cm-s^2)
p_i	Vapor pressure of component i (atm)
p_i^0	Vapor pressure of its pure component i (atm)
P_r	Pressure at a radial distance r from the vapor extraction well (atm)
P_{RI}	Pressure at the radius of influence (atm)
PV	Present value
P_w	Absolute pressure at an extraction well (atm or g/cm-s^2)
q	Specific discharge (L/day)
Q	Volumetric discharge/pumping rate (m^3/day, ft^3/min)
Q/H	Flow rate (cm^3/s) per unit thickness of a well screen (cm)
r	Radius of well/well casing/well screen/steam influence (m)
R	Ideal gas constant (0.082 atm-L/mol-K)
R	Retardation factor (unitless)
R	Groundwater recharge (m^3/m^2-day)
R	Radius of well screen plus sand pack/gravel envelope (m)
R	Stripping factor (unitless)
$R_{acceptable}$	Acceptable removal rate (kg/day)
Re	Reynolds number (unitless)
R_e	The effective radial distance over which y is dissipated (m)
R_{est}	Estimated vapor removal rate (kg/day)
RfC	Inhalation reference concentration (μg/m^3)
RfD	Oral reference dose (mg/kg-d)
R_I	Radius of influence of a vapor extraction well (m)
R_w	Radius of a vapor extraction well (m)
s	Drawdown ($s = h_0 - h$) (m)
S	Sorbed phase concentration (mg/kg)
S	Aquifer storativity (unitless)
S	Saturation (%)
S	Weight percentage of sulfur (%)
S_a	Capacity (size) of equipment A
S_s	Specific storage (m^{-1})
t	Time (s)
$t_{0.37}$	Time required for the water level to fall to 37% of the initial change (s)
$t_{1/2}$	Half-life (s)
t_D	Dimensionless form of time (unitless)
T	Absolute temperature in Kelvin (K) or Rankine (R)
T	Aquifer transmissivity (m^2/day)

TPH	Total petroleum hydrocarbons (mg/kg)
u	Time parameter in pumping test (dimensionless); arithmetic mean
UF	Uncertainty factor (unitless)
UR	Unit risk from drinking water $(\mu g/L)^{-1}$ and from inhalation $(\mu g/m^3)^{-1}$
v	Darcy velocity/specific discharge/gas flow velocity (m/s)
v_c	Contaminant velocity (m/s)
v_k	Kinematic viscosity (m^2/d)
v_p	Groundwater (pore) velocity (m/s)
V	Volume (m^3)
w	Width of an aquifer (x direction) (m)
$W(u)$	Well function
x	Distance in the x direction or coordinate (m)
x_i	Mole fraction of the component i in the mixture
X	Amount of chemical to be removed (kg)
y	Width of the pumping region (capture zone) (m)
y	Distance in the y direction or coordinate (m)
y_o	Drawdown at time $t = 0$ (slug test) (m)
y_t	Drawdown at time t (slug test) (m)
Y	Specific yield (%)
Y_{max}	Maximum half-width of the capture zone (m)
Y_{max}^0	Maximum half-width of the capture zone at $x = 0$ (m)
z	Distance in the z direction or coordinate (m)
z	Packing height of a stripping tower (m)
Z	The potential head (the elevation head above mean sea level, MSL) (m)
∇	Gradient operator (m^{-1})
∂	Latin letter d (pronounced as dee) which denotes partial derivative
$\partial C/\partial t$	Partial derivative of concentration with respect to time (kg/m^3-s)

Greek Symbols

α	Dynamic dispersivity
α	Compressibility of an aquifer skeleton
α	A scale factor related to the inverse of air entry pressure (cm^{-1})
β	Compressibility of water
γ	Interfacial tension between oil and water (dyne/cm)
γ	Specific weight (kg/L)
Δ	Delta which denotes difference (change in) when precedes symbol
θ	Contact angle at the solid–water–NAPL interface
θ_a	Soil air content (volumetric) (L^3/L^3)
θ	Soil moisture content (volumetric) (L^3/L^3)
θ_r	Residual soil water content (L^3/L^3)
θ_s	Saturated soil water content (L^3/L^3)
λ_i	Molar heat of vaporization (kJ/mol)
μ	Dynamic viscosity (M/L-T; g/cm-s; dyne×s/cm^2)
π	3.14159
ρ	Density (g/cm^3 or kg/L)
σ	Cosolvency power (dimensionless); standard deviation
ψ	Tension, suction, or the pressure head
ϕ	Angle (radian)
ω	Tortuosity factor (unitless)

Superscripts

*	At saturation/equilibrium
′	Prime symbol (e.g. A′ generally denotes that it is related to or derived from A)
0	Distance/time at 0
0	Standard state
n	Grid in the finite difference time domain

Subscripts

a	Air
aq	Aqueous phase
b	Bulk soil
est	Estimated
g	Gas phase
i	ith chemical/node
i	In/influent/reactant/feed
i, j, k	Three-dimensional spatial grids along x-, y-, and z-coordinates
n	NAPL phase
o	Out/Effluent/Product
oct	Octanol phase
p	Soil particle
p	Previous (past) year
s	Soil (adsorbed) phase/saturated zone
t	Time
T	Total
v	Void
w	Water/well

About the Companion Website

This book is accompanied by a companion website:

www.wiley.com/go/Zhang/Remediation_1e

The website includes Solutions manual and Weblinks.

Scan this QR code to visit the companion website

Chapter 1

Sources and Types of Soil and Groundwater Contamination

LEARNING OBJECTIVES

1. Identify the major use categories of surface water vs. groundwater
2. Discuss the current groundwater quality and the factors affecting groundwater quality
3. Understand the interactions between surface water and groundwater
4. Identify the major soil and groundwater contamination sources
5. Understand the terms relating to soil and groundwater contamination, including Superfund sites, brownfields, RCRA facilities, and underground storage tanks
6. Locate Superfund sites, brownfields, and/or RCRA facilities in an area of interest
7. Gain a general overview of soil and groundwater remediation
8. Develop a general understanding of the sources and types of soil and groundwater contamination in the United States and other regions of interest
9. Identify Internet resources related to global issues of soil and groundwater contamination and remediation

This introductory chapter will examine pollution problems (or contaminant sources) related to contaminated soil and groundwater. We will first take a look at the use of groundwater in comparison with the use of surface water in the United States. We should get a sense of what the general groundwater quality is in the United States and what factors affect it (detailed in Chapters 2–3). This chapter lays a foundation of future chapters regarding the assessment of such pollution problems (e.g. Chapter 5), and the remediation technologies to deal with these problems (Chapters 6–12). While our focus in this chapter is on the environmental problems in the United States, we will briefly introduce the global perspective of soil and groundwater contamination. To raise the awareness of such environmental problems, readers are encouraged to search relevant literature to get a better sense of the soil and groundwater pollution problems that are of primary concern at the local, regional, national, or global level. This chapter is concluded with remarks on the unique challenges of soil and groundwater remediation and the framework of environmental remediation.

1.1 Uses of Surface Water vs. Groundwater

Surface water, i.e. the water on the Earth's surface, occurs as streams, lakes, and wetlands, as well as bays and oceans. Surface water also includes the solid forms of water (snow and ice). **Groundwater** is the water beneath the surface of the ground which is stored in soil pore spaces and in the fractures of rock formations. In the discussion below, we will use primarily the US

Soil and Groundwater Remediation: Fundamentals, Practices, and Sustainability, First Edition. Chunlong Zhang.

Companion website: www.wiley.com/go/Zhang/Remediation_1e

data to illustrate the withdrawals and uses of surface water and groundwater, and their spatial and temporal trends. Readers interested in the uses of surface vs. groundwater in a particular region of the United States or other countries should refer to available sources. In the United States, abundant and reliable data are readily available from the US Environmental Protection Agency (EPA) and the US Geological Survey (USGS).

In the United States, fresh surface water and fresh groundwater withdrawals, reported in billion gallons per day (Bgal/d), were 230 and 76, respectively. Among eight categories of water use, thermoelectric power, irrigation, and public supply are the three largest, accounting for 49, 31, and 11%, respectively. The combined total of the other five categories is less than 10% including industrial (4%), aquiculture (2%), mining (1%), domestic (1%), and livestock (<1%) (USGS 2010) (Figure 1.1). In this text, **public supply** refers to water withdrawals by public and private water suppliers that provide water to at least 25 people or have a minimum of 15 connections. Public-supply water is delivered to users for domestic, commercial, industrial, and public services. **Domestic water** use includes indoor uses (drinking, food preparations, washing clothes and dishes, and flushing toilets) and outdoor uses (watering lawns and gardens and washing cars) at residences. Domestic water is either self-supported (well or captured as rainwater in a container) or provided by public suppliers.

Since the estimated use of fresh surface water is over three times higher than that of fresh groundwater, it is no surprise that thermoelectric power, irrigation, public supply, industrial, and aquaculture use mostly fresh surface water. For fresh groundwater, however, about 67% withdrawn groundwater is used for irrigation followed by 18% for public supply. The percentage of the US population obtaining drinking water from public suppliers has increased steadily

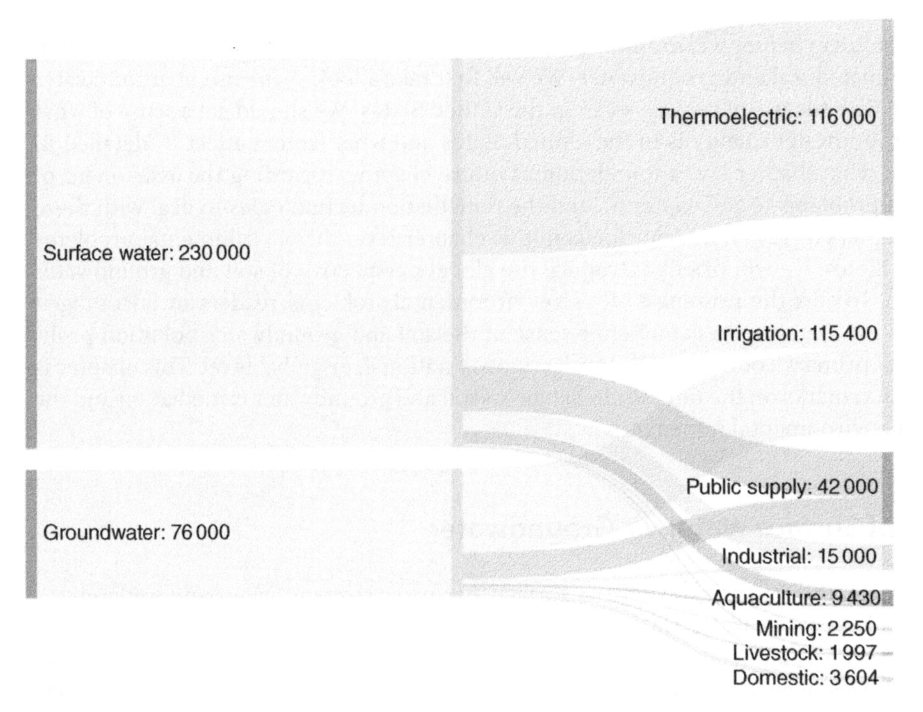

Figure 1.1 Sankey diagram showing the use of fresh surface water and fresh groundwater among eight categories of water uses in the United States. Data are in million gallons per day. *Source:* USGS (2010).

from 62% in 1950 to 86% in 2010 (USGS 2010). Most of the population providing their own household water obtained their supplies from groundwater sources. An interesting fact is that an estimated 43 million people in the United States, or 15% of the population, supplied their own water for domestic use. These self-supplied withdrawals totaled 3 Bgal/d, or about 1% of the estimated water withdrawals for all uses in 2005. Nearly all (98%) of these self-supplied withdrawals were from fresh groundwater (USGS 2009a).

On the global perspective, the United States ranks the third in its abstraction of the groundwater. The 10 countries that abstracted the largest quantities of groundwater, in a decreasing order, are India, China, the United States, Pakistan, Iran, Bangladesh, Mexico, Saudi Arabia, Indonesia, and Turkey (Figure 1.2). Approximately 72% of the assumed total global groundwater withdrawal takes place in these 10 countries (Margat and van der Gun 2013). In fact, the abstraction of groundwater is highly concentrated in a limited number of regions, especially in Asia. The abstraction is relatively modest in Africa, where only Egypt abstracts more than 5 km^3 per year, and in South America (except Brazil).

If we further examine the water uses among different states in the United States, they vary both spatially and temporally. The available groundwater resources and the uses of groundwater resources are not evenly distributed geographically in the United States as well as other countries with large geographical areas. For example, more than half of the fresh groundwater withdrawals in the United States occurred in six states. In California, Texas, Nebraska, Arkansas, and Idaho, most of the fresh groundwater withdrawals were for irrigation. In Florida, 52% of all fresh groundwater withdrawals were public supply whereas 34% was used for irrigation (USGS 2010). Figure 1.3a shows the spatial variations regarding the total fresh groundwater withdrawals and use categories in top 25 states with the most withdrawals, and Figure 1.3b shows the temporal variations through a comparison of surface water and groundwater withdrawals in parallel with population change from 1950 to 2010. The spatial variation is obvious with states that rely on groundwater for irrigation and public water supplies. On the temporal scale, both surface water and groundwater increased with population increase and then leveled off after the 1980s due primarily to an increase in water conservation.

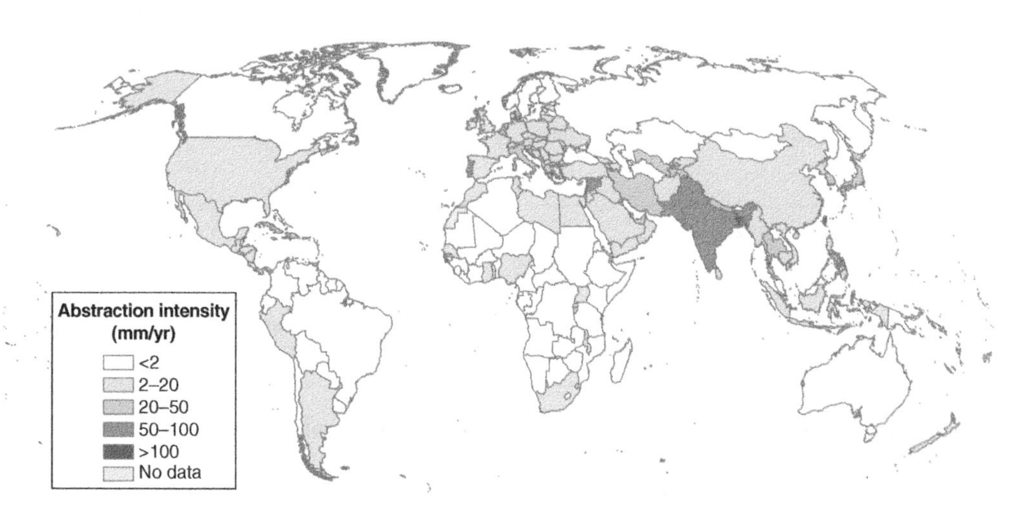

Figure 1.2 Intensity of groundwater abstraction by country for the year 2010, in mm/yr (1 mm/yr = 1000 m^3/yr/km^2). *Source:* © CRC Press, Reproduced with Permission.

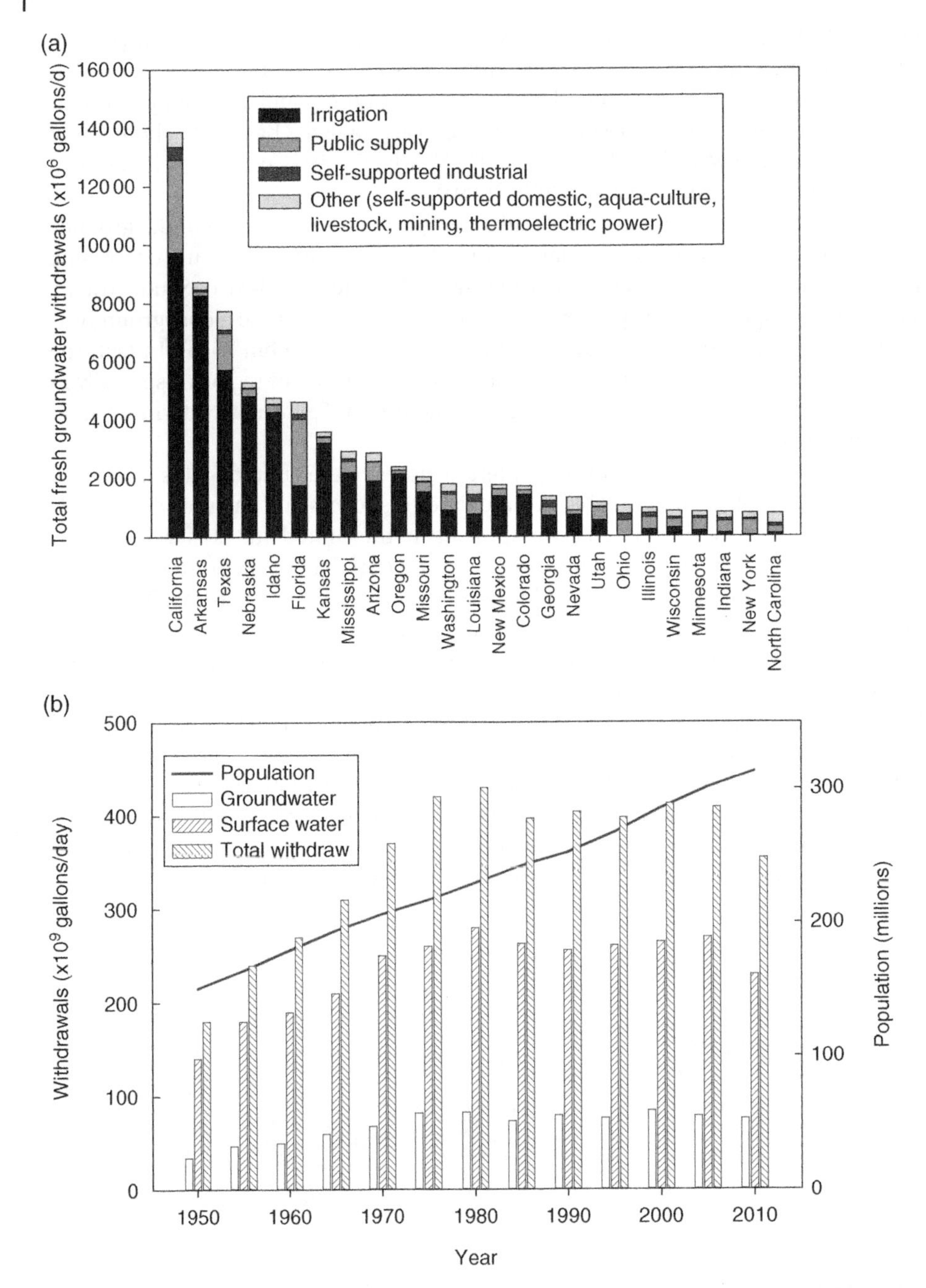

Figure 1.3 (a) Total water withdrawals and use categories by top 25 states, (b) Temporal trends in the population and fresh groundwater and surface water withdrawals from 1950 to 2010 (USGS 2010).

1.2 Groundwater Quantity vs. Groundwater Quality

Whereas the quantity of groundwater resource in its spatial and temporal distribution is important, the quality of groundwater is equally important. Groundwater is a renewable resource, yet it is subjective to overuse and contamination. Large-scale pumping of groundwater can cause large drawdowns (long-term drops in groundwater levels), which can further cause serious economic, health, and environmental impacts. Wells become dry and must be drilled deeper, thereby increasing the risk of elevated arsenic, radium, salts, and other naturally occurring substances. Overpumping can result in **subsidence** and dry up nearby streams, lakes, and wetlands that are essential to the natural habitat.

The USGS periodically collected groundwater quality data from domestic wells (private wells for household drinking water). From the monitoring data collected during the period 1991–2004 from wells located within major hydrogeologic settings of 30 regionally extensive aquifers (USGS 2009b), the following major conclusions can be made:

- About 23% of wells had at least one contaminant present at concentrations greater than a Maximum Contaminant Level (MCL) or Health-Based Screening Level (HBSL).
- Many of these contaminants frequently exceeding human-health benchmarks were mostly natural occurring, including radon, several trace elements (arsenic, uranium, strontium, and manganese), fluoride, and gross alpha- and beta-particle radioactivity. Contaminants from human activities include nitrate and fecal indicator bacteria.
- Certain contaminants present a geographic pattern such that their occurrences are in certain locations or regions rather than national. An example is the higher radon concentrations in crystalline-rock aquifers located in the northeast, the central and southern Appalachians, and Colorado. Also, nitrate was found more frequently in agricultural areas, rather than in other land-use settings at concentrations greater than the MCL.
- Concentrations of certain contaminants were also related to geochemical conditions. For example, in addition to showing regional patterns of occurrence, uranium concentrations were correlated with concentrations of dissolved oxygen. Also, relatively high concentrations of iron and manganese occurred everywhere, but were inversely correlated with dissolved oxygen concentrations.
- **Anthropogenic organic compounds** were frequently detected at low concentrations, using typical analytical detection limits of 0.001–0.1 μg/L, but were seldom present at concentrations greater than MCLs or HBSLs.
- The most frequently detected anthropogenic organic compounds were atrazine, its degradation product deethylatrazine, and volatile organic compounds (VOCs), such as chloroform, methyl *tert*-butyl ether, perchloroethene, and dichlorofluoromethane (Figure 1.4). The wide variety of compounds detected, including herbicides, solvents, disinfection by-products, gasoline hydrocarbons and oxygenates, refrigerants, and fumigants, reflect the diverse industrial, agricultural, and domestic sources that can affect the quality of source water to domestic wells.
- Seven of 168 organic compounds were present in samples at concentrations greater than MCLs or HBSLs, each in less than 1% of wells. These were diazinon, dibromochloropropane, dinoseb, dieldrin, ethylene dibromide, perchloroethene, and trichloroethene.
- Several combinations of organic compounds in mixtures with possible health effects were identified – specifically, atrazine and deethylatrazine, atrazine or simazine with nitrate, and perchloroethene and three other solvents – but combined concentrations either were less than the health benchmarks or no benchmarks were available for the mixtures. These co-occurrences may be a potential concern for human health, but the long-term cumulative effects of low concentrations of multiple contaminants on human health currently are unknown.

From the USGS survey, it is thus clear that most groundwater in the United States is generally considered of good quality and safe to drink (USEPA 2002). It is primarily the large number and widespread presence of potential sources of contaminants that make the groundwater quality

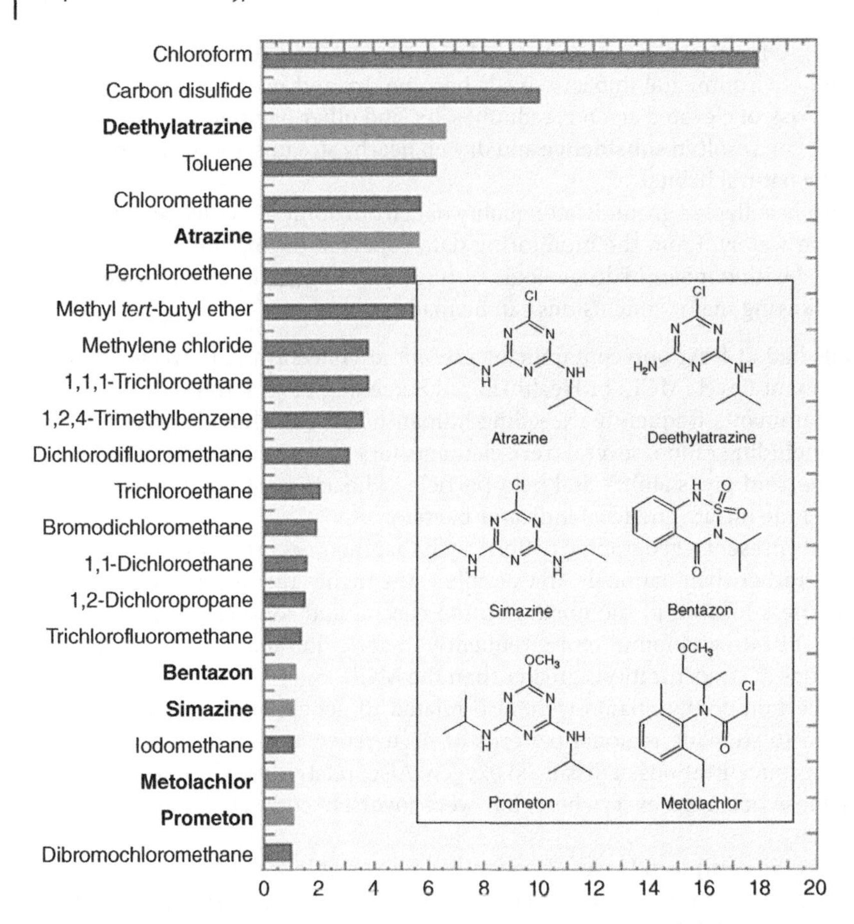

Figure 1.4 Detection frequency (%) for frequently occurring organic compounds in samples collected from domestic wells for the National Ambient Water Quality Assessment (NAWQA) Program in aquifer studies, 1991–2004. Six pesticides (with structures shown in the insert) and 17 volatile organic compounds were detected in more than 1% of wells at concentrations greater than a common threshold of 0.02 μg/L. Concentrations greater than 0.02 μg/L are shown here to eliminate the effects of different analytical detection limits among compounds (USGS 2009a).

variable both spatially and temporally. Degraded groundwater quality occurs in various contaminated sites such as industrial, agricultural, mining, domestic, and military operations. The remediation of these sites with anthropogenic sources of contaminants in soil and groundwater will be the focus of this book.

1.3 Major Factors Affecting Groundwater Quality

Factors affecting groundwater quality are contaminant- and site-specific, but they generally can be grouped into several categories. These include, but are not limited to, source factors, interactions with surface water (see Box 1.1), fate and transport factors including land use, groundwater age, degree of confinement, redox status, and geochemical conditions. We will present some details in the subsequent two chapters when we discuss contaminant fate and transport (Chapter 2) and hydrogeology (Chapter 3). The following are excerpts from a USGS

Box 1.1 Interactions between surface water and groundwater

Imagine a shallow domestic well for a drinking water supply is located in the west bank of a river, and the crop land with intensive pesticide (atrazine) applications is on the east bank. What caused the pesticides to show up in this domestic well in the wet season when runoff frequently occurred? In a subsequent dry season without recent pesticide application, why was this pesticide detected in the river water? The answer to this observation might be related to the frequent interactions between the surface water and groundwater.

Surface water and groundwater are not separate entities and can be considered as a single integrated water resource. The use and contamination of either water resources affects the quantity and quality of the other. Nearly all surface water features (streams, lakes, reservoirs, wetlands, and estuaries) interact with groundwater. These interactions take many forms. In many situations, surface water bodies gain water and chemicals from groundwater systems, or they can become a source of groundwater recharge and cause changes in groundwater quality. As a result, water withdrawal from streams can deplete groundwater. Conversely, pumping groundwater can deplete water in streams, lakes, or wetlands. Pollution of surface water can cause degradation of groundwater quality and conversely pollution of groundwater can degrade surface water.

The movement of water between groundwater and surface water provides a major pathway for chemical transfer between terrestrial and aquatic systems. From the surface water quality standpoint, the effect of contaminated groundwater is particularly relevant because much of the groundwater contamination occurs in the shallow aquifers that are directly connected to surface water. In some cases, contaminated groundwater can be a major and potentially long-term contributor of contaminants to the surface water of streams and lakes. Streams interact with groundwater in three basic ways: streams gain water from inflow of groundwater through the streambed (gaining stream), they lose water to groundwater by outflow through the streambed (losing stream), or they do both, gaining in some reaches and losing in other reaches. Lakes and wetlands can receive groundwater inflow throughout their entire bed, have outflow throughout their entire bed, or have both inflow and outflow at different localities (USGS 1998).

Observing and measuring the hydraulically connected surface water and groundwater is challenging, but it is essential for us to better remediate contaminated sites. An improved understanding of such interactions will also help us to evaluate applications of waste-discharge permits and to protect or restore biological resources in surface water.

study (USGS 2009b) regarding the factors affecting anthropogenic nitrate, pesticides, and VOCs. In a decreasing order of importance in the United States, these groundwater contaminants are nitrate, pesticides, VOCs, petroleum products, metals, brine, synthetic organic compounds, coliform bacteria, radioactive materials, other agricultural contaminants, arsenic (As), fluorides, and other inorganic compounds (USEPA 2002). Findings regarding the top three groundwater contaminants are as follows.

- *Nitrate:* Agricultural land use and the total nitrogen input from atmospheric deposition, fertilizer, manure, and septic systems were the source factors with the strongest positive correlation with nitrate concentrations. Groundwater age was a significant factor affecting nitrate concentration. Water recharged prior to 1953 had significantly lower median concentrations of nitrate than water recharged in 1953 or later. In addition, water from wells in confined aquifers and semi-confined/mixed confined aquifers had statistically significant lower nitrate concentrations than water from wells in unconfined aquifers. Dissolved oxygen concentration had a strong positive correlation with nitrate concentration, indicating that redox processes in the aquifer play an important role in the fate and transport of nitrate.

- *Pesticides:* All the frequently detected pesticides had positive correlations with agricultural land use and the application rate of the pesticide, if it was available. Atrazine, simazine, and prometon had a positive correlation with urban land use. The detections of pesticides were related to groundwater age and degree of confinement; samples of younger water and water from unconfined aquifers had a statistically significant higher rate of pesticide detection than samples of older water and water from aquifers with mixed confinement or confined aquifers. Dissolved oxygen (DO) concentration had a positive correlation with pesticide concentration, likely because well-oxygenated waters are associated with young aquifers. Dissolved organic carbon (DOC) had an inverse correlation with pesticide concentrations in some aquifers, and a positive correlation in others. In the cases where the correlation was positive, it is likely that the correlation was a result of land use and not concentration of dissolved organic carbon.
- *VOCs:* Detection frequencies for VOCs were slightly higher for samples in areas with urban or mixed land use than in areas with agricultural or undeveloped land use, but did not vary significantly among confinement categories or redox-status categories. All detected VOCs differed in their detection frequencies among the carbonate aquifers studied and these differences are likely related to various natural and anthropogenic factors, including hydrogeology, water chemistry, and land use.

1.4 Soil and Groundwater Contaminant Sources in the United States

In the United States, environmental awareness started since the first Earth Day in 1970 and by the establishment of the USEPA by President Nixon in the 1970s. For example, Love Canal was one of the most notorious scenarios that occurred in the history (see Box 1.2). It started in the 1940s and 1950s when Hooker Chemical Company used a number of disposal areas within the city for the disposal of over 100 000 tons of hazardous petrochemical wastes. Wastes were placed in the abandoned Love Canal, in a large unlined pit on Hook's properties. By the mid- 1970s, chemicals at the Love Canal had migrated from the sites – causing all kinds of injury and loss of property value. At that time, there was no relevant environmental law against such disposal.

The improper disposal and illegal dumping incidents described above are just part of the reason for the nationwide soil and groundwater contamination. The major milestone as a result of these pollution incidents was the 1984 report to the congress entitling "Protecting the Nation's Groundwater from Contamination," which listed 30 potential sources of groundwater contamination from the survey of National Water Quality Inventory. The schematic in Figure 1.5 illustrates that the sources of soil and groundwater contamination are numerous and are as diverse as human activities.

These major sources of soil and groundwater contamination in United States are shown in Figure 1.6. Underground storage tanks were most frequently identified among the top five contamination sources of concern (marked by 34 states). Twenty-five states characterized municipal landfills among the top 5 sources of concerns while 23 states ranked agricultural activity among the top 5. Abandoned hazardous waste sites (ranked by 21 states) and septic tanks (ranked by 20 states) ranked fourth and fifth, separately, among the top 5 sources of concern.

The following descriptions focus on contaminated sites and waste facilities based on cleanup market sectors, including Superfund sites on the National Priority List, Resource Conservation and Recovery Act (RCRA) facilities, underground storage tanks (USTs), the Department of Defense (DoD), the Department of Energy (DoE), states and private parties (including brownfields).

Box 1.2 The Love Canal Tragedy

Love Canal was a neighborhood in Niagara Falls, New York. Hooker Chemical (now Occidental Petroleum Corporation) had sold the site to the Niagara Falls School Board in 1953 for $1, with a deed explicitly detailing the presence of the waste and including a liability limitation clause about the contamination. The construction efforts of housing development, combined with particularly heavy rainstorms, released the chemical waste, leading to a public health emergency and an urban planning scandal. The dumpsite was discovered and investigated by the local newspaper, the *Niagara Falls Gazette*, from 1976 through the evacuation in 1978. Reporter Michael H. Brown first raised concerns about potential health problems in July 1978. In the mid-1970s Love Canal became the subject of national and international attention after it was revealed in the press. What was in there? How bad was it? What has been done?

Extensive testing and investigations revealed that the site had formerly been used to bury 21 000 tons of toxic wastes by Hooker Chemical. Materials included caustics, alkalines, fatty acids, and chlorinated hydrocarbons from the manufacturing of dyes, perfumes, and solvents for rubber and synthetic resins. It has been calculated that 248 separate chemicals, including 60 kg of dioxin, have been unearthed from the canal.

In 1979, the EPA announced the result of blood tests that showed high white blood cell counts, a precursor to leukemia, and chromosome damage in Love Canal residents. In fact, 33% of the residents had undergone chromosomal damage (in a typical population, chromosomal damage affects 1% of people). Ten years after the incident, New York State Health Department Commissioner David Axelrod stated that the Love Canal would long be remembered as a "national symbol of a failure to exercise a sense of concern for future generations." The Love Canal incident was especially significant as a situation where the inhabitants "overflowed into the wastes instead of the other way around."

The government relocated more than 800 families and reimbursed them for their homes. In 1994, the Federal District ruled that Hooker/Occidental had been negligent, but not reckless, in its handling of the waste and sale of the land to the Niagara Falls School Board. Occidental Petroleum was sued by the EPA and in 1995 agreed to pay $129 million in restitution. Residents' lawsuits were also settled in the years following the Love Canal disaster. Today, houses in the residential areas on the east and west sides of the canal have been demolished. All that remains on the west side are abandoned residential streets. Some older east side residents, whose houses stand alone in the demolished neighborhood, chose to stay. It was estimated that less than 90 of the original 900 families opted to remain. They were willing to remain as long as they were guaranteed that their homes were in a relatively safe area. On 4 June 1980, the Love Canal Area Revitalization Agency (LCARA) was founded to restore the area. The area north of Love Canal became known as Black Creek Village. LCARA wanted to resell 300 homes that had been originally bought by New York when the residents were relocated. These homes were farther away from where the chemicals had been dumped. The most toxic area (16 acres [65 000 m^2]) has been reburied with a thick plastic liner, clay, and dirt. A 2.4-m high barbed wire fence was constructed around this area. The Love Canal disaster is one of the most appalling environmental tragedies in the American history.

More detailed information about the Love Canal story can be located on Wiley's website for this book (see Audiovisual Aids).

1.4.1 Superfund Sites and Brownfields

As recognized by the USEPA, a **Superfund site** is an uncontrolled or abandoned place where hazardous waste is located. A site might have been abandoned because the owner has claimed bankruptcy, sold the properties, and the responsible parties may be in debate. Superfund sites originated from a regulatory act in 1980 termed "The Comprehensive Environmental Response,

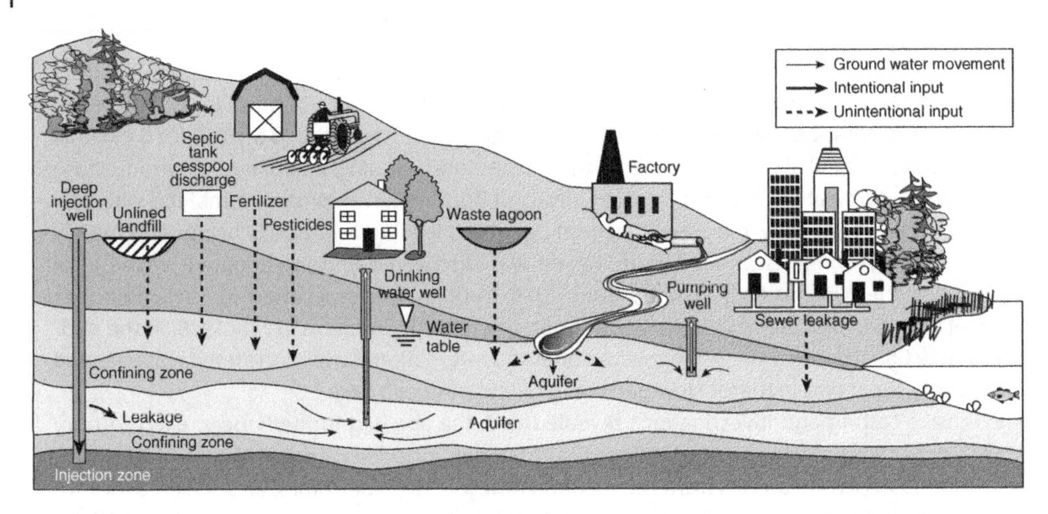

Figure 1.5 Sources of groundwater contamination. Note that the list of the groundwater contamination sources in the diagram is not exhaustive. Other potential sources may include underground storage tank, sewage treatment plant, spill, illegal dumping, and urban runoff (USEPA 2000).

Compensation, and Liability Act (CERCLA)" or commonly referred to as the Superfund Act. The Superfund Act requires that different remedial actions be evaluated for sites on the **National Priorities List** (NPL) and defines the criteria by which to evaluate options. The Superfund was originally a \$1.6 billion Superfund Trust Fund, and the EPA has deposited \$1.09–1.25 billion each fiscal year for Superfund remediation (Pichtel 2000; Congressional Research Service 2016). Superfund money is used to finance cleanups of abandoned sites for which no potentially responsible parties (PRPs) are immediately recognized. However, this pool of money is certainly not enough to remediate all sites.

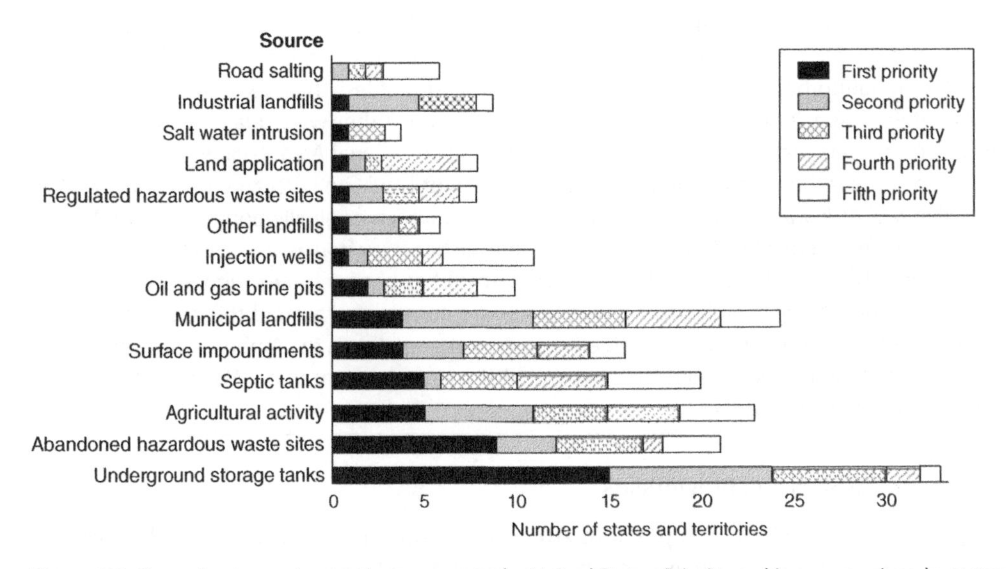

Figure 1.6 Groundwater contamination sources in the United States. Priority rankings were given by states and territories using score 1–5 (1 being the highest priority) for the various contamination sources (USEPA 1990).

Superfund sites are listed on the NPL upon the completion of a Hazard Ranking System (HRS) screening, public solicitation of comments about the proposed site, and after all comments have been addressed. There are approximately 36 814 potential hazardous sites in the United States. As of May 2018, there are a total of 1343 sites listed on the NPL. There is an estimated $27 million or 20 million m^3 (26 million yd^3) average per site to be cleaned. A full list of the NPL sites in the United States can be found at http://www.epa.gov/superfund/sites/index.htm. From the distribution of Superfund sites in the United States (Figure 1.7), NJ, CA, PA, NY, MI, FL, IL, TX, and WA are among those states with the most Superfund sites. Box 1.3 provides some more specifics on the common contaminated matrices (groundwater, soil, sediment, sludge, etc.) and frequent contaminants in the United States.

The Superfund is a unique system in the United States, and it was argued that there already is partially a Superfund system in Europe contained in the Environmental Liability Directive 2004/35/CE. In the future, more comprehensive regulatory framework will be promulgated to deal with soil contamination in European countries (Schirmeisen 2005).

Brownfields (BFs) are those abandoned, under-used commercial sites where viable development is hindered by a real or perceived contamination problem. Its name was derived from its counterpart of rural green fields. Brownfields routinely are associated with distressed urban areas, particularly central cities and inner suburbs that once were heavily industrialized, but since have been vacated. A small percentage of brownfield sites may be contaminated to the degree that they are candidates for Superfund sites under the CERCLA regulation (see Box 1.4 for a comparison of Superfund site and brownfield site).

There are an estimated 450 000 brownfield sites in the United States, 362 000 in Germany, 200 000 in France, and 100 000 in England (Oliver et al. 2005). Common examples of brownfields

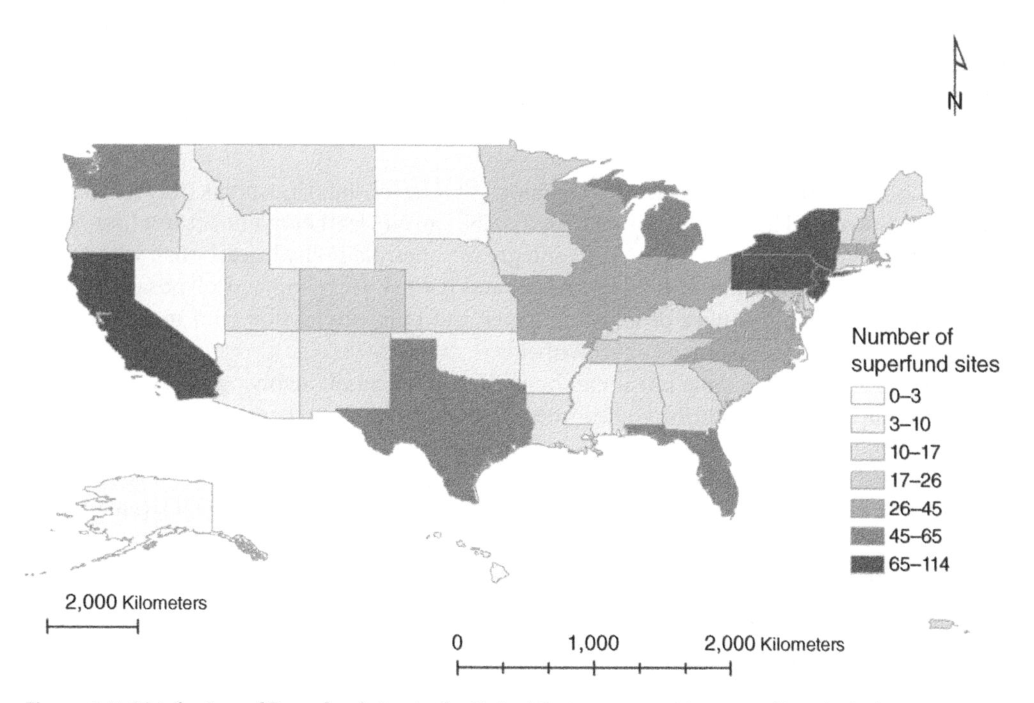

Figure 1.7 Distribution of Superfund sites in the United States grouped by states (Data include 1343 Superfund sites in the National Priority List. Courtesy of Marc Mokrech, Environmental Institute of Houston. Data source: http://www.epa.gov/superfund/.

Box 1.3 Superfund sites: How many? what matrices? what contaminants?

The number of Superfund sites in the NPL changes over time. Every year we delete and add more Superfund sites to the NPL. If we divide the total Superfund sites by 50 states, this is about 25 per state. Interested readers are referred to the US EPA website to find the existing Superfund sites in the area of concern. For example, at the time of this writing, there were 55 Superfund sites listed in the NPL for the State of Texas. From the distribution map, the Houston-Galveston area locates the most Superfund sites in the state. A total of 28 sites in the 8-county region were either listed on the EPA's National Priorities List or the Texas State Superfund List. In fact, Harris County is among the top 10 counties in the United States having the most Superfund sites. Additionally, the region generates 4.5 million tons of solid waste each year. It was estimated that the region's 21 landfills will reach full capacity in the near future based on 2000 disposal rates (Citizen's Environmental Coalition 2001).

The frequencies of contaminated matrices at NPL sites with **Record of Decisions** (RODs, a public document that explains which cleanup alternatives will be used to clean up a Superfund site) are reported to be groundwater (76%), soil (72%), sediment (22%), and sludge (12%) (USEPA 2004). The importance of contaminated "soil" and "groundwater" is apparent by comparing it with other environmental matrices.

In regard to the frequencies of major contaminants groups at NPL sites with RODs, the data from 944 NPL sites show that the number of site has: VOCs + SVOCs (semivolatile organic compounds) + metals = 41%, VOCs + SVOCs = 25%, either VOCs, SVOCs, or metals = 25%. Other contaminants of concerns include radioactive elements, nonmetallic inorganic, or unspecified organics or inorganics, which account for approximately 10%. Further analyses for the most common contaminant at NPL sites in the United States revealed the predominance of VOCs, SVOCs, and metals. In the group of VOCs, the statistics are: trichloroethylene (TCE) (50%), benzene (43%), toluene (36%), tetrachloroethylene (PCE) (36%), and vinyl chloride (VC) (29%). Polychlorinated biphenyl (PCBs) in the group of SVOCs is (36%), whereas metals are in a decreasing order of Pb (47%), As (41%), Cr (37%), Cd (32%), Ni (29%), Zn (29%) (USEPA 2004).

and their associated contaminants are: dry cleaning (VOCs, solvents), hospitals (formaldehyde, radionuclides, solvents, Hg), electroplating (Cd, Cr, CN, Cu, Ni), leather manufacturer (benzene, toluene), petroleum refinery (BTEX, fuel, oil and greases, petroleum hydrocarbons), wood preserving (PCP, creosote, arsenic, Cu, PCB, PAHs), and battery recycling and disposal (Pb, Cd, acids). In Europe, mining, timber processing, paper and pulp production, and iron and steel industries can be the biggest sources of brownfields.

The issue of BFs sounds simple but it can be really complicated, involving many interested parties and groups, such as realtors, environmental firms, politicians, lawyers, insurance companies, government agencies, investors, and the general public. BFs represent unique opportunities and challenges for these different groups of people. Much progress has been made for the cleanup of this nation's brownfields, for example, Houston leads the nation in reusing its brownfields. Currently, the Houston brownfields pilot has been assisting in the reuse of more than 550 acres of brownfields whose redevelopment costs a total of $463 million.

1.4.2 RCRA Facilities and Underground Storage Tanks

RCRA facilities or **solid waste management units** (SWMUs) are defined as "any unit which has been used for the treatment, storage, or disposal of solid waste at any time." These units include storage tanks, waste piles, drain fields, waste treatment units, and surface impoundment. Major

Box 1.4 Superfund sites vs. Brownfield sites

A Superfund site is an uncontrolled or abandoned place where hazardous waste is located, possibly affecting local ecosystems or people. Sites are listed on the National Priorities List (NPL) upon completion of a Hazard Ranking System (HRS) screening and public solicitation of comments. If a site is listed in the NPL, it receives Superfund monies.

In the US Superfund programs, there are federal and state levels of Superfund programs. In the state of Texas, for example, sites that receive an HRS score of 5.0 or greater may be eligible for listing in the state Superfund registry as state Superfund sites. However, a site that receives an HRS score of 28.5 or greater is also eligible for consideration on the National Priorities List as a federal Superfund site. Sites with an HRS score of 28.5 or greater that the EPA has determined are not of NPL caliber may then be proposed to the state Superfund registry. The state Superfund registry, established by the 69th Texas Legislature in 1985 and administered by the Texas Commission on Environmental Quality (TCEQ), lists those abandoned or inactive sites that have serious contamination but do not qualify for the federal program, and therefore are cleaned up under the state program. The state must comply with federal guidelines in administering the state Superfund program, but EPA approval of state Superfund actions is not required.

The term brownfield site means "real property, the expansion, redevelopment, or reuse of which may be complicated by the presence or potential presence of a hazardous substance, pollutant, or contaminant." The term brownfield site does not include a facility that is listed on the National Priorities List or is proposed for listing as a Superfund site. It is therefore apparent that Superfund sites are a top priority and are in the spotlight more often than brownfields.

The law termed CERCLA (see Chapter 4) regulates the cleanup of both Superfund sites and brownfield sites. The CERCLA holds all owners and operators of a brownfield site liable for contamination, regardless of whether or not they actually contributed to the contamination. This liability scheme has been criticized, and the Small Business Liability Relief and Brownfields Revitalization Act (SBLR&BRA) was promulgated in 2002 by aiming to provide liability protection for prospective purchasers, contiguous property owners, and innocent landowners and authorizes the funding for state and local programs that access and clean up brownfields.

types of SWMUs are for land disposal (landfills, waste piles, surface impoundments, land treatment units), treatment and storage (container storage and accumulation areas, tanks and tank connections), incineration, and miscellaneous uses such as boilers and furnaces.

The RCRA facilities are distributed unevenly in the United States, with CA and TX accounting for 20% of the RCRA facilities in the United States. The frequencies of contaminant media are in the decreasing order of groundwater, soil, sludge, surface water, sediment, air, and debris. The top 10 groundwater contaminants projected above action levels at the US RCRA facilities, ranked by the % facilities with each contaminant, are chromium (47%), benzene (30%), methylene chloride (23%), arsenic (20%), lead (20%), tetrachloroethylene (18%), trichloroethylene (17%), naphthalene (14%), 1,2,2-trichloroethane (11%), and toluene (10%) (USEPA 1993, 1994a).

Underground storage tanks (USTs) are defined as "any one or combination of tanks (including underground pipes connected thereto) that is used to contain an accumulation of regulated substances, and the volume of which ... is 10% or more beneath the surface of the ground" according to RCRA § 9001(1). Several types of tanks are excluded from UST, namely, "home heating oil tanks, septic tanks, otherwise regulated pipelines, USTs with less than 4 m^3 (1,100 gallon) capacity used for noncommercial purposes, and tanks with fuel oil used for heating commercial and industrial establishments."

The leaking of USTs are typically related to aging of tanks in use. Many federally regulated active and closed tanks are older than 25 years of service. The types of chemicals being stored were shown to be gasoline (62%), diesel fuel (20%), used oils (3%), kerosene (3%), heating oil (3%), hazardous material (3%), empty tanks (2%), and other (5%) (Tremblay et al. 1995). As a result of the large number of leaking gasoline storage tanks, gas service stations are now required to have UST Compliance Tags to be posted at all UST facilities after the inspection in certain states in the United States, and an owner should have an easy access to print out leak detection reports.

1.4.3 DoD/DoE Sites

The Department of Defense (DoD) sites and installation are distributed among the Army, Navy, Air Force, the Defense Logistics Agency (DLA), and the Formerly Used Defense Sites (FUDs). The % of DoD sites needing cleanup was estimated to be in the decreasing order of groundwater (71%), soil (67%), surface water (19%), and sediment (6%) (USDoD 1996). Most common types of DoD sites needing cleanup are: USTs, spill area, landfill, surface disposal area, storage area, disposal pit/dry well, unexploded munitions ordnance area, contaminated groundwater, fire/crash training area, and surface impoundment/lagoon. The major groups of contaminants include some common VOCs, SVOCs, metals, as well as some specific contaminants of concern such as explosives/propellants (TNT, 2,4-DNT, 2,6-DNT, RDX, HMX) and radioactive metals.

High concentrations of explosives in contaminated sites are particularly relevant in the United States as well as other countries in eastern and western Europe, Russia, China, and Japan. These sources include ammunition plants, testing facilities, military zones, and battlefields. In the United States, for example, extensive explosives contamination at active US military installations include 0.57 million m^3 (750 000+ cubic yards) of soil/sediments, 3.8 million m^3 (10 billion gallons) of contaminated groundwater, and a total current remediation spending of $2.66 billion.

The Department of Energy (DoE) sites commonly have radioactive contaminants (e.g. uranium, tritium, thorium, radium, plutonium) in addition to metals and organic contaminants described previously (USDoE 1995).

Table 1.1 is a summary of the estimated number of contaminated sites and cost in the United States. As can be seen, there are also numerous civilian federal sites, potential state program sites, and private sites that need evaluation and remediation. Because of the relatively large number of contaminated sites and facilities in DoE, DoD, and RCRA compared to that of the Superfund sites listed in NPL, the total cost for the remediation of Superfund sites is actually much smaller than these other segments in the United States.

1.5 Contaminated Soil and Groundwater: A Global Perspective

During prehistoric times, perhaps CO_2, SO_2, H_2S, NO_x, and other toxic gases from volcanoes were the only form of chemicals disturbing the earth. For as long as humans have lived on this planet, pollution has been present. Our ancestors then started to use forest fire (consequently gases) to deliberately capture animals. Another example is when the Romans discovered the thermally insulating properties of asbestos. Such minerals were subsequently mined and still used extensively today. In the 1800s, industrial pollution occurred as the result of the use of dyes and organic chemicals from coal tar industries. Pollution increased dramatically in the 1900s, due to the use of chemicals and generation of waste from the production of iron/steel, lead, petroleum refineries, and from the use of radium and chromium. During World War II (1930s–1940s), chemicals such as chlorinated solvents, pesticides, polymers, plastics, paints,

Table 1.1 Estimated number and cost of contaminated sites to be remediated in the United States (USEPA 2004).

Contamination segment	Number of sites	Cost ($ billion US dollar)	Comments
Superfund			
• Current Sites	456	19.4	Assuming 23–49 sites per year will be added to the NPL
• Projected Sites	280	12.7	
• Subtotal	736	32.1	
RCRA corrective action	3800	44.5	Cost does not include long-term monitoring and groundwater treatment
Underground storage tanks (USTs)	125 000	15.6	Include 35 000 sites already identified plus 90 000 projected sites
DoD	6400	33.2	Out of 30 000 identified sites, 24 000 sites have remediation completed or planned
DoE	5000	35.0	Cost does not include long-term stewardship for nuclear test sites
Civilian Federal Agencies	3000+	18.5	Does not include 8000–31 000 abandoned mine sites with potential cost of $18–51 billion
States	150 000	30.0	23 000 sites already identified plus 127 000 potential sites projected in the next 30 years
Total	294 000	208.9	Exclude sites where cleanup work has begun or is complete

and wood preservatives came into play. For example, plastics were very commonly used during the wars (e.g. bomber nose cones and parachutes), and organic pesticides like DDT replaced inorganic counterparts. The 1970s was the worst in terms of environmental quality in the developed countries, not only due to the air (smog) and water pollution, but also due to the soil and groundwater contamination.

Various techniques of soil and groundwater remediation originated from the rapid industrialization that occurred in the North American and European countries. The 1980s were a decade of rapid development of new environmental legislatures and commercialization of remediation techniques in the United States, Canada, Germany, England, and France. Since the 1990s, remediation techniques in these countries have been further refined for better cost-efficiency, low risk, and sustainability. Table 1.2 outlines the rough estimates of contaminated sites and remediation costs in various countries. A detailed account of remediation markets is not available.

As can be seen, the total costs to remediate all the contaminated sources discussed above are in the range of hundreds of billions of US dollars primarily for the contaminated soil and groundwater in the United States alone. In Europe, potentially polluting activities have occurred at about three million sites, and over 8% (nearly 250 000 sites) need to be remediated. Table 1.2 also reveals the rapid market growth for the remediation of contaminated sites in developing countries such as China, Latin America, and Africa. In China, for example, 90% of the groundwater has a certain degree of contamination and 60% has the severe groundwater contamination. Approximately 65% of the total irrigated agricultural land has heavy metal pollution (cadmium, lead, nickel, mercury) with an estimated cleanup cost of $1.2 trillion (Xu et al. 2017). In response to this growing concern, new environmental regulations have been promulgated and remediation demonstration projects have been implemented in China in the recent years.

Table 1.2 The global remediation market for contaminated soil and groundwater.

Countries/regions	Number of contaminated sites	Current market values	Estimated future market
United States	294 000	$6–8 billion/yr USD (30% of global market)	$209 billion USD in 30–35 yr
Canada	30 000	$250–500 million USD in 2005	$3.5 billion USD in 10 yr
Western Europe	600 000–2 500 000	$60 billion USD, timeframe unspecified	0.5–1.5% GDP
England	100 000	$8 billion USD	NA
Australia	160 000	>$3 billion/yr	NA
China	300 000–600 000	$3–6 billion USD	8% annual growth
Japan	500 000+	$1.20 billion USD, time frame unspecified	$3 billion USD by 2010
Asian Regions	3 000 000+	Unassessed	Unassessed
Latin America and Africa	N.A.	$9.7 billion USD	4.5% annual growth

Sources: Singh et al. (2009), Naidu and Birke (2015), European Environment Agency (2015), USEPA (2004). All money is converted to US dollars based on the 2018 rate.

1.6 Soil and Groundwater Remediation

The discussions we have made thus far in this chapter clearly indicate the importance of groundwater as a source of water supply for many domestic, industrial, and agricultural purposes. The importance of clean soil is self-evident because soil is essential for the terrestrial food production. Unfortunately, in the meantime, precious soil and groundwater resources have been threatened by various human activities during the last several decades. To mitigate this problem, soil and groundwater remediation, or environmental remediation in general, comes into play. Below is a brief overview regarding the unique challenges of soil and groundwater remediation, and the scope of this interdisciplinary field.

1.6.1 Unique Challenges Relative to Air and Surface Water Pollution

The complexity of soil and groundwater distinguishes it from other environmental matrices such as air, surface water, wastewater, and solid wastes. The term **remediation** is generally associated with sites contaminated with soil and groundwater, in conjunction with "treatment" or "disposal" techniques used for other contamination issues. The particular challenges for the remediation of contaminated soil and groundwater are summarized as follows:

- Unlike air or surface water, soil and groundwater contamination cannot be easily visualized. The detection of flow paths and contaminant movement can take a long time, thus delaying the remediation activities.
- Unlike contaminants in air or surface water, contaminants in groundwater cannot be readily degraded or diluted; therefore, contaminants will be accumulated overtime to potentially exceed the protection levels in a local or regional scale. Soil and groundwater contamination can thus be perceived as an irreversible process if an engineering measure is not implemented in the contaminated site.

- Remediation duration is typically long and cost is high. This relates in part to the slow groundwater flow and the difficulty in getting access to contaminants in complex geological formations.

1.6.2 Scope of Environmental Remediation

The approaches to achieve soil and groundwater remediation will be site specific, but the following technical activities will be addressed in the event of needed environmental remediation:

- What is the contaminant?
- Where did it come from and where is it going?
- How bad is it?
- How much do we need to remove?
- How can we remove it?
- How much will it cost?
- How do we know we are succeeding?
- How do we know when to stop?
- What legislation governs the remediation process?

The scope of this text will address the above issues. Logically, we will (i) first look at chemical and physical properties of common contaminants, aquifer characteristics, and movement of substances in soil and groundwater; then (ii) evaluate remedial technologies technically feasible given site-specific regulatory and economic constraints. By the order of appearance, this textbook will cover the following aspects:

- Types of contaminants and their physicochemical and biological characteristics (Chapter 2)
- Hydrogeological conditions of the aquifer soil and groundwater (Chapter 3)
- Environmental regulations, costs, and acceptable risk (Chapter 4)
- Location, use, and physical features of a site (Chapter 5)
- Technical feasibility of site-specific remedial actions (Chapters 6–12)
- Mathematical modeling of groundwater flow and contaminant transport (Chapter 13)

The field of soil and groundwater remediation has evolved rapidly, thanks to the environmental regulations and the knowledge we have acquired through innovative research and technology development over the last four decades. This text will present the current state of our understanding toward the principles and applications of environmental remediation. Sustainability, an increasingly important component of green remediation, will also be integrated into the discussion of site assessment and remediation.

Bibliography

Citizen's Environmental Coalition (2001). 2001 Environmental Resource Guide, Houston, Texas.

Congressional Research Service (2016). Environmental Protection Agency (EPA): FY2016 Appropriations, 43pp.

European Environment Agency (2015). Progress in Management of Contaminated Sites, 33pp.

Margat, J. and van der Gun, J. (2013). *Groundwater Around the World: A Geographic Synopsis.* Boca Raton, FL: CRC Press.

Naidu, R. and Birke, V. (2015). *Permeable Reactive Barrier: Sustainable Reactive Barrier.* Boca Raton, FL: CRC Press.

NICOLE Brownfield Working Group (2011). Environmental Liability Transfer in Europe: Divestment of Contaminated Land for Brownfield Regeneration.
Oliver, L., Ferber, U., Grimski, D. et al. (2005). The scale and nature of European Brownfield. In: *CABERNET. Proceedings of CABERNET 2005: The International Conference on Managing Urban Land* (ed. L. Oliver, K. Millar, D. Grimski, et al.), 274–281. Nottingham: Land Quality Press.
Pichtel, J. (2000). *Fundamentals of Site Remediation*. Rockville, MD: Government Institutes.
Schirmeisen, A. (2005). Is there or if not could there be a European SUPERFUND of some kind or other? Master's Thesis. University of the West of England, Bristol.
Singh, A., Kuhad, R.C., and Ward, O.P. (2009). Biological remediation of soil: an overview of global market and available technologies. In: *Advances in Applied Bioremediation* (ed. A. Singh), 1–20. Berlin: Springer.
Tremblay, D.L., Tulis, D., Kostecki, P., and Ewald, K. (1995). Innovation Skyrockets at 50,000 LUST Sites, Soil and Groundwater Cleanup, 6–13.
US Congress (1984). Protecting the Nation's Groundwater from Contamination, OTA-O-233, October 1984.
USDoD (1996). Restoration Management Information System, November 1996.
USDoE (1995). Estimating the Cold War Mortgage: The Baseline Environmental Report, DOE/EM-2032, March 1995.
USEPA (1990). National Water Quality Inventory, 1988 Report to Congress, EPA 440-4-90-003.
USEPA (1993). Draft Regulatory Impact Analysis for the Final Rulemaking on Corrective Action for Solid Waste Management Units Proposed Methodology for Analysis, March 1993.
USEPA (1994a). Analysis of Facility Corrective Action Data.
USEPA (1994b). RCRA corrective action plan. *OSWER Directive* 9902: 3–2A.
USEPA (2000). The Quality of Our Nation's Waters. A Summary of the National Water Quality Inventory: 1998 Report to Congress. EPA 841-S-00-001.
USEPA (2002). Drinking Water from Household Wells: Washington, D.C., US Environmental Protection Agency, Office of Water, EPA 816-K-02-003, 19pp.
USEPA (2004). Cleaning up the Nation's Waste Sites: Markets and Technology Trends, 2004 Edition (EPA 542-R-04-015).
USEPA (2007). Innovative Treatment Technology: Annual Status Report (12e), EPA 542-R-96-012, September 2007.
USEPA (2009). National Water Quality Inventory, 2009 Report to Congress. EPA 841-R-08-001.
USGS (1998). Ground Water and Surface Water: A Single Resource, Denver, Colorado
USGS (2006). Microbial Quality of the Nation's Ground-Water Resource, 1993–2004.
USGS (2009a). Quality of Water from Domestic Wells in Principal Aquifers of the United States, 1991–2004.
USGS (2009b). Factors Affecting Water Quality in Selected Carbonate Aquifers in the United States, 1993–2005.
USGS (2010). Estimated Use of Water in the United States in 2010.
Xu, C., Yang, W., Zhu, L. et al. (2017). Remediation of polluted soil in China: policy and technology bottlenecks. *Environ. Sci. Technol.* 51 (24): 14027–14029.

Questions and Problems

1 Gather information for the city or state of your concern regarding the relative % of surface water vs. groundwater, and their respective uses as categorized by the USGS.

2 Define the difference between "domestic" and "public supply" as the categorized water uses.

3 Describe the potential natural and anthropogenic sources in groundwater for: (a) nitrate, (b) radon, (c) radioactivity.

4 Describe the potential natural and anthropogenic sources in groundwater for: (a) uranium, (b) atrazine, (c) *Escherichia coli*.

5 What region of the US aquifer typically has high concentrations of radon in groundwater?

6 According to the USGS survey, what are the most frequently detected pesticides in the groundwater of the United States?

7 Are septic tanks considered to be USTs? Why or why not?

8 Gather information and write a short report about the contamination and remediation status of: (a) Superfund sites; (b) Brownfield sites; and/or (c) Underground storage tanks in the city or state of your concern.

9 Locate a brownfield site in your proximity and survey its past activities and associated contaminants.

10 In the United States, why is the cost for the remediation of Superfund sites much less than that in contaminated sites in the DoE and DoD? What are the unique contaminants of concern in the DoE and DoD sites?

11 Describe two primary factors affecting the concentrations of (a) pesticides, (b) nitrate in groundwater.

12 Describe two primary factors affecting the concentrations of (a) methyl *tert*-butyl ether (MTBE), and (b) total coliform in groundwater.

13 In Figure 1.3b, withdrawn fresh surface water and fresh groundwater appear to level off since 1985, even though the population continues to grow. Discuss what could be the reasons behind this?

14 Explain the difference between federal and state Superfund sites. Discuss how they are evaluated in your state.

15 Explain the difference between Superfund sites and brownfield sites in regard to their listings, applicable regulations, funding sources, etc.

16 Discuss why the remediation of groundwater in a contaminated site should consider the water quality of a nearby river.

17 In an agricultural region with extensive fertilizer applications, nitrate was detected in several domestic wells and the peak concentrations of nitrate in wells showed up sooner for wells closer to a river compared to wells farther away. Explain why.

Chapter 2

Subsurface Contaminant Fate and Transport

LEARNING OBJECTIVES

1. Relate the common contaminants in soil and groundwater to their structural features
2. Relate the common contaminants in soil and groundwater to their fate and transport processes
3. Determine the types of chemical and biological reactions (hydrolysis, redox, and biodegradation) that will alter the chemical structure
4. Use of solubility product (K_{sp}) to estimate aqueous solubility of inorganic metals in precipitation–dissolution reactions
5. Understand the factors that affect the aqueous solubility of organic contaminants and relate solubility to octanol/water partition coefficient (K_{ow})
6. Use density to differentiate light nonaqueous phase liquids (LNAPLs) and dense nonaqueous phase liquids (DNAPLs) and relate their movement in soil and groundwater
7. Differentiate various terms related to biodegradation including biotransformation, mineralization, aerobic, and anaerobic biodegradation
8. Use ideal gas law to estimate vapor phase concentration for vapor in equilibrium with its pure state
9. Relate Henry's law constant (H) to two common groups of soil and groundwater contaminants, i.e. volatile organic compounds (VOCs) and semivolatile organic compounds (SVOCs)
10. Differentiate the major sorption mechanisms between inorganic and organic contaminants and relate them respectively to cation exchange capacity and adsorption coefficients (K_d)
11. Use physicochemical properties (Appendix C) to determine the major fate and transport processes of a given contaminant
12. Relate the hydrological processes to contaminant fate and transport including advection, dispersion, and diffusion and their defining equations including Darcy's law and Fick's first law

The selection of remediation technology to a large extent depends on the type of chemical contaminants, their behavior (fate and transport), and the characteristics of soil and groundwater (hydrogeology). This chapter will first introduce some chemical contaminants that are frequently detected in contaminated soil and groundwater in the United States and worldwide. Chemical fate and transport, which is essentially the topic of environmental chemistry, is the focus of this chapter. This chapter will then introduce different types of "alterations" as well as "movement" of chemicals in soil and groundwater. Abiotic and biotic processes such as hydrolysis, oxidation–reduction, and biodegradation often alter the forms of inorganic species or the structure of organic compounds. Such alterations may lead to desired breakdown into nontoxic by-products or unwanted formation of more toxic daughter compounds. The movement of contaminants, crucial for the better monitoring and more effective cleanup of contaminated soil and groundwater, can be generally divided into the **interphase** (between phases) and **intraphase** (within a phase) processes. They include chemical exchange between

Soil and Groundwater Remediation: Fundamentals, Practices, and Sustainability, First Edition. Chunlong Zhang.

Companion website: www.wiley.com/go/Zhang/Remediation_1e

phases (soil-air-water) and chemical transport within a homogeneous environmental medium (e.g. groundwater). Collectively, they include volatilization, precipitation–dissolution, solubilization, adsorption–desorption, ion exchange, advection, and diffusion/dispersions of various types. More details about biotic processes will be provided in Chapter 9. Certain abiotic processes such as acid–base, complexation, and ion exchange, while important for some contaminants, are omitted in this discussion. An overview of these important fate and transport processes in soil and groundwater is listed in Table 2.1.

Table 2.1 A summary of important fate and transport processes in soil and groundwater.

Fate and transport processes[a]	Inorganic contaminants (ionic metals)	Organic contaminants (hydrophobic/neutral)	Process type[b]	Will be discussed in Section
Hydrolysis	Yes	Yes	A/B	2.2.1
Oxidation/reduction	Yes	Yes	A/B	2.2.2, 10.1.1, 12.1.2
Biodegradation	Maybe[c]	Yes	B	2.2.3, 9.1.3, 10.2.1
Complexation	Yes	Maybe[d]	A	2.3.2.1
Volatilization	Maybe[e]	Yes	IR	2.3.1, 8.2, 10.2.1
Precipitation/dissolution	Yes	Maybe[f]	IR	2.3.2, 7.3.2, 8.2.3, 11.1.2
Adsorption/desorption	Yes	Yes	IR	2.3.3, 7.3.2, 8.2.2, 10.2.1
Ion exchange	Yes	Maybe[g]	IR	2.3.3, 7.2.2
Advection	Yes	Yes	IA	2.4.1, 3.3, 8.2.4, 13.2, 13.3
Diffusion/dispersion	Yes	Yes	IA	2.4.2, 7.3.2, 8.2.3

[a] Certain physicochemical and biological processes typically not important in soil and groundwater are excluded in this table. These processes include, but are not limited to, photolysis, absorption, gravimetric settling, filtration, runoff, erosion, biological uptake (bio-concentration), condensation, and meteorological transport.
[b] A = Abiotic process; B = Biotic process; IA = Intraphase chemical movement; IR = Interphase chemical transport.
[c] Certain metal can be biotransformed through changes in species, but are not biodegraded.
[d] Not important unless the ionic organic compounds are chelating agents.
[e] Metals are not volatile with the exception of few organic metallic compounds such as Hg and Pb.
[f] Hydrophobic organic compounds are present in free phase liquid, hence solubilization may be important.
[g] Only important for ionic organics.

2.1 Frequent Soil and Groundwater Contaminants

To date there are an estimated 7 million chemicals and an approximate 100 000 of these chemicals have already been released into the environment. The chemicals commonly considered as important environmental pollutants, however, are likely less than a few hundred. Among them, fewer are routinely measured in environmental laboratories, and a smaller portion of the chemicals are “priority” pollutants listed by various organizations and countries based on selected factors such as quantity, persistence, bioaccumulation, mobility, toxicity, and other adverse biological effects. In Chapter 1, we have introduced some frequently detected compounds that are of both natural and **anthropogenic** (related to human activities) origins. The discussions that follow focus on organic compounds of primarily anthropogenic sources with frequent detection in contaminated soil and groundwater.

2.1.1 Aliphatic and Aromatic Hydrocarbons

Hydrocarbons consist solely of carbon and hydrogen. They can be either **aliphatic** (carbon atoms form open chains) or **aromatic** (carbon atoms form benzene ring) hydrocarbons. The carbons of aliphatic hydrocarbons with more than one atom can be bonded by single bonds (alkane), double bonds (alkenes), or triple bonds (alkynes) (Figure 2.1). **Alkanes** (or paraffins) are saturated hydrocarbons whereas both alkenes (or olefins) and alkynes are **unsaturated hydrocarbons**. Alkanes can be branched or cyclic (cycloalkanes) that have one or more rings of carbon atoms in the chemical structure of their molecules. Cycloalkanes are also called naphthenes. (Note that naphthenes are distinct from naphthalene which is an aromatic hydrocarbon.) Cyclic or unsaturated hydrocarbons such as naphthenes, olefins, and aromatics increase the octane rating of the gasoline whereas the saturated *n*-paraffins have the opposite effect. Alkanes are major components of petroleum products such as gasoline and fuel oil and may be analytically characterized as petroleum hydrocarbons or **total petroleum hydrocarbons** (TPHs). They are found as a result of oil spills or leaks from underground storage tanks. Within the same series of hydrocarbons (homologous series), as size increases, boiling point and density increase, and solubility and Henry's constant (Section 2.3.1) decrease. For example, hexene has a higher boiling point and density than its homolog pentene, whereas the aqueous solubility and the Henry's constant of hexane are lower than that of pentene.

Aromatic hydrocarbons of remediation importance include **BTEX series** (benzene, toluene, ethylbenzene, and xylene) and polycyclic aromatic hydrocarbons (PAHs) (Figure 2.2). BTEX are one-benzene ring compounds that are common components of gasoline and other petroleum products. The most common source of BTEX compounds are underground storage tanks. BTEX are very volatile, relatively water soluble, and lighter than water. Hence, BTEX are LNAPLs, a term used to denote light nonaqueous phase liquid (LNAPLs, see Section 2.3.2.1). Consequently, when BTEX enter groundwater, they can remain in water as a dissolved phase or float along the groundwater surface as a free phase.

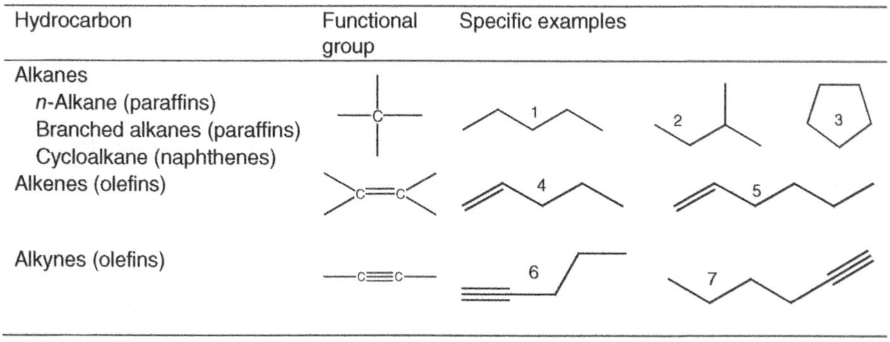

Hydrocarbon	Functional group	Specific examples		
Alkanes *n*-Alkane (paraffins) Branched alkanes (paraffins) Cycloalkane (naphthenes)	C	1	2	3
Alkenes (olefins)	C=C	4	5	
Alkynes (olefins)	C≡C	6	7	

Figure 2.1 Examples of alkanes, alkenes, alkynes, and cycloalkanes present in gasoline and fuel oils: pentane (1), isopentane (2), cyclopentane (3), pentene (4), hexene (5), pentyne (6), and hexyne (7).

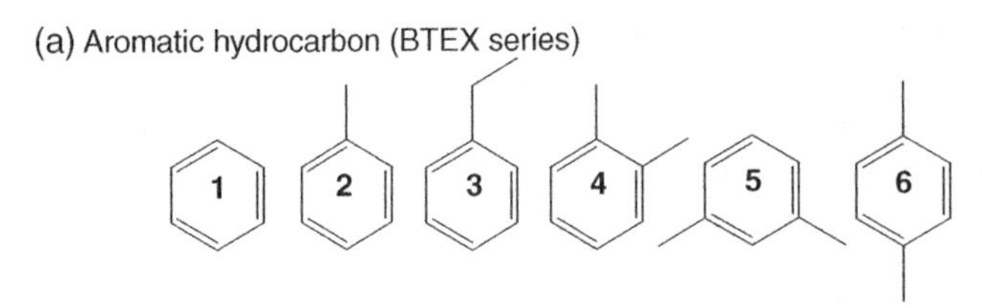

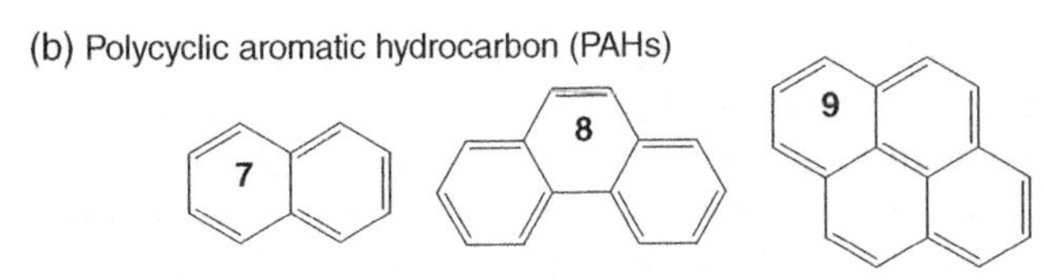

Figure 2.2 Common aromatic hydrocarbons in contaminated soil and groundwater. (a) Aromatic hydrocarbon (BTEX series): benzene (1), toluene (2), ethylbenzene (3), *o*-xylene (4), *m*-xylene (5), *p*-xylene (6), (b) Polycyclic aromatic hydrocarbon (PAHs): naphthalene (7), phenanthrene (8), and pyrene (9).

Polycyclic aromatic hydrocarbons (PAHs) have two or more fused aromatic rings in structure. Being the products of incomplete combustion, PAHs can be found in soils in industrial areas and in areas associated with the uses of petroleum, creosote, coal tar, and wood tar. PAHs generally have low volatility, water solubility, reactivity, and many are carcinogenic. Three PAHs with two to three benzene rings (namely naphthalene, phenanthrene, and pyrene) are shown in Figure 2.2. There are an estimated 30 000 PAHs, but only 16 common PAHs are listed under the USEPA **priority pollutants**.

Petroleum hydrocarbons from refinery industries are diverse in their compositions. Crude oil is a complex mixture of thousands of compounds, most of which are hydrocarbons. On average, crude oils contain 84.5% carbon, 13% hydrogen, 1.5% sulfur, 0.5% nitrogen, and 0.5% oxygen. A typical crude oil might consist of about 25% alkanes (paraffins), 50% cycloalkanes (naphthenes), 17% aromatics including PAHs, and 8% asphaltics which are molecules of very high molecular weight with more than 40 carbons (Fetter 1993).

The refined petroleum components are fractions of hydrocarbons with different carbon atoms and boiling points such as gasoline, kerosene, lubricating oils, paraffin, asphalt, and coke (a solid fuel made by heating coal in the absence of air). **Gasoline** is a complex mixture of as many as several hundred chemical components with carbon atoms ranging between C4 and C14. Gasoline has an average content of 10.8–29.6% *n*-alkane, 18.8–59.5% branched alkane, 3.2–13.7% cycloalkane, 5.5–13.5% olefins (unsaturated C_nH_{2n}), and 19.3–40.9% mono-aromatics (BTEX). Of the total gasoline, the BTEX contains 0.9–4.4% benzene, 4.0–6.5% toluene, 5.6–8.8% *m*-xylene, and 1.2–1.4% ethylbenzene (Pichtel 2007). Petroleum-derived **diesel** is composed of about 75% saturated hydrocarbons (primarily paraffins and cycloparaffins) and 25% aromatic hydrocarbons (including naphthalenes and alkylbenzenes).

2.1.2 Halogenated Aliphatic Hydrocarbons

Halogenated hydrocarbons refer to hydrocarbons that contain halogen atoms, i.e. fluorine (F), chlorine (Cl), bromine (Br), or iodine (I). Since most soil and groundwater contaminants contain chlorine, they are **chlorinated aliphatic hydrocarbons** (CAHs) or **chlorinated aromatic hydrocarbons.** Chlorinated aliphatic hydrocarbons frequently detected in soil and groundwater are chlorinated methane, ethane, or ethene. They, respectively, have one carbon with one or more Cl; two single-bonded carbons with Cl; and two double-bonded carbons with one to four Cl atoms (Figure 2.3). Trichloroethene (TCE) and tetrachloroethylenes (PCE, also called perchloroethene) are the two most commonly used solvents for degreasers in electroplating and commercial dry cleaning. Chlorinated aliphatic hydrocarbons are highly

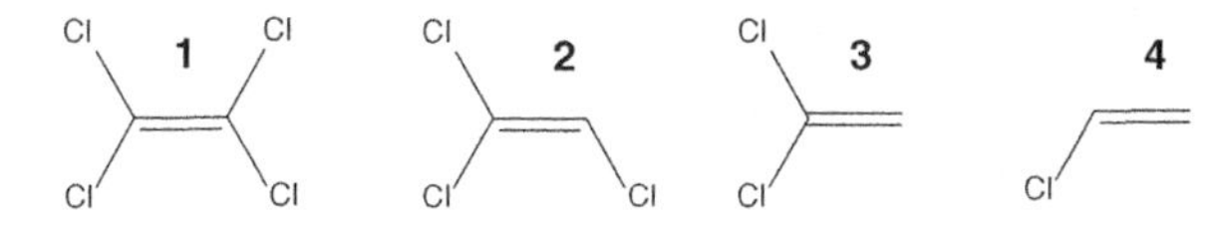

Figure 2.3 Important chlorinated aliphatic hydrocarbons in contaminated soil and groundwater: tetrachloroethylene (1), trichloroethene (2), dichloroethene (3), and vinyl chloride (4).

volatile, stable, and nonflammable. Compounds in this group have low water solubility and density greater than water (hence DNAPLs). They are subjective to **reductive halogenation** by certain anaerobic bacteria, hence they can be broken down into compounds that are not quite so harmful using anaerobic bioremediation techniques.

2.1.3 Halogenated Aromatic Hydrocarbons

Halogenated aromatic hydrocarbons, or particularly, chlorinated aromatic hydrocarbons include chlorinated benzenes, DDTs, polychlorinated biphenyls, and dioxins. **Chlorinated benzenes** (CBs), including monochloro, dichloro, trichloro, and hexachlorobenzene, are frequently found in leachate at many abandoned waste sites. Pentachlorophenol (PCP), also a CB, is frequently found in wood preserving sites. **Polychlorinated biphenyls** (PCBs) have two benzene rings joined by a single bond (Figure 2.4). PCBs are a series of 209 compounds (called **congeners**) with 1–10 Cl atoms in different positions. With the commercial name arochlor, PCBs were commonly used as dielectric fluid in transformers until the ban in the 1970s.

PCBs have a low vapor pressure, a very low water solubility, and a high octanol/water partition coefficient. Once PCBs are in groundwater they will not be very soluble and instead will bind tightly with the organic content of soil. Their low volatility means that not many PCBs

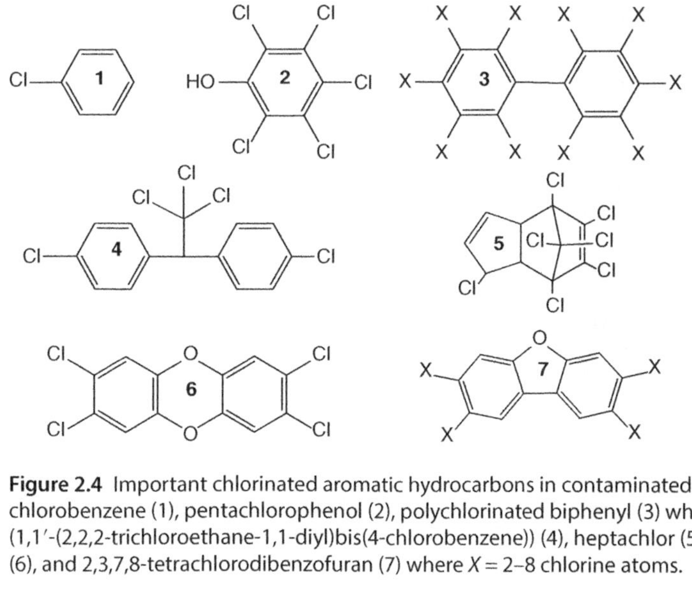

Figure 2.4 Important chlorinated aromatic hydrocarbons in contaminated soil and groundwater: chlorobenzene (1), pentachlorophenol (2), polychlorinated biphenyl (3) where $X = 1–10$ chlorine atoms, DDT (1,1′-(2,2,2-trichloroethane-1,1-diyl)bis(4-chlorobenzene)) (4), heptachlor (5), 2,3,7,8-tetrachlorodibenzodioxin (6), and 2,3,7,8-tetrachlorodibenzofuran (7) where $X = 2–8$ chlorine atoms.

will go into the gas phase and fill the pores, so vapor extraction remediation techniques will not work. In addition, the strong bonding between Cl and C atoms makes PCB a very stable compound resistant to both chemical and microbial attacks.

Dioxin or dioxin-like chlorinated compounds represent the most toxic contaminant in groundwater. Dioxins are not intentionally manufactured, but instead are the by-products of combustion of chlorine-containing compounds. Dioxins are very toxic; like PCBs, dioxin and dioxin-like chemicals are also very resistant to biodegradation.

2.1.4 Nitrogen-containing Organic Compounds

Nitrogen-containing organic compounds of environmental significance have a variety of structurally distinct features and functional groups. Figure 2.5a lists several important nitrogen-containing functional groups, including amine, hydroxyl amine, nitro, nitroso, nitrosamine, nitrile, azo, and amide.

Among the example compounds shown in Figure 2.5b, aminobenzene (also named as **aniline**) consists of a phenyl group attached to an amino group. It is the precursor of many industrial

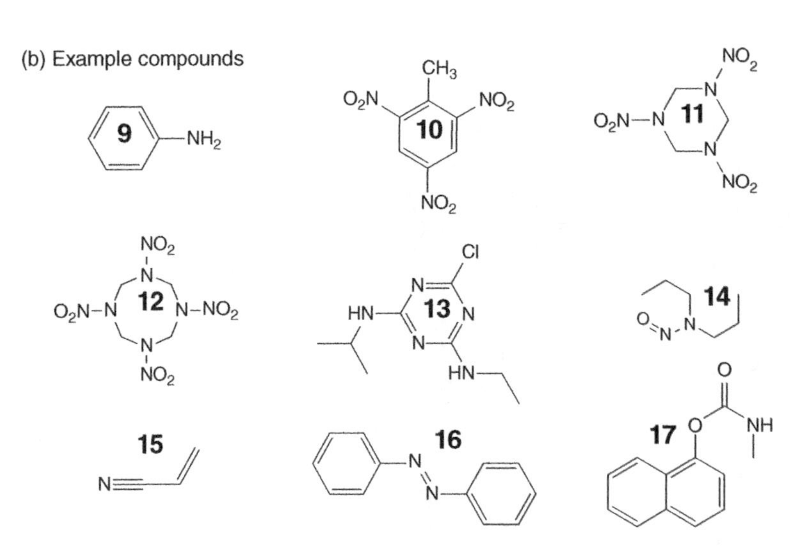

Figure 2.5 (a) Functionality of nitrogen-containing compounds in contaminated soil and groundwater: amine (1) (it is primary amine if $R_1 = R_2 = H$, secondary amine if $R_2 = H$ and $R_3 \neq H$, and tertiary amine if $R_2 \neq H$ and $R_3 \neq H$), hydroxylamine (2), nitro (3), nitroso (4), nitrosamine (5), nitrile (6), azo (7), and amide (8). (b) Examples of nitrogen-containing compounds: aminobenzene (9), 2,4,6-tronitrotoluene (TNT) (10), RDX (11), HMX (12), atrazine (13), di-*n*-propylnitrosamine (14), acrylonitrile (15), azobenzene (16), and 1-naphthyl-methyl carbamate (17).

chemicals such as dyes, pharmaceuticals, pesticides, and antioxidants. Three nitro (NO_2) compounds are given in Figure 2.5b, including 2,4,6-trinitrotoluene (TNT), 1,3,5-trinitroperhydro-1,3,5-triazine (RDX), and octahydro-1,3,5,7-tetranitro-1,3,5,7-tetrazocine(HMX). TNT is aromatic whereas RDX and HMX are nonaromatic heterocyclic compounds with alternating C and N atoms in the ring structure. These nitro compounds are ideal explosive chemicals since NO_2 (an electron acceptor) can act as a strong oxidant and quickly release energy when it is reduced from its +3 oxidation state. TNT, RDX, and HMX are well-known munitions and explosive compounds. Microbial reduction of these nitro compounds can result in more toxic nitroso (R—N=O) and hydroxylamine (R—NHOH) metabolic intermediates.

Nitrosamines are by-products of industrial operations and food and alcoholic beverage processing; many are carcinogenic. **Nitriles** with the cyano group (—C≡N) are the potential products from the combustion of degraded biomass. **Amides** consist of a carbonyl group (—C=O) bounded to a nitrogen atom. The nitrogen atom is either bonded to two hydrogen atoms (—$CONH_2$) or bonded to one or two aliphatic or aromatic groups (—$CONR_1R_2$). They are frequently detected in animal wastes and wastewater treatment facilities. Numerous herbicides contain amide groups. For example, the widely used herbicides and insecticide **carbamates** (HO—CO—NH_2) exhibit amide functionality. Azo group (—N=N—) is a strong light-absorbing chromophore (a color-bearing functional group of atoms) but it does not occur naturally. Dyes with azo structure represent the largest group of synthetic dyes, accounting for over 50% of all synthetic dyes.

A noteworthy nitrogen-containing compound is the widely detected **atrazine**, or 2-chloro-4-(ethylamino)-6-(isopropyl amino)-s-triazine in groundwater. This organic compound consists of a six-sided ring with alternating carbon and nitrogen atoms. It is a widely used herbicide in the United States as well as in other countries. Its use is controversial due to widespread contamination in drinking water (see Chapter 1) and its associations with birth defects and menstrual problems when consumed by humans at concentrations below government standards. Although it has been banned in the European Union, it is still one of the most widely used herbicides in the world.

2.1.5 Oxygenated Organic Compounds

These include a structurally diverse group of compounds such as alcohol/phenol (R—OH), ether (R_1—O—R_2), aldehyde (R—CHO), ketone (R_1—CO—R_2), and carboxylic acids (R—COOH) (Figure 2.6). Since oxygen atoms could allow participation in H-bonding, oxygenated organic compounds associated with these functional groups are usually very water soluble.

Among the chemicals listed in Figure 2.6, cresols are methyl phenols which have three isomers (*o*-, *m*-, and *p*-cresol). They are used in coal-tar refining and wood preservation. In our previous discussion (Figure 2.4), PCP was listed as a chlorinated aromatic hydrocarbon, but it also belongs to chlorinated phenols. In addition to phenols with methyl and chlorinated functional groups, there are also nitrophenols and many of these nitrophenols are precursors of pesticides and explosives. 4-Nonylphenol is the microbial degradation by-product of nonionic surfactants, which are used in large quantities. Since nonylphenols are frequently detected in the environment, their suspected endocrine disruption potential has recently raised concern.

Certain esters are ubiquitous in the environment. For example, phthalate esters are present in many cosmetic products such as nail polish, antiperspirant, lotions, hair spray, and shampoo. Phthalates are used as plasticizers to improve the flexibility of various plastics, and many esters from both natural and synthetic sources are used in flavorings, perfumes, solvents, and paints. Many pesticides are carboxylic acid esters. An example is benzadox shown in Figure 2.6,

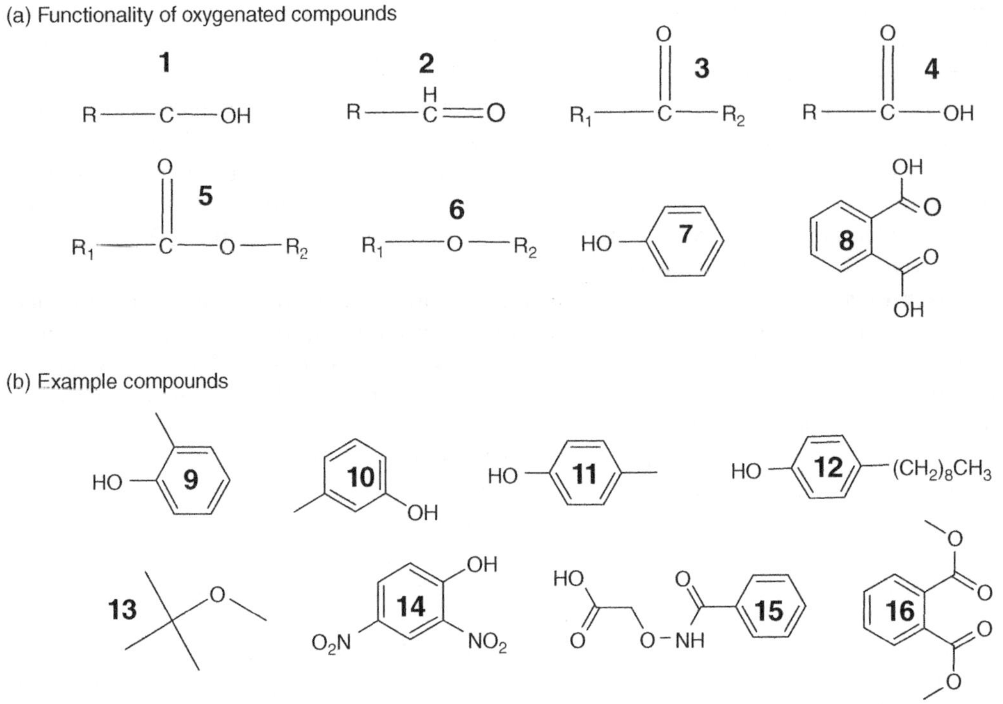

(a) Functionality of oxygenated compounds

(b) Example compounds

Figure 2.6 (a) Functionality of oxygen-containing compounds: alcohol (1), aldehyde (2), ketone (3), acid (4), ester (5), ether (6), phenol (7), and phthalate (9) (b) Examples of oxygenated compounds in contaminated soil and groundwater: *o*-cresol (9), *m*-cresol (10), *p*-cresol (11), 4-nonylphenol (12), MTBE (13), 2,4-dinitrophenol (14), 2-[(benzoylamino)oxy]acetic acid (benzadox, 15), and dimethyl phthalate (16).

which is soluble in water and is used as an herbicide to control kochia (a drought-resistant forage crop for sheep and cattle) in sugar beets.

Methyl tertiary butyl ether (MTBE) is a noteworthy oxygen-containing groundwater contaminant. After leaded gasoline was banned by the USEPA in 1973, MTBE became a gasoline additive in some US cities; some gasoline contains up to 15% of MTBE. As an ether, MTBE is fairly water soluble, consequently it is not very volatile when it is dissolved in water. Since the ether bond is relatively strong, its resistance to biodegradation adds challenges in remediation.

2.1.6 Sulfur- and Phosphorus-containing Organic Compounds

Many currently used pesticides contain sulfur and/or phosphorus. Esters and thioesters of phosphonic, phosphoric, and thiophosphoric acids are used in plasticizers, flame retardants, and pesticides including insecticides and acaricides. Figure 2.7 shows three common pesticides in this group, including adicarb (a systemic insecticide and nematicide), malathion (insecticide), and parathion (insecticide and acaricide). Compared to the legacy DDT pesticide (see Figure 2.4), these S- and P-containing pesticides are more acutely toxic, but much less persistent. Because of the weaker C=S and P=S bonds than the C—Cl bond, many S- and P-containing pesticides are subject to significant biotic hydrolysis. Therefore, these pesticides are not persistent in the environment.

Two additional S- and P-containing compounds with other uses are also shown in Figure 2.7. Triphenyl phosphate is a plasticizer as well as a fire retardant, and linear alkylbenzene sulphonates (LAS) is the largest-volume synthetic surfactant (detergent). LAS have hydrophilic sulfonic acids and hydrophobic aromatics. This amphiphilic character also makes aromatic

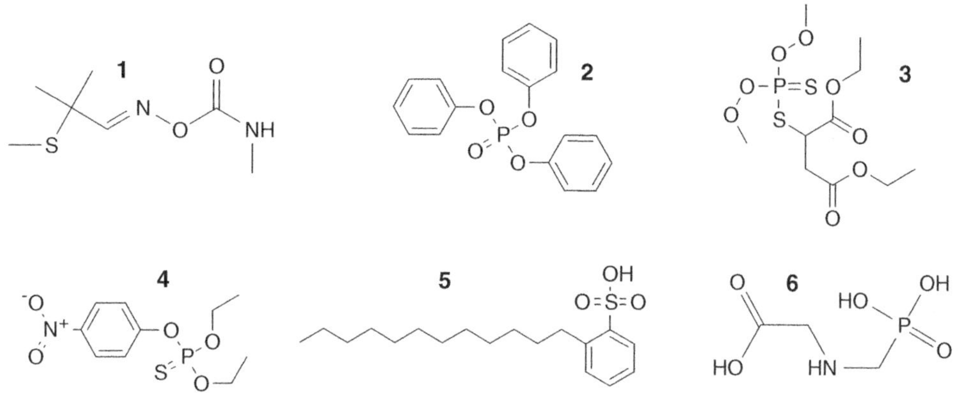

Figure 2.7 Examples of S- and P-containing pesticides. Adicarb (2-Methyl-2-(methylthio)propanal *O*-(*N*-methylcarbamoyl)oxime)) (1), triphenylphosphate (2), malathion (diethyl 2-[[dimethoxyphosphorothioyl] sulfanyl]butanedioate) (3), parathion (*o*, *o*-Diethyl *o*-[4-nitrophenyl]phosphorothioate) (4), dodecylbenzene sulfonic acid (5), and glyphosate (*N*-phosphonomethylglycine) (6).

sulfonates popular among surfactants, anionic azo dyes, and fluorescent and whitening agents. This group of anionic surfactants has relatively low cost, good performance, and more importantly, it is biodegradable because of its straight chain structure. However, because of their large quantity, some of these sulfonic acids (particularly those with branched structure) can contribute to groundwater contamination when they percolate to groundwater.

2.1.7 Inorganic Nonmetals, Metals, and Radionuclides

Inorganic nonmetals include selenium (Se), arsenic (As), cyanide (CN), nitrate (NO_3^-), nitrite (NO_2^-), and asbestos (an airborne fiber that can cause chronic lung disease by inhalation). The presence of these nonmetals in soil and groundwater can be from natural sources (such as As) or human activities (e.g. NO_3^-). Of the 118 elements in the periodic table, around 90 are metals. **Metals** of environmental concerns are typically the **heavy metals**, which include, but are not limited to, copper (Cu), zinc (Zn), lead (Pb), cadmium (Cd), nickel (Ni), mercury (Hg), and chromium (Cr). Se and As are **metalloids** although they are often included as heavy metals because of their toxicity. Among these heavy metals frequently detected in soil and groundwater, Hg, Pb, and As may become volatile such as AsH_3 and $As(CH_3)_2$ for arsenic; Hg^0, $Hg(CH_3)_2$ for mercury; and $Pb(C_2H_5)_4$ for lead. Their fate, transport, and toxicity are greatly influenced by pH and Eh conditions (see Section 2.2.2). Metals and metalloids may be present in several oxidation states (**species**); they vary in toxicity, solubility, and reactivity. As such, it is generally the species of metals that are important in remediation, rather than the total concentration of all the species. Regardless of frequent change in their oxidation states, metals cannot be destroyed, meaning that incineration is not a solution for remediation. This also implies that bioremediation and air stripping (to be introduced in future chapters) are not applicable for the total cleanup of metal and metalloid contamination.

The remediation of radionuclides in contaminated soil and groundwater is of particular challenge. **Radionuclides** include radioactive elements such as cesium (Cs), cobalt (Co), krypton (Kr), plutonium (Pu), radium (Ra), ruthenium (Ru), strontium (Sr), thorium (Th), and uranium (U). These are common pollutants in uranium mining, nuclear facilities, and many of the DoE sites. In a survey of the radioactively contaminated Superfund sites in the United States, Ra represents the most prevalent radionuclide, followed by U, Th, and radon (Rn) (USEPA 2007). Due to their long half-lives (e.g. in the order of several thousands to hundreds of millions of years), the safe storage and disposal are technically more demanding compared to other inorganic and organic contaminants.

What we have discussed above are the contaminants of the current concern in most of the contaminated sites. Many of these contaminants are called legacy contaminants such as DDT, PCBs, and PAHs. **Legacy contaminants** are the result of historical contributions; they are often persistent in the environment. What we have not discussed is the group of contaminants termed "emerging contaminants." A partial list of emerging contaminants is given in Box 2.1.

Box 2.1 Emerging soil and groundwater contaminants

The USEPA defines an **emerging contaminant** (EC) as a chemical or material characterized by a perceived, potential, or real threat to human health or the environment by a lack of published health standards. A contaminant also may be "emerging" because of the discovery of a new source or a new pathway to humans.

The USEPA published the emerging contaminant technical fact sheets (www.epa.gov/fedfac), which provide brief summaries of contaminants that present unique issues and challenges to the environmental community. Each fact sheet provides a brief summary of the contaminant's physical and chemical properties, environmental and health impacts, existing federal and state guidelines, and detection and treatment methods. Some of these listed emerging contaminants and other contaminants of concerns are as follows:

- Nanomaterials (NMs): NMs are small-sized substances that have structural components smaller than 1 micrometer (μm). They are increasingly used in a wide range of scientific, environmental, industrial, and medicinal applications. NMs are metal oxides, metals, and carbon nanotubes, including titanium silver, zinc, and aluminum.
- Perfluorooctanesulfonate (PFOS) and perfluorooctanoic acid (PFOA): A group of fully fluorinated, human-made compounds that are used as surface-active agents in a variety of products, such as fire-fighting foams, coating additives, and cleaning products.
- Polybrominateddiphenyl ethers (PBDE) and polybrominated biphenyls (PBB): PBDEs and PBBs are human-made brominated hydrocarbons that serve as flame retardants for electrical equipment, electronic devices, furniture, textiles, and other household products. PBBs have been banned in the United States since 1973, but PBDEs have been in widespread use in the United States since the 1970s.
- Perchlorate (ClO_4^-): It is a naturally occurring and human-made anion that has been detected at high concentrations at sites historically involved in the manufacture, maintenance, use, and disposal of ammunition and rocket fuel.
- Tungsten (W): It is a naturally occurring element that is a common contaminant at industrial sites that use the metal and Department of Defense (DoD) sites involved in the manufacture, storage, and use of tungsten-based ammunition.
- Pharmaceuticals, personal care products (PPCPs), and hormones: These chemicals have become emerging contaminants of concerns because of the widespread presence (e.g. 80% streams in the United States) and potential health effects on humans and animals, although the concentrations are typically low (ppb to ppt). PPCPs escape from municipal wastewater treatment plants, whereas antibiotic and hormonal compounds may originate from feedlot or dairy farm waste stream treatment.

2.2 Abiotic and Biotic Chemical Fate Processes

The three processes described below are hydrolysis, oxidation–reduction (redox), and biodegradation. Hydrolysis and redox can be both chemical (abiotic) and biotic, whereas biodegradation processes have the direct involvement of bacteria or plants.

2.2.1 Hydrolysis

Hydrolysis is a reaction in which a water molecule (or hydroxide) substitutes for an atom or group of atoms in another molecule, i.e. splitting of a bond and the addition of the hydrogen cation (H^+) or the hydroxide anion (OH^-) from the water. Hydrolysis can take place for both inorganic compounds (metals) and organic compounds.

Hydrolysis for Inorganics: In an aqueous solution, the positively charged metal (M) ions behave as Lewis acids, that is, the metal ions draw electrons density from the O—H bond in the water. When the —OH bond breaks, an aqueous proton is released producing an acidic solution.

$$\left[\mathrm{M}\left[\mathrm{H_2O}\right]_n\right]^{z+} + \mathrm{H}-OH \rightleftharpoons \left[\mathrm{M}\left[\mathrm{H_2O}\right]_{n-1}\left[\mathrm{OH}\right]\right]^{(z-1)+} + \mathrm{H_3O^+}$$

Thus, many inorganic contaminants exist in more than one ionic and molecular forms **(species)** as a result of hydrolysis in groundwater. For example, the concentration of dissolved Pb in water is represented by the concentrations of Pb^{2+} and various hydrolyzed species:

$$Pb(\mathrm{total}) = Pb^{2+} + Pb(OH)^+ + Pb(OH)_2 + Pb(OH)_3^- + \cdots$$

In most cases, it is the concentration of certain species rather than the total metal concentration that determines the fate and transport processes as well as their biological effects.

Hydrolysis for Organics: Not all environmental organic compounds are susceptible to hydrolysis. In fact, many compounds frequently detected in contaminated soil and groundwater are generally resistant to hydrolysis, including alkanes, BTEX, PAHs, PCBs, aromatic nitro compounds, ethers (e.g. MTBE), phenols, ketones, and carboxylic acids (Table 2.2). On the other hand, compounds with alkyl halides, carboxylic acid esters, carbamates (pesticides), amides, amines, and phosphoric acid esters are among the frequent groundwater contaminants potentially susceptible to hydrolysis. Hydrolysis generally produces more polar products and often (but not always) less toxic products.

Several examples of hydrolysis reactions are given in the following list according to the major hydrolysis mechanisms (Schwarzenbach et al. 2017). Note that hydrolysis can be either abiotic or biotic with the biotic hydrolysis being biologically mediated through hydrolase enzyme in bacterial cells.

Table 2.2 Organic functional groups in relation to hydrolysis (von Lyman et al. 1990).

a) Chemicals generally resistant to hydrolysis	
Alkanes, alkenes, alkynes	Aromatic amines
Benzenes/biphenyls	Alcohols/phenols
Polycyclic aromatic hydrocarbons (PAHs)	Glycols
Heterocyclic polycyclic aromatic hydrocarbons	Ethers
Halogenated aromatics/PCBs	Aldehydes
Dieldrin/aldrin and related halogenated hydrocarbon pesticides	Ketones
Aromatic nitro compounds	Carboxylic acids/sulfonic acids
b) Chemicals potentially susceptible to hydrolysis	
Alkyl halides	Nitriles
Amides/amides	Phosphonic acid esters
Carbamates	Phosphoric acid esters
Carboxylic acid esters	Sulfonic acid esters
Epoxides	Sulfuric acid esters

a) **Nucleophilic substitution** at saturated carbon atoms: An environmentally relevant example of nucleophilic substitution is the hydrolysis of methyl bromide, CH_3Br, under basic conditions. The attacking nucleophile (nucleus-liking and, hence, an electron-rich species) is the OH^-, and the leaving group is Br^-. Nucleophilic substitution commonly occurs at a saturated aliphatic carbon or at (less often) a saturated aromatic or other unsaturated carbon center.

$$CH_3Br + H_2O \rightarrow CH_3OH(\text{methanol}) + H^+ + Br^-$$

b) **β-Elimination**: Cl at the β-carbon position is eliminated. An example is the conversion of 1,1,2,2-tetrachloroethane ($Cl_2HC{-}CHCl_2$) into trichloroethene ($Cl_2C{=}CHCl$).

$$Cl_2HC-CHCl_2 + OH^- \rightarrow Cl_2C=CHCl + Cl^- + H_2O$$

c) **Ester hydrolysis**: Hydrolysis of an ester results in a carboxylic acid (or its salt) and alcohol by the action of water, dilute acid, or dilute alkali. Under alkaline condition, the reactions are one way rather than reversible.

$$(C_2H_5O)_2-\underset{\underset{S}{\parallel}}{P}-O-C_6H_4-NO_2 + HO^- \longrightarrow (C_2H_5O)_2-\underset{\underset{S}{\parallel}}{P}-O^- + HO-C_6H_4-NO_2$$

Parathion | *o,o*-Diethyl thiophosphoric acid | 4-Nitrophenol

d) **Carbamate hydrolysis**: Many carbamate pesticides can be hydrolyzed, in particular, base catalysis dominates over the environmental pHs.

$$H_3CNH-\underset{\underset{O}{\parallel}}{C}-O-(\text{2,2-dimethyl-2,3-dihydrobenzofuran-7-yl}) + H_2O \longrightarrow H_2N-CH_3 + CO_2 + HO-(\text{2,2-dimethyl-2,3-dihydrobenzofuran-7-yl})$$

Carbofuran | Methylamine | 2,3-Dihydro-3,3-dimethyl-7-benzofuranol

Hydrolysis is generally assumed to be the first order; however, the kinetic rates vary depending on the types of chemicals and the environmental conditions. As such, it is hard to assign an approximate range of half-life for the compound groups listed in Table 2.2. For example, carbamates have a half-life range from seconds to 10^5 years depending on the types of compounds and pH conditions. If the half-life is in several decades, natural attenuation of contaminants through hydrolysis would be negligible.

2.2.2 Oxidation and Reduction

Oxidation is a type of chemical reaction that occurs when a chemical loses (donates) electron(s) (i.e. the oxidation number of an atom becomes larger). **Reduction** occurs when a chemical gains electrons (i.e. the oxidation number becomes smaller). Thus, elemental iron (Fe) becomes oxidized because its oxidation number is increased from zero to +2. As shown below, molecular oxygen (O_2) is the oxidizing agent because oxygen gains electron from Fe, and the oxidation number of oxygen (O) is reduced from zero to −2.

$$2Fe(s) \rightarrow 2Fe^{2+}(aq) + 4e^-$$

$$O_2 + 4H^+ + 4e^- \rightarrow 2H_2O$$

$$2Fe(s) + O_2 + 4H^+ \rightarrow 2Fe^{2+}(aq) + 2H_2O$$

It is important to note that oxidation and reduction are complimentary, because electrons must be balanced, i.e. the number of moles of electrons lost from a chemical (reducing agent, e.g. Fe) will be equal to the moles of electrons gained by another chemical (oxidizing agent, e.g. O_2). This electron-transfer reaction is collectively called **redox**, implying that both a reducing agent (reductant) and an oxidizing agent (oxidant) must be present for a redox reaction to occur.

Redox for Inorganics: For inorganic compounds, redox reactions in the abiotic and biotic systems can be easily visualized by examining the oxidation numbers of the atoms involving changes in electron transfer. Carbon, nitrogen, iron, and sulfur are the common atoms with variable oxidation numbers. For example, nitrogen can either donate to or gain various numbers of electrons from another atom, thereby having various oxidation numbers from −3 to +5: −3 (NH_4^+), 0 (N_2), +2 (NO), +4 (NO_2), +3 (NO_2^-), and +5 (NO_3^-). Redox reactions of atoms in such inorganic species are a strong function of pH and Eh, which are the main factors in controlling the species and therefore the toxicity of inorganic chemicals. Here, Eh is the **reduction potential** with a unit in volts (V), or millivolts (mV), which is a measure of the tendency of a chemical species to acquire electrons during the reduction. Each species has its own intrinsic reduction potential; the more positive the potential, the greater the species' affinity for electrons and tendency to be reduced. In remediation and risk assessment, the toxicities of inorganic metals and metalloids are more species-dependent than their corresponding total concentrations. For example, hexavalent chromium, Cr(VI), is more toxic than Cr(III), whereas As(III) is more toxic than As(V). Example 2.1 illustrates the speciation of Cr under various pH and Eh conditions and its implication to remediation.

Redox for Organics: For organic compounds, the changes in oxidation number in redox reactions are less apparent than for the redox of inorganic species. In the example given below, the oxidation number of one carbon atom in DDT changes from +3 to +1, and the oxidation numbers of all other atoms (H, Cl, and the remaining carbon atoms) remain the same. Hence, the conversion of DDT to DDD requires a total of two electrons to be transferred from an electron donor to DDT. As we will learn more in the future, this important redox reaction is termed a **reductive dechlorination** to denote the occurrence of both the reduction reaction and the removal of chlorine atom.

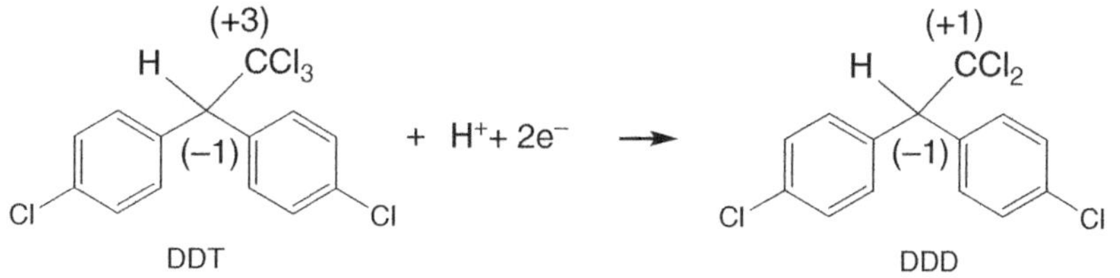

DDT [(+3) CCl_3, H, (−1), Cl, Cl] $+\ H^+ + 2e^- \longrightarrow$ DDD [(+1) CCl_2, H, (−1), Cl, Cl]

Many halogen-, nitrogen-, and sulfur-containing organic compounds (e.g. halogenated hydrocarbons and nitroaromatic compounds) are susceptible to redox reaction. Important examples of redox reactions in environmental remediation include combustion, ozonation, aerobic biodegradation, and reductive dechlorination. Like hydrolysis, many redox reactions of importance in environmental remediation are biologically mediated, meaning that they are catalyzed by microorganisms. In groundwater systems, biologically mediated redox reactions are generally far more important than abiotic redox.

Example 2.1 Use of Eh-pH diagram in determining predominant species and implication to the remediation of chromium in groundwater

The use of Cr(VI) in wood preservation with chromated copper arsenate (CCA) solution, metal plating facilities, paint manufacturing, leather tanning, and other industrial applications has the potential to introduce high concentrations of oxidized hexavalent chromium [Cr(VI)] to the environment. Chromium has become the second most common metal found at sites with signed Records of Decision (USEPA 2000). Chromium is present in Cr(VI) state and persists in anionic form as chromate, which is acutely toxic and very mobile in groundwater. On the other hand, its reduced form of Cr(III) is the most common for naturally occurring chromium, which is largely immobile in the environment. Use the Eh-pH diagram (Figure 2.8) to develop a general strategy for the remediation of Cr(VI) contaminated groundwater.

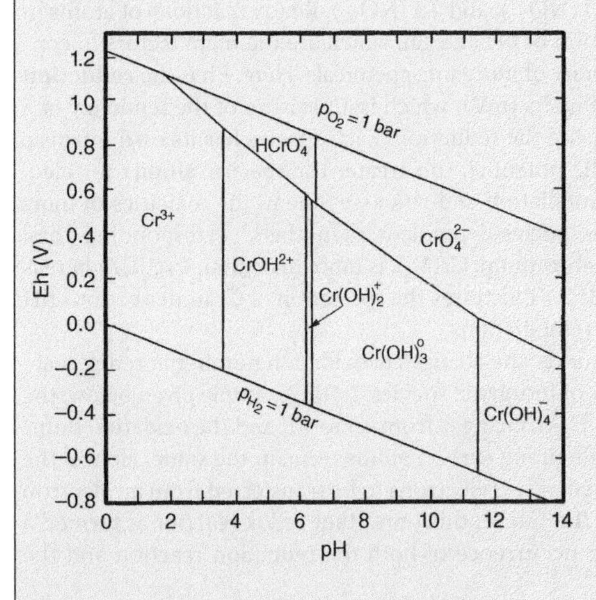

Figure 2.8 Eh-pH diagram of chromium. *Source:* Reproduced with permission from Palmer and Wittbrodt (1991).

Solution

Eh-pH diagrams indicate predominant species at equilibrium for specified Eh and pH ranges. The above Eh-pH diagram shows that under strong oxidizing conditions (i.e. high Eh value), chromium is present in its anionic Cr(VI) state as chromate ($HCrO_4^-$ and CrO_4^{2-}). Under reducing conditions (i.e. low Eh value), however, the trivalent Cr(III) is the most thermodynamically stable oxidation state. Cr(III) predominates as ionic (i.e. Cr^{3+}) at pH values less than 3.0. At pH values above 3.5, hydrolysis of Cr(III) yields trivalent chromium hydroxyl species $CrOH^{+2}$, $Cr(OH)_2^+$, $Cr(OH)_3^o$, and $Cr(OH)_4^-$. $Cr(OH)_3$ is the only solid species in the form of an amorphous precipitate. As such, the existence of the $Cr(OH)_3$ species as the primary precipitated product from Cr(VI) reduction is paramount to the viability of *in situ* treatment.

Since the total chromium concentration will not be changed regardless of changes in species distribution, remediation will be attempted to change the chromium to nontoxic and immobile species. The basic approach is to create a reducing environment such that the toxic and mobile Cr(VI) can be reduced into nontoxic Cr(III) in its stable form of [$Cr(OH)_3$]. Sulfur compounds such as sulfide (S^{2-}) and sulfite (SO_3^{2-}) in aquifer can reduce Cr(VI). For sulfides to reduce Cr(VI), Fe(II) must be present to act as a catalyst. Thus, in aquifer systems where sulfite is in excess, the reduction of Cr(VI) may occur as follows (Palmer and Wittbrodt 1991):

$$6H^+ + 2HCrO_4^- + 4HSO_3^-(\text{excess}) \rightarrow 2Cr^{3+} + 2SO_4^{2-} + S_2O_6^{2-} + 6H_2O$$

The metabisulfite ($S_2O_6^{2-}$) formed by the above reaction can then reduce Fe(III) to Fe(II), if it is present. In the presence of excess Cr(VI), the reduction to Cr(III) by sulfite follows the reaction:

$$5H^+ + 2HCrO_4^-(\text{excess}) + 3HSO_3^- \rightarrow 2Cr^{3+} + 3SO_4^{2-} + 5H_2O$$

Therefore, the addition of sodium metabisulfite ($Na_2S_2O_6$) as a reducing agent should reduce Cr(VI) to Cr(III) *in situ*, provided there are sufficient iron and manganese oxide adsorption sites within the aquifer treatment zone to which Cr(III) can be further precipitated. This process can be regarded as the geochemical fixation of chromium (USEPA 2000). Similarly, sodium dithionite ($Na_2S_2O_4$) can be added to reduce Fe(III) to Fe(II), which in turn reduces Cr(VI) to Cr(III). Fe is naturally occurring in groundwater, but it can also be added in the form of elemental iron (see Chapter 12).

2.2.3 Biodegradation

Biodegradation is the biological degradation of organic compounds through enzymatic processes, mostly by bacteria and less frequently used with fungi in various bioremediation systems. Two major processes are the aerobic and anaerobic biodegradation in the presence and absence of molecular oxygen, respectively. Biodegradation processes can partially break down a parent compound (a process termed **biotransformation**), or completely mineralized it into simple products such as SO_4^{2-}, NO_3^-, CH_4, CO_2, and H_2O. **Mineralization** is an ideal process in which organic contaminants are completely detoxified to form mainly methane under anaerobic condition, or CO_2 under aerobic condition. Transformation can also be used as a strategy in bioremediation. However, transformation of certain organic compounds sometimes can lead to the formation of more toxic products than the parent compound, which is a process that needs to be avoided in the bioremediation process. An example of biotransformation and mineralization is given in Box 2.2. Details of biodegradation will be discussed in Chapter 9. While we discuss bioremediation, it is important to note its use for the alteration or destruction of the structure of organic compounds. For inorganic metals and metalloids, biological processes can only change the species (oxidation numbers), because metals cannot be destroyed.

2.3 Interphase Chemical Transport

For our convenience in the discussion below, we use **interphase** chemical transport to refer to physicochemical processes between phases (air, water, solid), and **intraphase** to refer to the physicochemical processes within a single phase. For example, volatilization is an interphase process from liquid to gas, whereas advection is an intraphase process within the liquid (i.e. groundwater).

Box 2.2 Biodegradation of 2,4-DNT: Biotransformation versus mineralization

Bacteria have various strategies (pathways) to metabolize contaminants in the environment. Take 2,4-dinitrotoluene (2,4-DNT) as an example, this nitroaromatic compound frequently detected at sites of former ammunition plants can be either biotransformed or mineralized depending on the types of bacteria and the environmental conditions.

Under anaerobic condition, certain bacteria can reductively transform 2,4-DNT into 2,4-diaminotoluene via the formation of an intermediate product, 2-hydroxylamino-4-nitrotoluene, which is more toxic than 2,4-DNT (Hughes et al. 1999). This implies that anaerobically, this bacterium cannot be used for the cleanup of 2,4-DNT contaminated sites.

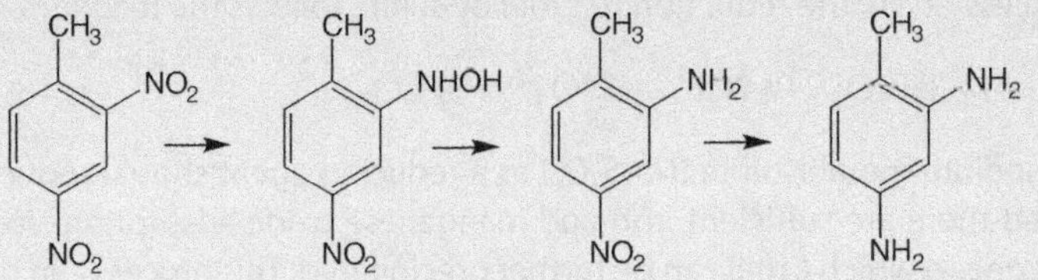

Interestingly, certain aerobic bacteria can use 2,4-DNT as the sole source of nitrogen and carbon and mineralize 2,4-DNT into NO_2^-, CO_2, and H_2O (Nishino et al. 2000). Mineralization is thus preferred because the parent compound has been fully degraded into CO_2, H_2O, and other innocuous compounds. The pilot-scale removal of 2,4-DNT and 2,6-DNT from contaminated soils has been demonstrated (Zhang et al. 2000).

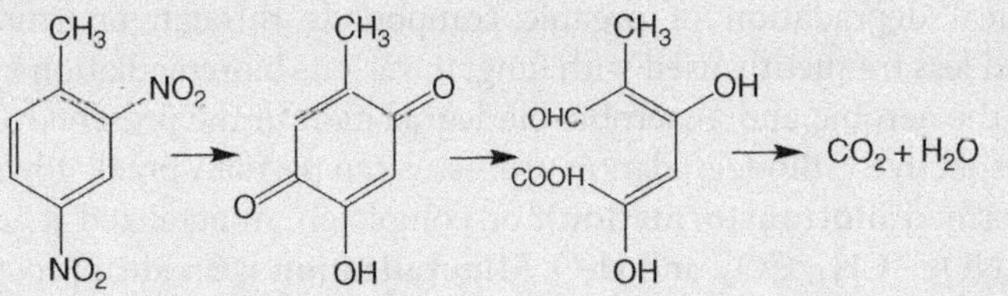

2.3.1 Volatilization

Volatilization (also known as vaporization) is conversion of a chemical from a liquid or solid state to a gaseous or vapor state by the application of heat, by reducing pressure, or by a combination of these processes. This term is often interchangeably misused with evaporation, which occurs at the surface of a liquid that requires energy to release the molecules from the liquid into the gas. The volatility of a pure compound (not in its aqueous solution) is measured by its vapor pressure (P) at the equilibrium condition. Thus, a compound with a higher vapor pressure (P) at a given temperature tends to be more volatile. For a pure compound, the equilibrium vapor phase concentration can be estimated using the **ideal gas law**:

$$PV = nRT \qquad \text{Eq. (2.1)}$$

where P = vapor pressure in standard atmosphere (atm), V = volume in liter (L), n = the number of moles of compound in its gaseous phase, R = ideal gas constant (0.082 atm-L/mol-K), and T = absolute temperature in K. From Eq. (2.1), the concentration in mass per unit volume (C)

can be converted from the concentration in mole per unit volume (n/V) using molecular weight (MW in g/mol):

$$C\left(\frac{mg}{\mathrm{m}^3}\right) = 10^6 \frac{PMW}{RT} \qquad \text{Eq. (2.2)}$$

where the coefficient 10^6 is used to convert from g/L to mg/m^3. Atmospheric pollutants are commonly expressed in mg/m^3, which can be further converted into ppm (volume ratio, ppm$_v$). Both mg/m^{3}and ppm$_v$ units are used in standards of regulatory agencies, such as EPA, OSHA, NIOSH, and ACGIH. Example 2.2 shows the use of Eq. (2.2) and illustrates the relationship between vapor pressure and volatility.

Example 2.2 Estimate vapor phase concentration from chemical spills

Atrazine is a pesticide having a vapor pressure of 4×10^{-10} atm at 25 °C, which is about nine orders of magnitude lower than that of trichloroethelene (TCE) used in dry cleaning (vapor pressure = 0.13 atm at 25 °C). Calculate the vapor phase concentrations of these two compounds if the vapors are in equilibrium with their pure states at 25 °C.

Solution

The molecular weight of atrazine and TCE are 215.7 and 131.39 g/mol, respectively (Appendix C). According to Eq. (2.2), we have:

$$C = 10^6 \frac{4\times10^{-10}\, atm \times 215.7 \dfrac{\mathrm{g}}{mol}}{0.082 \dfrac{atm\times \mathrm{L}}{mol\times \mathrm{K}} \times (273+25)\mathrm{K}} = 3.53\times10^{-3} \frac{mg}{\mathrm{m}^3} \text{ for atrazine, and}$$

$$C = 10^6 \frac{0.13\mathrm{atm}\times 131.39 \dfrac{\mathrm{g}}{\mathrm{mol}}}{0.082 \dfrac{\mathrm{atm}\times \mathrm{L}}{\mathrm{mol}\times \mathrm{K}} \times (273+25)\mathrm{K}} = 699000 \frac{\mathrm{mg}}{\mathrm{m}^3} \text{ for TCE}$$

With the calculated vapor phase concentrations, one can immediately know the volatile and hazardous nature of TCE if we compare these calculated concentrations with their health standards. For example, the ACGIH has assigned atrazine a threshold limit value (TLV) of 5 mg/m^3 in air as an 8-hour time-weighted-average (TWA) for a normal 8-hour workday and a 40-hour workweek. The corresponding ACGIH standard for TCE is 270 mg/m^3. Since these calculated concentrations are from their pure states, these values should represent the atmospheric concentrations immediately adjacent to a chemical spill rather than from their dissolved phase in groundwater. In other words, Eq. (2.2) cannot be used to calculate the concentration of TCE vapor in equilibrium with TCE dissolved in groundwater, but it applies to the scenario when vapor in pores of surface soil (vadose zone) is at equilibrium with pure TCE spilled. The above calculated vapor phase concentrations also imply that remediation technologies relying on the removal of vapor phase can be used very effectively for TCE, but not for atrazine.

For the most part, when we are concerned about the volatility of a compound dissolved in water, it is important to know that the Henry's law constant (H), rather than the vapor pressure, controls the volatility of contaminants from water. Although vapor pressure can still provide a measure of the volatility, the correct parameter to relate a chemical's volatility is the Henry's law constant. Henry's law constant is related to equilibrium vapor pressure (P) and water solubility (S) by:

$$H = \frac{P}{C_{aq}} \qquad \text{Eq. (2.3)}$$

Henry's law constant defined above has a unit of atm/(mg/L), atm/M, or atm/(mol/m^3) depending on the unit to express water solubility, i.e. mass per unit volume (mg/L), molarity specified with the symbol M (mol/L), or moles per cubic meter (mol/m^3), respectively. Another common unit is the dimensionless Henry's law constant as defined by:

$$H = \frac{C_g}{C_{aq}} \qquad \text{Eq. (2.4)}$$

where C_g is the vapor phase concentration in mg/L and aqueous phase concentration (C_{aq}) in mg/L. Since Henry's law constant has many different forms with various units, sometimes it could be very confusing. The choice of units depends on the specific disciplines and the convenience of a given circumstance. For remediation professionals, skill in the correct conversion of its value from one unit to another is important in the calculation and design of remedial systems (see Box 2.3, Table 2.3).

A chemical with a lower Henry's constant will have lower volatility, and generally has a lower vapor pressure, higher water solubility, and lower adsorptive capacity in soil. A compound with lower Henrys' constant is less likely to volatilize from groundwater into air among soil particles. Such compounds will be less susceptible to treatment using vapor-based technologies.

2.3.2 Solubilization, Precipitation, and Dissolution

Aqueous solubility (or simply **solubility**) is a measure of the maximum amount of chemical that can be dissolved in water. A chemical with a high aqueous solubility will disperse more in groundwater than a chemical with a low aqueous solubility. Since water is a very polar molecule (with an electronegative oxygen and two positive hydrogen atoms), water is an excellent solvent for polar compounds. However, it is a very poor solvent for most organic contaminants with a characteristic hydrophobic nature. Solubility depends on temperature and pressure, but the major factors in controlling a compound's solubility is very different between inorganic and organic compounds, which are described separately in the following section.

2.3.2.1 Solubility and Solubility Product for Inorganic Compounds

Precipitation is the formation of an insoluble precipitate from the reaction of two dissolved substances. The equilibrium between precipitation and its reversed **dissolution** reaction controls the solubility of many inorganic metals in contaminated groundwater system. Metals in contaminated groundwater can be removed by the precipitation reactions, and metals immobilized in contaminated soils can also be solubilized by the addition of some chelating agents followed by the subsequent removal of dissolved metals in groundwater. A chelating agent is a substance whose molecules can form several bonds to a single metal ion. An example of

Box 2.3 Henry's law constant: How volatile is volatile?

The volatility of a chemical increases with increasing Henry's law constant, but how volatile is volatile? The answer is not straightforward, as there is no clear dividing line to differentiate chemicals into various classes of volatilities. However, there is a general consensus regarding the relative volatility of most environmental contaminants. For example, we refer benzene-series compounds as volatile, chlorinated pesticides, phenols, PAHs, and PCBs as semivolatile, and most inorganic metal compounds as nonvolatile. The following guidelines (Table 2.3) by Mackay and Yuen (1980) are useful for evaluating the volatility of organic solutes in water at ambient temperature (25 °C).

Henry's law constant can vary several orders of magnitudes for various compounds, and it is also highly temperature dependent. The many forms of units add to the difficulty in using literature reported value of Henry's constants to perform engineering calculation. Readers who want to become proficient in unit conversion of Henry's constants should consult environmental chemistry texts (e.g. Hemondand Fechner-Levy 2014). Online unit conversions are also available for practitioners such as the USEPA's On-line Tools for Site Assessment Calculation (http://www.epa.gov/athens/learn2model/part-two/onsite/henryslaw.html). For example, the commonly used dimensionless Henry's constants are provided in Appendix C (concentration ratio based dimensionless *H*), these can be converted into three other units (atm·m^3/mol; atm/mol fraction; mole fraction based dimensionless *H*) using this online tool when needed.

The concept of volatility is important from the practical standpoint of environmental monitoring and remediation. Volatile organic compounds can be readily removed by volatility-based remediation techniques such as soil vapor extraction (Chapter 8). For environmental organic compounds classified as semivolatile organic compounds, their removal becomes harder by these techniques. With the exception of few metal compounds including mercury (Hg), lead (Pb), and arsenic (As), most metals are generally not considered volatile and therefore cannot be removed through vapor-based remediation.

Table 2.3 How volatile is volatile using Henry's law constant (*H*).

Volatility	Volatilization is	If
Highly volatile	Rapid	$H > 10^{-3}$ atm·m^3/mol (dimensionless $H > 4.09 \times 10^{-2}$)
Volatile	Significant	$10^{-3} < H < 10^{-5}$ atm·m^3/mol (4.09×10^{-4} < dimensionless $H < 4.09 \times 10^{-2}$)
Semivolatile	Slow	$10^{-5} < H < 3 \times 10^{-7}$ atm·m^3/mol (1.23×10^{-4} < dimensionless $H < 4.09 \times 10^{-4}$)
Nonvolatile	Negligible	$H < 3 \times 10^{-7}$ atm·m^3/mol (dimensionless $H < 1.23 \times 10^{-4}$)

precipitation is the removal of Fe(II) in groundwater, first by aeration (O_2) to oxidize Fe(II) to Fe(III), and its subsequent removal of Fe(III) by precipitation through the reaction with hydroxide.

$$Fe^{2+} + 2H^+ + \frac{1}{2}O_2 \rightarrow Fe^{3+} + H_2O$$

$$Fe^{3+} + 3OH^- \rightarrow Fe(OH)_3(s)\downarrow$$

The solubility of an inorganic precipitate can be calculated via **solubility product constant** (K_{sp}). For example, the K_{sp} for $Fe(OH)_3$ in the above precipitation can be expressed by:

$$K_{sp} = [Fe^{3+}][OH^-]^3$$

where [ ] denotes the molar concentration of given ions. If pH and K_{sp} are given, the concentration of Fe^{3+}, i.e. the solubility of $Fe(OH)_3$, can be determined. To do so, we can deduce a general formula to calculate the solubility of any inorganic compound from its K_{sp}.

For an inorganic compound M_xA_y with their respective oxidation numbers $+a$ for the metal (M) and $-b$ for its anion (A), the general reaction describing its precipitation-dissolution equilibrium can be written as:

$$M_xA_y(s) \rightleftarrows xM^{a+}(aq) + yA^{b-}(aq)$$

where x and y are the stoichiometric coefficients for M and A, respectively. Thus, $CaCO_3$ has $x = 1$ and $y = 1$, whereas $Ca_3(PO_4)_2$ has $x = 3$ and $y = 2$. The equilibrium **solubility product constant** (K_{sp}) can be written as follows:

$$K_{sp} = [M^{a+}]^x\,[A^{b-}]^y \qquad \text{Eq. (2.5)}$$

If we assume M_xA_y has "S" moles of M_xA_y that dissolves in 1 L of solution (i.e. its solubility is "S"), we can set up the "ICE chart" as commonly used in chemistry textbook:

	M_xA_y (s)	M^{a+} (aq)	yA^{b-} (aq)
Initial concentration (I)	All solid	0	0
Change in concentration (C)	$-S$ dissolves	$+xS$	$+yS$
Equilibrium concentration (E)	Less solid	xS	yS

We then substitute the equilibrium concentrations into Eq. (2.5):

$$K_{sp} = [xS]^x\,[yS]^y = (x^x y^y)S^{x+y} \qquad \text{Eq. (2.6)}$$

Solving Eq. (2.6) for S, we can get the general equation for solubility:

$$S = \left(\frac{K_{sp}}{x^x y^y}\right)^{1/(x+y)} \qquad \text{Eq. (2.7)}$$

Since the K_{sp} value can be obtained from a reference source, the solubility (S) of any inorganic precipitate can be readily calculated using Eq. (2.7) (see Example 2.3).

Example 2.3 Calculate solubilities of metallic compounds

Given the solubility products at 25 °C for lead (II) carbonate and cadmium (II) phosphate of 7.4×10^{-14} and 2.53×10^{-33}, respectively, calculate the solubility of these two compounds. Report solubility in the units of M and mg/L. Molecular weight: $PbCO_3$ = 267.21; $Cd_3(PO_4)_2$ = 527.18; Atomic weight: Pb = 207.2; Cd = 112.41. MW: $PbCO_3$ = 267.21; $Cd_3(PO_4)_2$ = 527.18. Atomic weight: Pb = 207.2; Cd = 112.41.

Solution:

a) For lead carbonate, $PbCO_3(s) \leftrightarrows Pb^{2+}(aq) + CO_3^{2-}(aq)$, $x = y = 1$ and $a = b = 2$. Hence, we have its solubility (Eq. 2.7):

$$S = (K_{sp})^{1/2} = \sqrt{(7.4\times10^{-14})} = 2.72\times10^{-7}\,\frac{\text{mol}}{\text{L}}$$

b) For cadmium phosphate, $Cd_3(PO_4)_2(s) \leftrightarrows 3\,Cd^{2+}(aq) + 2\,PO_4^{3-}(aq)$, $x = 3$, $y = 2$, $a = 2$, and $b = 3$.

$$S = \left(\frac{K_{sp}}{x^x y^y}\right)^{1/(x+y)} = \left(\frac{K_{sp}}{108}\right)^{1/5} = \sqrt[5]{\frac{(2.53\times10^{-33})}{108}} = 1.19\times10^{-7}\,\frac{\text{mol}}{\text{L}}$$

The molar concentrations can be further converted into mg/L by multiplying it with the molecular weight and then 1000 to convert from g/L to mg/L. Hence, we have the solubility in mg/L for:

$$S\,(\text{lead carbonate}) = 2.72\times10^{-7}\times267.21\times1000 = 0.0727\,\frac{\text{mg}}{\text{L}}\ (\text{i.e. } 72.7\text{ ppb})$$

$$S\,(\text{cadmium phosphate}) = 1.19\times10^{-7}\times527.18\times1000 = 0.0625\,\frac{\text{mg}}{\text{L}}\ (\text{i.e. } 62.5\text{ ppb})$$

Since the free forms of lead (Pb^{2+}) and cadmium (Cd^{2+}) are the toxic species, these concentrations can be calculated as:

$$Pb^{2+} = 2.72\times10^{-7}\times207.2\times1000 = 0.056\,\frac{\text{mg}}{\text{L}}\ (\text{i.e. } 56\text{ ppb})$$

$$Cd^{2+} = 1.19\times10^{-7}\times112.41\times1000 = 0.013\,\frac{\text{mg}}{\text{L}}\ (\text{i.e. } 13\text{ ppb})$$

These results clearly reveal the low solubility of these metal precipitates with low values of K_{sp}. Even at such low concentrations, Cd^{2+} has exceeded the MCL of 0.005 mg/L of the USEPA's drinking water standard. For Pb^{2+}, the USEPA standard states that if more than 10% of the drinking water samples exceeded the action level of 0.015 mg/L, then it requires the treatment system to control the corrosion.

The solution pH has a significant effect on the solubility of many metals. pH can either increase or decrease solubility depending on the specific metal of concern. Another factor to increase solubility is the use of organic chelating compounds through complexation/chelation. The increased solubility of metals in ground water usually enhances their removal (remediation) from the immobilized metal species to their soluble species.

2.3.2.2 Solubility and K_{ow} for Organic Compounds

For organic compounds, aqueous solubility is generally determined by its structure, mainly its polar and nonpolar functional groups. Compounds with polar groups such as alcohol (OH), acid (COOH), and aldehyde (CHO) can enhance water solubility, whereas nonpolar groups such as alkyl (CH_3, CH_3CH_2) and halogens (Cl, F, B, I) will reduce the solubility in water. Hence, the water solubility is in an increasing order of toluene (methyl benzene) < benzene < benzoic acid. In the same series of organic compounds, solubility will decrease as compound size increases. In environmental remediation, solubility of organic compounds can be greatly increased by using cosolvent (ethanol) through cosolvation or surfactants through micellar solubilization. Chapter 11 has the detailed description of micellar solubilization as the primary mechanism for the soil washing remediation technique.

A practically useful guideline regarding a chemical's solubility is that a compound with a higher water solubility is more likely to be more mobile in the groundwater, less likely to be adsorbed onto soil, less likely to be bioaccumulative, and more likely to be biodegraded (because the chemical must be dissolved to become bioavailable). Conversely, a chemical with a lower water solubility is more likely to be immobilized via adsorption, less mobile in groundwater, more bioaccumulative, persistent, and less prone to biodegradation. In fact, the latter is characteristic of many of the hydrophobic environmental organic compounds that represent the challenges in soil and groundwater remediation.

In Section 2.1, a chemical is designated as "very soluble" or "not soluble." The differentiation is sometimes arbitrary, since there is no clear dividing line to tell whether a chemical has a low or high solubility. A guideline perhaps useful for layman as well as practitioners is as follows (Ney 1998):

- **Low aqueous solubility:** <10 mg/L
- **Medium aqueous solubility:** 10–1000 mg/L
- **High aqueous solubility:** >1000 mg/L

An important parameter related to aqueous solubility of organic compounds is hydrophobicity which is typically measured by K_{ow} or its log-scale ($\log K_{ow}$). The **octanol–water partitioning coefficient** (K_{ow}) is experimentally determined by measuring the ratio of equilibrium concentrations of a chemical in octanol (C_{oct}) to that in water (C_w) at a 50 : 50 octanol:water volume ratio. Octanol is a good surrogate to mimic natural organic compounds.

$$K_{ow} = \frac{C_{oct}}{C_w} \qquad \text{Eq. (2.8)}$$

A chemical with a higher K_{ow} is a more hydrophobic or less polar compound and has a lower aqueous concentration in the water phase of the mixture than in the octanol phase. A chemical with a higher K_{ow} will have a greater potential for sorption in soil, a greater affinity to bioaccumulate in the food chain. Hence, most hydrophobic organic contaminants (e.g. PAHs, PCBs) tend to have a lower mobility in soil and groundwater. Hydrophobic compounds (with a high K_{ow}) have a higher affinity toward nonpolar compounds such as animal fat and organic compounds such as humus in soil. This is the reason why most hydrophobic organic compounds are accumulated in animal fat tissues and why soils particularly high in organic content will retain organic contaminants. A sandy soil will not retain as much organic contaminant as an agricultural soil high in organic content.

Like aqueous solubility, we can also group compounds' hydrophobicity based on K_{ow} values, and relate them to the fate and transport of chemicals for practical purpose. Since in the measurement of K_{ow}, an equal volume of octanol and water is used, the K_{ow} greater than unity implies the contaminant prefers the octanol phase. The K_{ow} values of most organic compounds are greater than 1 (Appendix C), mostly greater than 100 and they can be as high as 10^8. Because of this wide range, the logarithmic K_{ow} is commonly used. The practically useful guideline is given in Table 2.4 (Ney 1998):

2.3.3 Sorption and Desorption

Sorption is the transfer of a substance from a solution or a gaseous phase to a solid phase. Sorption is the term used to generally refer to two often indistinguishable processes: adsorption and absorption. **Adsorption** is the attraction to a solid surface, whereas **absorption** is the

Table 2.4 Aqueous solubility and octanol–water partitioning coefficient (K_{ow}).

Solubility	If	It is indicative of
High	$K_{ow} < 500$ ($\log K_{ow} < 2.7$)	high mobility, little to no bioaccumulation or accumulation, and degradation by microbes, plants, and animals
Medium	$500 < K_{ow} < 1000$ ($2.7 < \log K_{ow} < 3.0$)	the chemical that can go either way of low or high K_{ow} ($\log K_{ow}$)
Low	$K_{ow} > 1000$ ($\log K_{ow} > 3.0$)	immobility, nonbiodegradability, and a chemical that is bioaccumulative, accumulative, persistent, and sorbed in soil

penetration into the bulk of a solid (or liquid). Adsorption depends on properties of both contaminant (sorbate) and solid (sorbent). Between hydrophobic and ionic/polar contaminants, the adsorption mechanisms are vastly different. For solids, the principal factors of sorption include the homogeneity, permeability, porosity, surface area, surface charge, organic carbon content (for sorption of hydrophobic organics), and cation exchange capacity (CEC) (for ionic species).

Sorption mechanisms can be described as "like adsorbs like" (i.e. polar species on polar surfaces, and nonpolar species on nonpolar surfaces). Soil grains in aquifer are heterogeneous composite containing minerals and natural organic matter (e.g. humic acids). The mineral surfaces are dominated by polar or ionic functional groups capable of interacting with polar or ionic contaminants. The natural organic matter is generally made of hydrophobic molecules that tend to exclude water and other polar molecules, but retain nonpolar/hydrophobic molecules.

Sorption for Inorganics: For inorganic contaminants (metals in ionic forms), surface charge (measured as cation exchange capacity) determines the ability to attract and hold cations to negatively charged functional groups on the solid. Thus, clay or other soil particles with negatively charged surface will have strong sorption toward positively charged cations. **Cation exchange capacity** (CEC) is the maximum quantity of total cations that a soil is capable of holding, at a given pH value, available for exchange with the soil solution. CEC is expressed as milliequivalent of hydrogen per 100 g of dry soil (meq/100 g).

Sorption for Organics: For neutral organic contaminants, the organic carbon content of soil is the main factor in controlling the adsorption of hydrophobic contaminants. There are two parameters that measure sorption characteristics, including the adsorption coefficient (K_d) and the organic carbon normalized sorption coefficient (K_{oc}). The **adsorption coefficient** (partition coefficient) is defined by:

$$K_d = \frac{C_s\left(\frac{mg}{kg}\right)}{C_w\left(\frac{mg}{L}\right)} \qquad \text{Eq. (2.9)}$$

where C_s is mg of adsorbed contaminant per kg of soil and C_w is the aqueous phase concentration at equilibrium with a unit of mg/L. Note that K_d has a unit of L/kg. K_d is a good indicator for sorption of low-concentration environmental chemicals, and hydrophobic adsorption is the primary mechanism. Its value can be readily acquired through an isotherm which is a plot of amount of sorbed contaminant (C_s) vs. the equilibrium concentration in the aqueous phase (C_w). In a simplified case, a linear isotherm is assumed at the given temperature, but there are other nonlinear isotherm models (e.g. Freundlish and Langmuir; see Sparks 2003; Schwarzenbach et al. 2017). The disadvantage is that K_d value varies with aqueous concentration (C_w) if the compound presents a nonlinear isotherm.

For organic contaminants sorbed by soil organic matter, an empirical formula between K_d and K_{ow} is often used when experimental value of K_d is not available, such as (Schwarzenbach and Westall 1981):

$$\log K_d = 0.72 \log K_{ow} + \log f_{oc} + 0.49 \quad \text{Eq. (2.10)}$$

The above equation indicates that $\log K_d$ increases as both K_{ow} (hydrophobicity of sorbate, i.e. the compound of interest) and f_{oc} (organic content in sorbent, i.e. soil) increase. Thus, the higher the K_{ow} (i.e. the more the hydrophobic the compound), the higher the K_d value. It should be noted that the empirical equations such as Eq. (2.10) are only applicable to certain conditions. For instance, Eq. (2.10) is valid for the adsorption of nonpolar compounds for sorbents containing more than 0.1% organic carbon ($f_{oc} > 0.001$). Since K_d is strongly dependent on the soil organic matter content, a second parameter K_{oc} is used to define the compound-specific sorption value. Thus, **organic carbon normalized partition coefficient** is calculated as follows:

$$K_{oc} = \frac{K_d}{f_{oc}} \quad \text{Eq. (2.11)}$$

where K_{oc} is mg of contaminant per kg of organic carbon, and f_{oc} = % organic carbon in soil or other sorbents such as suspended solids or sediment (if f_{oc} = 5%, then f_{oc} = 0.05 in Eq. 2.11). Similar to K_d in Eq. (2.10), an empirical formula was also obtained between $\log K_{oc}$ and $\log K_{ow}$ for the same group of compounds and adsorbents:

$$\log K_{oc} = 0.72 \log K_{ow} + 0.49 \; (R^2 = 0.95) \quad \text{Eq. (2.12)}$$

More empirical expressions can be found from the literature (e.g. Chiou 2002), so that K_d and K_{oc} can be estimated using commonly known values of K_{ow}.

The reverse of the sorption process is termed **desorption** in which sorbed contaminants are disassociated from soil and return to the aqueous or gaseous phase. We generally assume the reversibility of the desorption process, but research has in fact indicated the slow desorption process which provides insight as to why the long cleanup time is required in many of the contaminated sites in the United States.

To further understand the sorption process, let's examine the K_d value in a more quantitative manner. Our effort here is to derive a useful equation to predict what fraction (f_w) of contaminant will remain in water of the aquifer materials from given K_d value as well as other needed parameters for soil particles. The value of f_w is important because only this dissolved fraction is susceptible to be removed by, for example, pump-and-treat remediation techniques introduced in Chapter 7.

The fraction of contaminant remained in soil water (f_w) is:

$$f_w = \frac{\text{contaminant mass in water}}{\text{contaminant mass in water} + \text{contaminant mass in soil}} = \frac{C_w V_w}{C_w V_w + C_s M_s} \quad \text{Eq. (2.13)}$$

Note that in the denominator, the mass of soil (M_s) instead of volume of soil is used, because the concentration unit for contaminant in soil is always reported in mass of contaminant per unit mass of soil (e.g. mg/kg). Rearranging Eq. (2.13) and by substituting sorption coefficient K_d

for C_s/C_w term, and then by substituting the product of its bulk density (ρ) and soil volume (V_s) for soil mass, we have:

$$f_w = \frac{1}{1+\frac{C_s M_s}{C_w V_w}} = \frac{1}{1+K_d \frac{M_s}{V_w}} = \frac{1}{1+K_d \frac{V_s \rho}{V_w}} \quad \text{Eq. (2.14)}$$

In order to substitute some commonly measured parameter for the M_s/V_w term, we use soil porosity (n) defined as:

$$n = \frac{\text{pore volume of water}(V_w)}{\text{pore volume of water }(V_w) + \text{volume of solid}(V_s)} = \frac{1}{1+\frac{V_s}{V_w}} \quad \text{Eq. (2.15)}$$

Rearranging Eq. (2.15) to get V_s/V_w term and substituting it into Eq. (2.14), we have:

$$f_w = \frac{1}{1+K_d \rho \left(\frac{1}{n}-1\right)} \quad \text{Eq. (2.16)}$$

The above equation can be used to calculate the percent of contaminant in soil water if the K_d value, soil bulk density (ρ), and porosity (n) are given (see Example 2.4). Note that the commonly used unit of K_d is L/kg which is the same as mL/g, and the unit of ρ is g/mL (or more commonly written as mg/cm^3).Therefore, the units for the product of K_d and ρ will cancel out. This results in a dimensionless number consistent with the dimensionless f_w. Example 2.4 illustrates the use of Eq. (2.16).

Example 2.4 Calculate the fraction of contaminant present in aqueous phase (groundwater) using K_d

Gasoline contains both MTBE (as the additive) and phenanthrene (as a minor component of the heavy fraction). The K_d value for MTBE was reported to be 0.56 L/kg in a silty soil with $f_{oc} = 0.05$ (i.e. 5% soil organic matter normalized to organic carbon) and 0.112 L/kg for a sandy soil with f_{oc} of 0.01 (Jacobs et al. 2000). Phenanthrene has a K_d value of 300 L/kg in a soil containing soil organic carbon of 0.0126 (Chiou et al. 1998). The porosity of the aquifer is assumed to be 30% and soil bulk density is assumed to be 1.8 g/cm^3 for comparison purpose. (a) Calculate f_w values for MTBE in both soils; (b) Calculate f_w for phenanthrene; (c) Schematically show how f_w changes with the changes in K_d, assuming the average porosity (n) of 0.30 and bulk density of 1.8 g/cm^3.

Solution

a) For MTBE in a silty soil: $f_w = \frac{1}{1+0.56\times 1.8\times\left(\frac{1}{0.3}-1\right)} = 29.8\%$ and MTBE in a sandy soil: $f_w = \frac{1}{1+0.112\times 1.8\times\left(\frac{1}{0.3}-1\right)} = 68.0\%$

b) For phenanthrene: $f_w = \dfrac{1}{1+300\times1.8\times\left(\dfrac{1}{0.3}-1\right)} = 0.079\%$

This calculation clearly shows that MTBE spilled from a gasoline tank will be transported to a very long distance from the source because of the lack of adsorption to soil. Phenanthrene, on the other hand, is very less likely to be dissolved in groundwater and dispersed far away from the source.

Substituting the given values of porosity ($n = 0.30$) and bulk density ($\rho = 1.8\,g/cm^3$) in Eq. (2.16), we can obtain the plot of f_w versus K_d (Figure 2.9). This plot clearly shows the relationship between f_w and K_d and its remediation implication. For example, when K_d exceeds 10, almost all organic contaminants will be retained by soils, making them very difficult to be removed. Generally, the higher the K_d value for a specific contaminant, the more mass will be adsorbed to the soil and the slower the contaminant will travel compared to groundwater (i.e. it will have a higher retardation factor). A plume containing a contaminant with a very high K_d value may even appear that the contaminant plume is not moving at all in a low hydraulic gradient.

Box 2.4 summarizes some practically useful guidelines on contaminant properties. This is followed by an example (Example 2.5) showing the use of these basic properties to deduce the relevant fate and transport processes (Table 2.5).

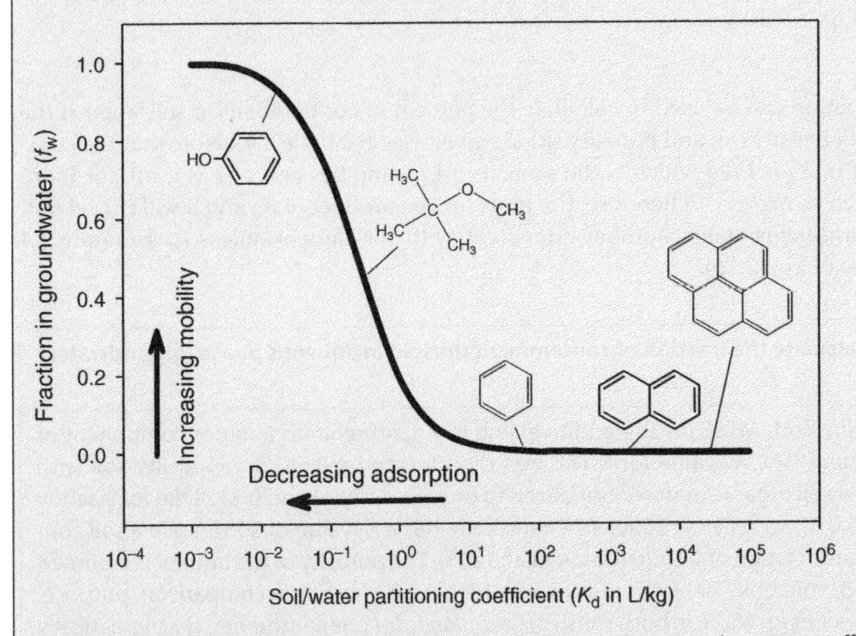

Figure 2.9 Fraction of contaminant present in groundwater as a function of sorption coefficient (K_d) with examples of soil and groundwater contaminants.

Box 2.4 Practically useful guidelines for chemical properties

Table 2.5 gives the useful guidelines for vapor pressure, solubility, and K_{ow} to provide readers with some general ideas as to tell where a given chemical falls in any category. The guidelines for Henry's law constant are given in Box 2.3, and the same for K_{oc} is not given for brevity. It should be noted that these guidelines are approximate and not scientifically sound. There are always some exceptions, but having the sense of fate and transport processes for a given set of properties will help understand the remediation technologies that we will discuss in future chapters.

Table 2.5 Vapor pressure, water solubility, and log K_{ow} vs. chemical fate.

Fate and transport	Vapor pressure		
	Low **$<10^{-6}$ mmHg** **(1.36×10^{-9} atm)**	**Medium** **10^{-6}–10^{-2} mmHg** **(1.36×10^{-9}–1.36×10^{-5} atm)**	**High** **10^{-2} mmHg** **(1.36×10^{-5})**
Volatility	Low	Medium	High
Solubility	High	Medium	Low
Persistence	Yes	Maybe	Negligible
Adsorption	High	Maybe	Low
Bioaccumulation	Yes	Maybe	Negligible
	Water solubility		
	Low **<10 mg/L**	**Medium** **10–1000 mg/L**	**High** **>1000 mg/L**
Mobility	Negligible	Either way	Yes
Persistence	Yes	Either way	Negligible
Adsorption	Yes	Either way	Negligible
Bioaccumulation	Yes	Either way	Negligible
Biodegradation	Maybe	Either way	Yes
	K_{ow}		
	Low **$K_{ow} < 500$** **($\log K_{ow} < 2.7$)**	**Medium** **K_{ow} 500–1000** **($\log K_{ow}$ 2.7–3.0)**	**High** **$K_{ow} > 1000$** **($\log K_{ow} > 3.0$)**
Solubility	High	Medium	Low
Persistence	Negligible	Either way	Yes
Adsorption	Negligible	Either way	Yes
Bioaccumulation	Negligible	Either way	Yes
Biodegradation	Yes	Either way	No, too slowly

Source: Adapted from Ney (1998).

Example 2.5 Use of practical guidelines to tell contaminant fate and transport

(a) Compound A: water solubility = 0.0017 mg/L, $K_{oc} = 238\,000$, $K_{ow} = 960\,000$, BCF = 61 600, where **bioconcentration factor** (BCF) is calculated by ratio of the equilibrium concentration in fish tissue to that in the aqueous phase. (b) Compound B: water solubility = 150–200 mg/L, $K_{ow} = 758$, vapor pressure at 20 °C = 14 Torr (1 Torr = 1 mm of mercury). Deduce the basic fate and transport characteristics of these two compounds in terms of soil adsorption, runoff with soil, leaching from soil, potential food chain contamination, and biodegradability.

Solution

a) Compound A: The low water solubility generally indicates that this chemical should adsorb to soil, runoff with soil particles, and bioaccumulate, and it should not be leached and should not be biodegraded. The high K_{oc} value indicates that this chemical should adsorb to soil very strongly. The high K_{ow} value indicates that this chemical should bioaccumulate and could cause food-chain contamination, and this is supported by the BCF data (BCF = 61 600 in flowing water). This chemical is actually dichlorodiphenyltrichloroethane or DDT.

b) Compound B: The medium water solubility indicates that this chemical could go either way in regard to leaching, runoff, adsorption, bioaccumulation, and biodegradation. The medium value of K_{ow} indicates that this chemical is volatile, and it could present inhalation problems with the parent and potential transformation products. Although it may bioaccumulate, it is quite unlikely because it should volatilize prior to bioaccumulation. The volatility also prevents residues from occurring in food, but there could be some in water if aquatic contamination occurred. This chemical is indeed tetrachloroethylene (PCE), a common chlorinated hydrocarbon solvent used in dry cleaning.

At the conclusion of our discussions in this section, readers may find Appendix C very useful in relating physicochemical properties (specific gravity, solubility, $\log K_{ow}$, vapor pressure, and Henry's constant) to the fate of contaminant of interest. Interested readers should spend time by examining the physicochemical properties listed for compounds of different categories, and develop an understanding of their basic fate and transport characteristics introduced above.

One particular notion relevant to groundwater remediation for data in Appendix C is the grouping of chemicals based on solubility and density. For water immiscible liquid (compounds of low aqueous solubility), **nonaqueous phase liquid** (NAPLs) is a term used to denote a free standing phase either below or above the groundwater table. Just imagine when vegetable oil is mixed with water, after standing, a free phase oil will be separated from water to form two immiscible phases: oil and water. If the nonaqueous phase has a density less than that of water (1 g/mL), this is called **lighter nonaqueous phase liquids** (LNAPLs) and conversely it is called **dense nonaqueous phase liquids** (DNAPLs). These two types of NAPLs are not defined by the specific chemical but by their densities relative to that of water. The differentiation between these two groups of chemicals is very important from the groundwater remediation standpoint. For example, benzene has a density of 0.8 g/mL, and TCE has a density of 1.46 g/mL (Appendix C). Benzene would float on top of groundwater table while TCE would sink below the groundwater table, and continue to sink until it reaches the bottom of an aquifer. Therefore, benzene can be referred to as a LNAPL or floater, and TCE is a DNAPL or sinker.

LNAPLs include BTEX series compounds, whereas DNPALs include chlorinated solvents such as TCE and PCE, PAHs, and PCBs. Alkanes can be LNAPL (short chain) or DNAPL (long chain and heavy fractions). The implication of such differentiation is apparent, because it is generally much easier to remove dissolved contaminants than free phase LNAPLs or DNPALs, and it is much easier to remove LNAPLs than DNAPLs. Unlike dissolved contaminants, NAPLs, particularly DNPALs, represent one of the most challenging group of contaminants in groundwater remediation.

2.4 Intraphase Chemical Movement

The transport of contaminants in the groundwater is also affected by different hydrological processes. These hydrodynamic processes include advection, dispersion, and diffusion. Each process alone or altogether will impact contaminant movement along with groundwater flow.

Since these three processes are generally physical, they will physically relocate (move) chemicals. They will not chemically change the structures of the contaminants.

2.4.1 Advection

Advection is the movement of chemical through the natural movement of bulk fluid of groundwater, or forced movement by pumping. Water flows in the direction of higher energy (potential head) to the lower potential head. **Head** (h) is a term in fluid dynamics used to relate the energy in an incompressible fluid to the height of an equivalent static column of that fluid. For example, the total head (energy) at a given point in a fluid is the energy associated with the movement of the fluid (the velocity head), plus energy from pressure in the fluid (pressure head), plus energy from the height of the fluid relative to an arbitrary datum (elevation head due to gravity). Head is expressed in units of height such as meters or feet. The flow rate of groundwater can be determined by **Darcy's law**:

$$v = -\frac{K}{n}\frac{\mathrm{d}h}{\mathrm{d}l} \quad \text{Eq. (2.17)}$$

where v = average linear velocity of groundwater flow (m/s), K = hydraulic conductivity (m/s), n = porosity, and $\mathrm{d}h/\mathrm{d}l$ (unit: m/m) is the hydraulic gradient which is the difference in hydraulic head ($\mathrm{d}h$) along with a distance of $\mathrm{d}l$. The **hydraulic head** (h) is a specific measurement of liquid pressure above a reference datum. In an aquifer, it can be calculated from the depth to water in a specialized water well (piezometric well), and given information of the piezometer's elevation and screen depth. The hydraulic head difference between two locations along the flow path ($\mathrm{d}h$) in a distance of $\mathrm{d}l$ can be used to calculate the **hydraulic gradient** ($\mathrm{d}h/\mathrm{d}l$). The minus sign (−) in Darcy's law is needed because derivatives are taken in the direction of groundwater flow, i.e. lower total head minus higher total head. The minus sign assures a positive velocity from high to low total head. We will define the hydraulic conductivity and illustrate the use of Darcy's law in Chapter 3.

Keep in mind that contaminants dissolved in groundwater are not moved at the same velocity as the bulk groundwater flow due to adsorption or other mechanisms. Hence a retardation factor (R, unitless) is employed to reflect the slowdown for the movement of contaminant in the plume. The **retardation factor** can be defined as the ratio of groundwater flow velocity (v) to the contaminant flow velocity (v_c). It can be measured from K_d value by:

$$R = \frac{v}{v_c} = 1 + K_d\frac{\rho}{n} \quad \text{Eq. (2.18)}$$

where v = average groundwater velocity (m/s) from Darcy's law (Eq. 2.17), v_c = velocity of contaminant at the point where the concentration is half of the initial concentration, ρ = bulk density of soil (g/cm^3), n = soil porosity, and K_d = soil–water partition coefficient (L/kg). So what exactly does the retardation mean? If, for example, we have R = 10 for a contaminant in aquifer, then it implies that the groundwater is moving 10 times faster than the organic contaminant. The effects of retardation factor on the movement of contaminated plume in groundwater are schematically shown in Figure 2.10.

2.4.2 Dispersion and Diffusion

We now visualize when a chemical dye is moving through a soil column with water flowing at a velocity of 5 cm/h. If the chemical is known to have a retardation factor of 10, then this chemical is moving at a velocity of 0.5 cm/h. Will all molecules of this chemical elute from the column at

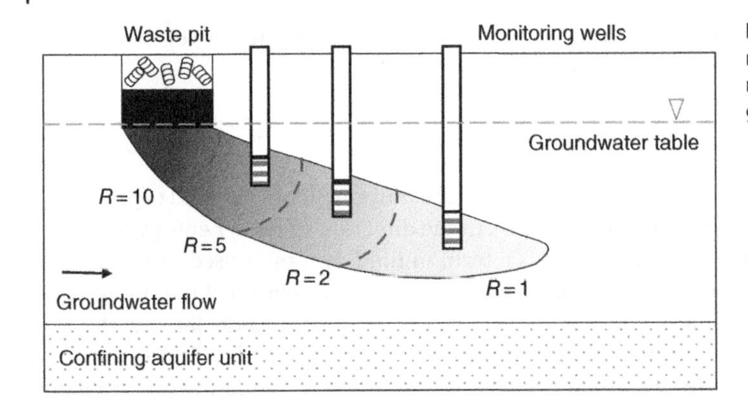

Figure 2.10 Effects of retardation factor (R) on the movement of contaminant in groundwater.

the same velocity? Or will these molecules be eluted from the column at the same instantaneous time? The answer is apparently no, and the reason is the two other hydrodynamic processes which make the "spreading" of chemical movement.

Dispersion is the "spreading" of chemicals in the direction of flow (longitudinal dispersion) or in the direction normal to the flow (transverse dispersion). In both directions, dispersion can be categorized in two ways, mechanical dispersion and hydrodynamic dispersion. The **mechanical dispersion** results from the "mixing" of contaminated water with uncontaminated water in the flow path, it does not include molecular diffusion, whereas **hydrodynamic dispersion** denotes mechanical dispersion plus diffusion. All dispersion processes occur at both a microscopic (soil pore-scale) and macroscopic level (macroscopic or field scale), hence we have micro-dispersion and macro-dispersion, respectively. It was reported that the pore-scale dispersion measured in the laboratory is on the order of centimeters, whereas macro-dispersion measured in the field is on the order of meters (Fetter 2001).

At the microscopic soil pore level, three basic causes of longitudinal dispersion are described in Figure 2.11: (a) Fluid that travels through larger pores will travel faster than fluid moving in smaller pores; (b) Some fluid parcels will travel through longer pathways than other fluid parcels; (c) As a fluid parcel moves through a porous medium, it moves faster through the center of the pore than along the edges due to friction. The lateral dispersion results from the splitting of flow paths which could even occur in laminar flow predominant in groundwater.

At the larger macroscopic scale such as field-scale, dispersion occurs as a result of heterogeneity of aquifer, for example due to the variations in hydraulic conductivity from one location to another location. The local differences in hydraulic conductivity can cause local flow

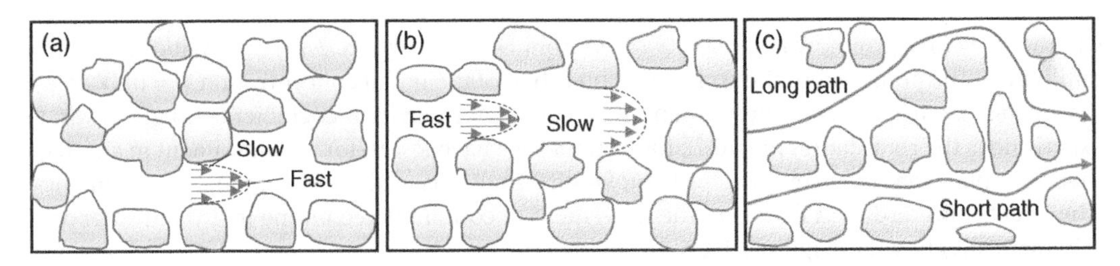

Figure 2.11 Three factors causing pore-scale longitudinal dispersion: (a) In a single pore (micro-pore scale), the flow velocity profile across a pore shows zero velocity at the edge of the soil particle and a maximum velocity at the center of the pore; (b) At the micro-pore scale, since the volume of water passing through a given pore is conserved, water flows at a faster velocity in a smaller pore; (c) Water takes different paths through porous media, causing some water travel longer in a longer path.

directions to be distorted and the actual directions of flow will diverge from the directions predicted on the basis of homogeneous aquifer.

The hydrodynamic dispersion (D_h) contributes from both the mechanical dispersion (αV) and molecular diffusion (D). Since it is very hard to distinguish these two processes in practice, we combine them as follows:

$$D_h = \alpha\, v + D. \quad \text{Eq. (2.19)}$$

where mechanical dispersion is proportional to the groundwater velocity (v) with a proportion factor termed dynamic dispersivity (α); D denotes the molecular diffusion coefficient. The diffusion of a chemical through groundwater is due to the concentration gradient. A chemical will diffuse even though the groundwater is not flowing. Diffusion can be described by **Fick's first law** under the steady-state condition:

$$J = -D\frac{dC}{dx} \quad \text{Eq. (2.20)}$$

where J = mass flux of contaminant per unit area per unit time (mg/m^2-s), D is the molecular diffusion coefficient (m^2/s), C = contaminant concentration (mg/L), x = distance along the gradient (m). Like the Darcy's law, the negative sign indicates that the movement is from greater to lesser concentrations and the mass flux (J) is always positive.

Taking together from the above discussion about advection and dispersion, a contaminant plume can be qualitatively depicted for the plume from an instantaneous source in a 1-D soil column or 2-D aquifer cell (Figure 2.12a and c), or a plume from a continuous source in a 1-D soil column or 2-D aquifer cell (Figure 2.12b and d). Because of the hydrodynamic dispersion, the concentration of a solute will decrease with distance from the source. In a 2-D aquifer cell, the contaminant will spread in the direction of groundwater movement more than in the direction perpendicular to the flow, because the longitudinal dispersivity is greater than the lateral dispersivity. A continuous source will yield a plume, whereas a one-time point source such as from a spill will yield a slug that grows with time as it moves down the groundwater flow path. The greater the heterogeneity (variations) in both micro-scale and macro-scale, the more dispersion will be seen in the plume.

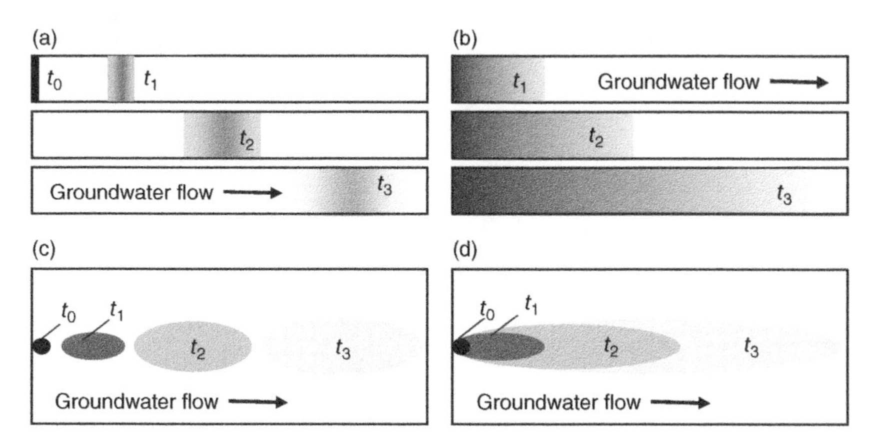

Figure 2.12 The development of a contaminant plume from (a) an instantaneous source in a one-dimensional soil column, (b) a continuous source in a one-dimensional soil column, (c) an instantaneous source in a two-dimensional aquifer cell, (d) a continuous source in a two-dimensional aquifer cell. The gradient darkness indicates contaminant concentrations.

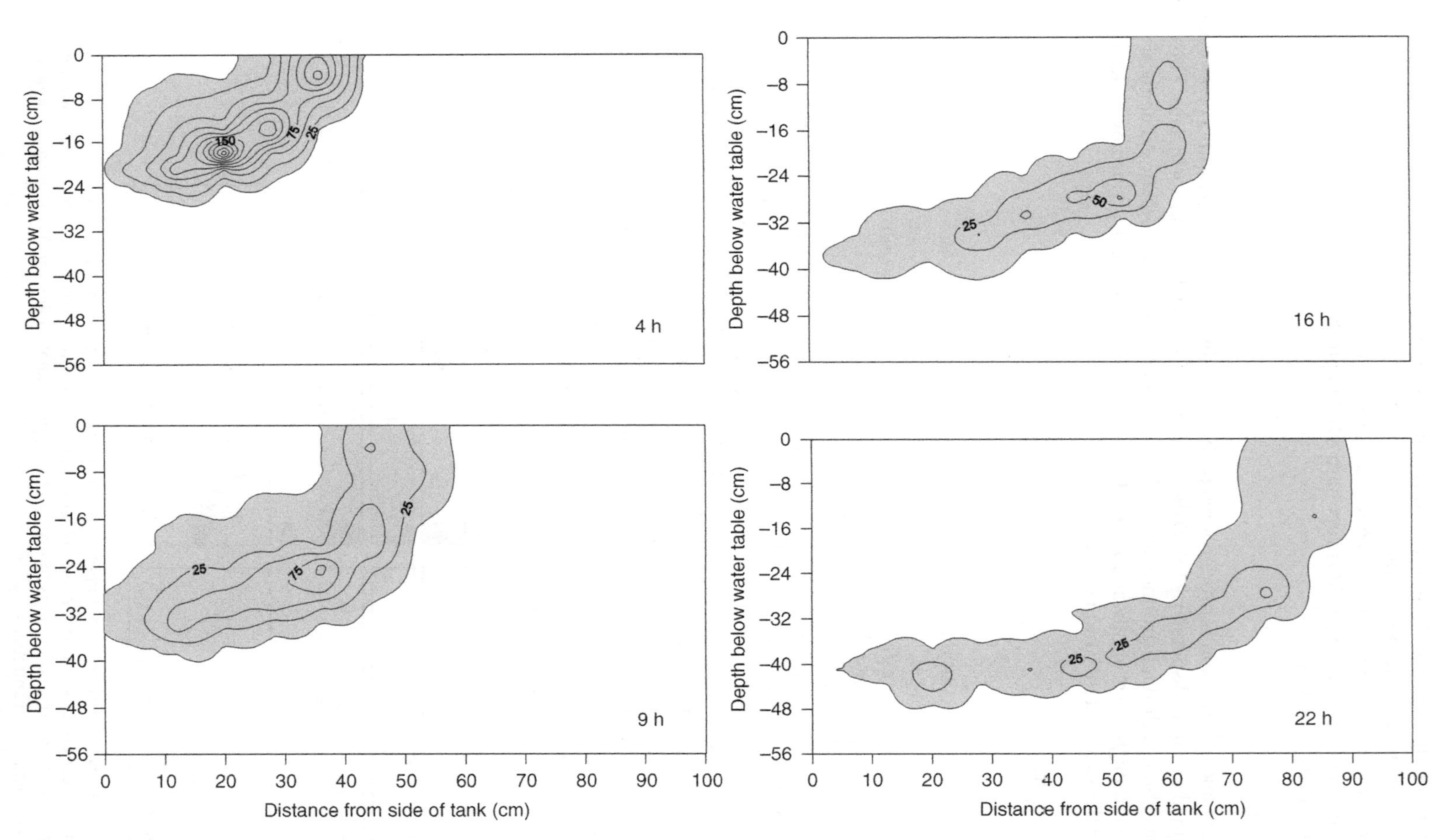

Figure 2.13 The observed benzene plume (unit of isoline: mg/L) at 4, 9, 16, and 22 hours after tracer injection. *Source:* Choi et al. (2005), Reprinted with permission from John Wiley & Sons.

Figure 2.13 is an actual plume of benzene observed at 4, 9, 16, and 22 hours following the injection of benzene in an aquifer model system made of polycarbonate materials. The benzene plume shows a banded shape propagating from the upper-left corner to the lower-right corner of the aquifer model (Choi et al. 2005). As the plume travels over time, the peak concentrations of the plume decrease rapidly due to the combined processes of advection, dispersion, volatilization, and sorption, which we have discussed previously.

Throughout this chapter, we have learned the complex contaminant fate and transport through various physical, chemical, biological, and hydrological processes. Since field data are expensive and difficult to obtain, contaminant fate and transport models can be used in soil and groundwater remediation. These models are useful to predict the potential contaminant concentrations at temporal and spatial points and to preliminarily define the media cleanup standards required to prevent exceedance of applicable human and environmental health limits set by the applicable regulatory agency. A proper model should answer the question of how long the unsafe contaminant concentration levels will persist and how far the contaminant plume will migrate. Typically, models input chemical properties data and hydrogeological parameters such as hydraulic conductivity, hydraulic gradient, porosity, and fate and transport process parameters for advection, dispersion, sorption, and biodegradation. In other words, the model predication of the plume will be based on the fate processes of the contaminant as well as the advective and dispersion transport in the aquifer. The framework of mathematical models for groundwater flow as well as chemical fate and transport processes in soil and groundwater will be detailed in Chapter 13.

Bibliography

Bedient, P.B., Rifai, H.S., and Newell, C.J. (1999). *Ground Water Contamination: Transport and Remediation*, 2e, Chapter 7. Upper Saddle River, NJ: Prentice Hall PTR.

Benjamin, M.M. (2014). *Water Chemistry*, 2e. New York, NY: McGraw Hill.

Chiou, C.T. (2002). *Partition and Adsorption of Organic Contaminants in Environmental Systems*, 257 pp. Wiley.

Chiou, C.T., McGroddy, S.E., and Kile, D. (1998). Partition characteristics of polycyclic aromatic hydrocarbons on soils and sediment. *Environ. Sci. Technol.* 32 (2): 264–269.

Choi, J.-W., Ha, H.-C., Kim, S.-B., and Kim, D.-J. (2005). Analysis of benzene transport in a two-dimensional aquifer model. *Hydrol. Process.* 19: 2481–2489.

Evangelou, V.P. (1998). *Environmental Soil and Water Chemistry*, 272–283. Wiley.

Fetter, C.W. (1993). *Contaminant Hydrogeology*. New York: Macmillan Publishing Company.

Fetter, C.W. (2001). *Applied Hydrogeology*, 4e, Chapter 10. Upper Saddle River, NJ: Prentice Hall.

Grasso, D. (1993). *Hazardous Waste Site Remediation – Source Control*. Lewis Publishers.

Hemond, H.F. and Fechner-Levy, E.J. (2014). *Chemical Fate and Transport in the Environment*, 3e. Academic Press.

Hughes, J.B., Wang, C.Y., and Zhang, C. (1999). Anaerobic biotransformation of 2,4-dinitrotoluene and 2,6-dinitrotoluene by *Clostridium acetobutylicum*: a pathway through dihydroxylamino-intermediates. *Environ. Sci. Technol.* 33 (7): 1065–1070.

Jacobs, J., Guertin, J., and Herron, C. (2000). *MTBE: Effects on Soil and Groundwater Resource*, 264 pp. Lewis Publishers.

Larson, R.A. and Weber, E.J. (1994). *Reaction Mechanisms in Environmental Organic Chemistry*, 217–221. CRC Press Inc.

Lehr, J., Hyman, M., Gass, T.E., and Seevers, W.J. (2002). *Handbook of Complex Environmental Remediation Problems*. McGraw-Hill.

Logan, B.E. (2012). *Environmental Transport Processes*, 2e. Wiley.
Loudon, G.M. (2015). *Organic Chemistry*, 6e. Oxford University Press.
von Lyman, W.J., Reehl, W.F., and Rosenblatt, D.H. (1990). *Handbook of Chemical Property Estimation Methods*, 3e, 960 pp. Washington, DC: American Chemical Society.
Mackay, D. and Yuen, T.K. (1980). Volatilization rates of organic contaminants from rivers. *Water Qual. Res. J.* 15 (1): 83–201.
Ney, R.E. Jr. (1998). *Fate and Transport of Organic Chemicals in the Environment: A Practical Guide*. Rockville, MD: Government Institute, Inc.
Nishino, S.F., Paoli, G.C., and Spain, J.C. (2000). Aerobic degradation of dinitrotoluenes and pathway for bacterial degradation of 2,6-dinitrotoluene. *Appl. Environ. Microbiol.* 66 (5): 2139–2147.
Palmer, C.D. and Wittbrodt, P.R. (1991). Processes affecting the remediation of chromium-contaminated sites. *Environ. Health Perspect.* 92: 25–40.
Piatt, J.J., Backhus, D.A., Capel, P.D., and Eisenreich, S.J. (1996). Temperature-dependent sorption of naphthalene, phenanthrene, and pyrene to low organic carbon aquifer sediments. *Environ. Sci. Technol.* 30 (3): 751–760.
Pichtel, J. (2007). *Fundamentals of Site Remediation*, 2e. Government Institute.
Schwarzenbach, R.P., Gschwend, P.M., and Imboden, D.M. (2017). *Environmental Organic Chemistry*, 3e, Chapter 2. New York: Wiley.
Schwarzenbach, R.P. and Westall, J. (1981). Transport of nonpolar compounds from surface water to groundwater: laboratory sorption study. *Environ. Sci. Technol.* 15 (11): 1360–1367.
Sparks, D.L. (2003). *Environmental Soil Chemistry*, 2e. Elsevier Science.
USEPA (2000). *In Situ* Treatment of Soil and Groundwater Contaminated with Chromium, 84 pp. Technical Resource Guide, EPA/625/R-00/005.
USEPA (2007). Technology Reference Guide for Radioactively Contaminated Media, EPA 402-R-07-004.
Walther, J.V. (2009). *Essentials of Geochemistry*, 2e. Sudbury, Massachusetts: Jones and Bartlett Publishers.
Zhang, C., Hughes, J.B., Nishino, S.F., and Spain, J. (2000). Slurry-phase biological treatment of 2,4-dinitrotoluene and 2,6-dinitrotoluene: role of bioaugmentation and effects of high dinitrotoluene concentrations. *Environ. Sci. Technol.* 34 (13): 2810–2816.

Questions and Problems

1 Describe why the remediation TCE from a leaking underground storage tank could be more challenging than BTEXs from a leaking gasoline tank?

2 Describe the basic chemical components of (a) gasoline, (b) diesel, and (c) crude oil.

3 Give an example of pesticide in each of the following groups and identify the major functionality: (a) N-containing pesticide, (b) *o*-containing pesticide, (c) S-containing pesticide, and (d) P-containing pesticide.

4 Explain why currently used pesticides such as carbamate and organo-phosphorus pesticides are not persistent in the environment compared to historically used chlorinated pesticides such as DDT.

5 Describe the susceptibility of chlorinated aliphatic hydrocarbons (chlorinated solvents) to abiotic or biotic (a) hydrolysis, (b) redox.

6 Describe what common environmental organic contaminants are susceptible to: (a) hydrolysis, (b) dehalogenation, and (c) redox.

7 What are the primary factors in determining the aqueous solubility of: (a) inorganic precipitates of metals and (b) organic compounds.

8 In soil and groundwater remediation, what can we do to improve the aqueous solubility and mobility of: (a) inorganic precipitates of metals, (b) organic hydrophobic compounds.

9 Describe the difference among: (a) biodegradation, (b) biotransformation, and (c) mineralization.

10 PCE can be transformed into TCE by certain anaerobic bacteria: $CCl_2{=}CCl_2 \rightarrow CHCl{=}CCl_2$. Indicate (a) the atom that undergoes oxidation number change, (b) the change of oxidation number from PCE to TCE, and (c) whether this is a reduction or oxidation half reaction?

11 Describe (a) how the vapor pressure and Henry's law constant are used differently to rank the volatility of a series of compounds. (b) whether it is possible to have a compound high in vapor pressure but very low in Henry's law constant.

12 Describe how the following functional groups attached to benzene ring affect the aqueous solubility of an aromatic hydrocarbon: (a) $CH_3{-}CH_2$ (b) Cl, (c) OH, and (d) COOH.

13 Rank the relative aqueous solubility among the following compounds: chlorobenzene, *m*-dichlorobenzene, 2,4-6-trichlorobenzene, benzoic acid, and benzene.

14 Rank the relative hydrophobicity (or order of $\log K_{ow}$) among the following compounds: benzene, benzoic acid, chlorobenzene, naphthalene, phenanthrene, and pyrene.

15 The K_{sp} values of $CaCO_3$ and $CaSO_4$ are 6.0×10^{-9} and 4.93×10^{-5}, respectively. Which one has the higher solubility in water?

16 Calculate the aqueous solubility of (a) CdS ($K_{sp} = 1 \times 10^{-27}$) and (b) $Ca_3(PO_4)_2$ ($K_{sp} = 2.07 \times 10^{-33}$). Report solubility in mg/L.

17 Derive an equation relating aqueous solubility (S) to the solution pH for $Cd(OH)_2$. The K_{sp} value of this compound is 7.2×10^{-15} at 25 °C.

18 Briefly comment the following chemical or group of chemicals with regard to their volatility, aqueous solubility, hydrophobicity, and bioaccumulation potential: (a) BTEXs, (b) Chlorinated solvents (e.g. PCE, TCE), (c) PAHs, and (d) Phenols.

19 Briefly predict the properties and/or fate of the following two chemicals based on the data provided in the table below.

Properties	Contaminant A	Contaminant B
Water solubility (mg/L)	0.0038 (25 °C)	1.1 (25 °C)
K_{ow} (unitless)	1 096 478	3.98
Vapor pressure (Torr)	5×10^{-9} (20 °C)	2660 (25 °C)

20 In a batch study, the K_d values reported by Piatt et al. (1996) (*Environ. Sci. Technol.*, 30:751–760) for naphthalene and pyrene are 1.6 and 8.3 mL/g, respectively. The soil of concern has an organic carbon content of 2%, porosity of 25%, and bulk density of 1.75 g/mL.

a. What fraction of the compound will be present in water phase?
b. What is the organic carbon normalized partition coefficient (K_{oc}) value?
c. What is the traveling velocity of the chemical relative to groundwater?

21 Based on the information given in this chapter, what remediation technology can/cannot be used for the following contaminant: (a) metals (b) pyrene (c) dioxin (d) dichloroethane. You might consider air stripping, pump-and-treat, bioremediation, and incineration.

22 The isotherms in the figure below were obtained by Chiou et al. for naphthalene at room temperature (reprinted with permission from Chiou et al. 1998. Copyright [1998] American Chemical Society). (a) Without using the raw data, schematically determine the K_d values for the four soil/sediment samples; (b) Which soil/sediment has the highest sorption potential for naphthalene, why? (c) If the organic carbon contents for Lake Michigan sediment, Marlette soil, Woodburn soil, and Mississippi River sediment were 0.0402, 0.018, 0.0126, and 0.0040, respectively, what are the K_{oc} values for each soil/sediment? (d) Comment why K_d values for naphthalene vary among four sorbents but not K_{oc}?

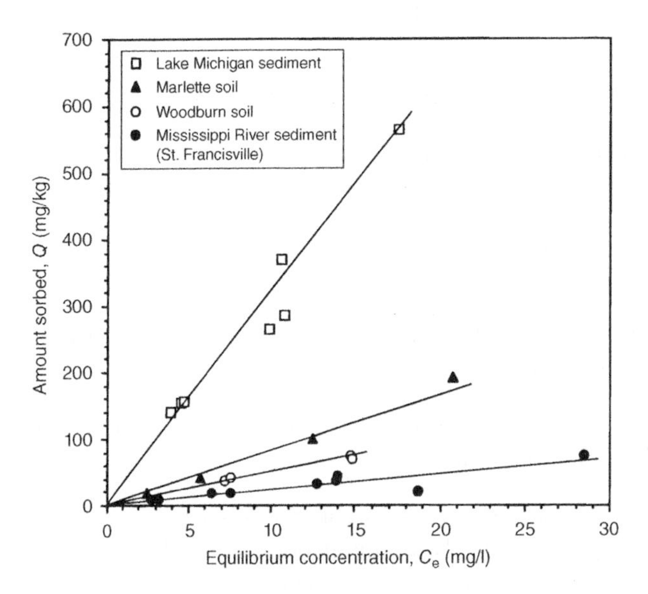

23 Using data provided in Example 2.4, calculate the retardation factors (R) for (a) MTBE in the silty soil, (b) MTBE in the sandy soil, and (c) phenanthrene in the specified soil.

24 Reconstruct Figure 2.9 using Excel to delineate the relationship between f_w and K_d, using porosity (n) value of 0.40 and soil bulk density (ρ) of 2.0 g/cm^3.

25 Reconstruct Figure 2.9, but instead for the plot of f_w, construct a plot between the fraction remained in soil f_s and K_d, assuming the same soil porosity (n) of 0.3 and soil bulk density (ρ) of 1.8 g/cm^3. Note that $f_s = 1 - f_w$.

26 What physicochemical and biological properties make the remediation of MTBE in groundwater especially challenging?

27 Describe the differences among (a) mechanical dispersion, (b) hydrodynamic dispersion, and (c) molecular diffusion.

28 How important is molecular diffusion compared to other chemical movement processes in groundwater? Under what hydrogeological condition, molecular diffusion becomes negligible?

29 Identify the functional group(s) for the pesticides given in Figure 1.4 in Chapter 1.

Chapter 3

Soil and Groundwater Hydrology

LEARNING OBJECTIVES

1. Understand the inorganic and organic constituents of soils and relate minerals and organic matters in soils to soil remediation
2. Articulate the structure of common soil silicates, soil organic composition and the calculation of relevant properties including porosity and density of various types
3. Understand the vertical distribution of aquifers (vadose zone and saturated zone), groundwater wells of various types, and terminologies associated with groundwater wells
4. Define various hydrogeological parameters and perform calculation of porosity, specific yield, hydraulic conductivity, transmissivity, and storativity
5. Use Darcy's law to calculate Darcy's velocity, and differentiate it from groundwater flow velocity (pore velocity) as well as contaminant flow velocity
6. Differentiate flows between vadose zone and saturated zone, and relate this to contaminant fate and transport
7. Understand the complexity of hydraulic conductivity and pressure head as a function of soil moisture contents in unsaturated soil
8. Perform calculation of flow to wells in steady-state confined aquifer regarding drawdown, radius of influence, and capture zone
9. Perform calculation of flow to wells in steady-state unconfined aquifer regarding drawdown and radius of influence
10. Describe how does the flow of dense nonaqueous phase liquids (DNAPLs) differ from that of light nonaqueous phase liquids (LNAPLs) and understand the challenge of DNAPLs remediation under various hydrogeological conditions

The knowledge about soil and groundwater is essential to the understanding of fate and transport of contaminants in the subsurface to evaluate various remediation approaches. The traditional discipline of soil science covers diverse areas of subjects including soil formation, classification, properties (physical, chemical, and biological), and their relation to the use and management of soils as a natural resource. Vastly different from but related to soil science, groundwater hydrology (i.e. hydrogeology) is the area of geology that deals with the distribution and movement of groundwater in soils and rocks of the Earth's crust (known as aquifers). Relevant to soil and groundwater remediation, the most important aspects of soil and hydrogeology are the movement and transport of water and the associated solutes (contaminants) in soil and groundwater. This chapter will briefly provide readers with basic understanding of soil and aquifer properties and principles of water movement. Hydrogeological techniques and methods to measure aquifer properties, also an essential part of hydrogeology, will be introduced in Chapter 5 as an integrated site investigation and subsurface characterization at a

Soil and Groundwater Remediation: Fundamentals, Practices, and Sustainability, First Edition. Chunlong Zhang.

Companion website: www.wiley.com/go/Zhang/Remediation_1e

contaminated site. In discussing principles of water movement, mathematical equations are essential (see Chapter 13); however, these equations are kept minimal in this chapter. For a detailed learning of the soil and groundwater and their governing equations, readers are referred to those classical texts such as Bear (1972), Ellis and Mellor (1995), Fetter (2001), Greenland and Hayes (1978), McBride (1994), and Sposito (2008).

3.1 Soil Composition and Properties

Soils, rocks, and pure minerals are part of the **lithosphere** of the terrestrial planet (lithosphere includes the Earth's crust and uppermost part of the mantle). The terrestrial land area makes up 29% of the total area of the Earth's surface, of which, forest, pasture and meadow, ice-covered land, and arable (farming suitable) land account for 40.9, 31.5, 17.2 and 14.8%, respectively (van Loon and Duffy 2017). Soil is an important natural resource that supports agriculture, forestry, as well as mineral deposits for various uses. It is also a dumping ground for humans to dispose of nonhazardous and hazardous wastes. The section below briefly describes various constituents of soil phases (gas, liquid, solid), as well as physical, chemical, and mineralogical properties relevant to soil and groundwater remediation.

3.1.1 Constituents of Soils

Soils consist of three phases, soil gas, soil liquid, and solid inorganic and organic materials (Figure 3.1). Soil gases and liquids are enclosed in the pores among solid particles. Since all pores below the groundwater table (the saturated zone) are filled with water, the soil gas phase is not present in the deep aquifer. In surface soils (0–15 cm), the amount of soil gas is inversely related to the amount of water. For example, all soil pores can be filled by air in arid soils, whereas soil air can be completely depleted in saturated soils following recent precipitation.

The composition of **soil gas** should be approximate to the typical composition of atmospheric air, i.e. 78.08% N_2, 20.95% O_2, 0.93% argon (Ar), 0.031% CO_2 and other minute gases (Sharma and Reddy 2004). In biologically active soils, O_2 contents decrease and CO_2 contents increase with soil depth. For contaminated soils, contaminants can be present in vapor form in soil pores.

Soil liquids are composed of water and dissolved solutes, including gas (e.g. dissolved oxygen), nutrients, and trace contaminants. For volatile gases, the concentrations of dissolved gases (or solutes) can be approximated by applying Henry's law constant (Section 2.3.1). In extreme cases, such as chemical spills, soil pores can be devoid of gas and completely replaced by a pure solvent. Water in soil pores can be adsorbed tightly by soil particles through various mechanisms, or can be free flowing in the form of free water molecules.

The fraction of soil particles that is **soil solid** deserve more attention not only because of their importance as a structural backbone of soils but also because of their physicochemical interactions that impact contaminant movement. Soil particles have a small amount of organic compounds in the range from <1 to 5% by weight to soil mass. Much higher contents of organic matter occur only in soils derived from peat (>20%) and surface soils in the forest areas (>10%). In contrast, desert soils are almost purely inorganic. Even though the proportion of soil organic matter is small for most soils, it plays a disproportionately significant role in the contaminant fate and transport through sorption of organic contaminants, and chelation reactions with metals and metalloids (**chelation** is the formation of two or more coordinate bonds with organic ligands). Organic matter content of soils is usually determined indirectly by converting the analytically determined organic carbon percentage (i.e. f_{oc} in Eq. 2.11, Section 2.3.3) to percent organic matter by a multiplication factor of 1.724 (Ranney 1969), based on the assumption that soil organic matter contains 58% carbon (i.e. 1/0.58 = 1.724).

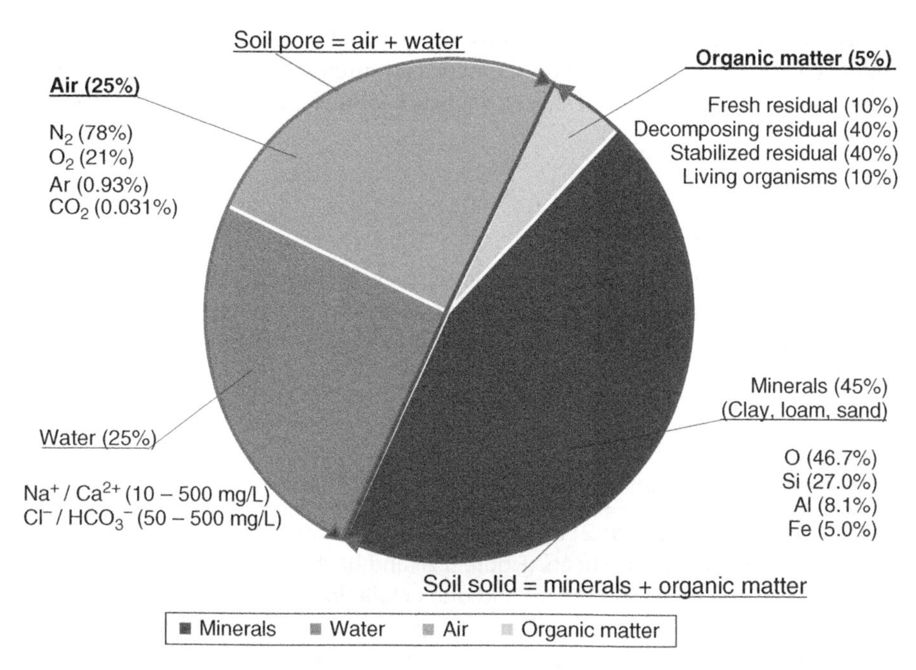

Figure 3.1 Pie chart showing the average composition of various soil components.

The primary sources of **soil organic matter** are the root exudates (a fluid that seeps out from root) from growing plants and decomposition of plant tissues such as leaves and branches that have fallen on the surface. The biomass of soil microorganisms such as bacteria, fungi, actinomycetes, and protozoa also make a small contribution to the total soil organic matter. Carbohydrates, proteins, hemicellulose, cellulose, lipids, and lignins from plant tissues collectively result in the formation of a degradation-resistant soil humic substance, called **humus**. Humus can be further classified based on solubility by first extracting with a strong base and then further acidified with an acid: (a) **humin** (a portion of soil organic compounds that are insoluble and nonextractable at all pHs), (b) **humic acid** (fraction that precipitated from the acidified extract), (c) **fulvic acid** (fraction that remains in the acidified solution). Humic substances are not a single compound, but they are macromolecules with molecular weight ranging from few hundreds for fulvic acid, to tens of thousands for humic acid and humin. They are important to contaminant fate and transport because of their acid–base, sorptive, and complexing properties. The elemental composition of humic substances is generally within the range of 45–55% C; 30–45% O; 3–6% H; 1–5% N, and 0–1% S.

Since more than 95% of the solid is inorganic minerals, the **elemental composition** of soil is dominated by these **soil inorganic constituents**. On average, soil is mainly made up of oxygen (46.7%), silicon (27%), aluminum (8.1%) and iron (5.0%). O, Si, and Al occur as constituents of minerals and as oxides. These **minerals** include carbonate minerals and silicate minerals. Carbonate minerals are mainly calcite ($CaCO_3$) and dolomite [$CaMg(CO_3)_2$] derived from limestone. **Silicate minerals (phyllosilicates**) are derived from various chemical weathering processes of igneous and metamorphic rocks. In coarse-grained soils, these silicate minerals include quartz (SiO_2), feldspar, and mica. In quartz and feldspar, silica (Si) and aluminum (Al), respectively, are arranged in their three-dimensional **tetrahedron sheets**, so named because each Si or Al atom is surrounded by four O atoms in a tetrahedral configuration. Mica is made of one octahedral layer sandwiched between two tetrahedral layers. In **octahedral sheets**, each Al atom is surrounded by six O atoms in its octahedral configuration. In fine-grained soils, silicate minerals are predominately clay minerals formed with various combinations of tetrahedral and octahedral sheets, such as kaolinite, illite, and montmorillonite (see Box 3.1, Figure 3.2).

3.1.2 Soil Physical and Chemical Properties

In this section, physical properties (including particle size, soil texture, density, and porosity) are introduced. For chemical properties, we will focus on cation exchange capacity (CEC). Additional geological and engineering characterizations of soil will be briefly described in Site Characterization in Chapter 5.

Box 3.1 Soil clays: Essential for soil fertility and useful as landfill liners?

Clay minerals can be classified as 1:1 or 2:1, depending on the combinations of tetrahedral silicate sheets and octahedral hydroxide sheets (Figure 3.2a and b). A 1:1 clay such as kaolinite and serpentine consists of one tetrahedral sheet and one octahedral sheet. A 2:1 clay such as talc, vermiculite, and montmorillonite consists of an octahedral sheet sandwiched between two tetrahedral sheets. The distance between two layers in kaolinite is about 1 nm. In montmorillonite, the distance between individual layers is larger (~1.4 nm), allowing hydrated cations, such as K^+, Na^+, Ca^{2+}, or NH_4^+, to readily exchange. Clays generally have negative charges which originate from the substitution of divalent (+2) ions (such as Fe) for Al^{3+} in the octahedral sheet, trivalent (+3) ions (such as Al^{3+}) for Si atoms in the tetrahedral layers, and more importantly Fe(II) or Mg(II) for Al(III) in montemorillonite. These **isomorphous substitutions** occur between ions of similar sizes, which do not result in any significant change in crystal structure but leads to a charge imbalance. The considerable net negative charge must be compensated by cations such as K^+, Na^+, or NH_4^+, leading to the retention of water and nutrients – a process related to the concept of cation exchange capacity (CEC).

The capacity of a soil to retain water and nutrients is also strongly related to particle size (surface area). For example, the fine particles of a clay soil can hold water and nutrient molecules more tightly than coarser particles of a sandy soil. Since clay soils hold more water and nutrients than sandy ones, they also hold it more tightly in smaller pores, so that water in clay is less readily taken up by plants, or less likely to flow toward a well. For these reasons, sandy soils require more frequent irrigation of comparatively small amounts of water, whereas clay soils are usually irrigated with larger amounts at longer intervals for plant growth. Nutrients in clay soils are also not readily leached by irrigated water or rain.

Relevant to remediation, clays are commonly used in the liner system to protect the groundwater from the contamination of landfill leachate. The above-mentioned features of clays suggest that groundwater flow in clays is very slow (more details in Section 3.3). Clay liners are constructed as a simple liner that is 2–5 ft thick. In composite and double liners, the compacted clay layers are usually between 2 and 5 ft thick, depending on the characteristics of the underlying geology and the type of liner to be installed. (Regulations specify that the clay used can only allow water to penetrate at a rate of less than 1.2 in. or 3.0 cm/yr.) The disadvantage of clay liners is their reduced effectiveness by fractures induced by freeze–thaw cycles, drying out, and the presence of some chemicals. These fractures will facilitate the leachate flow through the clay liners to the underlying groundwater. More clay can be added to the liners to safeguard the groundwater in the event of some loss of effectiveness in part of fractured clay layer.

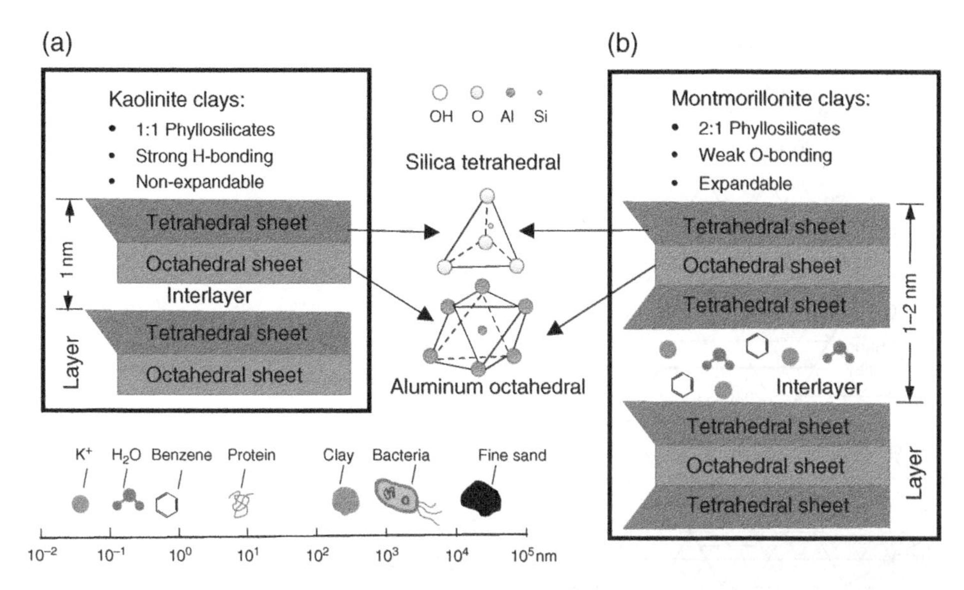

Figure 3.2 Basic units of clays (silicate tetrahedral and aluminum octahedral) and two types of clays (a) kaolinite clays (1 : 1 phyllosilicates), and (b) montmorillonite clays (2 : 1 phyllosilicates). The relative sizes of cations (K^+), water, organic compounds, and bacteria are also compared for their potential fit to the interlayers of the expandable clays.

The International Society of Soil Science defines soil as material with particle size less than 2.0 mm. By this convention, gravels with size >2.0 mm are nonsoil. Soil particles in their increasing sizes are: clay (<2 μm), silt (2–20 μm), fine sand (20–200 μm), and coarse sand (200 μm–2.0 mm). Due to the small size, clays are colloidal in nature with large surface area available for the interactions with contaminants through ion exchange and adsorption reactions. A **colloid** is a microscopically small substance having a particle size in the range of 1 nm–1 μm.

There are two soil classification systems used by remediation professionals. The Unified Soil Classification System (USCS) is based on grain sizes as well as other geological and engineering properties including shear stress and water-holding plasticity, whereas the traditional USDA method is based solely on grain size. It is important to use the consistent method for the same site for consistency, and accurately describe soil samples. In the USDA method (Figure 3.3), **soil texture** is a collective term used to classify 12 different soils according to the proportions of clay, silt, and sand. The nomenclature of soil texture is based on the triangular diagram shown in Figure 3.3. To use Figure 3.3, consider a case for a soil with 20% clay, 70% silt, and 10% sand. First, draw a line at the 20% on the clay axis in parallel to the sand axis, then draw a second line at the 70% silt axis in parallel to the clay axis. The intersection point is in the region known as the "silt loam" and the soil is so named. This is different from soil termed "loam," which contains approximately 40% sand, 40% silt, and 20% clay.

Density and porosity of soils (or aquifers in general) are two related terms. Soils and aquifers composed of materials such as clay, sand, gravel, or fractured rock are collectively called the **porous medium. Porosity** (n) is defined as the ratio of the void volume (i.e. the pore space, V_v) of the soil to the total (bulk) soil volume (V_b), which is the percentage of soil volume available to contain water.

$$n = \frac{\text{void volume}(V_v)}{\text{bulk volume}(V_t)} = \frac{V_b - V_p}{V_b} = 1 - \frac{V_p}{V_b} \qquad \text{Eq. (3.1)}$$

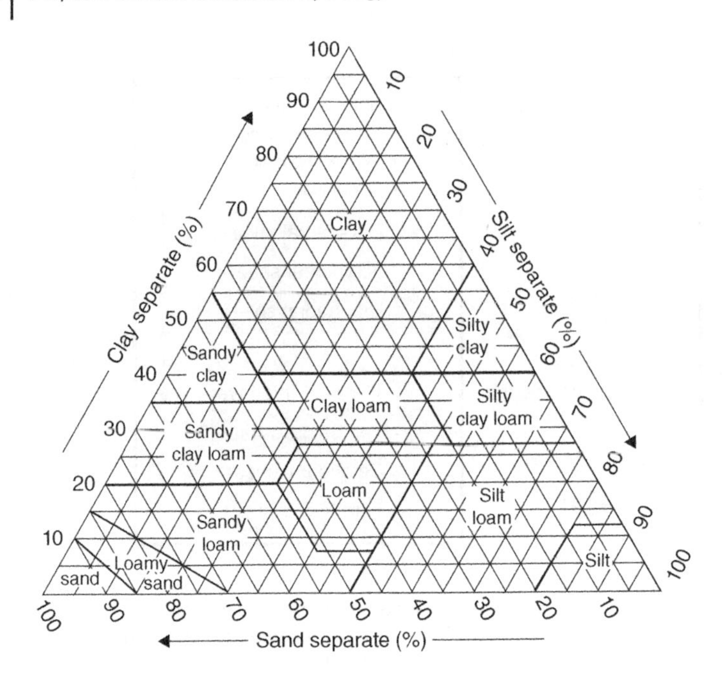

Figure 3.3 Soil texture triangle diagram. *Source:* USGS, 2018, www.nrcs.usda.gov.

where V_b is the bulk soil volume, and V_p is the soil particle volume ($V_v = V_b - V_p$). For soil with a given mass, the volume ratio is inversely related to the density ratio. Hence, if we use ρ_p and ρ_b to denote the particle density and the bulk density, then Eq. (3.1) can be written as:

$$n = 1 - \frac{\rho_b}{\rho_p} \qquad \text{Eq. (3.2)}$$

The density of soil particles (void excluded) depends on the types of materials, and varies from <1 g/mL for organic matter, >5 g/mL for some metal oxides, and >7 g/mL for metal sulfide. The average soil **particle density** is in the range of 2.5–2.8 g/mL. Soil **bulk density** includes the pore spaces between particles, so its value is smaller than soil particle density with an average of 1.2–1.8 g/mL. For example, a soil with a particle density of 2.65 g/mL and bulk density of 1.75 g/mL will have a porosity of 1–1.75/2.65 = 0.34. If n = 0.34, the aquifer can contain water up to 34% of the total volume. If water in all pores is replaced by NAPLs, then the maximal volume of a NAPLs pool that can be removed by pumping is theoretically also 34%.

For practical purposes, the average porosity of an aquifer could be in the range of 30–40% (Table 3.1). However, they vary with soil type, grain size, and packing manner. For example, peat soil may have a porosity of 92% whereas fractured rocks may have a porosity of only 10%. Soils with a uniform particle size have a larger porosity compared to soil containing a mixture of grain sizes because many of the larger pores can be occupied by smaller particles (Figure 3.4). Well-sorted (even size) grain materials have a higher porosity than similarly sized, poorly sorted materials, since some smaller grains can fill the pores where they should be occupied by water. Smaller size grains can drastically reduce porosity and hydraulic conductivity, even though it takes only a small fraction of the total volume of the material.

Effective porosity (n_e) is the portion of the total porosity available for fluid flow. Effective porosity is measured by allowing a liquid to drain under gravity into the pore spaces. The values of effective porosity (n_e) are smaller than the total porosity (n) because water in very small or discontinued pores cannot drain through under gravity. The difference between effective

Table 3.1 Range of porosity for common aquifer materials and their relation to effective porosity (specific yield) and hydraulic conductivity.

Soil type	Porosity (%)	Effective porosity (%)	Hydraulic conductivity[a] m/day	Hydraulic conductivity[a] ft/day
Clay	45	3	4×10^{-6}	1.3×10^{-5}
Sand	34	25	0.4	1.3
Gravel	25	22	40	130
Gravel and sand	20	16	4	13
Sandstone	15	8	0.04	0.13
Limestone, shale	5	2	4×10^{-4}	1.3×10^{-3}
Quartzite, granite	1	0.5	4×10^{-5}	1.3×10^{-4}

[a] These are only representative values for illustration purpose, since the hydraulic conductivity of the same geologic material may differ from aquifer to aquifer.

porosity and porosity is small for coarser sediments but very large for the soils with a high proportion of silt or clay. For example, soils that produce pumpable groundwater include gravel, sand, and silt. The amount of groundwater in clay can be abundant because of the large porosity. Unfortunately, it is often not possible to pump most water from a clay formation because of the commonly low effective porosity. This explains why clays with a high porosity (n) are poor in groundwater yield (Table 3.1). In bedrock, abundant water can generally be found in a highly fractured rock. A few unfractured rock types can also support abundant water, such as sandstone, limestone, and dolomite.

The last important soil parameter to be defined is the cation exchange capacity (CEC). As described previously, clay particles are predominantly negatively charged (anions), and have the ability to hold cations from being "leached" or washed away. The adsorbed cations are subject to replacement by other cations in a rapid and reversible process called "cation exchange." The **cation exchange capacity (CEC)** of a soil is a measurement of the negative charge per unit mass of soil, or the amount of cations in an exchangeable form. CEC is expressed as **milli-equivalents (meq)** of cations per 100 g of soil on a dry basis. For example, a meq of K^+ is the same as 1 mmol of K ions that has approximately 6.02×10^{20} positively charged ions (Avogadro's number = 6.02×10^{23}). A meq of Ca^{2+} is equivalent to 0.5 mmol of Ca^{2+} that contains approximately 3.01×10^{20} positively charged ions because each Ca^{2+} ion carries two positive charges. The greater the clay and organic matter content, the greater the CEC should be. A sandy soil with low organic matter may have CEC of 10 meq/100 g soil, but an organic-rich

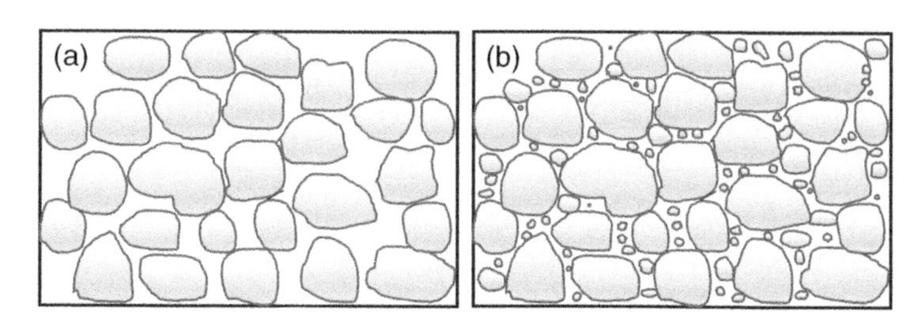

Figure 3.4 Porosity as a function of particle size and the homogeneity of a particle mixture: (a) well-sorted soil particles having high porosity; (b) poorly-sorted soil particles having low porosity.

silty clay may have CEC as high as 40 meq/100 g soil. Cation exchange is an important mechanism for plant nutrient retention and supply in soils and for the adsorption of cationic contaminants (e.g. metals). Sandy soils with low CEC usually have limited adsorption capability and are therefore more susceptible to cationic contaminants in groundwater.

3.2 Basic Concepts of Aquifer and Wells

Groundwater is recharged from precipitation and surface water, and it is also discharged into the surface water depending on the water balance in a particular region. Unlike surface water, however, groundwater flow is slow and it is typically difficult to visualize and predict. Groundwater flow is the subject of hydrogeology that requires special knowledge and skills. In the discussion below, we introduce some basic terms essential to this field.

3.2.1 Vertical Distribution of Aquifer

Think about what happens when water and associated contaminants are introduced into soil surface and moved subsequently downward at its vertical scale. Water will encounter three vertical zones with hydrogeologically very different subsurface formations (Figure 3.5). The first is the **unsaturated zone** (also called **vadose zone**) that extends from the soil surface to the groundwater table. The vadose zone can range from a few feet for high water table conditions

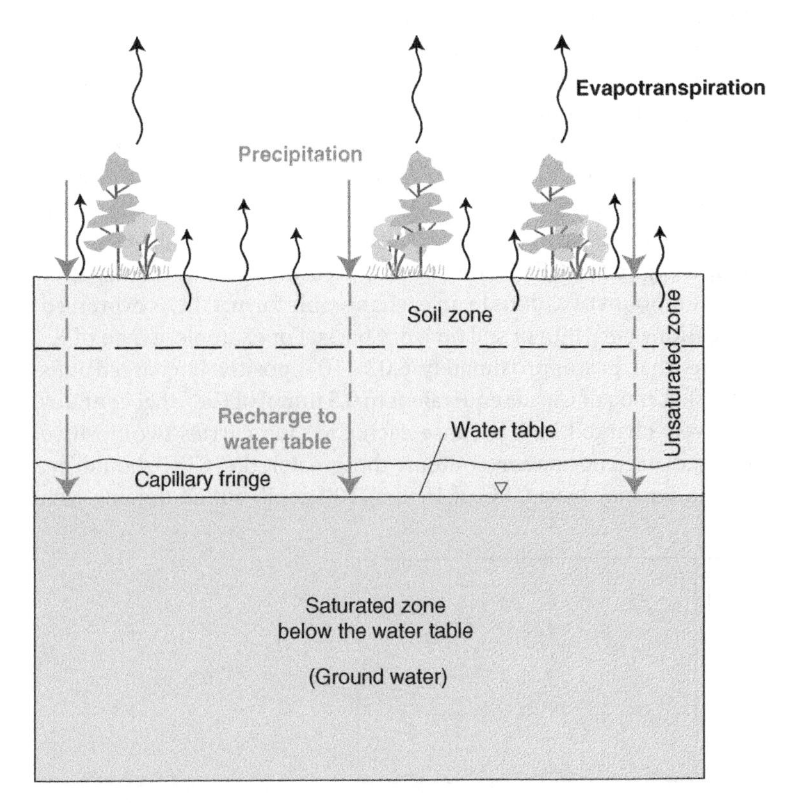

Figure 3.5 Vertical zones of aquifer: The unsaturated zone (vadoze zone), capillary fringe, groundwater table, and saturated zone. *Source:* USGS (1999).

to hundreds of the feet (meters) in arid regions. The second is a **capillary zone** (capillary fringe) which is a region where water is drawn somewhat above the groundwater table by capillary force due to surface tension. The capillary fringe sits just above the groundwater table in the vadose zone, and it can draw up groundwater from the confined aquifer beneath the groundwater table. The capillary rise ranges from 2.5 cm for fine gravel to more than 2 m for silt (Todd 1980). In very fine-grained soils, this capillary fringe can saturate the soil above the water table. The last is a **saturated zone**, which is located beneath the groundwater table but above the confined layer.

The **groundwater table** is the surface where the pressure head of the groundwater is equal to the atmospheric pressure (where gauge pressure = 0). It may be conveniently visualized as the "surface" of the subsurface materials that are "saturated" with groundwater. Here "saturated" means all of the voids of soil particles are filled with water. However, saturated conditions may extend above the water table as surface tension holds water in some pores below atmospheric pressure. Individual points on the water table are typically measured as the elevation that the water rises to in a well screened in the shallow groundwater. Most groundwater is present within 300 ft of the surface, but it can be as deep as 2000 ft (Moore 2002).

Note that water in the unsaturated zone cannot freely flow by gravity. Hence this part of the water cannot be used directly as the groundwater resource. Here, **aquifer** is the term used to define any geological unit capable of storing and supplying significant quantities of water, and can supply significant quantities to a well or spring. There are several other terms associated with aquifer. **Aquitard** is a geological layer of low permeability that can store groundwater and also transmit it slowly from one aquifer to another, also known as the **leaky confining layer**. Besides these, there are units called aquifuge and aquiclude, where **aquifuge** is an absolutely impermeable unit that will not transmit any water, and **aquiclude** is a formation that has very low hydraulic conductivity and hardly transmits water.

There are three types of aquifer: unconfined, confined, and perched aquifers (Figure 3.6). **Unconfined aquifers** are aquifers that have no confining layers between the water level and ground level. A **confining layer** is a layer of material that has little or no porosity. **Confined aquifers** are aquifers that have a confining layer between the water level and ground level. Again, the unconfined aquifer and confined aquifer are separated by the groundwater table where the pore water pressure is equal to atmospheric pressure. The third type, **perched aquifers** (not shown in Figure 3.6) are aquifers that have a confining layer below the groundwater, and sits above the main groundwater table.

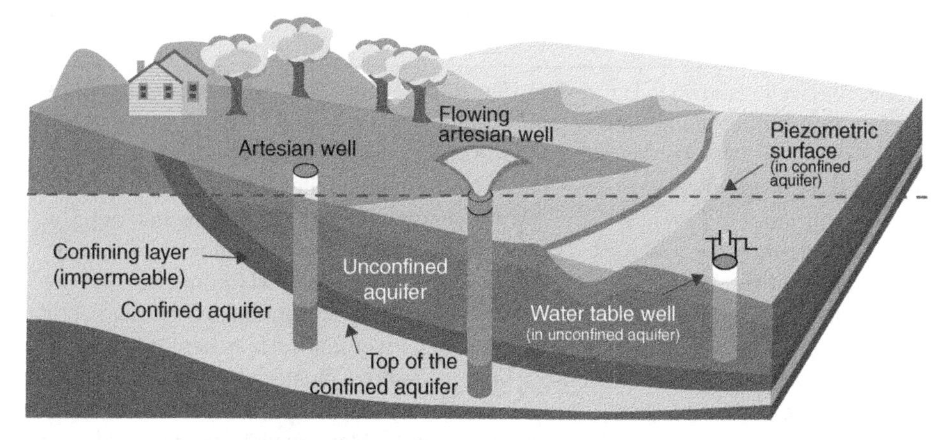

Figure 3.6 Schematic cross-section of unconfined and confined aquifers showing various types of wells. *Source:* Environment Canada, 2018.

3.2.2 Groundwater Well and Well Nomenclature

Groundwater wells have many variations depending on their uses. Wells can be used for the production of drinking water, extracting contaminated water, injecting water for hydraulic control, monitoring hydrogeological parameters, and collecting samples for chemical analysis. Production or pumping wells are large diameter (>15 cm in diameter) wells cased with metal, plastic, or concrete, and constructed for extracting water from the aquifer by a pump (if the well is not artesian). **Monitoring wells** or **piezometers** are often smaller diameter wells used to monitor the hydraulic head or sample the groundwater for chemical constituents. Piezometers can be installed over a very short section of an aquifer, but they can also be completed at multiple levels, allowing discrete samples to be collected or measurements to be made at different vertical elevations at the same location.

As can be seen in Box 3.2 and Figure 3.7, various wells can be named depending on whether the well is located in an unconfined or confined aquifer. A **water table well** is situated in an unconfined aquifer; its water table rises or declines in response to rainfall and changes in the stage of surface water to which the aquifer discharges. An **artesian well** draws water from a confined aquifer which is overlain by a relatively impermeable (confining) unit (aquitard). The water level in artesian wells stands at some height above the water table because of the pressure (artesian pressure) of the aquifer. The level at which it stands is the potentiometric (or pressure) surface of the aquifer. If the potentiometric surface is above the land surface, a flowing well or spring results and is referred to as a **flowing artesian well**. The **potentiometric surface** is an imaginary level to which water will rise in a well. In a confined aquifer, this surface is above the top of the aquifer unit; whereas, in an unconfined aquifer, it is the same as the groundwater table.

3.2.3 Hydrogeological Parameters

This section describes several important hydrogeological parameters, including specific yield, hydraulic conductivity, permeability, transmissivity and storativity.

3.2.3.1 Specific Yield and Specific Retention

Specific yield (Y), also known as the drainable porosity, is a hydrogeological parameter related to effective porosity. It is the ratio of the volume of water that drains from a saturated aquifer due to gravity to the total aquifer volume. This ratio is less than or equal to the effective

Box 3.2 Groundwater well nomenclature

A typical monitoring well is shown in Figure 3.7 to illustrate several terminologies one should be familiar with for groundwater monitoring and remediation.

The steel or PVC plastic pipe that extends from the surface to the screened zone is called the **casing**. The **well screen**, a portion of the well that is open to the aquifer, is typically a section of slotted pipe allowing water to flow into the well while coarse soil particles are screened out. The annular space around the screen, often filled with sand or gravel to limit the influx of fine aquifer materials, is termed the **filter pack**. Above the filter pack is an **annular seal** (grout), which is the seal around the annulus of the well casing (the void immediately surrounding the casing). It is usually cemented up to the surface with a low permeability material such as bentonite clay to protect leakage from above the screened interval or percolating rainwater. The centralizers keep casing centered in the wellbore during the well installation. The materials used to install a monitoring well are critical for the quality of monitoring data. Many environmental regulatory agencies therefore require that wells be installed by licensed drillers. Common monitoring wells are 2 and 4 inch in diameter and are installed in 6 and 10 inch diameter boreholes.

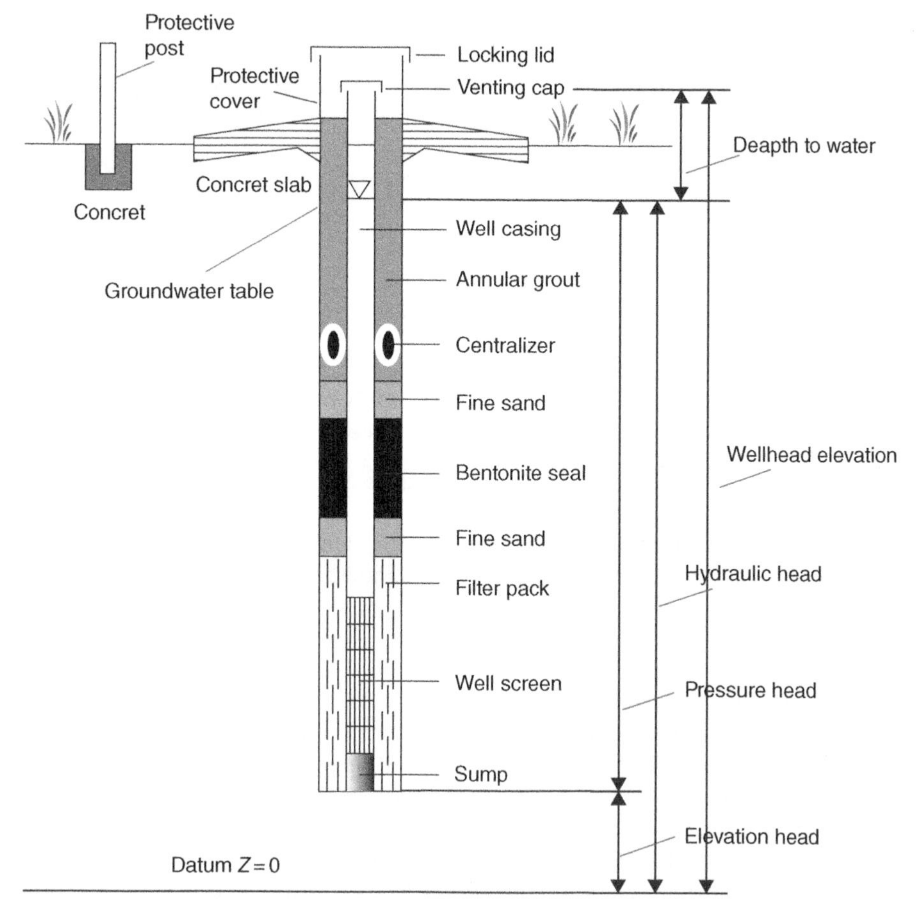

Figure 3.7 Schematic of a monitoring well. *Source:* Adapted from the Louisiana Department of Environmental Quality, 2000.

porosity, indicating that not all the water can be drained out of the aquifer under the forces of gravity. Opposite to the specific yield, **specific retention** (R) is the ratio of volume of water an aquifer material can retain against gravity drainage to the total aquifer volume. It is apparent that the total porosity value is equal to the sum of specific yield and specific retention.

$$n = \mathrm{Y} + \mathrm{R} \qquad \text{Eq. (3.3)}$$

Hence, for an aquifer with a unit volume of 1 m^3, if the drainable water is 0.2 m^3 and the retained water (not drainable) is 0.15 m^3, then the specific yield is 0.2 and the specific retention is 0.15. The total porosity for this aquifer is 0.35 or 35%. See Example 3.1 for the calculations related to these concepts. If we know the total volume of a contaminated aquifer (V_T), the volume of drainable contaminated groundwater (V) can be calculated with a given porosity and specific yield as follows:

$$\text{Groundwater volume}(\mathrm{V}) = \text{specific yield}(\mathrm{Y}) \times \text{total volume}(\mathrm{V_T}) \qquad \text{Eq. (3.4)}$$

Specific yield can be determined in the lab through a soil column test, or in the field through a pumping test. In the lab, a soil sample with a known volume is fully saturated in a soil column by slowly flooding from the bottom, allowing the air to escape upward. The water is then allowed to drain by gravity from the column and the volume is measured. The ratio of the volume of water drained to the volume of the soil column is the specific yield.

Example 3.1 Porosity, specific yield and specific retention

An aquifer has an average bulk density of 1.85 g/cm^3 and a soil particle density of 2.65 g/cm^3. The specific yield is 20%. How much water can be drained per cubic meter of aquifer? How much water can be retained per cubic meter of aquifer?

Solution

First, the porosity is calculated according to Eq. (3.2):

$$n = 1 - \frac{\rho_b}{\rho_p} = 1 - \frac{1.85}{2.65} = 0.30(30\%)$$

Volume of total drainable and retained water = 0.30 × 1 m^3 = 0.30 m^3. Since the specific yield is 20%, the volume of drainable water = 0.20 × 1 m^3 = 0.20 m^3. Hence the volume of water retained by unit volume of soil (1 m^3) is 0.3 − 0.2 = 0.1 m^3.

Specific yield and specific retention are related to grain size. Hence, if two soil samples have the same porosity but different grain sizes such as clay and sand, the sample with smaller grain sizes will have a lower specific yield. Specific yield will be very close to the total porosity for coarse porous media (Table 3.1).

3.2.3.2 Hydraulic Conductivity and Permeability

Hydraulic conductivity (K) and permeability (or intrinsic permeability, k) are two important parameters in the field of hydrogeology. They are often used interchangeably, yet the preference is commonly given to the use of hydraulic conductivity in groundwater hydrology and permeability used in the petroleum industry. Permeability is defined as the ability of an aquifer material to transmit a fluid; its value depends only on the porous media (i.e. soil, subsurface strata). Normally porosity and permeability correspond to each other, and a high porosity always means high permeability; however, this relationship is not applied for materials like clay and silt. Permeability is independent of the fluid (water) but "intrinsic" to porous media. The faster the fluid can travel through the aquifer, the higher the permeability is.

While **hydraulic conductivity** is also the measure of aquifer's ability to transmit water, it is dependent on both aquifer material (hence the intrinsic permeability, k) and the properties of fluid (water), as can be seen from the density and viscosity of water in the following equation:

$$K = k\frac{\rho g}{\mu} \qquad \text{Eq. (3.5)}$$

where ρ = density of water (M/L^3; g/cm^3 or kg/m^3), g = gravitational constant (L/T^2; 9.81 m/s^2), μ = dynamic viscosity of water (M/L·T). Equation (3.5) indicates that an aquifer's ability to transmit water increases for higher density of fluid (ρ) and is inversely related to the viscosity of the fluid (μ). The units of K and k are different as the dimensional analysis shows below:

$$K = k\left(\mathrm{L}^2\right)\frac{\rho\left(\frac{\mathrm{M}}{\mathrm{L}^3}\right)\times g\left(\frac{\mathrm{L}}{\mathrm{T}^2}\right)}{\mu\left(\frac{\mathrm{M}}{\mathrm{L}\times\mathrm{T}}\right)} = \frac{\mathrm{L}}{\mathrm{T}}$$

The unit for k is the square of length (L^2; m^2 or ft^2), and the unit for K is the velocity unit (L/T; m/s or ft/s) which is equivalent to L^3/L^2·T such as m^3/m^2·day or gallon per unit square foot of area per day (gal/ft^2·day). In a layman's term, hydraulic conductivity is the ease with which water is conducted through a porous material. In a mathematical term, hydraulic conductivity is the proportionality constant in Darcy's law (Eq. 2.17 in Chapter 2). Since the hydraulic gradient (dh/dl) and porosity (n) are both dimensionless, K has a velocity unit of L/T (m/s, ft/s, cm/s). It relates to the flow velocity, but it is not the same.

Since hydraulic conductivity is such an important hydrogeological parameter in relating water flow in an aquifer, a quantitative understanding of its value is essential. The greater the hydraulic conductivity, the less resistance to the groundwater flow. As Figure 3.8 shows, hydraulic conductivity can vary by several orders of magnitude. Whereas a sand aquifer has a K value of several meters per day (permeable), a clay aquifer may have a K value of only 10^{-6} m/day (0.365 mm/day) (impermeable). This agrees with what we have discussed in Box 3.1 as to why a clay liner can be used in landfill design to prevent leachate from contaminating underlying groundwater.

Note that in groundwater modeling, a homogeneous aquifer is often assumed. A **homogeneous aquifer** essentially has the same hydraulic conductivity in any area of the aquifer. If the hydraulic conductivity differs from one location to another, the aquifer is then said to be **heterogeneous**. Hydraulic conductivity may also differ in three different directions (x, y, and z). An **isotropic aquifer** has the same hydraulic conductivity in all directions ($K_x = K_y = K_z$), whereas an **anisotropic aquifer** has different hydraulic conductivities in different directions.

3.2.3.3 Transmissivity and Storativity

Transmissivity (T) is the rate at which water flows horizontally (in the flow direction along x-axis) through a unit width (e.g. w = 1 ft along the y-axis) of an aquifer under a unit hydraulic gradient (i.e. dh/dl = 1 ft/1 ft = 1). Hence,

$$T = \frac{Q}{w \times \frac{dh}{dl}} \qquad \text{Eq. (3.6)}$$

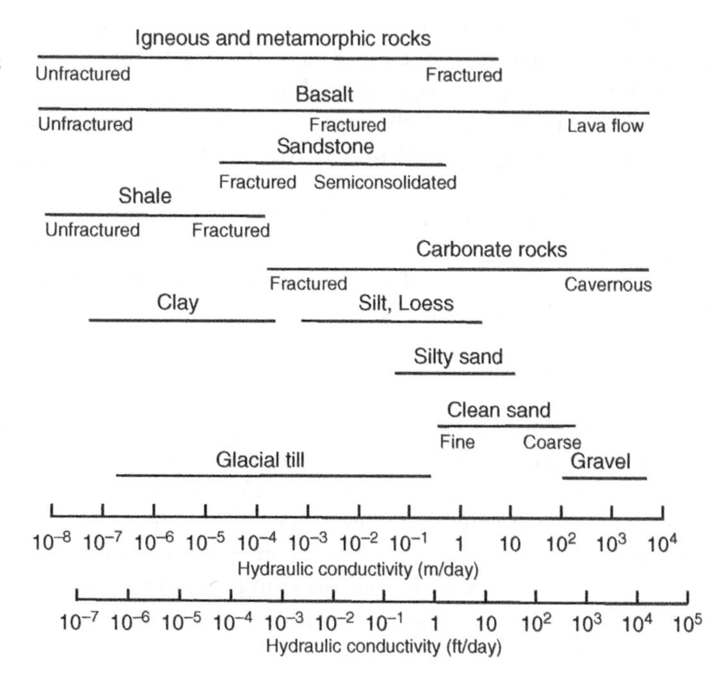

Figure 3.8 Hydraulic conductivity in relation to various aquifer materials. *Source:* Adapted from Heath (1983).

where Q is the volumetric water flow (L^3/T; m^3/day), and it is the product of flow velocity (v in L/T; m/s or ft/s) and cross-sectional area (A in L^2; m^2 or ft^2) perpendicular to the flow path:

$$Q = v \times A \qquad \text{Eq. (3.7)}$$

We now substitute the Darcy's velocity ($v = K\,dh/dl$) and the cross-sectional area perpendicular to the flow direction (x-direction) ($A = b \times w$) into the above equation, and rearrange:

$$Q = v \times A = \left(K\frac{dh}{dl}\right)(b \times w) = (K \times b) \times w\frac{dh}{dl} \qquad \text{Eq. (3.8)}$$

where w = width of aquifer perpendicular to the flow direction (y-direction) and b = saturated aquifer depth (z-direction). Substituting Eq. (3.8) into (3.6) and rearranging the above equation, we get:

$$T = \frac{Q}{w \times \frac{dh}{dl}} = \frac{(K \times b) \times w\frac{dh}{dl}}{w \times \frac{dh}{dl}} = K \times b \qquad \text{Eq. (3.9)}$$

The above equation also indicates that transmissivity (T) is indeed the rate of water transmitted (Q in volume per unit time) per unit width of aquifer (w) per unit hydraulic gradient (dh/dl). When aquifer width (w) and hydraulic gradient (dh/dl) are both in unity (i.e. $w = 1$ ft, $dh/dl = 1$), T is numerically equal to the volumetric flow rate Q. Thus, transmissivity is a measure of how much water can be transmitted horizontally, such as to a pumping well. Transmissivity is also equal to the product of hydraulic conductivity (K) and the saturated thickness of the aquifer (b) (Eq. 3.9).

Transmissivity has a unit of L^2/T (m^2/day, ft^2/day), or it is better perceived as $L^3/L{\cdot}T$ (gal/day × ft) to reflect the above-mentioned definition. It is directly proportional to the aquifer thickness (b). For a confined aquifer, T remains constant since the saturated thickness remains constant. The aquifer thickness of an unconfined aquifer is from the base of the aquifer (or the top of the aquitard) to the water table. If the water table fluctuates, the transmissivity of the unconfined aquifer will change accordingly.

Storativity or **storage coefficient** (S) is a dimensionless number. It is defined as the volume of water that an aquifer will release from or take into storage per unit surface area of the aquifer (not flow cross-section area) per unit change in hydraulic head (not unit hydraulic gradient as defined for transmissivity). It is a measure of the ability of an aquifer to release groundwater from storage, due to a unit decline in hydraulic head. In both confined and unconfined aquifers, the volume of groundwater drained (V) due to a lowering of hydraulic head (Δh) can be calculated as:

$$V = S \times A' \times \Delta h \qquad \text{Eq. (3.10)}$$

where A' is the surface area overlying the drained aquifer. As a dimensionless quantity, storativity ranges between 0 and the effective porosity (n_e) of the aquifer. For a confined aquifer, storativity is in the range of 0.00005–0.005, and for an unconfined aquifer its range is 0.07–0.25. This order of magnitude difference reflects the greater storage associated with draining pore spaces in the unconfined aquifer compared to confined aquifers which release very little water. As a consequence of their low storage, confined aquifers respond quickly to recharge events or pumping, whereas in unconfined aquifers, longer pumping periods are required to achieve the same drawdown.

3.3 Groundwater Movement

Groundwater flows from regions of higher energy to lower energy. The energy of groundwater takes the forms of kinetic energy (due to the motion of fluid velocity), gravitational potential energy (due to the elevation from a reference datum), and pressure energy (due to the weight of overlying water and rocks). Box 3.3 below illustrates the law of conservation of energy and associated concepts when it is applied to groundwater system.

Box 3.3 Bernoulli's equation: What is head and hydraulic head?

Bernoulli's equation is an important equation in fluid dynamics. This equation was named after the Swiss scientist Daniel Bernoulli who published this principle in his book *Hydrodynamica* in 1738. It can be derived from the principle of conservation of energy. Bernoulli's equation states that, in a steady flow, the sum of all forms of energy in a fluid along a streamline is the same at all points on that streamline. Mathematically, we have the energy per unit weight of fluid:

$$H = Z + \frac{P}{\rho g} + \frac{v^2}{2g} = \text{constant}$$

where H = total head (energy), Z = the elevation of fluid above a reference plane, with the positive direction pointing upward (opposite direction to the gravitational acceleration), P = pressure of the fluid, g = the acceleration due to gravity (9.80 m/s^2), ρ = density of the fluid, v = fluid flow velocity at a point on a streamline. Bernoulli's equation states that the total energy at a given point in a fluid is the energy from the height of the fluid relative to an arbitrary datum (elevation head due to gravity), plus energy from pressure in the fluid (pressure head), plus the energy associated with the movement of the fluid (the velocity head).

On several instances in Chapters 2 and 3, we introduced the concept of head, pressure head, and hydraulic head. We now can use the Bernoulli's equation to calculate these three components of energies by assuming some representative values: P = 1200 N/m^2, v = 5×10^{-6} m/s for a fluid (density = 1.00 × 103 kg/m^3) at 0.5 m above a reference elevation. Note that 1 N = 1 kg × m/s^2.

$$H = 0.5\,\text{m} + \frac{1200\,\frac{\text{N}}{\text{m}^2}}{1.00\times10^3\,\frac{kg}{\text{m}^3}\times 9.80\,\frac{\text{m}}{\text{s}^2}} + \frac{\left(5\times10^{-6}\,\frac{\text{m}}{\text{s}}\right)^2}{2\times9.80\,\frac{\text{m}}{\text{s}^2}} = 0.5\,\text{m} + 0.12\,\text{m} + 1.28\times10^{-12}\,\text{m} = 0.62\,\text{m}$$

The above calculation indicates the following: (i) Three energy terms can be expressed in the unit of height of an equivalent static column of water (L; m or ft), thus energies are the "head," i.e. the elevation head (Z), the pressure head ($p/\rho g$), and the kinetic (velocity) head ($v^2/2g$). (ii) The velocity of groundwater flow in porous media is so small that the kinetic energy (i.e. 1.28×10^{-12} m) can be safely ignored. This means that the elevation and pressure are the two driving forces for groundwater flow in porous media.

Take a groundwater well for example. As schematically shown in Figure 3.7, since the velocity head is considered zero, the total hydraulic head is then the sum of elevation head of a well screen and the pressure head. The total hydraulic head can be calculated from the elevation head of the wellhead and depth to water (DTW) in a specialized water well (piezometric well) as follows:

Hydraulic head = Wellhead elevation − Depth to water (DTW)

Again referring to Figure 3.7, if DTW is 5 ft and wellhead elevation is 35 ft, then the hydraulic head is 35 − 5 = 30 ft. The hydraulic head is the distance (m or ft) from the groundwater table (water level in the well) to the reference datum. This distance is equivalent to the energy the well water has in reference to the geodetic datum. The geodetic datum is the reference datum with zero elevation that is used to measure height (altitude) or depth above and below the mean sea level. It becomes clear that this simplified energy unit of m or ft., rather than the common unit for energy such as Joules (J), adds a lot of convenience in hydrogeological calculations.

As further explained below, the pressure energy (head) in the saturated aquifer vastly differs from that in the unsaturated aquifer. Thus in the following section, we discuss the flow of water in these two aquifer zones separately.

3.3.1 Flow in Saturated Zone

One of the striking differences for groundwater flow in a saturated zone compared to the flow in an unsaturated zone is the ability of water to flow freely under gravity. This free flow is driven by the hydraulic head (dh) or more accurately the dimensionless hydraulic gradient along two adjacent points or wells (dh/dl) in the flow direction, as stated by Darcy's law we introduced in Chapter 2:

$$v = -K\frac{dh}{dl}$$

Darcy's law states that the velocity increases as the hydraulic gradient increases with a proportionality constant. This constant is the hydraulic conductivity we defined in the previous section. It is very important not to confuse **Darcy's velocity** (v) with two other velocity terms, i.e. the groundwater flow velocity and the contaminant movement velocity along the groundwater flow path. The different names of these three velocity terms and their formulas are listed in Table 3.2 for comparison, followed by Example 3.2 to calculate these different velocities. Note that Eq. (2.18) in Chapter 2 was referred to calculate the velocity of contaminant (v_c) using the retardation factor R. The calculations shown in Example 3.2 should help differentiate these three velocities.

Darcy's velocity is neither the velocity of groundwater flow, nor the velocity of contaminant movement (v_c). In fact, the groundwater velocity, or the **pore velocity** (v_p), will be larger than Darcy's velocity because water can only flow through connected pores.

Table 3.2 Darcy's velocity, groundwater velocity, and contaminant velocity.

Velocity terms	Other names	Governing equation	
Darcy's velocity (v)	Specific discharge Effective velocity	$v = -K\frac{dh}{dl}$	
Groundwater velocity (v_p)	Pore velocity Seepage velocity Linear velocity Flow front velocity	$v_p = \frac{v}{n} = -\frac{K}{n}\frac{dh}{dl}$	Eq. (3.11)
Contaminant velocity (v_c)	None	$v_c = \frac{v_p}{R} = \frac{v_p}{1 + K_d\frac{\rho}{n}}$	Eq. (3.12)

Example 3.2 Darcy's velocity, seepage velocity and contaminant velocity

A confined aquifer 20 m thick has two monitoring wells spaced 500 m apart along the direction of groundwater flow. The difference in water levels between the two wells is 2 m. The hydraulic conductivity is 50 m/day. Assume the aquifer has a porosity of 35% and bulk density of 1.8 g/cm^3. (a) Estimate Darcy's velocity and the volumetric flow rate per meter of aquifer width perpendicular to the flow. (b) Estimate the seepage velocity (actual velocity of nonsorbing chemical). (c) Estimate the traveling time from the upgradient to the downgradient well. (d) If methyl tertiary-butyl ether (MTBE) (K_d = 0.2 L/kg) has contaminated the upgradient well, how much time it will take for MTBE to reach the downgradient well? (e) What if this contaminant is trichloroethene (TCE) with a K_d value of 5.2 L/kg?

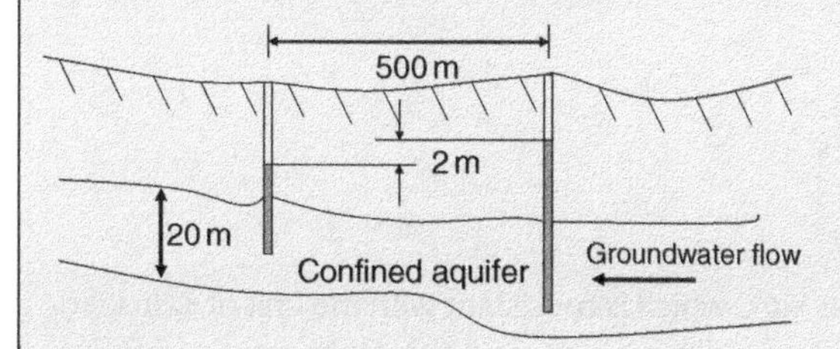

Solution:

a) The gradient = dh/dl = −2 m/500 m = −0.004. Using Darcy's law, we have Darcy's velocity (v) and the rate of flow (Q):

$$v = -K\frac{dh}{dl} = -50\frac{\text{m}}{day}(-0.004) = 0.2\frac{\text{m}}{day}$$

$$Q = v \times A = 0.2\frac{\text{m}}{day} \times (1\,\text{m} \times 20\,\text{m}) = 4\frac{\text{m}^3}{day}$$

where 1 and 20 m in the above calculation denote the assumed unit aquifer width and the actual depth of the aquifer, respectively.

b) The seepage velocity or the actual velocity for the nonsorbing chemical is:

$$v_p = \frac{v}{n} = \frac{0.2\frac{\text{m}}{day}}{0.35} = 0.57\frac{\text{m}}{day}$$

c) The traveling time from the upgradient to the downgradient well: t = 500 m/(0.57 m/day) = 877 days = 2.4 years!

d) The retardation factor (R), velocity, and traveling time for MTBE:

$$R = 1 + K_d\frac{\rho}{n} = 1 + 0.2\frac{\text{L}}{kg}\frac{1.8\frac{\text{g}}{cm^3}}{0.35} = 2.03$$

$$v_c = \frac{v_p}{R} = \frac{0.57\frac{\text{m}}{day}}{2.03} = 0.28\frac{\text{m}}{day}$$

$$t = \frac{L}{v_c} = \frac{500\,\text{m}}{0.28\dfrac{\text{m}}{day}} = 1786\,\text{days} = 4.9\,\text{years}$$

(e) The retardation factor (R), velocity, and travel time for TCE:

$$R = 1 + K_d \frac{\rho}{n} = 1 + 5.2\frac{\text{L}}{kg}\frac{1.8\dfrac{\text{g}}{cm^3}}{0.35} = 27.7$$

$$v_c = \frac{v_p}{R} = \frac{0.57\dfrac{\text{m}}{day}}{27.7} = 0.0206\frac{\text{m}}{day}$$

$$t = \frac{L}{v_c} = \frac{500\,\text{m}}{0.0206\dfrac{\text{m}}{day}} = 24272\,\text{days} = 66\,\text{years}$$

Note that the unit of ρ in g/cm^3 is the same as kg/L, which is consistent with the unit of K_d in L/kg. The above example indicates that the groundwater flow is slow and the movement of sorbing contaminant is even slower. Unlike contaminants in surface water, it may take many years of pumping efforts to clean up sites that could be contaminated decades ago! In fact the hydraulic conductivity of 50 m/day (164 ft/day) assumed in this problem is for an aquifer made of well sorted sand and gravel (Table 3.2). Groundwater velocity is generally in the order of 1 ft/day to 1 ft/year (30 cm/day to 30 cm/year) throughout the United States, depending on the hydraulic conductivity and the gradient of groundwater system. Hence it is possible that the well water we drink today perhaps is the water recharged from the surface two centuries ago!!!

If we want to know the travel time for groundwater to reach a specific location (e.g. a domestic water supply well), then the time required = distance/v_p. Similarly, if we want to calculate the time required for a specific contaminant to travel a certain distance, then the time required = distance/v_c. Eq. (3.12) indicates that when $R = 1$ (i.e. nonsorbing or conservative chemicals, such as NaCl), the contaminant will travel at the same velocity as groundwater. For all the hydrophobic organic contaminants, the **contaminant velocity** will decrease as the K_d value (sorption) increases.

At this point, one should be cautious about the applicable conditions for Darcy's law. In using Darcy's law, one should ascertain if the flow is in the laminar flow region where a fluid flows in parallel layers without turbulence (disruption). Darcy's law applies when flow is laminar or the dimensionless **Reynolds number** (Re) < 1.

$$Re = \frac{v_p\, d}{v_k} \qquad \text{Eq. (3.13)}$$

where v_p = pore velocity of groundwater flow (L/T; m/day), d = mean soil particle diameter (L; m), v_k = kinematic viscosity (L^2/T; m^2/day). If we use d = 2 mm (0.002 m), and water viscosity at 10 °C = $1.31 \times 10^{-6}\ m^2/s$, then the Reynolds number (Eq. 3.13) for Example 3.2 is:

$$Re = \frac{v_p\, d}{v_k} = \frac{0.57\dfrac{\text{m}}{day} \times 0.002\text{m}}{1.31\times10^{-6}\dfrac{\text{m}^2}{\text{s}} \times \dfrac{24\times60\times60\,\text{s}}{\text{day}}} = 0.01(\text{dimensionless})$$

Therefore, it is a laminar flow ($Re < 1$) and Darcy's law is valid. Since groundwater flow is typically slow, the flow of groundwater is generally considered to be laminar. Only in a very coarse-grained aquifer, the groundwater flow can be turbulent.

3.3.2 Flow in Unsaturated Zone

As the name implies, the water (moisture) content in the unsaturated (vadose) zone is not saturated. In an unsaturated zone, the amount of water (hence the degree of saturation) depends on recent rainfall, infiltration, evapotranspiration by vegetation, and groundwater recharge through capillary force. For example, following rainfall or irrigation, the initially saturated surface soil can lose water over time due to evapotranspiration (Figure 3.9a). In shadow groundwater, dry surface soil can also gain water over time from groundwater in the saturated zone through capillary force (Figure 3.9b). Generally, the water content in the unsaturated zone increases toward the groundwater table. Since soil pores are only partially filled with water, the moisture content (θ) is less than the porosity (n) in the unsaturated zone.

Unlike water in a saturated zone, water in the unsaturated zone is held in place by surface forces, and the water in these partially filled and disconnected soil pores cannot flow freely under gravity. As a result, the fluid pressure P in the unsaturated zone is less than atmospheric pressure, so the pressure head ψ is less than zero (Figure 3.9c). This negative pressure can be better understood if one inserts a water-filled porous cup in the unsaturated soil. Because of the capillary suction, water (moisture) will be pulled out from the cup to saturated soil particles surrounding the cup. If a pressure-sensing device is connected to this porous cup, the increased vacuum inside the cup (or the negative capillary pressure) can be measured. This device is a **soil tensiometer**.

In a significant contrast to an unsaturated soil, soil pores below the groundwater table are filled with water, and the moisture content (θ) equals the porosity (n). The fluid pressure P is greater than atmospheric pressure, so the pressure head ψ (measured as gauge pressure) is greater than zero. The hydraulic head h must be measured with a **piezometer** in the saturated zone. To better understand this pressure change, it is of importance to note that pressure at the groundwater table is zero at transition from the unsaturated zone to the saturated zone.

Figure 3.9 reveals that both moisture content (θ) and pressure head (ψ) are functions of the soil depth. As a result, the hydraulic conductivity, $K(\theta)$, in the unsaturated zone is now a function of moisture content (θ) (Figure 3.9). This is in a significant contrast with constant K_s in the saturated zone, where the moisture is constant in the saturated zone and the hydraulic conductivity (K_s) is not a function of the pressure head (ψ). However, Darcy's law discussed in the previous section for saturated zone still applies in the unsaturated zone.

$$v = -K(\theta)\frac{dh}{dz} \qquad \text{Eq. (3.14)}$$

where v = Darcy's velocity; h = potential head = $z+\psi$, where z = depth below surface and ψ = tension, suction, or the pressure head; θ = volumetric moisture content; and $K(\theta)$ is the moisture-dependent unsaturated hydraulic conductivity. Equation (3.14) indicates that the water movement in the unsaturated zone is much more complicated to quantify mathematically than what is occurring in the saturated zone. The governing equations for the flow and transport of water and solute in the vadose zone is a nonlinear partial differential equation and its discussion will be presented in Chapter 13.

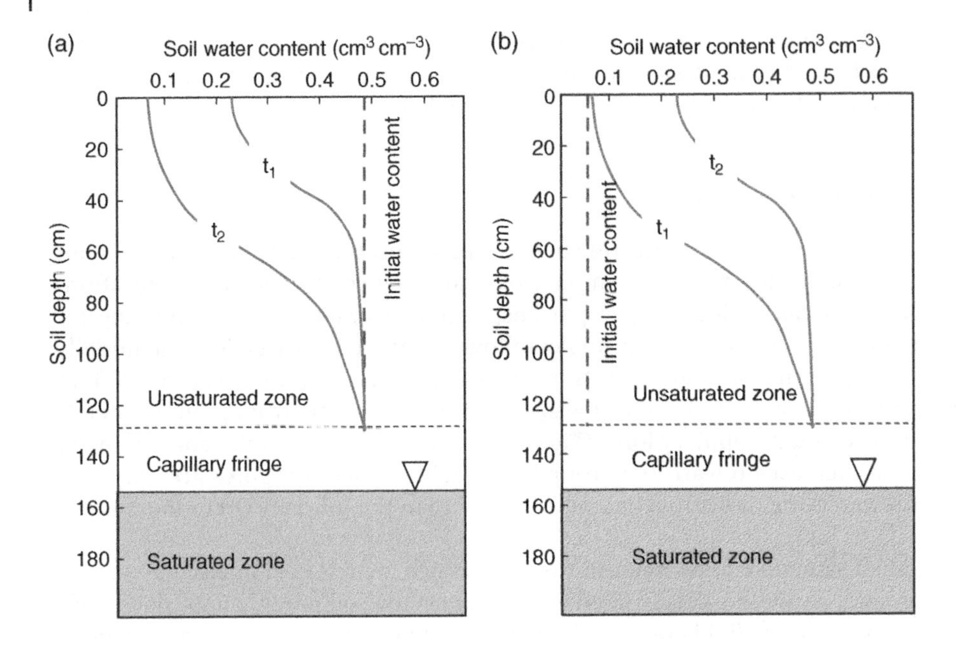

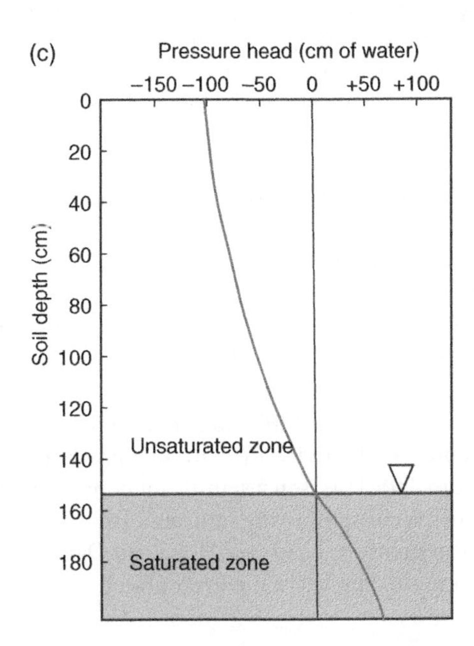

Figure 3.9 Soil moisture (θ) and pressure head (ψ) as a function of soil depth: (a) Change of soil moisture profile over time following rainfall or irrigation at two successive times ($t_1 < t_2$), (b) Change of soil moisture profile over time due to upward movement of groundwater through capillary force ($t_1 < t_2$), (c) a typical pressure head (ψ) as a function of soil depth from unsaturated zone to saturated zone.

Because of the importance in controlling water and contaminant movement in the unsaturated zone, it is noteworthy to further examine the relationship between soil moisture (θ) and $K(\theta)$ as well as ψ. As shown in Figure 3.10, the relationship between the pressure head (ψ) and volumetric water content (θ) for a given soil (commonly known as **the moisture**

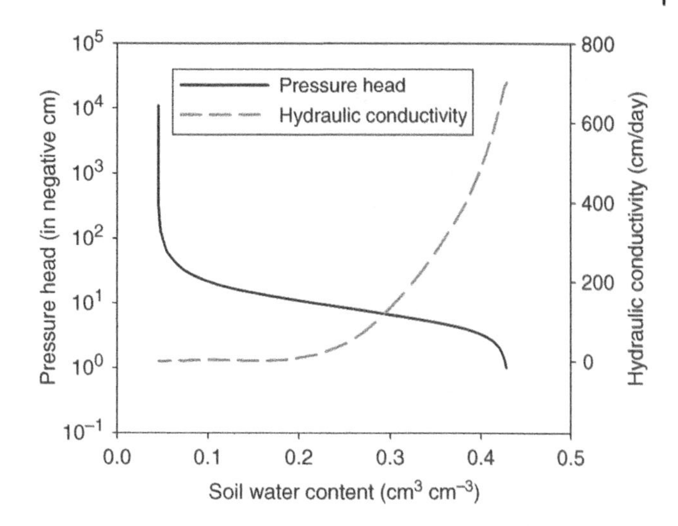

Figure 3.10 Pressure head (ψ) and unsaturated hydraulic conductivity (K) as a function of soil moisture content (θ). Graph generated using parameter values reported by Kosugi et al. (2002) for a sandy soil: $\theta_s = 0.43\,\text{cm}^3\,\text{cm}^{-3}$, $\theta_r = 0.045\,\text{cm}^3\,\text{cm}^{-3}$, $\alpha = 0.145\,\text{cm}^{-1}$, $n = 2.68$, $b = 1$, $K_s = 712.8\,\text{cm/day}$).

characteristic curve) is nonlinear. Van Genuchten (1980) provided the following curve-fitting empirical equation between ψ and θ:

$$\theta(\psi) = \theta_r + \frac{\theta_s - \theta_r}{[1 + (\alpha\psi)^n]^{1-(1/n)}} \qquad \text{Eq. (3.15)}$$

where $\theta(\psi)$ = volumetric moisture ($\text{cm}^3\,\text{cm}^{-3}$) at pressure head ψ, θ_r = residual soil water content ($\text{cm}^3\,\text{cm}^{-3}$), θ_s = saturated soil water content ($\text{cm}^3\,\text{cm}^{-3}$), α = a scaling factor related to the inverse of air entry pressure (cm^{-1}), and n = a dimensionless curve-shape parameter related to soil pore distribution. The example shown in Figure 3.9 is for a soil with saturated soil water content (porosity) θ_s of 0.43. When $\theta_s = 0.43\,\text{cm}^3\,\text{cm}^{-3}$, this soil is saturated and the pressure head is hence zero (the pressure is atmospheric). The line for air entry pressure is where significant volumes of air appear in the soil pores. At very high soil tension (i.e. a large negative pressure head or tension head), the curve becomes nearly vertical, reflecting residual water content held tightly to the soil grains ($\theta_r = 0.045\,\text{cm}^3\,\text{cm}^{-3}$). The hydraulic conductivity in an unsaturated soil is determined largely by the size (cross sectional area) of pathways for water transmission. It increases nonlinearly with water content to its saturated value as the water content increases to saturation.

The unsaturated hydraulic conductivity, $K(\theta)$, can be estimated from the saturated hydraulic conductivity (K_s) by the following equation (Campbell 1974):

$$K(\theta) = K_s \left(\frac{\theta}{\theta_s}\right)^{2b+3} \qquad \text{Eq. (3.16)}$$

where the parameter b can be estimated from the moisture characteristic curve, i.e. the slope of log(ψ) versus log (θ) curve, and its values were estimated to be in the range of 0.16–12.5 for five soils Campbell (1974) employed. These empirical relationships, although approximate, are useful in estimating these hard-to-measure parameters.

From the practical standpoint of vadose zone soil remediation, it is extremely important to further realize the different phases of contaminant present in the vadose zone. Contaminants in both saturated and unsaturated zones can be (a) dissolved in soil water, (b) sorbed to soil particles, and/or (c) nonaqueous phase liquid if a free phase exists such as from a recent oil spill. It is the additional vapor phase present in the vadose zone that differentiates it from the saturated zone. Therefore, the consideration of vapor behavior is crucial in the remediation of a contaminated vadose zone soil. The flow and transport of vapor in the vadose zone occurs in response to pressure and the concentration gradient. This will be illustrated in a quantitative detail in Section 8.2. As we will see, soil vapor extraction is used to remediate a shallow aquifer including a contaminated vadose zone.

3.3.3 Flow to Wells in a Steady-State Confined Aquifer

Our attempt here is to derive the governing equation to describe groundwater flow under steady-state condition in a confined aquifer. A steady-state refers to a constant hydraulic head over time, but hydraulic head can be variable in space by responding to the pumping of water from a well. We consider two water table observation wells with a distance of r_1 and r_2 away from the pumping well at a volumetric rate of Q (Figure 3.11). Applying Darcy's law to obtain Q at any radius r, we have:

$$Q = -KA\frac{dh}{dr} = -K\left(2\pi rb\right)\frac{dh}{dr} \qquad \text{Eq. (3.17)}$$

where A is the surface area of the side of cylinder with a radius r and height b (i.e. thickness of confined aquifer): $A = 2\pi rb$. Integrating Eq. (3.17) after separating variables (h and r) gives:

$$Q = 2\pi Kb\frac{h_2 - h_1}{\ln\frac{r_2}{r_1}} \qquad \text{Eq. (3.18)}$$

The solution (Eq. 3.18) to a steady-state radial flow to a pumping well is called the **Thiem equation**. Since transmissivity (T) is equal to the product of K and b (Eq. 3.9), we can rearrange Eq. (3.18) to estimate transmissivity:

$$T = Kb = \frac{Q}{2\pi\left(h_2 - h_1\right)}\ln\frac{r_2}{r_1} \qquad \text{Eq. (3.19)}$$

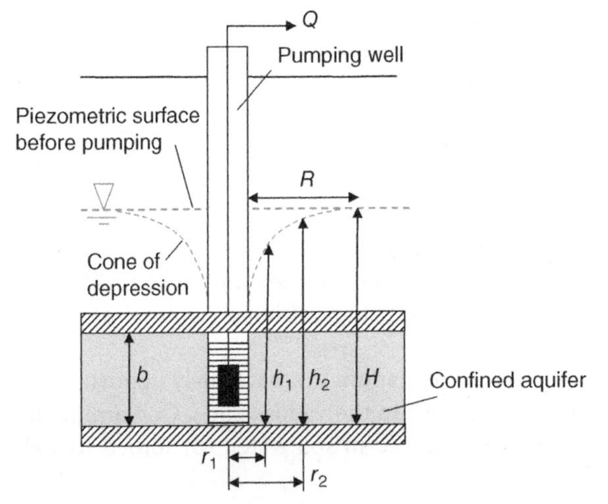

Figure 3.11 Cone of depression in a confined aquifer. The horizontal dashed line represents the water table before pumping and the curved line around the pump represents the cone of depression.

The above equation can be used to experimentally determine K or T in the field by obtaining the water levels at two observation wells along with the **cone of depression**. A special case for Eq. (3.18) is when we set the pumping well as one of the observation wells, i.e. with $h_1 = h_w$ at $r_1 = r_w$ (radius of the pumping well), we have:

$$Q = 2\pi Kb \frac{h_2 - h_w}{\ln \frac{r_2}{r_w}} \qquad \text{Eq. (3.20)}$$

If we rearrange the above equation and solve for T, we can obtain an equation similar to Eq. (3.19) (i.e. h_1 and r_1 replaced by h_w and r_w, respectively). Therefore, if we know the water tables (h_2) from only one well along with the cone of depression, we can determine the T value. Note that h_1, h_2 and h_w in the above equations are the water levels (elevation) expressed as the **mean sea level** (MSL). In many instances, the drawdown (s) is alternatively used in the above equations. **Drawdown** is the change in hydraulic head observed at a well in an aquifer, typically due to pumping a well as part of an aquifer test or well test. If the original static head is H, then the drawdown for the two observation wells at r_1 and r_2 distance away (Figure 3.11) are $s_1 = H - h_1$, and $s_2 = H - h_2$, respectively. For the term $h_2 - h_1$ in Eq. (3.19), it can be replaced by $(H - s_2) - (H - s_1) = s_1 - s_2$. Substitute this into Eq. (3.19) or (3.20), we can express Q and T as a function of drawdown in place of groundwater table. For example, Eq. (3.19) becomes:

$$T = \frac{Q}{2\pi (s_1 - s_2)} \ln \frac{r_2}{r_1} \qquad \text{Eq. (3.21)}$$

Another special case is when the drawdown in well at r_2 is zero (i.e. $s_2 = 0$), meaning pumping has no effect on the water level at this location; the corresponding r_2 is called the **radius of influence**, which can be derived from Eq. (3.21) as follows:

$$r_2 = r \times e^{(2\pi s\, T/Q)} \qquad \text{Eq. (3.22)}$$

The radius of influence is an important concept in pump-based groundwater and vapor treatment, because this is the measure of how far a pumping/vacuum well can reach to extract a contaminated plume. Example 3.3 shows the use of some of the equations discussed above for estimating hydrological parameters in a confined aquifer from steady-state drawdown data. More discussions relevant to this concept will be presented in Sections 3.3.4, 5.3.3, 7.2.1, and 8.2.4.

Example 3.3 Transmissivity of a confined aquifer

Estimate the transmissivity of a confined aquifer using the following drawdown data of a fully penetrating pumping well and two monitoring wells. Aquifer thickness = 10 m (33 ft); Well diameter = 4 in. (0.1 m); groundwater pumping rate = 25 gpm (0.0016 m^3/s or 0.06 ft^3/s); Steady-state drawdown for a monitoring well 1.83 m (6 ft) from the pumping well = 0.67 m (2.2 ft); Steady-state drawdown for a monitoring well 7.62 m (25 ft) from the pumping well = 0.52 m (1.7 ft).

Solution

Since the drawdown data (s) rather than the static head (h) are given, we can directly use Eq. (3.21). When SI units are used:

$$T = \frac{Q}{2\pi(s_1 - s_2)} \ln \frac{r_2}{r_1} = \frac{0.0016 \frac{m^3}{s}}{2\pi(0.67\,m - 0.52\,m)} \ln \frac{7.62\,m}{1.83\,m} = 0.0024 \frac{m^2}{s} \left(0.027 \frac{ft^2}{s}\right)$$

This calculated T is the same as 0.0024 m^3/m·s or 0.027 ft^3/ft·s with the unit of volume per unit width per unit time. In the US unit, one can directly plug in gpm to report the T value in the unit of gal/ft · s as follows.

$$T = \frac{Q}{2\pi(s_1 - s_2)} \ln\frac{r_2}{r_1} = \frac{25\,gpm}{2\pi\left(2.2\,\text{ft} - 1.7\,\text{ft}\right)} \ln\frac{25\,\text{ft}}{6\,\text{ft}} = 11.4\frac{gal}{\text{ft}\cdot\text{m}} = 0.19\frac{gal}{\text{ft}\cdot\text{s}}$$

3.3.4 Flow to Wells in a Steady-State Unconfined Aquifer

Figure 3.12 shows the cone of depression curve in a well located in an unconfined aquifer. Applying Darcy's law:

$$Q = -KA\frac{dh}{dr} = -K\left(2\pi rh\right)\frac{dh}{dr} \quad \text{Eq. (3.23)}$$

Unlike a confined aquifer with a constant aquifer thickness (b) (Eq. 3.17), the hydraulic head (h) is variable for an unconfined aquifer (Eq. 3.23). The integrated form of Eq. (3.23) will also be slightly different from Eq. (3.18):

$$Q = \pi K \frac{h_2^2 - h_1^2}{\ln\frac{r_2}{r_1}} \quad \text{Eq. (3.24)}$$

where h_1 and h_2 are groundwater elevations observed in wells at r_1 and r_2 distance away from the pumping well, respectively. Example 3.4 illustrates the use of Eq. (3.24) in unconfined aquifers. Similar to a confined aquifer, we can derive the radius of influence. We denote $h_2 = H$ for simplification, because the initial head H is always known. The radius of influence in an unconfined aquifer is:

$$r_2 = r_1 \times e^{\left[(\pi K/Q)\left(H^2 - h_1^2\right)\right]} \quad \text{Eq. (3.25)}$$

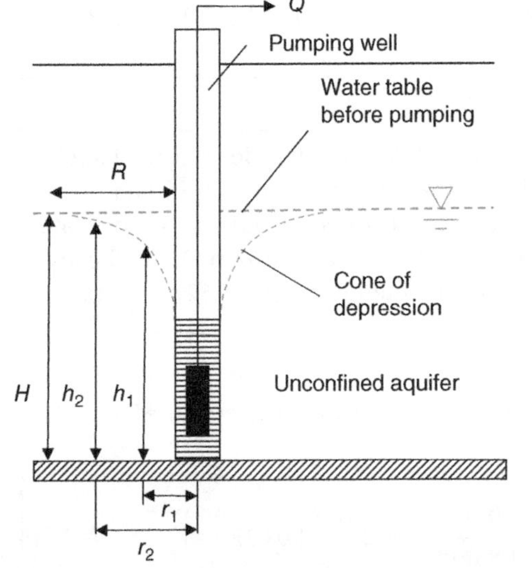

Figure 3.12 Cone of depression in unconfined aquifer.

Example 3.4 Hydraulic conductivity of a unconfined aquifer

A well discharges 100 gal per minutes (gpm) from an unconfined aquifer. Prior to pumping, the water table was 50 ft MSL (mean sea level). After a long time period the water table was recorded as 30 ft MSL in an observation well located 100 ft away and 40 ft MSL at another well located 1500 ft away. Determine the hydraulic conductivity of this aquifer in ft/s.

Solution

Rearrange Eq. (3.24), we have:

$$K = \frac{Q}{\pi\left(h_2^2 - h_1^2\right)} \ln\frac{r_2}{r_1} = \frac{100\frac{gal}{\min} \times \frac{1\min}{60\,\text{s}} \times \frac{1\,\text{ft}^3}{7.48\,gal}}{\pi\left(40^2 - 30^2\right)\text{ft}^2} \ln\frac{1500}{100} = 2.75\times10^{-4}\,\frac{\text{ft}}{\text{s}}$$

For a very thick unconfined aquifer (H and h_1 are large, $H \approx h_1$), $H^2 - h_1^{\,2} = (H - h_1)(H + h_1) \approx (H - h_1) \times 2H = 2sH$, where s is the drawdown in a well having a groundwater table of h_1. Substituting this into Eq. (3.25), we have a simplified equation for the radius of influence applicable for a thick unconfined aquifer:

$$r_2 = r_1 \times e^{\left[2\pi k s H / Q\right]} \qquad \text{Eq. (3.26)}$$

Related to the radius of influence, the concept of **capture zone** is introduced here. Capture zone is defined by the USEPA as the three-dimensional region that contributes to the groundwater extracted by one or more wells or drains. A capture zone is important to either avoid pumping contaminated groundwater if the well is used for drinking water purposes, or to remove (remediate) a contaminated plume if the well is used for remediation purposes. This concept is only briefly introduced here, interested readers should refer to references for more detailed derivation and calculation examples (e.g. Bedient et al. 1999; USEPA 2008; Hemond and Fechner-Levy 2014).

For one single extraction well as shown in Figure 3.13, the equation to calculate the width of the capture zone (y) shown in Figure 3.13 is:

$$y = \frac{Q}{\left(bv\right)} \qquad \text{Eq. (3.27)}$$

The above equation indicates that the width of the pumping region (y), or capture zone, increases as the pumping rate (Q) increases, and increases as the aquifer thickness (b) or Darcy's

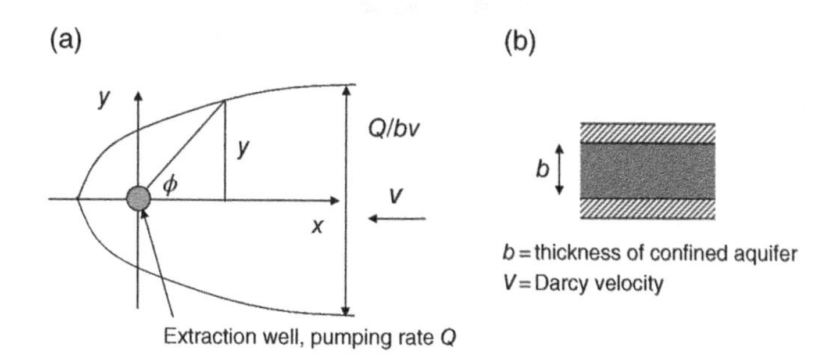

Figure 3.13 Capture zone curve: (a) Plan view, (b) Vertical view.

velocity (v) decreases. For multiple (n) optimally spaced wells arranged symmetrically along the y-axis, Eq. (3.27) should be modified as: $y = n\ Q/(b\ v)$.

Many mathematical equations have been introduced thus far to describe groundwater flow in both confined and unconfined aquifer systems. This might appear to be overwhelming for some to start with. A valid question for the majority of readers is how to choose an appropriate equation in solving a given practical problem.

At first, readers should comprehend what is behind each of these mathematical equations, the assumptions of these equations and the intended use. For example, Eqs. (3.17) through (3.22) apply only to flow in confined aquifers whereas Eqs. (3.23) through (3.26) apply only to unconfined aquifers. All these equations are derived from Darcy's law. If the acquisition of the hydraulic conductivity is attempted from geological testing in the field, the rearrangement of Eqs. (3.20) and (3.24) should give the right formula for us to determine K in a confined and unconfined aquifer, respectively. If the radius of influence is of concern, then Eq. (3.22) should be used for flow in confined aquifers, and Eq. (3.25) or (3.26) should be used in unconfined aquifers – the choice between these two depends on what parameters are given with the known values, e.g. hydraulic head (h) or drawdown (s).

3.3.5 Flow of Nonaqueous Phase Liquid

The movement of water immiscible (undissolved) liquid (e.g. oils, NAPLs) is more complicated than the movement of water and the chemical in its dissolved phase. A schematic description of NAPL's movement is given in Figure 3.14, the patterns of movement are mainly dependent on its density, i.e. whether it is in the form of LNAPLs or DNAPLs (see Chapter 2 for their definitions). The different plumes between LNAPLs (e.g. gasoline) and DNAPLs (like heavy fuels), as schematically represented in the figure, is self-explanatory. That is, the LNAPLs tend to float on the surface of the groundwater table, and the DNAPLs tend to sink and eventually sit on top of the unconfined aquifers. In contrast, the dissolved contaminants will travel with water.

The movement of NAPLs in the subsurface is a very complex process. As a result, it is very difficult to delineate NAPL's movement and remediate NAPLs and especially DNAPLs in a contaminated aquifer. Table 3.3 shows the ease of cleanup as a function of contaminant chemistry and hydrogeology at hazardous waste sites. It clearly indicates how difficult it is if we want

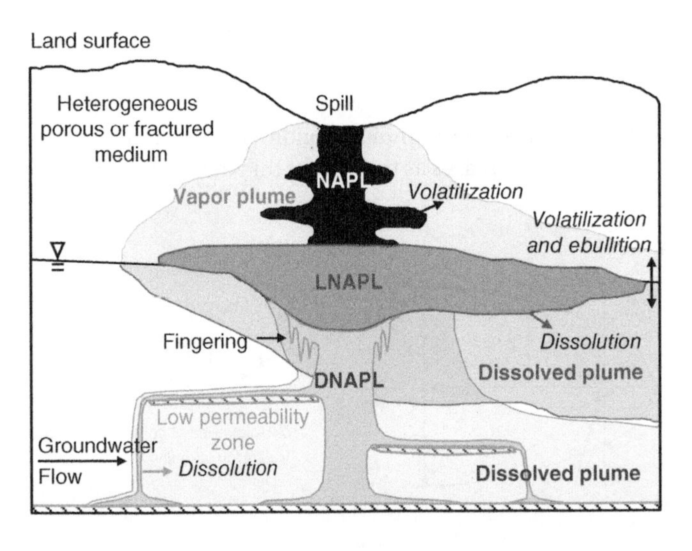

Figure 3.14 Comparison of LNAPL and DNAPL movement in aquifer. This vertical cross-section depicts the infiltration of a spilled LNAPL and DNAPL in the subsurface and the subsequent development of vapor and dissolved organic plumes. *Source:* Essaid et al. (2015). Reprinted with permission from John Wiley and Sons.

Table 3.3 Ease of remediation as a function of contaminant behavior and hydrogeological conditions (MacDonald and Kavanaugh 1994).[a]

	Contaminant chemistry					
Hydrogeology	Mobile, dissolved (degrades/volatilzes)	Mobile, dissolved	Sorbed, dissolved (degrades/volatilizes)	Sorbed, dissolved	Separated-phase LNAPL	Separated-phase DNAPL
Homogeneous, single layer	1	1–2	2	2–3	2–3	3
Homogeneous, multiple layers	1	1–2	2	2–3	2–3	3
Heterogeneous, single layer	2	2	3	3	3	4
Heterogeneous, multiple layers	2	2	3	3	3	4
Fractured	3	3	3	3	4	4

[a] Relative ease of cleanup: 1 = easiest; 4 = most difficult. The 1–4 scale used in this table should not be viewed as objective and fixed, but as a subjective, flexible method for evaluating sites. Other factors that influence the ease of cleanup, such as the contaminant mass at a site and the length since it was released, are not shown in this table.

to remediate, for example, DNAPLs in a fractured aquifer. The cleanup of DNAPLs represents the most challenging in environmental remediation.

As a concluding remark of this chapter, it is important to point out the relevance of computer models to simulate groundwater flow in complex aquifers. Groundwater flow modeling is a standard practice in the field of hydrology, through which the rate and direction of groundwater movement through aquifers in the subsurface can be predicted. The fate and transport models also require the development of a calibrated groundwater flow model or, at a minimum, an accurate determination of the velocity and direction of groundwater flow that has been based on field data. Regardless, the selection and proper use of a model must be based on a thorough understanding of the flow and solute transport processes we have discussed in Chapters 2 and 3, including typically the direction of flow, geometry of the aquifer, the heterogeneity or anisotropy of the aquifer materials, the contaminant transport mechanisms, and chemical reactions.

Bibliography

Bear, J. (1972). *Dynamics of Flows in Porous Media*. New York: Elsevier.

Bedient, P.B., Rifai, H.S., and Newell, C.J. (1999). Chapter 7: contaminant fate processes. In: *Ground Water Contamination: Transport and Remediation*, 2e. Upper Saddle River, NJ: Prentice Hall PTR.

Campbell, G.S. (1974). A simple method for determining unsaturated conductivity from moisture retention data. *Soil Sci.* 117 (6): 311–314.

Ellis, S. and Mellor, A. (1995). *Soils and Environment*. London: Routledge.

Essaid, H.I., Bekins, B.A., and Cozzarelli, I.M. (2015). Organic contaminant transport and fate in the subsurface: evolution of knowledge and understanding. *Water Resources Res.* 51: 4861–4902.

Fetter, C.W. (1993). *Contaminant Hydrogeology*, 2e. New York: Macmillan Publishing Company.

Fetter, C.W. (2001). Chapter 10: water quality and ground-water contamination. In: *Applied Hydrogeology*, 4e. Upper Saddle River, NJ: Prentice Hall.

Greenland, D.J. and Hayes, M.H.B. (1978). *The Chemistry of Soil Constituents*. Chichester: John Wiley and Sons.

Heath, R.C. (1983). *Basic Ground-Water Hydrology. US Geological Survey Water-Supply Paper 2200*, 86. Alexandria, Virginia: United States Government Printing Office.

Hemond, H.F. and Fechner-Levy, E.J. (2014). *Chemical Fate and Transport in the Environment*, 3e, Academic Press.

Knox, R.C., Sabatini, D.A., and Canter, L.W. (1993). *Subsurface Transport and Fate Processes*. Boca Raton, FL: Lewis Publishers.

Kosugi, K., Hopmans, J.W., and Dane, J.H. (2002). Parametric methods. In: *Methods of Soil Analysis. Part 4*, Physical Methods, Soil Science Society of America Book Series, no. 5. Washington, DC: Soil Science Society of America.

Leake, S.A. (1997). Modeling Ground-Water Flow with MODFLOW and Related Programs: US Geological Survey Fact Sheet 121–97, 4 pp.

Lee, C.C. and Lin, S.D. (2007). *Handbook of Environmental Engineering Calculations*, 2e. New York: McGraw-Hill.

van Genuchten, M.T. (1980). A closed-form equation for predicting the hydraulic conductivity of unsaturated soils. *Soil Sci. Soc. Am. J.* 44: 892–898.

van Loon, G.W. and Duffy, S.J. (2017). *Environmental Chemistry: A Global Perspective*, 4e. New York: Oxford University Press.

MacDonald, J.A. and Kavanaugh, M.C. (1994). Restoring contaminated groundwater: an achievable goal? *Environ. Sci. Technol.* 28 (8): 362A–368A.

Manahan, S.E. (2005). *Environmental Chemistry*, 6e. New York, NY: CRC Press.
McBride, M.B. (1994). *Environmental Chemistry of Soils*. New York: Oxford University Press.
Moore, J.E. (2002). *Field Hydrogeology: A Guide for Site Investigations and Report Preparation*. Boca Raton, FL: Lewis Publishers.
Ranney, R.W. (1969). An organic carbon-organic matter conversion equation for Pennsylvania surface soils. *Soil Sci. Soc. Am. Proc.* 33: 809–811.
Sharma, H.D. and Reddy, K.R. (2004). *Geoenvironmental Engineering: Site Remediation, Waste Containment, and Emerging Waste Management Technologies*. Hoboken, New Jersey: John Wiley & Sons.
Sposito, G. (2008). *The Surface Chemistry of Soils*, 2e. Oxford: Oxford University Press.
Todd, D.K. (1980). *Ground Water Hydrology*, 2e. New York, NY: John Wiley & Sons.
USEPA (2008). A Systematic Approach for Evaluation of Capture Zones at Pump and Treat Systems: Final Project Report, EPA 600/R-08/003 http://www.epa.gov/ord
USGS (1999). *Sustainability of Ground-Water Resources, US Geological Survey Circular 1186*. Colorado: Denver.

Questions and Problems

1. How are contaminants generally associated with three phases of soil (gas, liquid, solid)?

2. Define the compositional differences between soils of following groups: clay, silty clay, clay loam, silty clay loam.

3. Define the compositional differences between soils of the following groups: sandy clay, sandy clay loam, sandy loam, loamy sandy, and sandy.

4. What are the differences in general elemental compositions between soil organic matter and inorganic minerals?

5. If a soil sample is measured to have a total organic carbon of 1.2%, what is the approximate soil organic matter content?

6. Describe the structural differences between kaolinite and montmorillonite. Why montmorillonite can retain more water and cationic nutrients?

7. Explain why soil clays carry negative charge and explain how does this relate to the cation exchange capacity (CEC)?

8. Define the terms: humic substance, humin, humic acid, and fulvic acid.

9. Why does soil organic matter play a disproportionate role in contaminant fate and transport compared to soil inorganic constituents?

10. How is the soil named with a texture of 50% clay, 10% silt and 40% sand?

11. Describe the difference between (a) water table wells, (b) flowing artisan wells, and (c) nonflowing artisan wells.

12 Describe the difference among three major aquifer types (a) confined, (b) unconfined, and (c) perched aquifer.

13 Describe the functions of the following for a monitoring well: (a) casing, (b) screen, (c) annular seal.

14 Clays have much smaller grain size than silt and sand, they have porosities as high as 40–50%, but why are they generally considered impermeable?

15 A silt loam soil sample has particle density = 2.65 and bulk density = 1.5, what is the porosity?

16 An aquifer has a porosity of 0.42 and specific retention of 0.20, what is its specific yield?

17 If the depth to water (DTW) is 7 ft and the wellhead elevation is 50 ft, what is the hydraulic head of the groundwater table in the well?

18 If the porosity of a contaminated aquifer containing sands and gravel is 39% and the specific yield is 25%, how much groundwater in gallons can be drained for a leaking underground storage tank with an estimated volume of contaminated aquifer of 2000 ft × 1500 ft × 25 ft?

19 A sandstone aquifer ($H \times W \times L$ = 20 m × 1000 m × 5 km) has a hydraulic conductivity of 5×10^{-2} m/day. The head change over the length of 5 km is 5 m. Determine (a) the daily flow capacity of this aquifer, (b) the transmissivity of this aquifer.

20 The aquifer underneath an abandoned solvent recovery facility has a hydraulic conductivity of 1.5×10^{-1} cm/s. The hydraulic gradient (dh/dl) from site contour map shows the change of groundwater elevation from 150 to 145 ft at a distance of 250 ft apart. The contaminated site has a width of 500 ft and aquifer thickness of 35 ft. (a) calculate the groundwater flow rate in gallon per minutes (1 ft^3 = 7.48 gal). (b) calculate the groundwater flow velocity if the effective porosity can be assumed to be 30%.

21 Describe the major differences between flows in saturated and unsaturated aquifer.

22 Explain how these parameters affect the hydraulic conductivity in vadoze zone: (a) moisture content, (b) porosity, (c) soil air, (d) homogeneity of soil particle size.

23 A 130-ft long concrete settling tank is supported on a layer of silty sand, which has a hydraulic conductivity $K = 3.28 \times 10^{-6}$ ft/s. Below this layer lies a layer of clay which can be considered impervious. A construction trench is cut to facilitate repairs to the tank. Determine (a) the hydraulic gradient (dh/dl) of the groundwater flow into the trench, (b) the minimum pumping capacity (Q) in ft^3/s required to dewater the trench to the level of the clay layer.

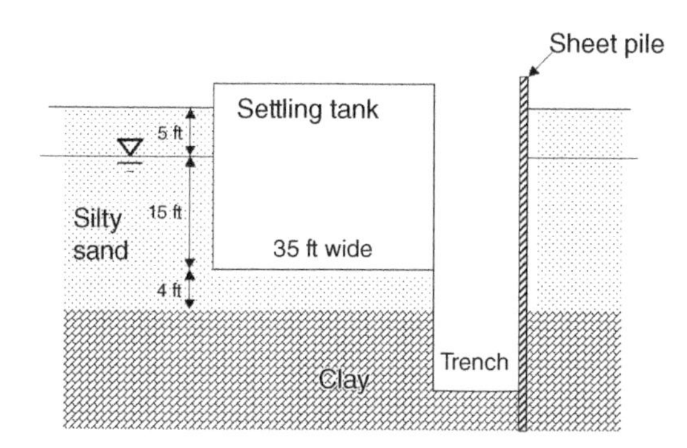

24 A well discharges 75 gpm from an unconfined aquifer. Its original water table was recorded as 35 ft. MSL. After a long time period the water table was recorded as 20 ft. MSL in an observation well located 75 ft away and 34 ft MSL at another observation well located 2000 ft away. Determine the hydraulic conductivity of this aquifer in ft/s.

25 A well is installed to pump water from a contaminated aquifer that is confined at 35 m thick. Two observation wells are installed 200 and 950 m apart from the pumping well. The pumping well constantly drains water at a rate of 0.35 m^3/min. When steady-state is achieved, the drawdowns measured in two wells are 7 and 3 m at the corresponding wells of 200 and 950 m away from the pumping well. Determine (a) transmissivity (T), and (b) hydraulic conductivity (K).

26 Explain the relative ease of the remediation for: (a) dissolved MTBE plume in a shallow homogeneous aquifer, (b) dissolved pentachlorophenol plume in a heterogeneous aquifer, (c) a pool of gasoline containing alkane and BTEX in a confined aquifer, (d) leaks of tetrachloroethylene containing water into a shallow aquifer, (e) a pool of PCBs from an old transformer facility leaked into an aquifer with fractured rocks.

27 Calculate the elevation head, pressure head, velocity head, and total hydraulic head for a groundwater having pressure $P = 1000\,N/m^2$, $v = 2.5 \times 10^{-6}\,m/s$, density $= 1.05 \times 10^3\,kg/m^3$ at 0.5 m above a reference elevation.

Chapter 4

Legal, Economical, and Risk Assessment Considerations

LEARNING OBJECTIVES

1. Appreciate the importance of three independent but related factors in the decision-making of soil and groundwater remediation: regulation, cost, and risk
2. Develop regulatory awareness for some important groundwater protection laws in the United States, including SDWA, RCRA, CERCLA, and Superfund Act
3. Identify resources of regulations in your own countries and regions of concern and compare them to the laws in the United States that drive the remediation of soil and groundwater
4. Understand the MCLs and MCLGs of contaminants in the current National Primary Drinking Water Regulations (NPDWR)
5. Understand the RCRA regulations related to municipal solid waste under Subtitle D and disposal of hazardous waste under Subtitle C
6. Understand the regulations relevant to Superfund sites and brownfield sites (CERCLA, Superfund Act) and the Small Business Liability Relief and Brownfields Revitalization Act (SBLR&BRA)
7. Develop a sense of cost estimate in environmental remediation systems in regard to cost components and basic approaches for cost estimate (e.g. unit cost, cost indices, scale-up)
8. Calculate the present value from interest (discount) rate and future value or vice versa and use cost estimates to evaluate remediation alternatives
9. Define how clean is clean based on available standards and guidelines including, but not limited to, background levels, detection limits, regulatory standards, criteria, and screening levels
10. Calculate risk for the exposure to carcinogenic compounds using cancer slope factor (CSF), and unit risk
11. Calculate risk for the exposure to noncarcinogenic compounds using reference dose/concentration (RfD, RfC) and chronic daily intake (CDI)
12. Determine risk-based cleanup levels for soil and groundwater for both carcinogenic and noncarcinogenic compounds

This chapter introduces three independent but all related topics to environmental remediation. Legal considerations, economic constraints, and risk factors are all important to the protection of groundwater resource and the cleanup of contaminated sites. First, environmental regulations are the driving force for the implementation of all remediation techniques. Secondly, the selected remediation technique optimal for a particular contaminated site should be technically sound but also cost-efficient at the same time. Thirdly, while the total removal of contaminants is not essential in many cases, an acceptable risk and cleanup goal (i.e. the issue of how clean is clean) should be clearly defined prior to all remediation efforts. The purpose of this chapter is to help readers understand these three issues fundamental to soil and groundwater remediation. The legal, economic, and risk assessment considerations present how nontechnical factors can become important in practice and how difficult it can be for the decision-making of environmental remediation.

Soil and Groundwater Remediation: Fundamentals, Practices, and Sustainability, First Edition. Chunlong Zhang.

Companion website: www.wiley.com/go/Zhang/Remediation_1e

4.1 Soil and Groundwater Protection Laws

The environmental regulations pertaining to soil and groundwater remediation in the United States have been well established since the 1970s. Hence, the relevant soil and groundwater laws in the United States will be introduced first in the following discussions. This will be followed by a brief overview of the environmental regulatory framework in other countries. Such a broad overview can be important since environmental pollution does not stop always at national borders and the issues have to be addressed sometimes by cross-border measures. The regulatory information outlined below should be used as a guide for readers to search appropriate sources on the details of specific regulatory provisions in the country and region of concern. Prior to reading these various environmental legislations, it would be beneficial to get familiar with several basic terms introduced in Box 4.1.

4.1.1 Relevant Soil and Groundwater Laws in the United States

Several important federal environmental statutes were enacted for the protection of soil and groundwater under the auspices of the USEPA during the period from the 1970s through 1980s. In the order of appearance, these are (Figure 4.1): Safe Drinking Water Act (SDWA) (1974),

Box 4.1 Basic terms relevant to environmental legislature

In the United States, federal environmental laws are passed by the US Congress. These laws define their goals but not technical details on how to achieve these goals. A law is first proposed by a member or members of Congress as a **bill**. If both houses, the House of Representatives and the Senate, approve a bill, it goes to the President to either approve or veto. If approved, the bill becomes the **law** or **act**, and the text of the act is known as a **public statute**. Once an act is passed, the House of Representatives standardizes the text of the law and publishes it in the **United States Code** (USC).

In the environmental arena, the USEPA is typically authorized by the Congress to develop full technical details on how to achieve the goals stated in environmental laws. These details are known as the **regulations**, which have the legal requirements in compliance with the law. During the development of regulations, the USEPA does the research and publishes the proposed regulation in the **Federal Register** for a specified period for public comments. Subsequently, the regulation is published in the **Code of Federal Regulations** (CFR), which becomes an official record of the regulation. All environmental regulations are printed in Title 40 of the 50 volume publications. In Title 40, each regulation has its parts and sections. For example, 40 CFR 280.230 is Part 280, Section 230 of Title 40, referring to "Operating an underground storage tank or underground storage tank system." From the practical standpoint, the distinction between laws and regulations might not be important.

Additionally, there are also technical guidelines and policies issued by the USEPA. **Policies** are put into place by the administrative branch, may or may not go through a public hearing process. While they do not have the force of law, they can greatly impact how cleanup programs are implemented. In the United States, states and local governments have made laws and regulations corresponding to the federal ones. If this is the case, the laws and regulations at the state and local levels should be stricter than the federal laws and regulations.

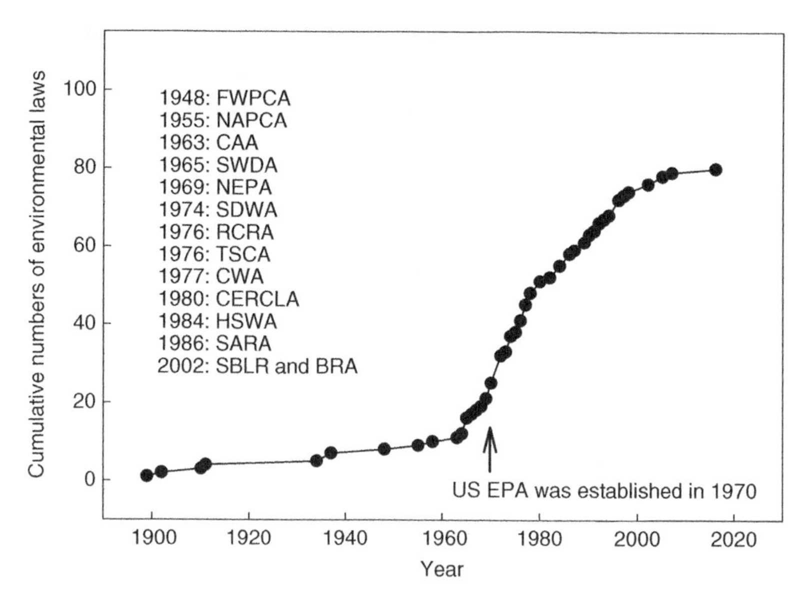

Figure 4.1 Cumulative numbers of federal environmental laws over time in the United States. Also shown are several critical drivers relevant to soil and groundwater remediation.

Resource Conservation and Recovery Act (RCRA, 1976), Comprehensive Environmental Response, Compensation and Liability Act (CERCLA, 1980), Hazardous and Solid Waste Amendment (HSWA, 1984), Superfund Amendment and Reauthorization Act (SARA, 1986), and Small Business Liability Relief and Brownfields Revitalization Act (SBLR&BRA, 2002). In addition to these federal statutes, each state in the United States plays an instrumental role in groundwater protection by implementing its own laws for groundwater supplies, allocations, and property rights related to groundwater protection (Bedient et al. 1999). A state may adopt environmental regulations that are more stringent than corresponding federal regulations on the same issue.

4.1.1.1 Safe Drinking Water Act

The **Safe Drinking Water Act** (SDWA) concerns the safety of public water supplies with two major initiatives: (i) the development of national drinking water standards and (ii) the implementation of a broad-scale groundwater protection program called the Underground Injection Control (UIC) program. Since the drinking water standard does not regulate sources of groundwater contamination, its impact on the prevention of groundwater contamination was minimal. The UIC program, on the other hand, regulates underground injection from various wells through state's permitting program in controlling underground injection.

The SDWA has mandatory water quality standards for drinking water contaminants. These enforceable standards called **maximum contaminant levels** (MCLs) are established to protect the public against consumption of drinking water contaminants that present a human health risk. An MCL is the maximum allowable concentration of a contaminant in drinking water delivered to the consumer. The SDWA also has a **maximum contaminant level goal** (MCLG) – the level of a contaminant in drinking water below which there is no known or expected health risk. MCLGs allow for a margin of safety and are nonenforceable public health goals. MCLs are commonly set as close to MCLGs as feasible using the best available treatment technology and taking cost into consideration. There are approximately 150 contaminants in the current **National Primary Drinking Water Regulations** (NPDWR), which fall into six categories: (a) microorganisms, (b) disinfectants, (c) disinfection by-products, (d) inorganic chemicals, (e) organic chemicals, and (f) radionuclides. MCLGs and MCLs for selected contaminants are provided in Table 4.1.

Table 4.1 National Primary Drinking Water Standards for selected contaminants.

Contaminants	MCLG (mg/L)	MCL (mg/L)	Contaminants	MCLG (mg/L)	MCL (mg/L)
Arsenic	0	0.01	Lead	0	0.015[a]
Atrazine	0.003	0.003	Mercury (inorganic)	0.002	0.002
B(a)P	0	0.0002	Nitrate-N	10	10
Barium	2	2	Nitrite-N	1	1
Benzene	0	0.005	PCB	0	0.0005
Bromate	0	0.01	Selenium	0.05	0.05
Cadmium	0.005	0.005	PCE	0	0.005
Chromium (total)	0.1	0.1	Toluene	1	1
Copper	1.3	1.3[a]	TCE	0	0.005
Ethylbenzene	0.7	0.7	Uranium	0	0.03
Fluoride	4	4	Xylenes (total)	10	10

[a] For copper and lead, these are action levels not MCLs. Lead and copper are regulated by a treatment technique that requires systems to control the corrosiveness of their water. If more than 10% of tap water samples exceed the action level (1.3 mg/L for copper and 0.015 mg/L for lead), water systems must take additional steps.

In addition, the EPA has established the **National Secondary Drinking Water Regulations** (NSDWR) that set nonmandatory water quality standards for 15 contaminants (color, odor, total dissolved solids, corrosivity, foaming agents, pH, Cl, F, sulfate, Al, Cu, Fe, Mn, Ag, and Zn). These **secondary maximum contaminant levels** (SMCLs) are not enforceable, but serve as guidelines to assist public water systems in managing their drinking water for aesthetic considerations (such as taste, color, and odor). At the SMCLs, these contaminants are not considered to present human health risk, but may present: (i) aesthetic effects – undesirable tastes or odors, (ii) cosmetic effects – effects which do not damage the human body but are still undesirable; and (iii) technical effects – damage to water equipment or reduced effectiveness of treatment for other contaminants.

The Safe Drinking Water Act includes a process (every six years) to identify and list "unregulated contaminants" called the **contaminant candidate list** (CCL). At least five or more contaminants can be called from this list for regulatory determinations. The EPA uses this list to prioritize research and data collection efforts to help determine whether new contaminants should be regulated. Many of these chemicals, termed emerging contaminants (see Box 2.1), are known or anticipated to occur in public water systems, such as pharmaceuticals, personal care products, and hormonal compounds.

4.1.1.2 Resource Conservation and Recovery Act

Initially passed in 1976, the Resource Conservation and Recovery Act (RCRA) has several subsequent amendments, including the most important Hazardous and Solid Waste Amendments of 1984 (see Section 4.1.1.4). RCRA was not promulgated by the USEPA until the late 1980s in response to several environmental tragedies including the Love Canal incident (Box 1.2). The RCRA regulates the disposal of hazardous wastes under **Subtitle C** and nonhazardous municipal solid wastes under **Subtitle D. Hazardous wastes** covered under Subtitle C fall into two categories: listed wastes and characteristic wastes. **Listed wastes** are wastes that the EPA has already determined to be hazardous, and **characteristic wastes** are hazardous if a waste is ignitable, corrosive, reactive, or toxic. Some hazardous wastes are exempt from RCRA regulations including sewage, wastewater discharges already regulated by the CWA (Clean Water Act), mining wastes, nuclear wastes, and municipal garbage. **Nonhazardous wastes** covered under Subtitle D include household waste, scrap metal, construction waste, and sludge from wastewater treatment plants.

The RCRA established standards for facilities that generate hazardous wastes, transporters of hazardous wastes, and facilities that treat, store, or dispose of hazardous wastes (termed the **TSD facilities**). Through a RCRA manifest program, it creates a "paper trail" to follow the waste from "cradle to grave." Each generator and transporter must have an EPA identification number and each TSD facility must be permitted. The generator is responsible for determining whether its waste is hazardous, and the transporter must follow the Department of Transportation regulations for hazardous waste transportation. The owner or operator of the TSD facility must keep a well-maintained manifest and operating record and submit a biennial report to the regulatory agency.

The RCRA also gave TSD facilities in existence prior to 19 November 1980 the "interim status," with which these facilities must file application for a final permit. The final permit identifies the terms and conditions under which the RCRA facility must operate. If they failed the final permit application, these facilities were deemed to undergo "closure" activities. During the interim status, the facility must also implement groundwater monitoring and a groundwater quality assessment program. However, this RCRA program does not require an immediate cleanup or corrective action even if groundwater contamination is found. When a TSD facility fails and faces a closure, a closure plan must be developed for the control, minimization, or the elimination of the wastes.

4.1.1.3 Comprehensive Environmental Response, Compensation and Liability Act

The Comprehensive Environmental Response, Compensation, and Liability Act (CERCLA), commonly known as the **Superfund Act**, was enacted by Congress in 1980. CERCLA (i) set prohibitions and requirements for closed and abandoned hazardous waste sites; (ii) provided for liability for parties responsible for releases of hazardous wastes at these sites; and (iii) established a trust fund for cleanup when no responsible party could be identified. By creating a tax on the chemical and petroleum industries, CERCLA provided a broad federal authority to respond directly to releases or threatened releases of hazardous substances from abandoned or uncontrolled hazardous waste sites. Congress directed the USEPA to establish a **National Priority List** (NPL) using a hazard ranking system (HRS) and guide the expenditure of these Superfund monies (see Box 1.4). Only NPL sites (a total of 1343 sites as of 2018) shall be eligible for Superfund-financed remedial activities.

The CERCLA procedure starts with the **remedial investigation** (RI) where data are compiled and site characterization is achieved. Then a feasibility study (FS) examines various remediation alternatives by a set of criteria such as long-term effectiveness, reduction of toxicity, mobility and volume through treatment, short-term effectiveness, implementability, and cost. After considering community input and acceptance, the final remedy shall be selected and documented in a **record of decision** (ROD). Following ROD, the remedial design/remedial action (RD/RA) stage takes place toward the cleanup of Superfund sites.

4.1.1.4 Hazardous and Solid Waste Amendment

This amendment of RCRA came after Congress' disappointment over the EPA's permission on the disposal of free liquid hazardous wastes in land disposal facilities. Passed in 1984, HSWA targeted the nation's groundwater contamination issue. HSWA has four major initiatives

(Bedient et al. 1999): (i) The ban on land disposal of hazardous waste was implemented. These banned activities include "... any placement of such hazardous waste in a landfill, surface impoundment, waste pile, injection well, land treatment facility, salt dome formation, salt bed formation, underground mine, or cave." (ii) The exemption for small generators was considerably reduced by changing the small quantity generator exemption from 1000 to 100 kg/month. (iii) Underground storage tanks become regulated for their leakage detection, analysis, and remediation. (iv) **Solid waste management units** or SWMU ("smoos") became regulated for the required testing of groundwater contamination. If a leakage is detected, a cleanup action must be undertaken.

4.1.1.5 Superfund Amendment and Reauthorization Act

The Superfund Amendment and Reauthorization Act added a provision to the CERCLA that protects "innocent" purchasers of contaminated property from the liability under CERCLA, a so-called **due diligence**. If the buyer of a property had "no reason to know" about the contamination at the time of property purchasing, then it will protect them from the liability. This new provision has fundamentally changed all parties involved in the transactions of real estates that could be related to the liability of property contamination. Lenders and buyers tend to examine whether there was "no reason to know" about contamination prior to the transaction of a property or land. This change created a nationwide consulting business, now called Phase I (mainly visual inspection), Phase II (mainly sampling and analysis), and Phase III (remediation actions).

Another new requirement was the reporting under SARA Title III. Facilities exceeding a certain size threshold in hazardous substances must report to the USEPA annually for their total amount of released hazardous wastes in both permitted and unpermitted facilities. This reporting requirement virtually affected all major industrial facilities in the United States.

4.1.1.6 Small Business Liability Relief and Brownfields Revitalization Act

The Small Business Liability Relief and Brownfields Revitalization Act (SBLR&BRA) was enacted in 2002 in an attempt to reform the strict regulatory approach of CERCLA and address the concerns over the environmental, economic and environmental justice associated with brownfields and their redevelopment (Collin 2003). It exempts persons from Superfund response cost liability at National Priorities List sites as generators and transporters if the person can demonstrate that (a) the total amount of the material containing hazardous substances was less than 110 gallons of liquid materials or 200 pounds of solid materials and (b) all or part of disposal, treatment, or transport occurred before 1 April 2001. The SBLR&BRA establishes $250 million per year for brownfields revitalization, including $200 million per year for brownfields assessment, cleanup, revolving loans, and environmental job training; and $50 million per year to assist State and Tribal response programs.

4.1.2 Framework of Environmental Laws in Other Countries

Similar to the Superfund Act in the United States, Canada has the Contaminated Sites Regulation. It has the provisions of "polluter pays" and the established fund for priority sites of national importance. The European countries have different regulations, but hold the same basic principles, such as the obligation to reach a "good chemical condition" of the groundwater, the precautionary principle, the "polluter pays" principle, and the sectoral scope of rules and regulations (specialty principle). The EU-wide environmental laws alone amount to well over 500 directives, regulations, and decisions.

The EU's Water Framework Directive (WFD) targets water policy aiming for rivers, lakes, groundwater, and coastal waters to be of "good quality." The EU Groundwater Directives is a "daughter" directive of the WFD, which include special measures for the prevention and control of groundwater pollution by determining common criteria for achieving "good chemical status" and for identifying quality trends. Table 4.2 lists E.U. countries that have overarching legislation on soil contamination and countries that lack overarching or comprehensive legislation on soil contamination. Nine out of 12 countries/regions have overarching legislations that specifically address soil contamination. The remaining three countries Finland, France, and Czech Republic cover contaminated soil through other legislations (Fraye and Visser 2016).

Table 4.2 Selected EU countries with/without overarching legislation on soil contamination.

Country/region	Legislation
Flanders – Belgium	Soil Remediation Decree, 1995
Brussels Region – Belgium	Ordinance pertaining to the management of contaminated soil, 2004
UK	Contaminated Land Regulation, 2000 (England and Scotland) and 2001 (Wales) or Waste and Contaminated Land Order, 1997 (N. Ireland)
Walloon Region – Belgium	Decree pertaining to the remediation of contaminated soil, 2004
The Netherlands	Soil Protection Act, 1987
Germany	Federal Soil Protection Law, 1998
Spain	Royal Decree, 2005
Italy	Ministerial Decree 471, 1999
Sweden	Environmental Code, 1998
Finland[a]	Environmental Protection Act, 86/200
France[a]	Law on Classified Installations, 2003/699
Czech Republic[a]	National Act on Waste, 2005

[a] Countries lacking overarching or comprehensive legislation on soil contamination.

4.2 Cost Constraints in Remediation

Cost is the deciding factor in the selection of proper remediation techniques among various alternatives. Many cost analysis methods are typically covered in engineering economics course (Box 4.2). Without going into detail, the intent here is to introduce the basic concepts with some hands-on calculation examples.

In the Superfund program, cost estimates are generally conducted as early as during the feasibility study (FS) phase. At the FS stage, the design for the remedial action is still conceptual, not detailed. Cost estimate is considered to be "screening of alternatives" at +100% to −50% accuracy, to "order-of-magnitude" at +50/−30% accuracy. For example, at the screening-level +100% to −50% accuracy for the actual cost of \$100 000, the estimated cost of an alternative is expected to be between \$200 000 and \$50 000. As a project progresses, the design becomes more complete and the cost estimate becomes more definitive, thus increasing the accuracy of the cost estimate to +15% to −10% (i.e. \$115 000–\$90 000) in the final project cost estimate (Figure 4.2). The following discussions only focus on "order-of-magnitude" cost estimation.

Box 4.2 Engineering economics in environmental remediation

Engineering economics is often a required course in undergraduate engineering curricula in the United States and other countries. In the United States, it is also part of two step exams (Engineering-In-Training, EIT; Professional Engineer, PE) dealing with various environmental engineering project cost estimates. In engineering economic analysis, the time value of money is central, and the cash flows are discounted using an interest rate. Instead of using mathematical formulas such as Eq. (4.3), these interest rates (discount rates) are readily available in many engineering handbooks. Software tools are also readily available to perform more complicated engineering economic analysis.

In soil and groundwater remediation, there are usually many possible remediation alternatives, including a "do nothing alternative" as a base for comparison. Costs of cleanup vary among sites, and can be very difficult to evaluate. For example, geologic factors such as soil permeability and contaminant transport are all site-specific. Economic analysis can be complicated by factors other than these **direct costs** (capital and labors), such as the **opportunity cost** of using resources to clean sites rather than build bridges, schools, and hospitals. These opportunity costs are particularly high in less developed nations since environmental problems compete with a host of other significant social issues. The concept of opportunity cost allows for a comparison of broad social priorities. Environmental remediation also creates **institutional costs** including administrative, monitoring, and enforcement expenditure. These vary across remediation alternatives and the institutional costs could become more burdensome in less developed economies due to limited resource.

A complete cost analysis shall also consider how the **benefits of remediation** arose from improved human and ecosystem health or the avoidance of future degradation in health or environmental quality. The benefits of cleanup are, therefore, directly linked to the exposure pathways associated with a site. In general, the benefits of cleanup are greatest when toxicity is high and there is a large exposed population or habitat. There are also social benefits associated with remediation such as the job opportunities and the improved land and property values after the completion of cleanup (Boyd 1999).

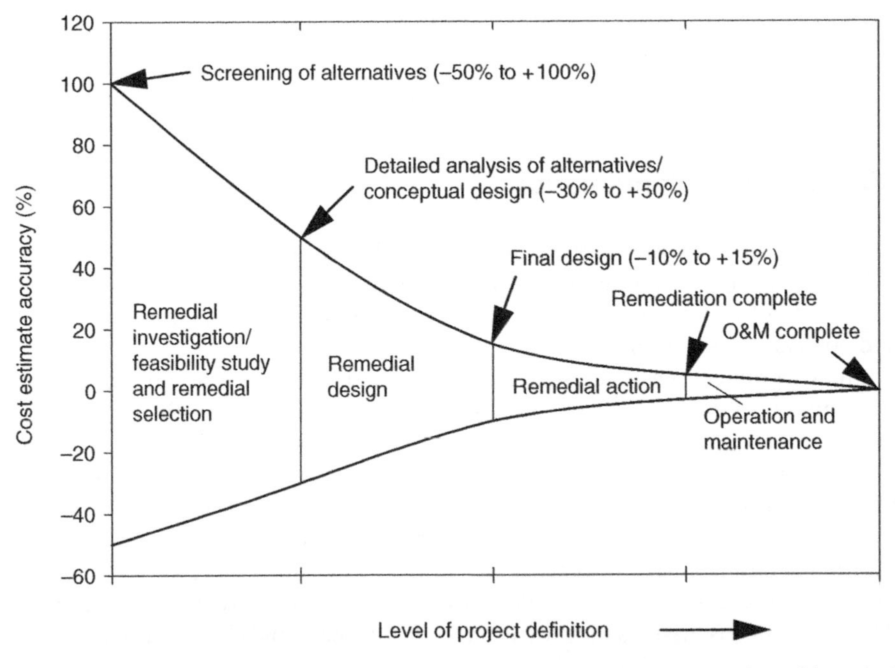

Figure 4.2 Expected accuracy of cost estimate along Superfund pipeline. *Source:* Adapted from the USEPA (2000).

4.2.1 Remediation Cost Elements

Capital cost: **Capital costs** are expenditures that are initially incurred to build or install remedial action (e.g. groundwater treatment system and related site work). They are exclusive of costs required to operate or maintain the action throughout its lifetime. Capital costs include all labor, equipment, and material costs, including contractor markups such as overhead and profit associated with activities such as mobilization/demobilization; monitoring; site work; installation of extraction, containment, or treatment systems; and disposal. Capital costs also include expenditures for professional/technical services that are necessary to support construction of the remedial action.

Annual O&M cost: **Operation and maintenance (O&M) costs** are those post-construction costs necessary to ensure the continued effectiveness of a remedial action. These costs are estimated mostly on an annual basis. Annual O&M costs include all labor, equipment, and material costs, including contractor markups such as overhead and profit associated with activities such as monitoring; operating and maintaining extraction, containment, or treatment systems; and disposal. Annual O&M costs also include expenditures for professional/technical services necessary to support O&M activities.

Periodic cost: **Periodic costs** are those costs that occur only once every few years (e.g. five-year reviews, equipment replacement) or expenditures that occur only once during the entire O&M period or remedial timeframe (e.g. site closeout, remedy failure/replacement). These costs may be either capital or O&M costs, but because of their periodic nature, it is more practical to consider them separately from other capital or O&M costs in the estimating process.

A check list for the above three cost categories is provided in Table 4.3. The checklists are not all-inclusive and, therefore, the listed cost elements should not be assumed to apply to every remedial alternative. Rather, the checklists can be used to identify applicable cost elements, which can be added to or modified as needed according to the site-specific conditions.

4.2.2 Basis for Remediation Cost Estimates

The basis for a screening-level cost estimate can include a variety of sources, including cost curves, generic unit costs, vendor information, standard cost estimating guides, historical cost data, and estimates for similar projects, as modified for the specific site. Both capital and O&M costs should be considered, where appropriate, at the screening level. Cost estimates developed during the detailed analysis phase are used to compare alternatives and support remedy selection. The types of costs that shall be assessed include the following: (a) Capital costs, including both direct and indirect costs; (b) Annual operations and maintenance costs; and (c) Net present value of capital and O&M costs.

Cost Estimate Based on Unit Cost: **Unit costs** have been reported, for example, by the USEPA, based on historical data of several remediation technologies (Figure 4.3). For example, the unit costs for pump-and-treat systems are based on the cost per 1000 gallons of groundwater treated per year, or per unit mass or volume of soil remediated for thermal desorption, bioventing, and soil vapor extraction. Figure 4.3 shows that pump-and-treat is less costly if the system is used for treating a larger amount of water compared to the same system if it were used for a smaller amount of water. These trends in unit costs may be useful as part of the broad assessment of technologies. Caution should be exercised when extrapolating such cost data for use in another site with very dissimilar contaminants, geological conditions, and regulatory requirements.

Cost Estimate Based on Cost Indices: If the costs for a past project are known, the prices can be converted to the current year by Eq. (4.1)

$$C = C_p \frac{I}{I_p} \qquad \text{Eq. (4.1)}$$

Table 4.3 Check list of cost elements for remediation project.

Capital cost	Annual O&M cost	Periodic cost
Construction Activities • Mobilization/Demobilization • Monitoring, Sampling, Testing, and Analysis • Site Work • Surface Water Collection or Containment • Groundwater Extraction or Containment • Gas/Vapor Collection or Control • Soil Excavation • Sediment/Sludge Removal or Containment • Demolition and Removal • Cap or Cover • On-Site Treatment (specify treatment technology) • Off-Site Treatment/Disposal • Contingency	O&M Activities • Monitoring, Sampling, Testing, and Analysis • Extraction, Containment, or Treatment Systems • Off-Site Treatment/Disposal • Contingency	Construction/O&M Activities • Remedy Failure or Replacement • Demobilization of On-Site Extraction, Containment, or Treatment Systems • Contingency
Professional/Technical Services • Project Management • Remedial Design • Construction Management	Professional/Technical Services • Project Management • Technical Support	Professional/Technical Services • Five Year Reviews • Groundwater Performance and Optimization Study – Remedial Action Report
Institutional Controls • Administrative Control • Legal Control	Institutional Controls • Administrative Control • Legal Control	Institutional Controls • Administrative Control • Legal Control

Source: Adapted from USEPA (2000).

where C and C_p are current year and previous year costs, respectively; I and I_p are current year and previous year index values, respectively. These **cost index** values are publicly available on the website of Bureau of Labor Statistics (www.bls.gov/cpi), or can be subscribed through Engineering News-Record, Chemical Engineering Cost Index, and Marshall and Swift Equipment Cost Indices for weekly to monthly data.

Costs Estimate Based on Scale-up: When treatment equipment is scaled up, its price can be determined by its relative size according to the six-tenths-factor rule (0.6 power rule):

$$C_a = C_b \left(\frac{S_a}{S_b} \right)^{0.6} \quad \text{Eq. (4.2)}$$

where C_a and C_b are costs for two different sized equipment A and B, respectively; and S_a and S_b are capacities (sizes) of equipment A and B, respectively. The exponential factor varies slightly for different types of equipment. This scaling method is applicable when the scaling is less than 10 folds (Peters et al. 2003; Holland et al. 2007).

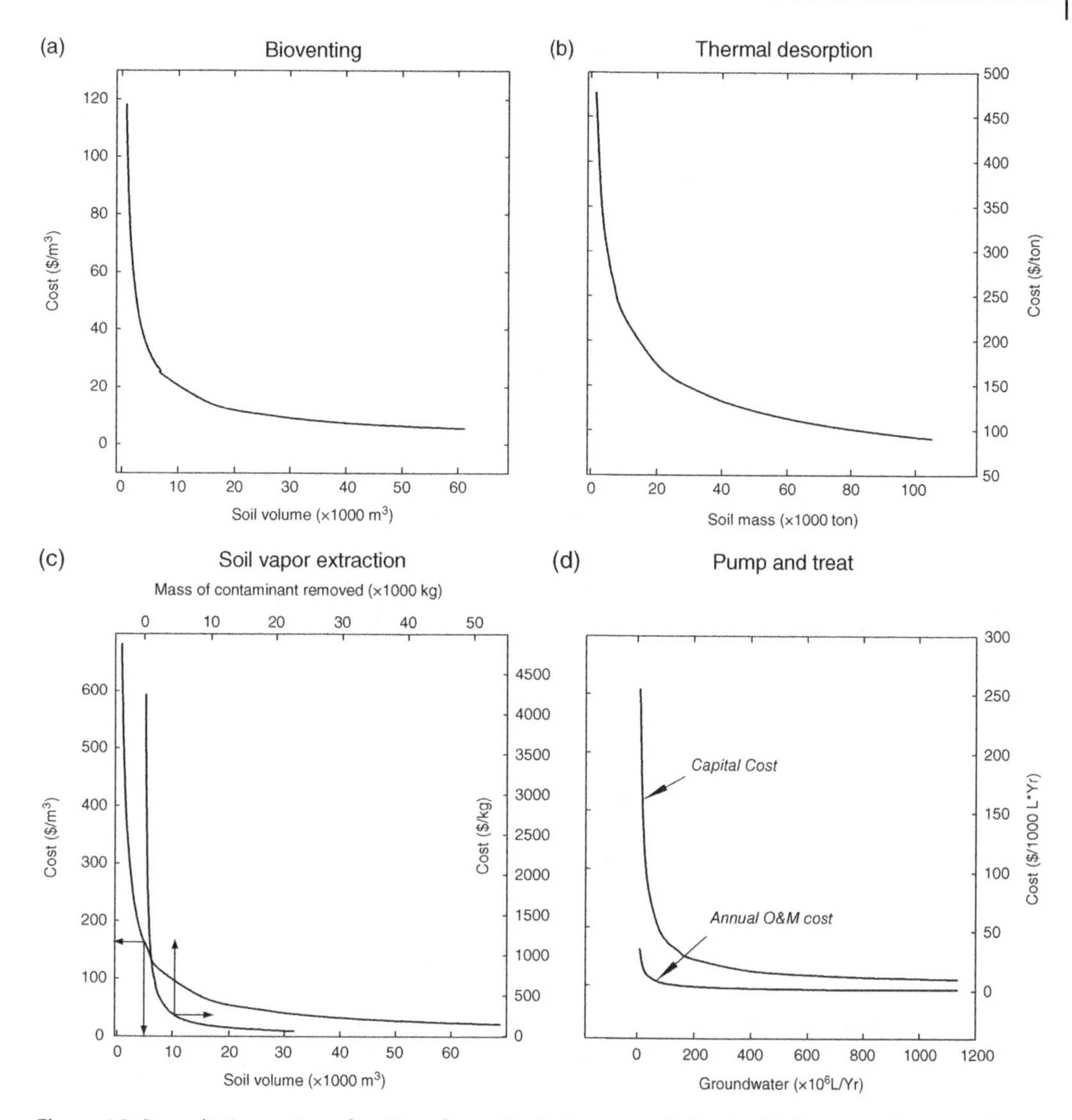

Figure 4.3 Remediation cost as a function of quantity (volume, mass) of treated soil or groundwater, or the mass of contaminant removed. Cost data from the USEPA (2001b) were adjusted for inflation rate to 2018.

Cost Estimate Based on Percent of Total Costs: Certain costs can be estimated using the percentage of pertinent total costs. For example, total direct capital cost can be used to estimate instrumentation and control, engineering and supervision, and general maintenance in the annual O&M. On the other hand, contingencies (incidental expenses) should occur at every stage of construction, hence the use of total indirect capital cost is inappropriate. **Contingencies** refer to costs, and are amounts that are held in reserve to deal with unforeseen circumstances.

4.2.3 Cost Comparisons among Remediation Alternatives

Remedial costs typically involve construction costs that are expended at the beginning of a project (e.g. capital costs) and subsequent costs that are required to implement and maintain the remedy (e.g. annual O&M costs, periodic costs). A single cost figure is needed in order to directly compare various remediation alternatives for screening and evaluation purposes. The **present value method** can be used to evaluate expenditures, either capital or O&M, which occur over the same life spans. Another method for comparison is the **annual cost (annual cash flow) method** especially applicable when remediation alternatives have different life spans. With this method, alternatives are compared by the costs or benefits per year. Both are the standard methods used in engineering economics. Details of these methods as well as other methods can be readily found in the texts of engineering economics.

In the present value method, this single cost number, referred to as the **present value** (PV), is the amount needed to be set aside at the initial point in time (base year) to assure that funds will be available in the future as they are needed, assuming certain economic conditions. The present value of a future payment is calculated using the following equation:

$$\mathrm{PV} = \frac{\mathrm{FV}_t}{(1+i)^t} \qquad \text{Eq. (4.3)}$$

where FV_t is the **future value** (payment) in year t ($t = 0$ for present or base year) and i is the interest (discount) rate. The USEPA recommends a **discount rate** of 7% be used in developing present value cost estimates for remedial action alternatives during the feasibility study. For example, suppose one needs to make a \$10 000 payment for a pump to be replaced in Year 6. Using an interest rate of 7%, the present value would be:

$$\mathrm{PV} = \frac{\$10\,000}{(1+0.07)^6} = \$10\,000 \times 0.666 = \$6\,660$$

Therefore, \$6660 would need to be set aside or invested in the present year (Year 0), at an interest or discount rate of 7%, in order to have \$10000 in Year 6. The number of 0.666 (i.e. $1/(1+i)^n$) is called the **annual discount factor**, which is typically tabulated for various years (t) and various interest rates (refer to Table 4.4 for an interest rate of 7%). The cost analysis using Eq. (4.3) and annual discount factors is illustrated in Table 4.5, Example 4.1.

Using the present value method, the costs of several hypothetical remedial alternatives can be calculated (Table 4.6). In this example, six remedial alternatives have varying amounts of initial capital costs, annual O&M costs, and years of analysis. The cost analysis depicted in Figure 4.4 using data from Table 4.6 clearly illustrates that Alternative F is the least costly alternative (lowest present value cost), although it has the second highest total cost in base year dollars. This is because much of its total costs are in the future, which become quite small after the discount rate is applied. The cost of Alternative C is less than that of alternative D, but its present value is higher, since it has large upfront capital costs. This example illustrates the

Table 4.4 Annual discount factors at a discount rate of 7%.

Year	Factor	Year	Factor	Year	Factor	Year	Factor	Year	Factor
1	0.935	7	0.623	13	0.415	35	0.0937	70	0.00877
2	0.873	8	0.582	14	0.388	40	0.0668	80	0.00446
3	0.818	9	0.544	15	0.362	45	0.0476	90	0.00227
4	0.763	10	0.508	20	0.258	50	0.0339	100	0.00115
5	0.713	11	0.475	25	0.184	55	0.0242	150	0.0000391
6	0.666	12	0.444	30	0.131	60	0.0173	200	0.00000133

Example 4.1 Estimate present value from capital, annual, and periodic costs

A remedial alternative has a construction cost of $1 800 000 in the present year (Year 0), annual O&M costs of $50 000 for 10 years, and periodic costs of $10 000 in Year 5 and $60 000 in Year 10. What is the present value for this remediation project?

Solution

Using the discount factors from Table 4.4 (i.e. 0.935, 0.873, 0.816) and placing these discount factors in column e of Table 4.5, we can calculate the present value each year for up to 10 years (column f). The present value for capital cost, annual M&O, and periodic cost is totaled at $2.18 million, as shown in Table 4.5.

Table 4.5 An example of present value calculation for a remedial alternative.

Year	Capital costs ($)	Annual O&M costs ($)	Periodical costs ($)	Total costs ($)	Discount factor at 7%	Total present costs at 7% ($)
	a	b	c	$d = a + b + c$	e	$f = d \times e$
0	1 800 000	0	0	1 800 000	1.000	1 800 000
1	0	50 000	0	50 000	0.935	46 750
2	0	50 000	0	50 000	0.873	43 650
3	0	50 000	0	50 000	0.816	40 800
4	0	50 000	0	50 000	0.763	38 150
5	0	50 000	10 000	60 000	0.713	42 780
6	0	50 000	0	50 000	0.666	33 300
7	0	50 000	0	50 000	0.623	31 150
8	0	50 000	0	50 000	0.582	29 100
9	0	50 000	0	50 000	0.544	27 200
10	0	50 000	60 000	110 000	0.508	55 880
Total	1 800 000	500 000				2 183 760

Table 4.6 Comparison of present value for six remediation alternatives.

Remedial alternative	Initial capital cost	Annual O&M cost	Period of analysis[a]	Total cost	Present value at 7%
Alternative A	0	0	0	0	0
Alternative B	3650	583	15	12 395	8960
Alternative C	10 800	548	30	27 240	17 600
Alternative D	2 850	696	50	37 650	12 455
Alternative E	5 500	230	80	23 900	8771
Alternative F	2 000	120	220	28 400	3714

Source: Adapted from USEPA (2000).
[a] In this example, the period of analysis is the same as project duration.

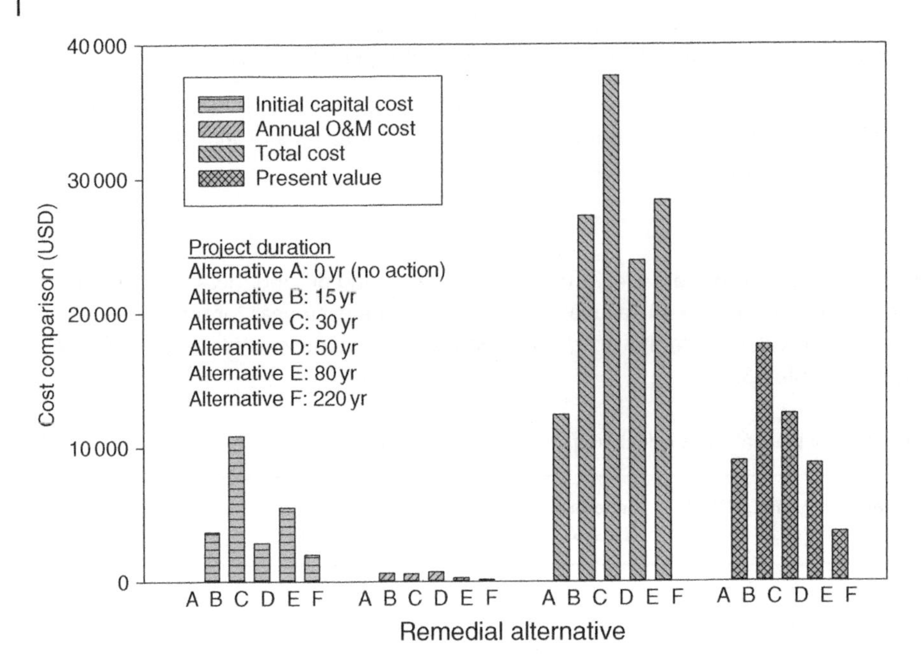

Figure 4.4 Cost comparison of six remedial alternatives.

effect of varying initial capital cost, annual O&M costs, and period of analysis on the present value cost of alternatives.

As an additional example, the present value of Alternative B in the amount of $8960 is calculated as follows using Excel spreadsheet (Table 4.7). Interested readers can do the same for other alternatives in Table 4.6. Note that, unlike Table 4.5, the periodic costs are excluded for illustration purpose. Also note that the present value method can compare different remediation alternatives well, even though the period of analysis (duration of remedial activity) is different. This is because all the cost items are adjusted to the present value of the Year 0.

4.3 Risk-based Remediation

Ideally, contaminated sites would be cleaned up completely for a total removal of all contaminants present. Unfortunately, and often times, this is not the case. The total removal may be too costly, time-consuming, and not always possible. Consequently, before we remediate a site, it is important to clearly define our goal as to how clean the site must be. When the appropriate level of cleanup is achieved, further remediation effort is not warranted.

4.3.1 How Clean Is Clean

The criteria used to determine how much cleanup must be done depend on several factors. These factors include site-specific current or future use of the land, the applicable regulatory standards, and the potential risks to humans and the ecosystem from the contaminants. Various numerical standards have been available to help decision-makers develop site-specific *Not to Exceed* (NTE) values, i.e. the concentrations that should not be exceeded at any single sampling

Table 4.7 An example of present value calculation for a remedial alternative B.

Year	Capital costs ($)	Annual O&M costs ($)	Total costs ($)	Discount factor at 7%	Total present costs at 7% ($)
	a	b	$c = a + b$	d	$e = c \times d$
0	3650	0	3650	1000	3650
1	0	583	583	0.935	545
2	0	583	583	0.873	509
3	0	583	583	0.816	476
4	0	583	583	0.763	445
5	0	583	583	0.713	416
6	0	583	583	0.666	388
7	0	583	583	0.623	363
8	0	583	583	0.582	339
9	0	583	583	0.544	317
10	0	583	583	0.508	296
11	0	583	583	0.475	277
12	0	583	583	0.444	259
13	0	583	583	0.415	242
14	0	583	583	0.388	226
15	0	583	583	0.362	211
Total	3650	8745			8960

point on site, or concentrations that should not be exceeded *on average* across a site (Sellers 1999). Considerations should be given to:

- Background levels in soils
- Analytical detection limit
- Regulatory standards, criteria, and screening levels
- Risk associated with human health
- Removal rate based on contaminant mass
- Best available technology

Background levels in soils: Many compounds considered to be environmental pollutants have some natural origin, i.e. they are not solely from human activities. These include many of the inorganic heavy metals and metalloids, such as cadmium, lead, mercury, chromium, and arsenic. These metals and metalloids are typically present in natural soils at the ppm (mg/kg) level in an unpolluted soil. For organic compounds, PAHs is the group of chemicals that have many natural sources, including forest fires and natural seepage. Therefore, the background levels of these contaminants in soil and sediment can be used as a cleanup level. Once a contaminated site has returned to its background level, remediation can be considered complete. The **background levels** in soils can be determined from local soil samples collected upwind or offsite away from the boundary of the contamination. Table 4.8 is a summary statistics for the background concentrations of Cd, Zn, Cu, Ni, and Pb in 3405 surface soil samples from agricultural areas of the United States. These soil samples were collected

Table 4.8 Background levels of metals and other soil parameters in 3045 surface soil samples from major agricultural areas of the United States.

	Cu	Zn	Pb	Cd	Ni	CEC	OC	pH
Statistics			mg/kg dry soil			cmol/kg	%	
US soils:								
Mean	29.6	56.5	12.3	0.265	23.9	26.3	4.18	6.26
±s	40.6	37.2	7.5	0.253	28.1	37.6	9.53	1.07
Minimum	<0.6	<3.0	<1.0	<0.010	0.7	0.6	0.09	3.9
5th percentile	3.8	8.0	4.0	0.036	4.1	2.4	0.36	4.7
50th percentile	18.5	53.0	11.0	0.20	18.2	14.0	1.05	6.1
95th percentile	94.9	126.0	23.0	0.78	56.8	135.0	33.3	8.1
Maximum	495	264.0	135.0	2.0	269.0	204.0	63.0	8.9
Detection limit	0.6	3.0	1.0	0.010	0.6	1.0	0.01	0.1
Percentage < DL	0.19	0.83	0.29	1.64	—	—	—	
World soils:								
Mean	30	50	10	0.06	40			
Minimum	2	10	2	0.01	5			
Maximum	100	300	200	0.70	500			

Source: Adapted from Holmgren et al. 1993.

at least 8 km downwind from any stack emitter (coal-fired electric generator, smelter, foundry, etc.), 200 m from US state highways, 100 m from rural roadways, 100 m from current, abandoned, or known obliterated building sites, and 50 m from field boundaries. These soils had no prior sludge applications.

Analytical detection limit: If a risk-based cleanup level (to be discussed in the following sections) is lower than the detection limit with the currently available state-of-the-art analytical methods, then the cleanup level can only be based on the detection limit. The **detection limit** is the lowest concentration detectable at a certain confidence level (e.g. 99%) that this concentration is significantly different from zero. Concentrations lower than the detection limit cannot be detected using state-of-the-art analytical instruments. For example, Table 4.8 shows that Cd concentrations in certain soil samples were below the detection limit of 0.01 mg/kg. It is thus clear that cleanup goal for Cd cannot be below 0.01 mg/kg.

Regulatory standards, criteria, and screening levels: Various regulatory standards, criteria, and screening levels have been developed by federal and state agencies for contaminants in soils, sediment, and groundwater. While criteria and screening levels have their technical ramifications, they are not mandatory cleanup levels. Only **regulatory standards** are enforceable. If, for example, an aquifer is used as a drinking water source, drinking water standards such as the USEPA's maximum contaminant levels in the Safe Drinking Water Act, or analogous state standards can be used. For surface water receiving groundwater discharge, groundwater cleanup standards may also be based on the potential risks from consumption of surface water or the risk to aquatic organisms. In this case, the ambient water quality criteria (AWQC) established under the Clean Water Act (CWA) should be adopted. Implementation for the protection of groundwater quality under the SDWA and CWA are mandatory.

Risk associated with human health: Risk-based assessments are becoming increasingly used on a voluntary basis (i.e. outside of the standard regulatory arena) to demonstrate the presence, absence, or extent of environmental or health-related concerns in specific exposure circumstances (Teaf et al. 2003). An example is the development of **soil screening levels** (SSLs), which are the risk-based concentrations derived from standardized equations combining exposure information assumptions with EPA toxicity data. SSLs are used in addition to or in lieu of promulgating standards or criteria. They are not national cleanup goals, so merely exceeding SSLs do not trigger the need for cleanup. However, where contaminant concentrations equal or exceed SSLs, further study or investigation is warranted. SSLs are used to streamline the evaluation and cleanup of soils by helping site managers eliminate areas, pathways, and/or chemicals of concern at the National Priority List sites.

The soil screening numbers developed by the California Environmental Protection Agency are the risk-based soil screening levels. Table 4.9 shows an example of soil screen numbers (mg/kg of dry soil) for selected nonvolatile chemicals. The State of California has also established the soil gas screening numbers (μg/L soil gas) for volatile chemicals (California EPA 2005). Details on the derivations of risk-based concentrations can also be found in several USEPA's guidelines (USEPA 1996a–c). The use of these numerical values for risk assessment is illustrated in the next sections.

Table 4.9 Soil screening numbers for nonvolatile chemicals based on total exposure to contaminated soil: Inhalation, ingestion, and dermal adsorption.

	Soil screen number (mg/kg)[a]			Soil screen number (mg/kg)[a]	
Contaminants	**Residential**	**Commercial/ industrial**	**Contaminants**	**Residential**	**Commercial/ industrial**
Ag	380 (nc)	4.8×10^3 (nc)	1,4-Dioxane	18 (ca)	64 (ca)
Aldrin	0.033 (ca)	0.13 (ca)	Dioxin	4.6×10^{-6} (ca)	1.9×10^{-5} (ca)
As[b]	0.07 (ca)	0.24 (ca)	Endrin	21 (nc)	230 (nc)
B(a)P	0.038 (ca)	0.13 (ca)	Fluoride	4.6×10^3 (nc)	5.7×10^4 (nc)
Cd	1.7 (ca)	7.5 (ca)	Hg	18 (nc)	180 (nc)
Cr(III)	10^5 (nc, max)	10^5 (nc, max)	Lindane	0.5 (ca)	2 (ca)
Cr(VI)	17 (ca)	37 (ca)	Ni	1.6×10^3 (nc)	1.6×10^4 (nc)
Cu	3.0×10^3 (nc)	3.8×10^4 (nc)	Pb	80 (nc)	320 (nc)
DDD	2.3 (ca)	9.0 (ca)	PCB	0.089 (ca)	0.3 (ca)
DDE	1.6 (ca)	6.3 (ca)	Perchlorate[c]	28 (nc)	350 (nc)
DDT	1.6 (ca)	6.3 (ca)	Se	380 (nc)	4.8×10^3 (nc)
Dieldrin	0.035 (ca)	0.13 (ca)	Zn	2.3×10^4 (nc)	1.0×10^5 (nc)

[a] ca = based on a carcinogenic potency factor; nc = based on chronic toxic effects other than cancer; max = based on the maximum concentration allowed, 100 000 mg/kg, and not toxicity.

[b] For anthropogenic As source only. In case of naturally occurring As, the concentrations may be far above the screening number. In this case, the agency with authority over remediation decisions should be consulted.

[c] While these concentrations are considered safe for exposure to perchlorate in soil, the potential for significant groundwater contamination may exist, since the safe drinking water level is 6 μg/L.

4.3.2 Estimate Environmental Risk from Carcinogenic Compounds

Risk can be considered as the chance of harmful effects to human health or to ecological systems resulting from exposure to an environmental stressor. Human health risks are grouped into two types, i.e. carcinogenic risks and noncarcinogenic risks. A carcinogenic chemical may present both risks, whereas a noncarcinogenic chemical will only cause noncarcinogenic risks. For example, benzene is known to be carcinogenic. It can lead to the risk of leukemia (cancer) as well as the noncarcinogenic risk such as the decreased lymphocyte count. Methyl chloride is not classifiable as to human carcinogenicity, it will present only noncarcinogenic risk such as cerebellar lesions. Carcinogenic compounds fall into four classes based on the **weight of evidence** (WOE) of their carcinogenicity (Table 4.10).

Carcinogens are treated as having "no threshold effect," i.e. any exposure to a cancer-causing substance will result in the initiation of cancer. The dose-response curves for carcinogens and noncarcinogens are considered different. The **cancer slope factor**, also known as the **potency factor** (PF), is the slope of the oral dose-response curve, which is the unit risk for a chronic daily intake of 1 mg/kg-day. Hence it has the unit of $(\text{mg/kg-day})^{-1}$.

For evaluating risks from contaminants in other environmental sources, dose-response measures are expressed as risk per concentration unit. These measures are called the **unit risk** for air (i.e. **inhalation unit risk**) and the unit risk for drinking water (oral). Since the exposure concentration units for air and drinking water are usually $\mu g/m^3$ and μg/L, respectively, unit risk has a unit of $(\mu g/m^3)^{-1}$ for inhalation and $(\mu g/L)^{-1}$ for drinking water. The values of CSF and unit risk for several carcinogens are illustrated in Table 4.11. These values are essential for cancer risk assessment.

A compound with a higher CSF will present a higher risk. For carcinogens, an acceptable risk is generally between 1×10^{-4} (1 in 10 000 cancer chance) and 1×10^{-6} (1 in a million cancer chance). The risk through oral ingestion is calculated as follows:

$$\text{Cancer Risk} = \text{CDI} \times \text{CSF} \qquad \text{Eq. (4.4)}$$

where CDI = **chronic daily intake** (mg/kg-day), i.e. mg of contaminant per unit body weight (BW) per day. The USEPA has set up the default assumptions for the determination of chronic daily intake from drinking water source for adult male, adult female, and child. The USEPA's assumption for body weight (BW) are 70, 50, and 10 kg for adult male, adult female, and child,

Table 4.10 USEPA weight of evidence of carcinogenicity.

Class	Weight of evidence (WOE)
Class A	*Known* human carcinogen (sufficient human evidence)
Class B	*Probable* human carcinogen. B1 = limited human evidence, or B2 = sufficient animal evidence with inadequate human evidence
Class C	*Possible* human carcinogen (limited animal evidence)
Class D	Not classifiable (no adequate human or animal evidence)
Class E	Evidence of noncarcinogenicity (no evidence for carcinogenicity in two animal species, or both human and animal studies)

Table 4.11 Cancer slope factor (CSF) and unit risk for cancer assessment of selected carcinogenic chemicals.[a]

Compounds	WOE Class	Oral CSF $(mg/kg\text{-}d)^{-1}$	Drinking water unit risk $(\mu g/L)^{-1}$	Inhalation unit risk $(\mu g/m^3)^{-1}$
Arsenic, inorganic	A	1.5	5.0×10^{-5}	4.3×10^{-3}
Benzene	A	1.5×10^{-2}–5.5×10^{-2}	4.4×10^{-7}–1.6×10^{-6}	2.2×10^{-6}–7.8×10^{-6}
B(a)P	B2	1.0	NA	6.0×10^{-4}
Bromate	B2	0.7	2.0×10^{-5}	NA
Chloroform	B2	NA	NA	2.3×10^{-5}
p,p'-DDT	B2	3.4×10^{-1}	9.7×10^{-6}	9.7×10^{-5}
PCB	B2	2.0	NA	1.0×10^{-4}
PCE	B2	2.1×10^{-3}	6.1×10^{-8}	2.6×10^{-7}
TCE	A	4.6×10^{-2}	NA	4.1×10^{-6}
Vinyl Chloride	A	7.2×10^{-1}	2.1×10^{-5}	4.4×10^{-6}

[a] NA = These values are not estimated or assessed under the current IRIS program. A complete list can be found in the USEPA's Integrated Risk Information System (IRIS) at http://www.epa.gov/iris/subst/index.html.

respectively, and the amount of water consumption per day is 2 L for adult and 1 L for child. CDI can be calculated as follows depending on the exposure pathways:

a) CDI from drinking water consumption:

$$CDI\left(\frac{mg}{kg\text{-}day}\right)=\frac{C\left(\frac{mg}{L}\right)\times IR\left(\frac{L}{day}\right)\times EF\left(\frac{350\,days}{yr}\right)\times ED(yr)}{BW(kg)\times AT\left(yr\times\frac{365\,days}{yr}\right)} \quad \text{Eq. (4.5)}$$

b) CDI from the inhalation of air:

$$CDI\left(\frac{mg}{kg\text{-}day}\right)=\frac{C\left(\frac{mg}{m^3}\right)\times IR\left(\frac{m^3}{day}\right)\times EF\left(\frac{350\,days}{yr}\right)\times ED(yr)}{BW(kg)\times AT\left(yr\times\frac{365\,days}{yr}\right)} \quad \text{Eq. (4.6)}$$

c) CDI from fugitive inhalation of dust/soil:

$$CDI\left(\frac{mg}{kg\text{-}day}\right)=\frac{C\left(\frac{mg}{kg}\right)\times IR\left(\frac{kg}{day}\right)\times ED\left(\frac{350\,days}{yr}\times yr\right)\times BR(\%)\times ABS(\%)}{BW(kg)\times AT\left(yr\times\frac{365\,days}{yr}\right)} \quad \text{Eq. (4.7)}$$

where C = contaminant concentration in water (mg/L), air (mg/m^3), or dust/soil (mg/kg), IR = intake (ingestion) rate, EF = exposure frequency, ED = exposure duration, BW = body weight, AT = averaging time (70 yr × 365 days/yr = 25 550 days), BR = breathing rate for dust (%), ABS = absorption rate for dust (%). Table 4.12 is a summary of the EPA default assumptions to calculate the CDI. Note that EF is 350 days by assuming 2 weeks of vacation per year. ED is 30 years assuming an individual resides in a house for only 30 years of his/her life. Note also that these defaults may be slightly different between literatures. For example, 21 years rather than 25 years was used for industrial setting, and 245 days was assumed for the same instead of 250

Table 4.12 USEPA standard default exposure factors.

Land use	Exposure pathway	IR	EF (d/yr)	ED (yr)	BW (kg)
Residential	Water	1 L/d (child) 2 L/d (adult)	350	30	70
	Air	20 m^3/d (total) 15 m^3/d (indoor)	350	30	70
	Soil/dust	200 mg/d (child) 100 mg/day (adult)	350	6 (child) 24 (adult)	15 (child) 70 (adult)
Industrial	Water	1 L/d	250	25	70
	Air	200 m^3/d (total)	250	25	70
	Soil/dust	250 mg/d	250	25	70

days (Mihelcic and Zimmerman 2014). One should check the most recent USEPA's update.

To calculate the cancer risk from unit risk (UR), the following equation is used:

$$\text{Cancer Risk} = \text{EC} \times \text{UR} \qquad \text{Eq. (4.8)}$$

where EC = exposure concentration (μg/L in water; μg/m^3 in air). The unit risk (UR) is considered to be the lifetime cancer risk estimated to result from continuous exposure to an agent at an exposure concentration of 1 μg/L in water or 1 μg/m^3 in air. If the drinking water contains 0.2 μg/L of bromate with a unit risk in drinking water of 2×10^{-5} $(\mu g/L)^{-1}$ (Table 4.11), the cancer risk can be calculated using Eq. (4.8):

$$\text{Cancer risk} = 0.2\,\frac{\mu g}{L} \times 2 \times 10^{-5}\left(\frac{\mu g}{L}\right)^{-1} = 4.0 \times 10^{-6} \quad \left(\text{e.g. 4 in 1 million}\right).$$

This suggests that four excess cancer risk cases are expected in one million people who are exposed to this bromate-containing drinking water. Example 4.2 illustrates the use of oral CSF to estimate cancer risk with Eq. (4.4).

Example 4.2 Calculation of cancer risk

PCBs from a leaking transformer have contaminated a community water system servicing 300 000 people. An average level of 10 μg/L was present in the water for a period of six months. (a) What is the individual lifetime risk of cancer from drinking the water? (b) What is the collective risk in the community as a result of the PCB-contaminated water? (Adapted from King 1999)

Solution

a) We first determine CDI using Eq. (4.5) assuming an adult male has the daily intake of water of 2 L/day, and the average body weight of 70 kg. Hence, we have:

$$CDI = \frac{10^{-2}\frac{mg}{L} \times 2\frac{L}{day} \times 6\,month \times \frac{30\,day}{month}}{70\,kg \times 70\,yr \times \frac{365\,day}{yr}} = 2.01 \times 10^{-6}\frac{mg}{kg \times day}$$

Applying Eq. (4.4) with the CSF value for PCBs of 2.0 (Table 4.11), the individual lifetime risk can be calculated as:

$$Risk = CDI \times CSF = 2.01 \times 10^{-6}\frac{mg}{kg \times day} \times 2.0\left(\frac{mg}{kg \times day}\right)^{-1} = 4.02 \times 10^{-6}$$

b) The collective risk is simply the product of the individual risk and the total population at risk. Assuming everyone drank the same source of contaminated water, this collective risk would be $300\,000 \times 4.02 \times 10^{-6} = 1.21$. This means only one cancer death as a result of this accidental release of PCBs.

The cancer risk of 4.02×10^{-6} calculated from the above example is equivalent to 1.2 cancer deaths among 1 000 000 people. It is interesting to note that this death rate is smaller than the estimated risk level of annual fatality rate of fires of 28 per 1 000 000 and 240 per 1 000 000 motor vehicle driving in the United States. To get an even better sense of the risk from exposure to carcinogenic compounds, some interesting data in Box 4.3 and Figure 4.5 can be referenced. Indeed the risk from contamination calculated from the above example is tolerable if one can tolerate X-ray or contaminated seafood consumption. But the exposure to contaminated drinking water from a public water supply can be a more sensitive issue.

Box 4.3 What is the cleanup risk tolerance? When to stop remediation?

A risk lower than 10^{-6} (1 in million) and 10^{-4} (1 in 10 000) is commonly perceived tolerable. Although it is a useful default for practitioners, this is indeed overly simplified. A definite answer to a tolerable risk is that it depends. For example, Clean Water Act 304(a) has criteria set as 10^{-6} risk level, and recommends that States and Tribes set criteria at 10^{-5} or 10^{-6}, and most highly exposed populations should not exceed 10^{-4} risk level. The USEPA is unlikely to approve statewide 10^{-4} risk level. In the Superfund sites, a risk range for excess cancer 10^{-4}–10^{-6} is applicable to a linear low-dose carcinogen. If the cancer risk is greater, an action must be taken. For nonlinear low-dose risk, a hazard index of greater than 1 triggers action.

One should, however, bear in mind that nothing is absolutely safe and almost everything is at risk. For a better sense of the risk, Figure 4.5 schematically compares the lifetime risks of cancer to individual exposed over 70 years. It is interesting to note that the cancer risk of drinking water contaminated with chlorinated compounds could even be much lower than the exposure to naturally occurring radioactive radon on average.

The order of magnitude for the environmental risk can be compared to other risks due to factors in our daily lives. For example, the annual fatality per 1 000 000 people at risk in an increasing order is: lighting (0.5), floods (0.6), fires (28), farming (360), coal mining (630), firefighting (800), smoking-cancer (1 200), smoking-all causes (3 000), and motorcycling (20 000). This implies each individual has a 0.5 chance in 1 000 000 (5 in 10 millions) to get a struck by lightning (Grasso 1993). While this risk can be ignored, the risk at 1 in 1 million is a very conservative number for risk assessment.

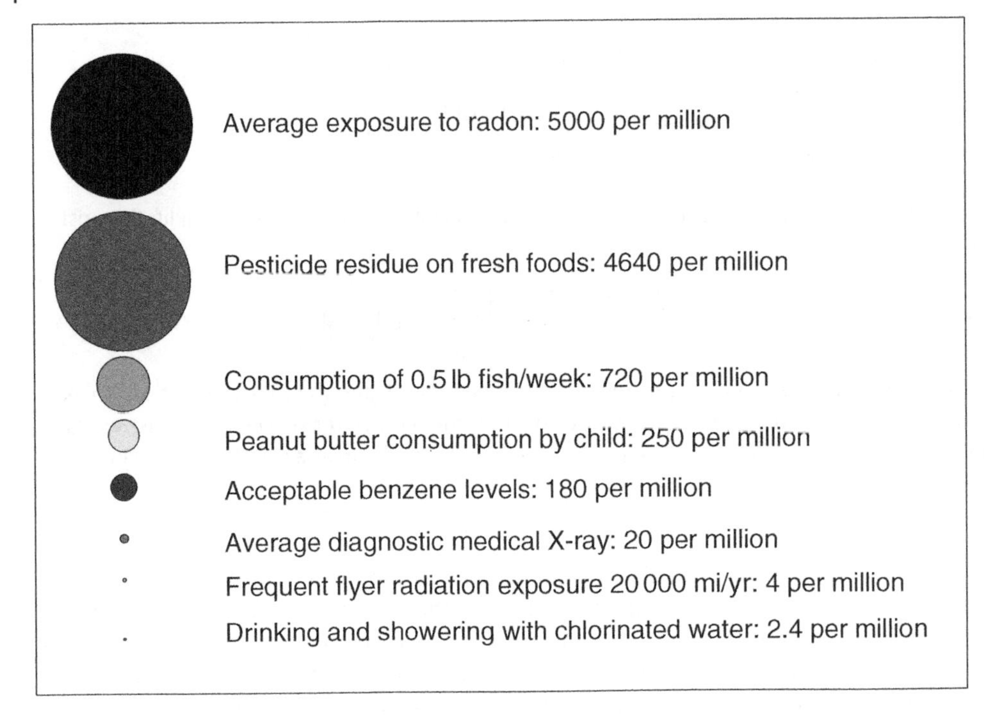

Figure 4.5 Lifetime risks of cancer for an individual exposed over 70 years.

4.3.3 Estimate Environmental Risk from Noncarcinogenic Compounds

For the risk assessment of noncarcinogenic chemicals, a **reference dose** (RfD), or **reference concentration** (RfC), is used. RfD is also used for carcinogenic chemicals for noncarcinogenic effects. Take benzene for an example, the exposure to carcinogenic benzene can also lead to numerous noncancerous health effects associated with functional aberration of vital systems in the body such as reproductive, immune, nervous, endocrine, cardiovascular, and respiratory systems. A RfD is an estimate of a daily exposure level that is likely to be without an appreciable risk of adverse effects during a lifetime. Like CSF, RfD values can also be found from the USEPA's Integrated Risk Information System (IRIS) at http://www.epa.gov/iris/subst/index.html (see Table 4.13 for RfD and RfC values of selected chemicals). For oral intake of noncarcinogens through drinking water or eating food, RfD is a "safe" reference oral dose reported in mg/kg-day. When inhalation exposure (from breathing) of noncarcinogens is concerned, RfC is used to denote the "safe" reference inhalation concentration ($\mu g/m^3$).

The values of RfD and RfC in Table 4.13 were derived from **no-observed-adverse-effect level** (NOAEL) in mg/kg-day. A NOAEL is an experimentally determined dose at which there was no statistically or biologically significant indication of the toxic effect of concern. An **uncertainty factor** (UF) is applied to convert NOAEL into RfD for the protection of sensitive subgroups. The conversion is as follows:

$$\text{RfD} = \frac{\text{NOAEL}}{\text{UF}} \qquad \text{Eq. (4.9)}$$

The "uncertainty" described above includes variations in susceptibility among the members of the human population (interindividual or intraspecies variability), uncertainty in extrapolating animal data to humans (interspecies uncertainty), uncertainty in extrapolating from data obtained in a study with less-than-life-time exposure (extrapolating from subchronic to

Table 4.13 RfD and RfC values for noncancer assessment of selected chemicals.[a]

Compounds	Oral RfD (mg/kg-d)	Inhalation RfC (μg/m³)
Arsenic, inorganic	3.0×10^{-4}	NA
Benzene	4.0×10^{-3}	3.0×10^{-2}
B(a)P	3.0×10^{-4}	2.0×10^{-3}
Bromate	4.0×10^{-3}	NA
Chloroform	1.0×10^{-2}	NA
p,p'-DDT	NA	NA
PCB	2.0	1.0×10^{-4}
PCE	6.0×10^{-3}	40
TCE	5.0×10^{-4}	20
Vinyl chloride	3.0×10^{-3}	100

[a] NA = These values are not estimated or assessed under the current IRIS program. A complete list can be found in the USEPA's Integrated Risk Information System (IRIS) at http://www.epa.gov/iris/subst/index.html.

chronic exposure), uncertainty in extrapolating from **lowest-observed-adverse-effect level** (LOAEL) rather than from NOAEL, and uncertainty associated with extrapolation when the database is incomplete.

The USEPA's prescribed UF value is to multiply 10 for interspecies variability (variation within population), 10 for intraspecies variability (animal to human), 10 for LOAEL in place of NOAEL, and 10 for subchronic when long-term human effects are not available (subchronic to chronic). The value of UF therefore can be from 10 to 1000.

Hazard quotient (HQ) is the ratio of dose (or concentration) to the reference dose (or reference concentration) for a single pathway of exposure (e.g. water, air):

$$\text{HQ}(\text{dimensionless}) = \frac{\text{CDI}\left(\dfrac{\text{mg}}{\text{kg-day}}\right)}{\text{RfD}\left(\dfrac{\text{mg}}{\text{kg-day}}\right)} \qquad \text{Eq. (4.10)}$$

Hazard index (HI) is a summation of more than one hazard quotients for multiple chemicals and/or multiple exposure pathways that an individual is exposed. If multiple contaminants are involved, one needs to calculate the sum of HQ for the same pathway (e.g. water, air):

$$\text{HI} = \sum(\text{HQ}_{\text{chemical A}} + \text{HQ}_{\text{chemical B}} + \cdots \qquad \text{Eq. (4.11)}$$

Exposures associated with a HQ of less than 0.2 and a hazard index value of less than 1.0 indicate that no adverse human health effects (noncancer) are expected to occur. If the HQ is greater than 0.2, or the HI is greater than 1, it is unsafe, and either the risk assessment should be refined or risk management measures should be taken. Example 4.3 illustrates the noncarcinogenic risk assessment using hazard quotient (Eq. 4.10).

4.3.4 Determine Risk-Based Cleanup Levels for Soil and Groundwater

Here we will describe the use of the equations in the previous section to derive the risk-based cleanup levels relevant to soil and groundwater remediation. Given the acceptable risk or the HI, the general calculation approaches are the same for carcinogenic and noncarcinogenic

Example 4.3 Calculation of noncancer risk

Chloroform can be formed from the disinfection process when chlorine (Cl_2) reacts with organic compounds in drinking water. If an average concentration of 80 μg/L chloroform was determined in the tap water, determine the lifetime noncarcinogenic risk of an adult receiving this tap water.

Solution

We first use the default values of IR, EF, ED, and BW from Table 4.12 and apply Eq. (4.5) to calculate the CDI:

$$\mathrm{CDI}\left(\frac{\mathrm{mg}}{\mathrm{kg\text{-}day}}\right)=\frac{\mathrm{C}\left(\frac{\mathrm{mg}}{\mathrm{L}}\right)\times\mathrm{IR}\left(\frac{\mathrm{L}}{\mathrm{day}}\right)\times\mathrm{EF}\left(\frac{350\,\mathrm{days}}{\mathrm{yr}}\right)\times\mathrm{ED}(\mathrm{yr})}{\mathrm{BW}(\mathrm{kg})\times\mathrm{AT}\left(\mathrm{yr}\times\frac{365\,\mathrm{days}}{\mathrm{yr}}\right)}$$

$$=\frac{0.08\left(\frac{\mathrm{mg}}{\mathrm{L}}\right)\times 2\left(\frac{\mathrm{L}}{\mathrm{day}}\right)\times 350\left(\frac{\mathrm{days}}{\mathrm{yr}}\right)\times 30(\mathrm{yr})}{70(\mathrm{kg})\times\mathrm{AT}\left(\mathrm{yr}\times\frac{365\,\mathrm{days}}{\mathrm{yr}}\right)}=0.066\frac{\mathrm{mg}}{\mathrm{kg\text{-}day}}$$

We then use the RfD value in Table 4.13 and apply Eq. (4.10) to estimate the risk in terms of hazard quotient:

$$\mathrm{HQ}(\mathrm{dimensionless})=\frac{\mathrm{CDI}\left(\frac{\mathrm{mg}}{\mathrm{kg\text{-}day}}\right)}{\mathrm{RfD}\left(\frac{\mathrm{mg}}{\mathrm{kg\text{-}day}}\right)}=\frac{0.066\left(\frac{\mathrm{mg}}{\mathrm{kg\text{-}day}}\right)}{1\times10^{-2}\left(\frac{\mathrm{mg}}{\mathrm{kg\text{-}day}}\right)}=6.58$$

Since HQ > 0.2, we conclude that the level of chloroform in this tap water will present a noncancer risk. For example, chloroform is known to exhibit moderate/marked fatty cyst formation in the liver and elevated level of serum glutamic pyruvic transaminase (SGPT), which is an enzyme released into blood when the liver or heart is damaged.

compounds. There are some slight variations among the different exposure routes, for example, oral intake from drinking water or contaminated food, inhalation from breathing air, and direct ingestion of soil and dust. These are illustrated in the following section.

4.3.4.1 Determining Maximum Concentration in Drinking Water and Air

First, for **carcinogens,** we use the previously given formula:

$$\mathrm{Risk}=\mathrm{CDI}\times\mathrm{CSF}=\frac{\mathrm{C}\times\mathrm{IR}\times\mathrm{EF}\times\mathrm{ED}}{\mathrm{BW}\times\mathrm{AT}}\times\mathrm{CSF}$$

By rearrange the above equation, we can solve for concentration (C):

$$C\left(\frac{\mathrm{mg}}{\mathrm{L}}\right)=\mathrm{risk}(\mathrm{acceptable})\times\frac{\mathrm{BW}\times\mathrm{AT}}{\mathrm{CSF}\times\mathrm{IR}\times\mathrm{EF}\times\mathrm{ED}} \qquad \text{Eq. (4.12)}$$

Second, for **noncarcinogens**, a similar formula can be derived for the calculation of maximum concentration:

$$C\left(\frac{\text{mg}}{\text{L}}\right) = \text{HI} \times \frac{\text{RfD} \times \text{BW} \times \text{AT}}{\text{IR} \times \text{EF} \times \text{ED}} \qquad \text{Eq. (4.13)}$$

Thus, if a tolerable risk can be specified, Eqs. (4.12) and (4.13) can be used for carcinogenic and noncarcinogenic effects, respectively, to estimate the safe concentration in drinking water and air (see Example 4.4).

Example 4.4 Determine risk-based maximum concentrations in drinking water

Determine the maximum concentration of benzene in drinking water, given the target excess individual lifetime cancer risk = 10^{-5} and CSF = 0.029 (mg/kg-day)$^{-1}$ for benzene.

Solution

Since we are dealing with the carcinogenic effects of benzene, Eq. (4.12) can be used. Substituting the values of risk and CSF and using default values for other parameters in Eq. (4.12), we have:

$$C\left(\frac{\text{mg}}{\text{kg}}\right) = 10^{-5} \times \frac{70\,\text{kg} \times 70\,\text{yr} \times 365\,\text{day/yr}}{0.029\frac{\text{kg-day}}{\text{mg}} \times 2\frac{\text{L}}{\text{day}} \times 350\frac{\text{day}}{\text{yr}} \times 30\,\text{yr}} = 0.03\frac{\text{mg}}{\text{L}}$$

Note that if the noncarcinogenic effect is of concern for benzene in this problem (e.g. effects on reproductive organs), we would use Eq. (4.12), but assuming the value of HI = 1 to calculate the maximum concentration of benzene at the safe level. The final cleanup level (maximum concentration) will be selected from whichever is more conservative from the risk standpoint.

4.3.4.2 Determining Allowable Soil Cleanup Level

Here we use Example 4.5 to show how a risk-based soil cleanup level can be determined following the direct oral ingestion of contaminated soil dust.

Example 4.5 Determine risk-based soil cleanup levels

Determine risk-based soil cleanup level of methylene chloride given the following conditions: target cancer risk < 10^{-6} and CSF = 0.0075 (mg/kg-day)$^{-1}$ for methylene chloride through oral ingestion of soil exposure pathway. Assume 80% retention rate and 90% absorption rate for dust. Use other USEPA default assumptions when needed.

Solution

Combining Eqs. (4.4) with (4.5), the maximum concentrations can be calculated as follows:

$$C\left(\frac{\text{mg}}{\text{kg}}\right) = 10^{-6} \times \frac{70\,\text{kg} \times 70\,\text{yr} \times 365\,\text{day/yr}}{0.0075\frac{\text{kg-day}}{\text{mg}_{\text{methylene chloride}}} \times \frac{100\,\text{mg}_{\text{soil}}}{\text{day}} \times 350\frac{\text{day}}{\text{yr}} \times 30\,\text{yr} \times 80\% \times 90\%}$$

$$= 0.00032\frac{\text{mg}}{\text{mg}_{\text{methylene chloride}}} = 315\frac{\text{mg}_{\text{methylene chloride}}}{\text{kg}_{\text{soil}}}$$

Therefore, 315 mg/kg is the maximum concentration that should be targeted to protect human from cancer risk under 1 in million (10^{-6}).

4.3.4.3 Risk Involving Multimedia

In a realistic scenario, the risk from the exposure through multiple sources (e.g. drinking water and fish) is often a concern. The total risk can be calculated by adding the risk in drinking water and the risk from eating contaminated fish. The procedure to calculate the risk-based water quality standard will generally be the same as that for the single phase (medium). Equilibrium partition constants (see Chapter 2) between the phases have to be assumed for multimedia risk assessment. In the case of fish and drinking water, the USEPA default for fish consumption of 6.5 g/day per person is assumed, and the following partition coefficient is introduced to relate the contaminant concentrations between water and fish:

$$\mathrm{BCF}\left(\frac{\mathrm{L}}{\mathrm{kg}}\right)=\frac{\text{concentration in fish}\left(\frac{\mathrm{mg}}{\mathrm{kg}}\right)}{\text{concentration in water}\left(\frac{\mathrm{mg}}{\mathrm{L}}\right)} \quad \text{Eq. (4.14)}$$

Hydrophobic contaminants have a very high bioaccumulation factor (or bioconcentration factor, BCF). Hence the risk associated with drinking water and eating fish for hydrophobic contaminants come mostly from contaminated "fish" rather than contaminated "drinking water," if an individual consumes a high amount of certain fish (such as shark, but shrimp, salmon, and catfish are normally much lower in contaminant accumulation). Example 4.6 illustrates the calculation of exposure doses from both drinking water and fish consumption. The combined risk can then be estimated from the doses.

Example 4.6 Calculate risks from contaminated drinking water and fish

Determine the predominant risk for people exposed to contaminated water and consuming contaminated fish. Given the chemical of concern has a BCF of 10^3 L/kg, aqueous concentration = 0.01 mg/L for a female (BW = 50 kg) who drinks 2 L/day, and eats 30 g fish/day.

Solution

$$\text{Dose from water}\left(\frac{\mathrm{mg}}{\mathrm{kg\text{-}day}}\right)=\frac{2\frac{\mathrm{L}}{\mathrm{d}}\times 0.01\frac{\mathrm{mg}}{\mathrm{L}}}{50\,\mathrm{kg}}=0.0004$$

$$\text{Dose from fish}\left(\frac{\mathrm{mg}}{\mathrm{kg\text{-}day}}\right)=\frac{0.01\frac{\mathrm{mg}}{\mathrm{L}}\times 1000\frac{\mathrm{L}}{\mathrm{kg}}\times 0.03\frac{\mathrm{kg}}{\mathrm{day}}}{50\,\mathrm{kg}}=0.006$$

The % dose from fish = 0.006/(0.006 + 0.0004) = 93.75%. It reveals that the major part of the risk (93.75%) comes from the consumption of contaminated fish insofar as the very bioaccumulative characteristic of the contaminant is concerned.

Before we conclude this section, let's look at a more complicated example (Example 4.7) by considering the risk involving multimedia, i.e. to determine the allowable risk-based soil cleanup level based on exposure of drinking groundwater.

Example 4.7 Determine risk-based soil cleanup level based on exposure to drinking water

The USEPA standard for the MCL of benzene is 0.005 mg/L in drinking water, what should be the cleanup standard for soil? Use an empirical number of 16 to convert benzene concentration in groundwater (0.005 mg/L) into pore water concentration.

Solution

To solve this problem, we need to keep the following in mind:

a) The bulk groundwater concentration is different from the soil pore water concentration. The groundwater concentration is the one that human will be exposed to by drinking it; however, the pore water concentration is the one we need to use for equilibrium calculation. This is because only pore water is in direct contact with soil water and soil air.
b) To convert groundwater concentration (0.005 mg/L) into pore water concentration, an empirical number of 16 is used for benzene. Thus pore water concentration:

$$C_{\text{porewater}} = 0.005\frac{\text{mg}}{\text{L}} \times 16 = 0.08\frac{\text{mg}}{\text{L}}$$

c) The next step is to back calculate the total benzene concentration in soil, which is the concentration adsorbed on soil, the concentration in soil water, plus the concentration in soil air. We now need to have the knowledge of how the contaminant partitions between soil and soil pore water (K_d), between soil pore water and soil air (Henry's law constant, H) that we have learned in Chapter 2. This gives us:

Allowable soil concentration

$$= \text{sorbed} + \text{dissolved in soil water} + \text{vapor in soil air}$$

$$= C_{\text{pore water}} \times K_d + \frac{C_{\text{pore water}}\theta_{\text{water}}}{\text{soil density}} + \frac{C_{\text{pore water}}\theta_{\text{air}} \times H}{\text{soil density}}$$

where θ_{water} and θ_{air} are % soil water and % soil air void (30 and 20% assumed in this problem, i.e. $\theta_{\text{water}} = 0.3$, and $\theta_{\text{air}} = 0.2$), soil density = 2.1 g/cm^3(i.e. 2.1 kg/L), H (dimensionless) = 0.1, K_d = 1.05 L/kg.

d) Substitute all the numbers in the above equation, we can obtain the allowable soil cleanup standard to be 0.096 mg/kg.

Allowable soil concentration

$$= 0.08\frac{\text{mg}}{\text{L}} \times \frac{1.05\text{L}}{\text{kg}} + \frac{0.08\frac{\text{mg}}{\text{L}} \times 0.3}{2.1\frac{\text{kg}}{\text{L}}} + \frac{0.08\frac{\text{mg}}{\text{L}} \times 0.2 \times 0.1}{2.1\frac{\text{kg}}{\text{L}}}$$

$$= 0.096\frac{\text{mg}}{\text{kg}}$$

Note that the unit of concentration (mg/L) × K_d (L/kg), or the unit of concentration/soil density (kg/L) results in the unit for contaminant concentrations in soil (mg/kg).

We conclude our discussions on risk assessment by pointing to sources of technical details in risk assessment developed by the ASTM (Box 4.4).

Box 4.4 The ASTM procedure for risk assessment

The American Society of Testing and Materials (ASTM) developed a guide to **risk-based corrective action (RBCA)**, which is a consistent decision-making process for the assessment and response to a petroleum release, based on the protection of human health and the environment (ASTM 1995). Sites with petroleum release vary greatly in terms of complexity, physical and chemical characteristics, and in the risk that they may pose to human health and the environment. The RBCA process recognizes this diversity, and uses a tiered approach where corrective action activities are tailored to site-specific conditions and risks. While the RBCA process is not limited to a particular class of compounds, this guide emphasizes the application of RBCA to petroleum product releases through the use of the examples. Ecological risk assessment, as discussed in this guide, is a qualitative evaluation of the actual or potential impacts to environmental (nonhuman) receptors.

The Tier 1 evaluation begins with a site assessment that usually includes a review of historical records of site activities and past release, even future use of site and surroundings. Tier 2 provides an option to determine the **site-specific target levels (SSTLs)** for chemicals of concern. In Tier 3, details for SSTLs of source areas and points of compliance are developed using statistical and fate and transport models with site-specific input parameters.

Bibliography

ASTM E1739-95 (2015). *Standard Guide for Risk-Based Corrective Action Applied at Petroleum Release Sites.* West Conshohocken, PA: ASTM (American Society for Testing and Materials).

Boyd, J. (1999). *Environmental Remediation Law and Economies in Transition.* Washington, DC: Resources for the Future.

Bedient, P.B., Rifai, H.S., and Newell, C.J. (1999). *Ground Water Contamination: Transport and Remediation*, 2e, Chapter 14. Upper Saddle River, NJ: Prentice Hall PTR.

California EPA (2005). Human-Exposure-Based Screening Numbers Developed to Aid Estimation of Cleanup Costs for Contaminated Soil.

Collin, F.P. (2003). The small business liability relief and brownfields revitalization act: a critique. *Duke Environ. Law Policy Forum* 132: 303–328.

Federal Remediation Technologies Roundtable (n.d.). Cost and Performance Case Studies. http://www.frtr.gov/costperf.htm (accessed December 2018).

Fraye, J. de Visser, E.-L. (2016). The interaction between soil and waste legislation in ten European Union countries. A Network for Industrially Contaminated Land in Europe (NICOLE).

Grasso, D. (1993). *Hazardous Waste Site Remediation: Source Control.* CRC Press.

Holland, F.A., Watson, F.A., and Wilkinson, J.K. (2007). Process economics. In: *Perry's Chemical Engineers' Handbook*, 8e (ed. R.H. Perry, D.W. Green and J.O. Maloney). New York: McGraw-Hill.

Holmgren, G.G.S., Meyer, W., Chaney, R.L., and Daniels, R.B. (1993). Cadmium, lead, zinc, copper, and nickel in agricultural soils of the United States of America. *J. Environ. Qual.* 22: 335–348.

King, W.C. (1999). *Environmental Engineering P.E. Examination Guide & Handbook*, 2e. American Academy of Environmental Engineers Publication.

Kingscott, J. and Weiman, R.J. (2002). Cost evaluation for selected remediation technologies. *Remediation* Spring: 99–116.

Lawal, Q., Gandhi, J., and Zhang, C. (2010). Direct injection, simple and robust analysis of trace-level bromate and bromide in drinking water by IC with suppressed conductivity detection. *J. Chromatogr. Sci.* 48: 537–543.

MacDonald, D., Ingersoll, C., and Berger, T. (2000). Development and evaluation of consensus-based sediment quality guidelines for freshwater ecosystems. *Arch. Environ. Contam. Toxicol.* 39 (1): 20–31.

Mihelcic, J.R. and Zimmerman, J.B. (2014). Chapter 6: environmental risk. In: *Environmental Engineering: Fundamentals, Sustainability, Design*, 2e, John Wiley & Sons.

Peters, M.S., Timmerhaus, K.D., and West, R. (2003). *Plant Design and Economics for Chemical Engineers*, 5e. New York: McGraw-Hill.

Sellers, K. (1999). *Fundamentals of Hazardous Waste Site Remediation*. Boca Raton, FL: Lewis Publishers.

Teaf, C.M., Covert, D.J., and Coleman, R.M. (2003). Risk assessment applications beyond baseline risks and cleanup goals. *Soil Sediment Contamination* 12 (4): 497–506.

USEPA (1985). *Remedial Action Costing Procedures Manual*, EPA/600/8-87/049. Washington, DC: USEPA.

USEPA (1996a). *Soil Screening Guidance: Fact Sheet*, EPA/540/F-95/041. Washington, DC: USEPA.

USEPA (1996b). *Soil Screening Guidance: User's Guide*, 2e, EPA/540/R-96/018. Washington, DC: USEPA.

USEPA (1996c). *Soil Screening Guidance: Technical Background Document*, 2e, EPA/540/R-96/128. Washington, DC: USEPA.

USEPA (1996d)). *The Role of Cost in the Superfund Remedy Selection Process. Quick Reference Fact Sheet*, EPA 540/F-96/018. Washington, DC: USEPA.

USEPA (1997). *Rules of Thumb for Superfund Remedy Selection*, EPA 540-R-97-013. Washington, DC: USEPA.

USEPA (2000). *A Guide to Developing and Documenting Cost Estimates During the Feasibility Study*, EPA540-R-00-002. Washington, DC: USEPA.

USEPA (2001a). *Cost Analyses for Selected Groundwater Cleanup Projects: Pump and Treat Systems and Permeable Reactive Barriers*, EPA 542-R-00-013. Washington, DC: USEPA.

USEPA (2001b). *Remediation Technology Cost Compendium – Year 2000*, EPA-542-R-01-009. Washington, DC: USEPA.

Questions and Problems

1 Identify soil and groundwater related laws in your area of concern (country, state/province, region).

2 Describe the difference between (a) Superfund sites and (b) RCRA facilities.

3 A real estate investor purchased an apartment complex several years ago for rent in a downtown area that was rebuilt from a previous brownfield without his knowledge. Some of tenants complained about health issues related to the apartment complex. Is he liable for this preexisting environmental issue in his property? What should he do prior to his purchase transaction?

4 How are Superfund sites listed in the National Priority List? Are all hazardous waste sites in the United States eligible for Superfund monies?

5 Describe the difference among (a) standards, (b) criteria, and (c) screening level guidelines.

6 Describe the difference (a) between MCLs and MCLGs and (b) between primary and secondary standards in the SDWA.

7 What is the present value for a purchase of (a) \$20 000 in 10 years from now and (b) \$20 000 in 15 years from now? Assume the same interest rate of 7%.

8 What is the present value for a purchase of \$20 000 in 10 years from now at an interest rate of (a) 4.5%, (b) 5.5%, (c) 6.5%, and (d) 7.5%?

9 A lab analytical instrument in a remediation consulting company has an initial capital cost of \$200 000, annual O&M costs of \$15 000, and a salvage value of \$25 000 at the end of its expected life of 10 years. What is its present value?

10 Use Excel spreadsheet to confirm the present values listed in Table 4.6 for Alternatives C, D, E, and F.

11 Use the present value method to select the more cost-effective remediation from two options. Assume a discount rate of 7%. (Hint: When converting annual O&M to present value, an alternative approach is to use tabulated value for the conversion, or use the formula $P = A\ [((1+i)^t - 1)/i\ (1+i)^t]$ to convert from annual O&M to the present value).

Option 1: Capital cost = \$500K, annual O&M = \$15K, salvage value = \$50K, life expectancy = 15 years.

Option 2: Capital cost = \$350K, annual O&M = \$10K, equipment replacement \$20K every 3 years, salvage value = \$5K, life expectancy = 15 years.

12 Based on the literature given in this chapter, make a list of the major cost elements needed for a common (a) pump-and-treat remediation and (b) soil vapor treatment.

13 When drinking water is disinfected with chlorine, an undesired by-product chloroform (CH_3Cl), is formed. Chloroform is a Class B2 probable carcinogen with a cancer slope factor of 6.1×10^{-3} $(mg/kg\text{-}day)^{-1}$. Suppose a 70-kg person drinks 2 L of water every day for 70 years with a chloroform concentration of 0.10 mg/L (the drinking water standard).

a. Find the upper-bound cancer risk for this individual.

b. If a city with 500 000 people in it also drinks the same amount of this contaminated water, how many extra cancers per year would be expected? Assume the standard 70-year lifetime.

c. Make comments by comparing the extra cancers per year caused by chloroform in the drinking water with the expected number of cancer deaths from all causes. The cancer death rate in the United States is 193 per 100 000 per year. Which one is higher?

14 Bromate (BrO_3^-) is a Class B2 carcinogen, and it has been detected in bottle water with prior ozonation treatment when the source water contains bromide (Br^-). In a survey on bottle water sold in grocery stores in the United States (Lawal et al. 2010), a highest concentration of bromate was detected to be 7.72 μg/L (99% confidence level of 0.32–2.58 μg/L). Using this highest concentration as the worst-case scenario, calculate the cancer risk for (a) an adult male and (b) an adult female who drinks this brand of bottle water. Bromate has a CSF of 0.7 $(mg/kg\text{-}day)^{-1}$. Assume intake rate (IR) = 2 L/day, exposure frequency (EF) = 350 days/yr, exposure duration (ED) = 30 years, body weight (BW) for an individual male = 70 kg, adult female = 50 kg, and average lifetime (AT) = 70 years or $70 \times 265 = 25\,550$ days.

15 Based on the data given in Example 4.6, calculate separately the actual risk for an adult female who (a) drinks water; (b) consumes fish, if the oral CSF of the contaminant of concern is 0.15 mg/kg-day.

16 EPA has classified TCE as a Group B2 carcinogen. In a dry cleaning establishment, its indoor air concentration was found to be at an average concentration of 50 μg/m^3. What will be the cancer risk from inhalation of TCE for an employee of 25 years' service? Make assumptions using EPA's default values whenever possible.

17 If we are only considering the noncarcinogenic effects of toluene from inhalation, is there any risk associated with a daily inhalation of air contaminated with 10 μg/m^3 for a resident living nearby an industrial complex? The IRIS database indicates that the inhalation RfD for toluene is 0.114 mg/kg-day.

18 Determine the maximum concentration of chloroform in drinking water, given the target excess individual lifetime cancer risk = 10^{-5} and CSF = 6.1×10^{-3} (mg/kg-day)$^{-1}$ for chloroform.

19 Explain the two different risk assessment procedures by the USEPA and the ASTM.

20 The analysis of a soil sample collected from a farm shows the following concentrations (dry basis) of several metals – Cu (15 mg/kg), Pb (25 mg/kg), Ni (10 mg/kg), Zn (30 mg/kg). The farmer who owns this piece of land was concerned about the presence of the metals, and some of these appear to be toxic. If you are approached by this farmer, what would be your professional comment?

21 Vinyl chloride can be frequently detected near landfills, wastewater treatment plants, and polyvinyl chloride (PVC) production and fabrication facilities. If an average concentration of 150 ppb$_v$ (380 μg/m^3) is detected, what would be its lifetime cancer risk for the inhalation of vinyl chloride from atmosphere? The inhalation unit risk of vinyl chloride = 4.4×10^{-6} (μg/m^3)$^{-1}$. Report in excess cancer risk per 1 000 000 population.

22 The average concentrations of PCE and TCE in groundwater monitoring wells within 5 mi of a Superfund site were 150 and 200 μg/L, respectively. This contaminated aquifer was the sole source of drinking water supply for local residents. Estimate the cancer risk using the unit risk of PCE and TCE in drinking water.

23 Estimate the cancer risk from the contaminated groundwater for residents within 5–10 mi from the same Superfund site in Question 22. The PCE and TCE concentrations were determined to be 0.5 and 25 μg/L, respectively.

Chapter 5

Site Characterization for Soil and Groundwater Remediation

LEARNING OBJECTIVES

1. Delineate the objectives of site characterization and identify the situation requiring site characterization (environmental site assessment, ESA)
2. Differentiate the purposes and components of Phase I and Phase II assessment
3. Describe the general scopes of geologic and soil characterization and associated methods using direct drilling and sampling approaches
4. Become acquainted with the indirect geophysical instrumental methods increasingly used in subsurface geological characterization
5. Describe the procedures of installing piezometers and monitoring wells, and techniques in well development and well purging
6. Comprehend the common methods in measuring hydraulic head, hydraulic gradient, and determine groundwater flow direction using the three-well approach
7. Understand the governing equations and visualize the procedures of slug test and pumping test (through YouTube links) in determining hydraulic conductivity, transmissivity, and storativity
8. Explain various groundwater sampling tools including bailer and various pumps
9. Know the use of lysimeters and the techniques for vadose zone soil gas sampling
10. Identify some common QA/QC issues in sampling groundwater for chemical characterization
11. Identify the sources of standard methods and instruments commonly used for groundwater chemical analysis
12. Identify general skill settings and elements required for hydrologists and environmental scientists to conduct ESA

In this chapter, we will discuss the "what" and "how" for the components of site characterization that are essential to soil and groundwater remediation. The purpose of site characterization is to formulate the problems in terms of contaminant migration and exposure pathways, to select the appropriate treatment technique or eliminate the unsuitable ones, and to provide feedback on remediation and site management. This chapter will cover general scopes and steps of site investigation and characterization, flow system characterization, and contaminant characterization. Whereas geologic and hydrogeologic site investigations define the flow conditions, soil and groundwater sampling and analysis delineate the concentration change over space and time as well as the migration of contaminants. A good knowledge of the contaminated site will help develop a better site-specific remediation scenario.

Soil and Groundwater Remediation: Fundamentals, Practices, and Sustainability, First Edition. Chunlong Zhang.

Companion website: www.wiley.com/go/Zhang/Remediation_1e

5.1 General Consideration of Site Characterization

Before we discuss the specific technique for site characterization, we need to define the objectives and general scopes of site characterization. Phase I and Phase II site assessment will be elaborated with an example of various scenarios to analyze whether Phase I and/or Phase II is required.

5.1.1 Objectives and Scopes of Site Characterization

The purposes of site characterization are several fold: (a) To document the presence or absence of regulated contaminants in the subsurface, (b) To map the fate and transport of contaminants, (c) To identify environmental receptors and evaluate potential threats to human health and the environment, (d) To evaluate feasible remedial actions, technologies, and costs. In other words, the questions to be answered during site characterization are: Where is the contaminant (nature and extent of contamination)? Where is it going (future migration and control)? What harm will it do (receptors and their risk)? How do we fix it (technical options for remediation)?

To address all of these questions, a variety of data is needed from the activities of site characterization. Such data can be roughly grouped into two types: chemical characterization data, and flow characterization data. For chemical characterization, both historical data and new sample analysis data on contaminant sources, extent and migration may be collected. For flow characterization, both geologic and hydrogeologic data are indispensable for a full site characterization. Geological data provides the geological framework for the conduit of water flow, whereas hydrogeologic data is essential to determine directions and rates of groundwater flow. The field or lab activities in acquiring such data are sometimes overlapped. For example, drillings for geologic samples can provide chemical information regarding contamination. In practice, site investigation and characterization are often conducted in a phased manner as will be discussed in the subsequent section.

Before any field and lab work commences, a site characterization goal must be clearly delineated to meet the site-specific conditions and regulatory requirement. Site characterization, or commonly termed as **environmental site assessment (ESA)**, is the required first step in response to Superfund Amendments and Reauthorization Act (SARA) (see Chapter 4). Transaction of properties in a much smaller setting than Superfund sites could also entail the requirement of environmental assessment. If CERCLA is not an applicable regulation, ASTM standards E1528-96 (Transaction Screen Process) may be applicable. SARA provides an incentive for environmental diligence in commercial property transactions through the so-called innocent purchaser or innocent landowner defense. The ASTM Practice E1527 (i.e. the **E1527 standard**) for Environmental Assessments in Real Estate Transactions (Phase I Environmental Site Assessment Process), issued by ASTM in 1993, is an applicable guideline for such hazardous waste sites. The EPA does not require that the ASTM standard protocols to be applied. However, this protocol is a generally accepted standard guideline.

Under the "no reason to know" provision, buyers have the incentives to hire environmental companies for assessment service to avoid the liability for and costs associated with the cleanup of a site that is later found to be contaminated with toxic materials. The "no reason to know" is not granted to the buyer for ignoring the assessment of the potential environmental issues associated with the acquired property. Approximately 75% of the clients of ESAs in the United States are reported to be the buyers of properties. The term **due diligence** is the applicable process that requires environmental characterization of a commercial real estate or other conditions, usually in connection with a commercial real estate transaction. The purpose of this practice is to define good commercial and customary practice in the United States for those who wish to conduct limited environmental investigation or audit prior to transactions. The due diligence is small yet part of the Phase I Environmental Site Assessment. For example, site assessments of some brownfields are conducted to facilitate the reuse of properties. If a contamination exists on-site, then the characteristics of contamination need to be determined. This includes the threat it poses, potential solutions for cleanup, and the cost of cleanup necessary to prepare the site for redevelopment.

The Standard Practice E1527 addresses **recognized environmental condition** (REC), which was defined as "The presence or likely presence of any hazardous substances or petroleum products on a property under conditions that indicate an existing release, a past release, or a material threat of a release of any hazardous substances or petroleum products into structures on the property or into the ground, groundwater, or surface water of the property." The E1527 standard also lists several environmental issues as being explicitly excluded from the Phase I ESA. The exclusions from this standard include (Alter 2012): asbestos-containing materials, radon, lead-based paint, lead in drinking water, wetlands, industrial hygiene, health and safety, indoor air quality, and high-voltage power lines.

5.1.2 Basic Steps: Phase I, II, and III Assessment

Three phases, termed Phase I, Phase II, and Phase III, are employed in environmental site assessment depending on the site-specific condition. Briefly, **Phase I site assessment**, also termed Phase I audits, is mostly literature review and site visit. **Phase II site assessment** involves further sampling and analysis. Phase III, also known as a remedial design and applications, is conducted for a remedial action. Since Phase I and II are truly the site "assessment," we will only focus on Phase I and II assessment in the following discussions.

5.1.2.1 Phase I Environmental Site Assessment

The purposes for Phase I ESA are: (a) To determine the potential liabilities for prior or current contamination of the site. (b) To determine whether surface or subsurface soil sampling or other sampling is required to fully assess any potential contamination. (c) To examine the applicable state and federal laws with respect to the property transfer if appropriate. (d) To provide a written analysis of the findings of interviews, record searches and field inspections in a form that will allow informed decisions to be made by the developers, lenders, and other interested parties. As shown in Figure 5.1, the components of Phase I assessment are as follows:

a) *Review records:* Records for site information include, but are not limited to, site maps and building plans, deed and title records, ownership and lease records, fire insurance maps, state and federal permitting records, prior audits/assessments, and compliance documents. Records for contaminant migration pathways include, for example, topography maps (information for surface drainage and possibility of contaminant transport), previous land topography (prior to addition of fill material or re-routing of streams or channels), aerial photographs (e.g. tidal or river – stage variations), soil and subsurface data (locations of underground pipeline or utilities), groundwater information (location of municipal, agricultural, or other registered wells; on-site production well or local groundwater pumping records), environmental and health records from state and local health departments, the Agency for Toxic Substances and Disease Registry (ATSDR).
b) *Conduct site visit:* On-site visit to physically observe regulated activities or structure and operations. Features to note include odors, stained soil and pavement, corrosive stains, storage tanks (USTs, UST vent pipes, and ASTs), drum containers, PCB transformers, septic systems, dry wells, pits, ponds, lagoons, waste piles, tank piping, and stressed vegetation.

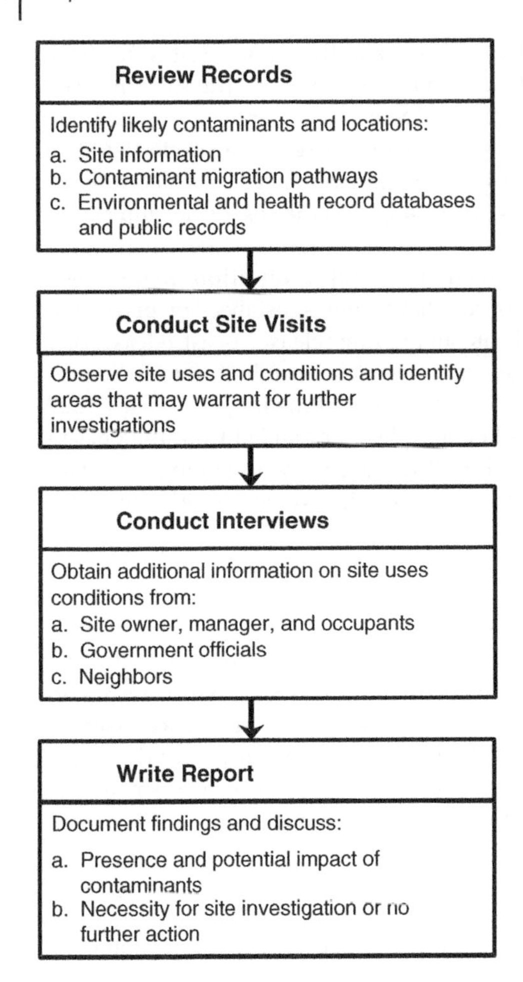

Figure 5.1 Flow chart for Phase I Site Assessment. *Source:* Adapted from the USEPA (2001).

c) *Conduct interviews:* Interview present and previous owners, occupants, neighbors, and individuals familiar with the site and property. Interview local and state officials and request historical information and documents. Investigation through others who may provide or assist with relevant information.
d) *Write report:* The obtained data and information is compiled in a report structured to meet requirements of the assessment objectives. This report should include scope of investigation, site depiction, user-supplied information, records evaluation, site reconnaissance, interview, findings of known or suspect environmental conditions, limitations such as lack of samples as qualifiers on report, recommendations and conclusions as to whether it is needed for site investigation and actions. Note that Phase I uses existing information and does not include sampling. It is the process intended to satisfy requirements of CERCLA innocent purchaser defense.

5.1.2.2 Phase II Environmental Site Assessment

The purposes of Phase II ESA are the follow-up when a Phase I Site Assessment identified conditions of environmental consequence. Many lenders will require a Phase II Environmental Site Assessment for property that has been known to have hazardous substances, such as dry cleaning operation, gas stations, and hazardous waste storage. Its primary purpose is to identify nature and extent of contamination in order to make informed business decisions and satisfy

Superfund innocent purchaser defense. Phase II ESA may include the Phase I portion of the assessment and also include soil, groundwater and surface water sampling in suspect areas. A Phase II assessment can also be implemented without the Phase I ESA if the property is known to have an existing contamination.

A flow chart for Phase II assessment recommended by the USEPA is given in Figure 5.2. Components of Phase II are: (a) To develop data quality objectives (DQO), (b) To establish risk-based screening levels (see Chapter 4), (c) To conduct assessment activities (site sampling and analysis), and (d) To present a report including sampling and analysis results, findings, and conclusions.

Subsurface soil and groundwater sample collections and analysis are the main components in Phase II assessment. Prior to sampling and analysis, a written **data quality objective (DQOs)** should be developed, by including problem statement, identification of decision and inputs to the decision, defining study boundaries, development of decision rule, and the specifications of limits on decision errors. The **screening levels** for contaminants can be established using EPA or state-level guidelines. The basic techniques and QA/QC protocols of sampling and analysis for soil and groundwater can be found elsewhere (Zhang 2007; Zhang et al. 2013). Additional details unique to site characterization will be presented in the remaining of this chapter.

The outcome of the Phase II Environmental Site Assessment is a report documenting findings from sampling and analysis (including field examination and method, sampling, analysis, geological and hydrological conditions, analytical data, and results) and a conclusion drawn from data pertaining to the project. If hazardous conditions are identified that may require remediation, the recommendations section of the report will discuss the general options available.

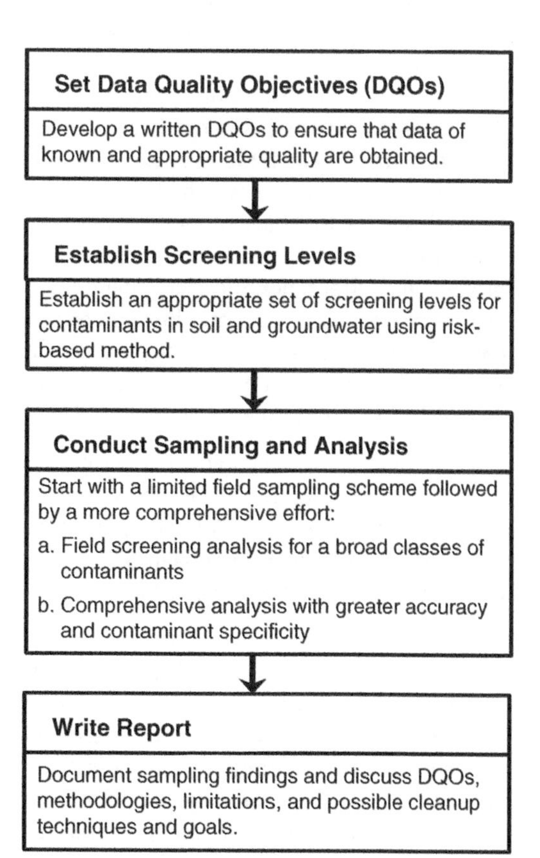

Set Data Quality Objectives (DQOs)
Develop a written DQOs to ensure that data of known and appropriate quality are obtained.
Establish Screening Levels
Establish an appropriate set of screening levels for contaminants in soil and groundwater using risk-based method.
Conduct Sampling and Analysis
Start with a limited field sampling scheme followed by a more comprehensive effort:
a. Field screening analysis for a broad classes of contaminants
b. Comprehensive analysis with greater accuracy and contaminant specificity
Write Report
Document sampling findings and discuss DQOs, methodologies, limitations, and possible cleanup techniques and goals.

Figure 5.2 Flow chart for Phase II Site Investigation. *Source:* Adapted from the USEPA (2001).

To conclude what we have discussed on three phases of site assessment, Phase I is required to obtain site information through literature and site visit, whereas Phase II is required only if further sampling and analysis is needed for decision-making. If Phase II shows potential risk and problems, then a Phase III is to further delineate a complex situation with a focus on remediation design and plan. Phase III assessment utilizes the most recent studies and modeling programs detailing alternative cleanup methods while taking into consideration the costs and logistics for the most cost-effective soil and groundwater cleanup strategy.

Box 5.1 provides some useful information for various stakeholders who might be involved in site assessment, ranging from property owners to environmental professionals. Example 5.1 is used to illustrate several situations where it is determined whether an environmental assessment is needed or not.

Box 5.1 Site assessment: Who does the work and what to expect?

As a stakeholder, whether you are a resident of a local community, a business owner, or a regulator, some basic knowledge of ESA is helpful and sometimes essential. For example, when a business owner considers purchasing a commercial or industrial property, he/she needs to hire someone to prepare ESA. This is analogous to a situation where one takes a used car to a mechanic before buying, or hire an inspector to inspect a house prior to purchasing. Phase I and Phase II Environmental Site Assessments (ESAs) are so developed to evaluate environmental issues at any site previously used for commercial purposes.

The applicable standards include ASTM E 1527-05 (Standard Practice for Environmental Site Assessments: Phase I Environmental Site Assessment Process), E 2247-08 (Standard Practice for Environmental Site Assessments: Phase I Environmental Site Assessment Process for Forestland or Rural Property), and E 1903-97R02 (Standard Guide for Environmental Site Assessments: Phase II Environmental Site Assessment Process). There are also other local, state, or federal regulations, beyond CERCLA, that have other site-assessment requirements.

Each Phase I or II ESA should be performed by a trained and experienced environmental professional. Oftentimes, these are environmental scientists who have been trained to integrate diverse disciplines. Many states have professional registrations which are applicable to the preparers of Phase I ESAs; for example, the state of California has a registration entitled "California Registered Environmental Assessor Class I or Class II." ASTM E 1527-05 set forth qualifications for an individual to perform Phase I ESAs:

- A current Professional Engineer's or Professional Geologist's license or registration from a state or US territory with three years equivalent full-time experience;
- Have a Baccalaureate or higher degree from an accredited institution of higher education in a discipline of engineering or science and five years equivalent full-time experience; or
- Have the equivalent of 10 years full-time experience.

A person not meeting one or more of these qualifications may assist in the conduct of a Phase I ESA if the individual is under the supervision or responsible charge of a person meeting the definition of an Environmental Professional when concluding such activities. Environmental firms with such capacity are often listed in Yellow Pages under Engineers – with possible subheadings of Consulting, Environmental, Geotechnical, or others. Hiring a firm with insufficient experience or training can only compound the risk if it fails to provide accurate data. A trained eye can have great value on a really complicated site.

Sufficient time should be given to allow environmental professionals to gather information and prepare for ESA. The ESA can be a useful document as part of the purchase agreement. Possible options might be:

- Require that the current landowner clean up the property prior to the sale.
- Reduce the cost of the property commensurate with the cost of remediation required.
- Pursue acquisition and cleanup alternatives that help control environmental liability for the property.

An experienced environmental professional performing the ESAs can help propose the most up-to-date remedial methods and provide reasonable cost estimates. All details about who pays for cleanup costs and criteria for "how clean is clean" should be included in the final contract with the seller.

For property owners, buyers, and other stakeholders, more basic information (such as hiring an environmental consultant and use of ESA results) can be found from: http://dnr.wi.gov/files/PDF/pubs/am/AM465.pdf. For environmental professionals, technical details on conducting Phase I and Phase II site assessment can be located in ASTM (2011, 2013).

Example 5.1 Determine whether Phase I and/or Phase II is (are) required

Given the following four situations, what action should each involved party take? (a) Consider an abandoned battery-manufacturing facility in a city of 25 000 people. The city is committed to rejuvenating the site and wants to find out how "bad" the contamination really is. (b) A big pharmaceutical company wants to acquire a medium-sized manufacturing facility to expand its production for a new drug, but wants to avoid environmental liability due to past operation. (c) A family-owned dry cleaning shop located in a metropolitan area has been in operation for decades. The owner wants to expand and modernize the operation for increased competition. The lender (banker) is approached by the owner for financing. (d) A private investor wants to purchase a pre-1978 building in a downtown area for commercial development. There are concerns about the lead paint and asbestos in this old building.

Solution

(a) The city needs to hire an environmental service company to perform a site investigation of historical data, and if needed, collect soil samples for potential lead contamination in soil. This might also be a brownfield with which assessment can be done to determine the feasibility of site redevelopment through various activities, including background investigations, site sampling and analysis, and evaluation of cleanup options and costs. (b) This pharmaceutical company should conduct a thorough (Phase I and perhaps Phase II) assessment for this medium-sized manufacturing facility not only to avoid potential liability, but also for a reduced purchasing price during the negotiation process for its acquisition. Ideally, a team with collective expertise in analytical chemistry, environmental engineering, statistics, economics, and public policy should be assembled for such an acquisition task. (c) The owner will have to hire an environmental professional to conduct a Phase I assessment as a part of the document for banker loan before the financing can be approved. (d) Asbestos or lead-based paint are not covered under the Phase I ESA, because these dangers do not subject an owner to the same broad liabilities under brownfield or Superfund. It is true that most financial organizations require certain environmental information be included with any loan eligible for purchase. The health risks associated with lead paint and asbestos are routinely evaluated by home inspectors and other professionals.

5.2 Soil and Geologic Characterization

In this section, we introduce several techniques that are commonly applied in the characterization of soil and geological formation. Geologic characterization may include the evaluation of many parameters related to stratigraphy, lithology, and structural geology. The methods employed can be grouped into direct sampling and drilling, indirect geophysical methods, and drive method, which are briefly introduced in this section. Direct soil sampling will be discussed in Section 5.4.1.

5.2.1 Stratigraphy, Lithology, and Structural Geology

Stratigraphy is a branch of geology concerned about the formation, composition, sequence, and correlation of stratified rocks and unconsolidated (loose) materials (e.g. clays, sands, silts, and gravels). Stratigraphic data can be obtained from the field through driller's logs, well cuttings, and/or corings, as well as the laboratory analysis of porosity, saturation with respect to a specific fluid component. Stratigraphic investigation will reveal aquifers or confining formations that will be likely to transport or prohibit the movement of water and contaminants.

Lithology concerns the physical character and composition of unconsolidated deposits or rocks. These include mineralogy, organic carbon content, grain size, grain shape, and packing. The first two items affect sorption, whereas the last three items affect water storage and flow. **Structural geology** includes studying and mapping features produced at the time of deposition and by movement after deposition. Structural features include folds, faults, joints, fractures, and interconnected voids (i.e. caves and lava tubes). Structural features affect groundwater transport, for example, joints and fractures can result in a preferential flow of groundwater.

Some of the geological characterizations can be done in the field, such as soil classification and hydraulic conductivity, location and groundwater surface elevation. Additional soil and geological samples are performed in the laboratory to determine, for example, soil moisture, organic matter content, grain size, density, cation exchange capacity, and Atterberg limits. The **Atterberg limits** measure the shear strength of fine-grained soils such as clays and silts because their shrinking and expanding volume changes in response to moisture changes. The Atterberg limits test is relatively inexpensive but the results can be useful for handling and processing of contaminated soils.

5.2.2 Direct Drilling Methods

Soil sampling methods can be broadly classified as hand-held and power-driven. The hand-held methods are typically used for near surface soil sampling within 2–3 m (Section 5.4.1), and the large drilling rig is capable of drilling 100 m (hundreds of feet). While the criteria for selection varies, hollow stem auger is by far the most commonly used method for well installation in unconsolidated deposit, and air rotary drilling is probably the most commonly used method in consolidated formation (e.g. solid rocks). Drilling is needed for sampling at deeper depths, which is commonly achieved with a power-operated drill rig mounted on a truck or an all-terrain vehicle. Among various drilling methods, the method of choice depends on the purpose of boring such as for lithologic sampling, soil samples for chemical analysis, or installation of monitoring wells.

Auger drilling method: The **solid stem auger** and the **hollow stem auger** are the two major auger drilling methods. With a continuous flight auger, the cutting bit loosens the soil and the auger flights carry the cuttings to the surface. The solid auger sampler (Figure 5.3a) is effective in cohesive soils above the saturated zone. It is not good for loose soil or below groundwater table and not suitable for monitoring well installation. The hollow stem auger can advance the boring to a "desired" sampling depth for precise sampling. It is the preferred environmental drilling method that is effective in various soil types and useful for the installation of groundwater monitoring wells through the inside of the auger.

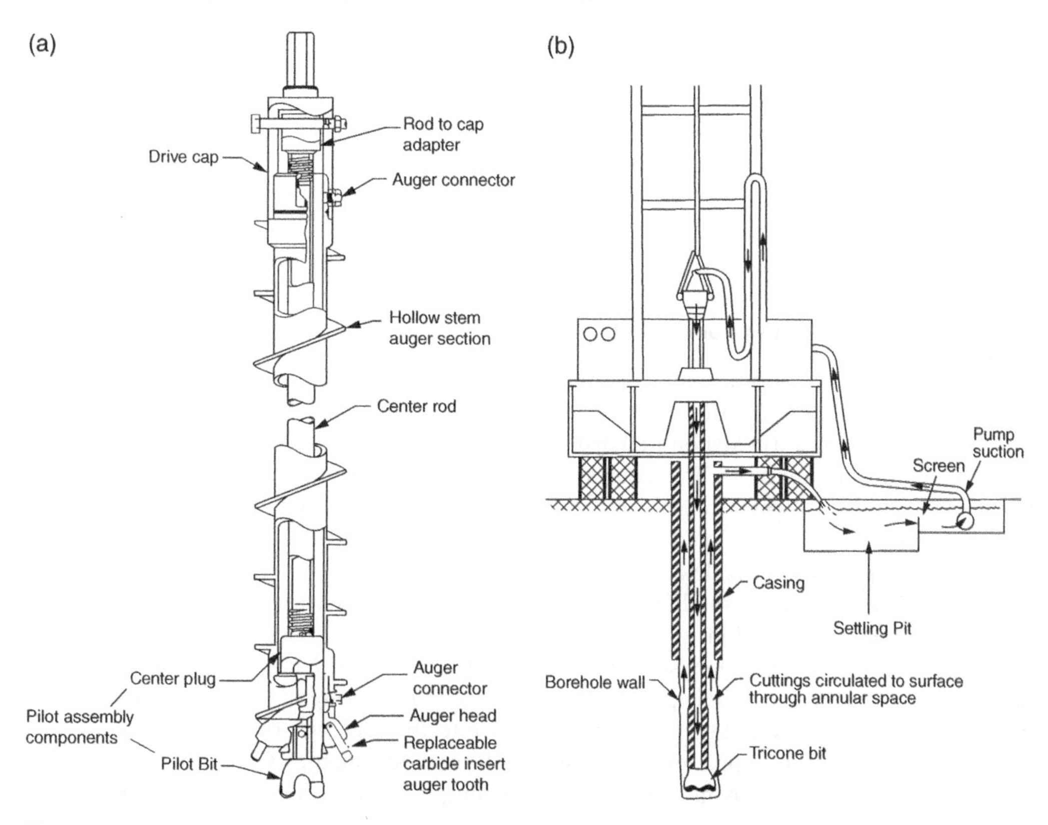

Figure 5.3 A conceptual comparison of (a) the hollow-stem auger, and (b) the rotary drilling. *Source:* Courtesy of Central Mine Equipment Company; National Water Well Association of Australia.

Rotary drilling method: With the **rotary drilling method** (Figure 5.3b), the cutting head is rotated in the presence of drilling fluid (known as mud). As the fluid is pumped down the hole, drill cuttings are circulated up the borehole to the surface. Boreholes stability is maintained through the use of a steel casing inserted in the borehole or through the use of the drilling mud. Wet rotary drilling is effective in site characterization when auger drilling is not practical such as in a deep aquifer, hard formation, and in flowing sand conditions, but requires addition of water and drilling mud.

Dry drilling with auger without the use of fluid is common in environmental site assessment to avoid any introduction of foreign materials. The wet rotary drilling is used to drill boreholes for installing piezometers or monitoring wells at a greater depth. Similar to wet rotary drilling, the air rotary drilling methods using compressed air are also frequently employed, particularly for the installation of piezometers or monitoring wells in consolidated materials, and for deeper unconsolidated materials. Table 5.1 compares hollow-stem auger and the direct-rotary drilling when both are used for the construction of monitoring wells.

Table 5.1 Auger and rotary drilling techniques: advantages and disadvantages for construction of monitoring wells.

Type	Advantages	Disadvantages
Auger	• Minimal damage to aquifer • No drilling fluids required • Auger flights act as temporary casing, stabilizing hole for well construction • Good technique for unconsolidated deposits	• Cannot be used in consolidated deposits • Limited to wells less than 150 ft in depth • May have to abandon holes if boulders are encountered
Rotary	• Quick and efficient method • Excellent for large and small diameter holes • No depth limitations • Can be used in consolidated (solid rock) and unconsolidated (loose) deposits	• Required drilling fluids which alter water chemistry • Results in a mud cake on the borehole wall, requiring additional well development, and potentially causing changes in chemistry • Loss of circulation can develop in fractured and high-permeability materials

5.2.3 Drive Method Using Cone Penetrometer

Cone penetrometry, or commonly known as the **cone penetrometer test (CPT)**, uses electronic strain gauges mounted in a steel cone-shaped probe, which are pushed at a constant rate into the subsurface by a truck-mounted hydraulic system (Figure 5.4). It is thus a drive method rather than a drilling method. The gauges measure the resistance and friction related to soil properties at various depths in real time. For example, high tip resistance and low friction indicate coarse-grained soils. In addition, electro-conductivity and laser-induced fluorescence can also be used to correlate hydrocarbons or presence of free-phase hydrocarbons in the subsurface. CPT can also be modified to enable pore pressure measurement to infer the permeability of geologic formation.

CPT differs from other geophysical methods (next section) in that it is destructive to the integrity of soil samples. CPT can provide useful information during the initial soil characterization, such as soil types and contaminants sensitive to the sensing probes. Primarily developed for geotechnical investigation, the direct push methods such as Geoprobe systems are increasingly used in environmental field for soil sampling and monitoring of contaminated sites.

5.2.4 Indirect Geophysical Methods

Instead of directly examining the geological soil and rock samples, subsurface geologic formation can also be indirectly measured by a variety of geophysical instruments. Such indirect measures can be based on various instrumental signals from surface (above ground), surface to borehole, borehole to borehole, and single borehole devices. Geophysical techniques have also become increasingly useful for estimating groundwater contamination and locating hazardous wastes (Evans and Schweitzer 1984). Table 5.2 provides a summary of these geophysical methods in regard to their governing principles and applications. Several geophysical techniques are nondestructive, because these surveys can provide subsurface information needed for site characterization without digging, probing, or drilling. These nondestructive subsurface survey tools can rapidly delineate variations in the subsurface conditions, making them ideal for an accurate and rapid site characterization. The use of geophysical signals, however, relies heavily on the data interpretation. Consequently, these methods are usually used in conjunction with the previously described direct methods to confirm the interpretation.

Figure 5.4 Typical cone penetrometer (USEPA 1993a).

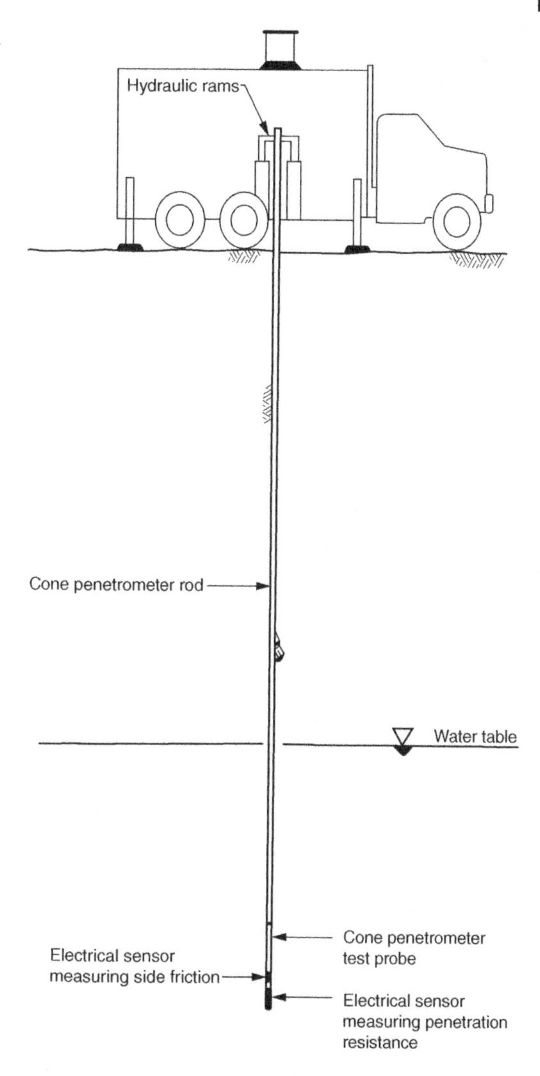

Electrical resistivity: In conventional resistivity survey, electrodes are inserted into soil for measuring current (I) and voltage difference (V). The measured electrical resistance ($R = V/I$) between the electrodes depends on the electrical properties of the geologic materials which, in turn, depend upon the resistivity (ρ) of the pore water, the amount of solutes in pore water, and soil texture. **Resistivity** (unit in Ω m) relates directly to resistance R (unit in Ω) by a coefficient determined by section area/length (m^2/m).

Freshwater has a resistivity of 10–200 Ω m, whereas an aquifer may have a resistivity in a wide range of 50–2000 Ω m. Rock materials and gravels are more resistive than silt and clay, because electrically charged surfaces of fine particles are better conductors (lower resistivity). Water is very conductive, and the elevated concentrations of dissolved chemicals in water will make it a better conductor (lower resistivity). Porosity and local stratigraphy, therefore, can be deduced from the resistivity measurements. Contaminant plumes with high concentrations of some solutes frequently appear as a highly conductive layer. Resistivity methods thus can be useful in identifying and mapping plumes of certain contaminants. Electrical resistivity can be used to provide detailed subsurface information to outline aquifer boundaries, depth to the water table and bedrock, and change in soil type or levels of contamination.

Table 5.2 Comparisons of common geophysical methods for site characterization.

Method and principles	Applications and limitations
Electrical resistivity: Electrodes measuring current and voltage	Mapping conductive or unconductive contaminant plumes; stratigraphy of aquifers. No depth limit, but commonly used at depths <90 m.
Electromagnetic conductivity: Measuring the conductivity after the induction of currents in response to alternating magnetic field	Mapping conductive or unconductive contaminant plumes and detect buried metallic objects such as waste drums. Depth up to 60 m (frequency domain) to 300 m (time domain)
Magnetometers: Measuring the magnetic field strength using nuclear magnetic resonance magnetometers	Especially useful to detect buried metal drums or other ferromagnetic (iron and steel) metal objects. No depth limit, but commonly used at depths <90 m.
Ground-penetrating radar (GPR): Measuring the reflection of radio waves transmitted into the ground	Detect buried objects (both plastic and metals); stratigraphy in conductive soils such as clays; depth to groundwater. Typically at depths <9–30 m.
Seismic reflection/refraction: Monitoring the sound wave by a geophone from an impact at the surface using a hammer or an explosive device	High resolution mapping of top of bedrock (stratigraphy). At depth >300 m (seismic reflection) to no limit (seismic refraction). Slow survey.

As an example of the electrical resistivity method, the background resistivity and its changes following air sparging are depicted in Figure 5.5. Prior to sparging, the resistivity values ranged 200–400 Ω m in the saturated zone, and reached up to 1600 Ω m in an unsaturated zone (LaBrecque et al. 1996). With air introduced into the saturated zone, resistivity increased by 500 Ω m above the background level.

Instead of using the electrode probes hammered into the ground, capacitively coupled resistivity is commercially available, which uses antenna to be hauled along the ground by a field crew or an all-terrain vehicle (Figure 5.6). This system uses a dipole array to measure resistivity. The detailed operational principle is beyond the scope of this text.

Electromagnetic conductivity (EM): **Electrical conductivity**, expressed as Siemen per minute (S/m), is the inverse of the electrical resistivity (e.g. 2 S/m = 0.5 Ω m). Since EM or terrain conductivity survey instrument measures the conductivity, in principle, it shares the same applications as the measurement of electrical resistivity discussed above. EM consists of a transmitter and receiver. The transmitter coil produces an alternating magnetic field that induces electrical currents within the ground. The induced currents vary with the electrical conductivity of the geological materials and alter the magnetic field of the transmitter. This alteration is detected by a receiver coil. Since these devices are generally carried near the earth surface, without the need of installing electrodes or geophones, the EM method is rapid and is likely to be more cost-effective than traditional resistivity methods. EM can be used to detect changes in subsurface conductivity related to contaminant plumes or buried metallic wastes such as drums. As such, it is used to offer a rapid, cost-effective method of mapping contaminant plumes.

Magnetometers: A **magnetometer** measures the magnetic strength in units of gammas or nanotesla (1 gamma = 1 nanotesla or nT). The Earth has an approximate 32 000 nT at 0 °C latitude and 0 °C longitude. Areas with large amounts of buried metal, such as steel drums, will have magnetic anomalies associated with them. For example, a profile over a single underground storage tank has a 1000-gamma (1000 nT) anomaly (Barrows and Rocchio 1990). The strength of the anomaly will vary with the amount and depth of the buried metal.

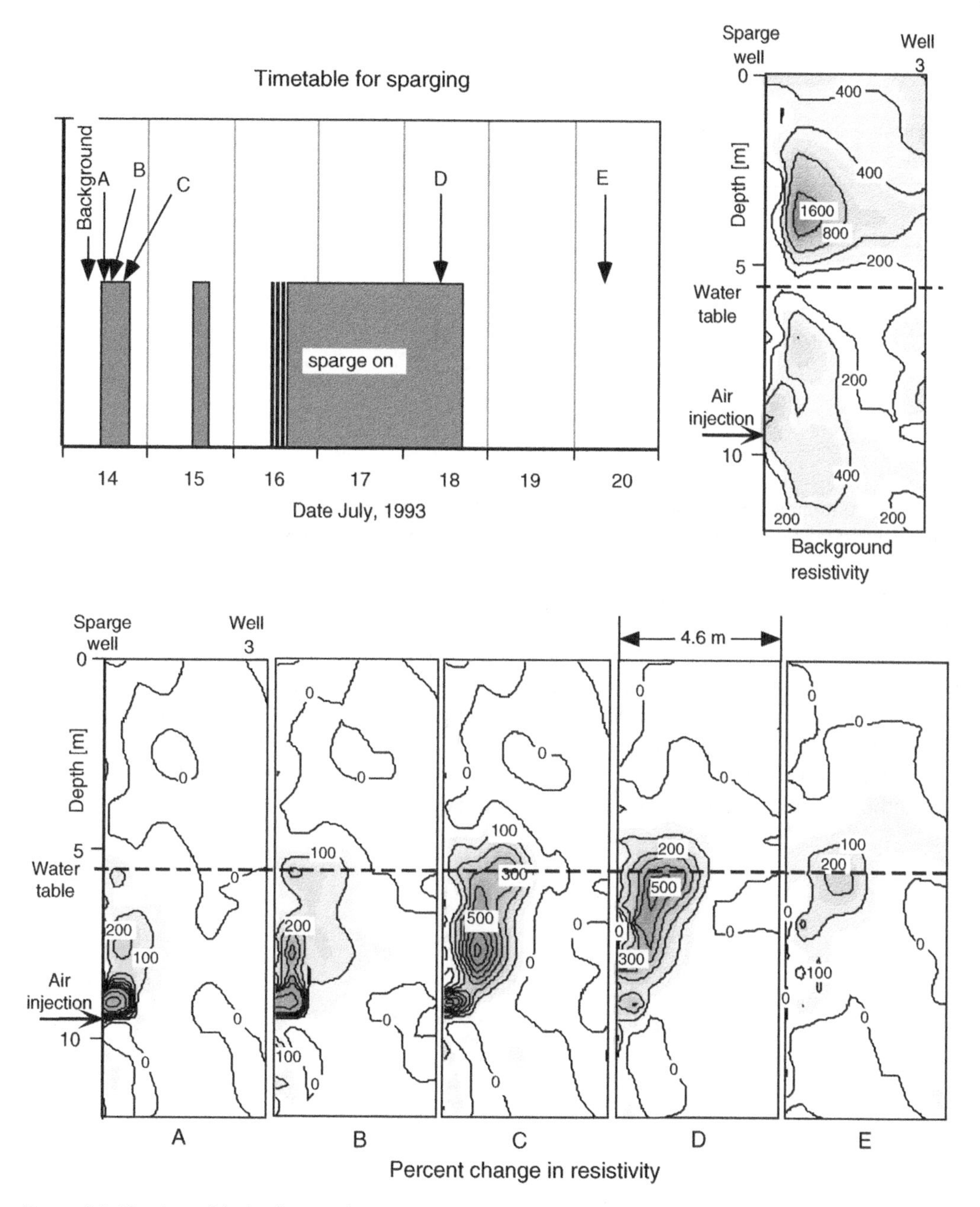

Figure 5.5 The timetable, background resistivity image, and percentage change in resistivity images for the air sparging experiment at Florence, Oregon. Percentage change in resistivity images are shown for (A) 20 minutes after sparging began, (B) 1 hour after sparging began, (C) 4 hours after sparging began, (D) after 48 hours of continuous sparging and (E) 18 hours after sparging stopped. *Source:* LaBrecque et al. (1996). Reprint with permission from IOP Publishing, Inc.

Since a handheld proton nuclear magnetic resonance magnetometer is frequently used, one person can rapidly perform a survey over a site of a few acres. The surveyor sets up a grid system and measures the magnetic field at each intersection of the grid. Such methods are especially useful for the identification of buried drums or other foreign ferromagnetic (iron

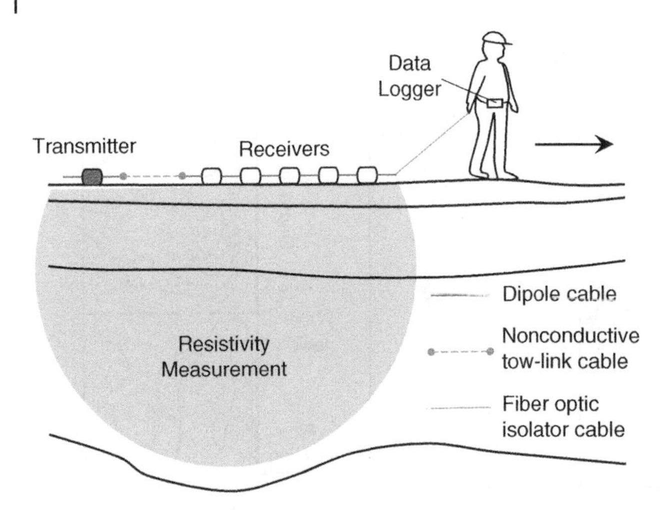

Figure 5.6 Schematic of a capacitively coupled array resistivity system (one transmitter and five receivers) towed by a field crew.

and steel) materials. As shown in Figure 5.7 a magnetometer successfully detected buried underground storage tanks underneath a municipal garage. These steel tanks were approximately 2000 L with a diameter of 1.4 m and a length of 2.1 m. The top of the USTs were approximately 1.4 m below ground surface.

Ground-penetrating radar (GPR): With **ground-penetrating radar** method, radio waves of 50 MHz to 2.5 GHz are transmitted into the ground and the reflected waves are monitored and analyzed (Figure 5.8). Reflections occur as a result of geologic variations in porosity and water content. The method is useful for determining stratigraphic variations and for locating buried objects such as gasoline tanks and steel drums. A GPR survey can be conducted over the entire survey area or only in those areas where a previous EM survey showed anomalous conditions. The method is only applicable for conductive soils such as clays.

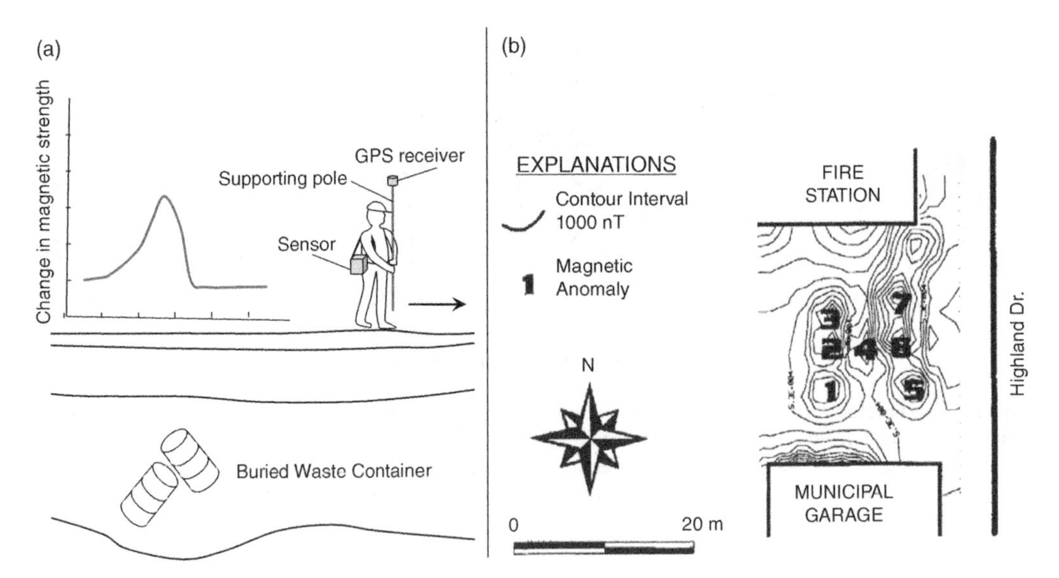

Figure 5.7 (a) Schematic showing the operational principle of a magnetometer, (b) Magnetic intensity contour map showing the magnetic anomalies in relation to the location of 12 UST subsequently removed at the municipal garage of the Village of Kohler, Sheboygan County, Wisconsin. *Source:* van Biersel et al. (2002). Reprinted with permission from John Wiley & Sons.

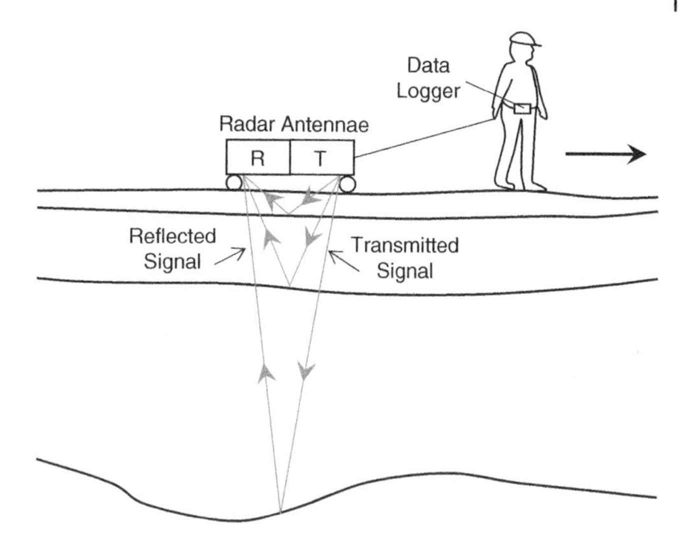

Figure 5.8 Schematic of a ground-penetrating radar (GPR) system towed by a field crew.

The resulting profile from a device illustrated in Figure 5.8 is a scan of the one-dimensional subsurface beneath the antenna. The GPR survey can also show the cross-sectional information in a two-dimensional pictorial form. To build up a two-dimensional section of the subsurface (a radargram), the antenna is traversed across the surface to collect a number of adjacent scans. An example of 100 MHz radar data collected over an unlined and capped portion of the landfill cells in eastern England is shown in Figure 5.9, where the circled portion (i.e. the vertical stripe approximately 2 m wide and 12 m in depth) of the radargram corresponds to the leachate from containment walls. Leachate increased the conductivity of the pore water, which led to the absorption of radar signal. GPR can also detect metal objects, service pipes/sewer lines, and other underground objects without invasive methods of boreholes and drilling.

Seismic reflection/refraction: The principles involved in **seismic reflection/refraction** are generally similar to GPR, except that GPR employs electromagnetic energy rather than acoustic (sound) energy. An impact is made at a particular point on the ground surface using a mechanical hammer or an explosive device, and the resulting sound waves are monitored by sensing devices (geophones) positioned at various distances from the impact source (Figure 5.10). In the refraction method, the travel time of refracted waves along an acoustic interface is measured. In the reflection method, the travel time of waves reflected off an interface is measured. The time of arrival of the sonic waves at the receiver site depends on velocity and density contrasts that occur as the wave passes through different stratigraphic layers. The received sound signal and travel

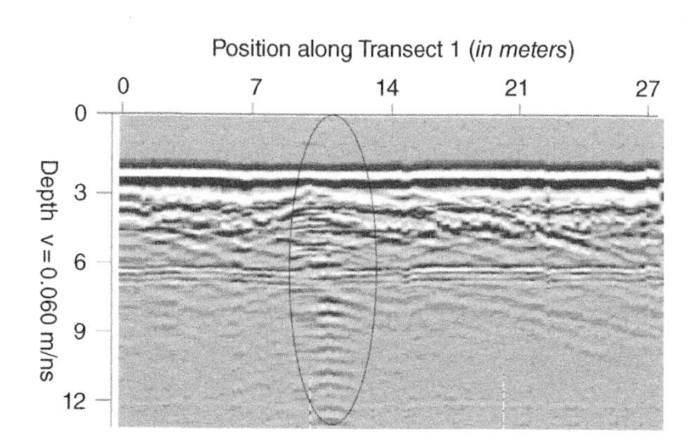

Figure 5.9 A radargram acquired from a ground-penetrating radar (GPR) operated at 100 MHz from a landfill site located in eastern England. The anomalous feature is circled. *Source:* Reprinted with permission from Splajt et al. (2003). Copyright (2003) American Chemical Society.

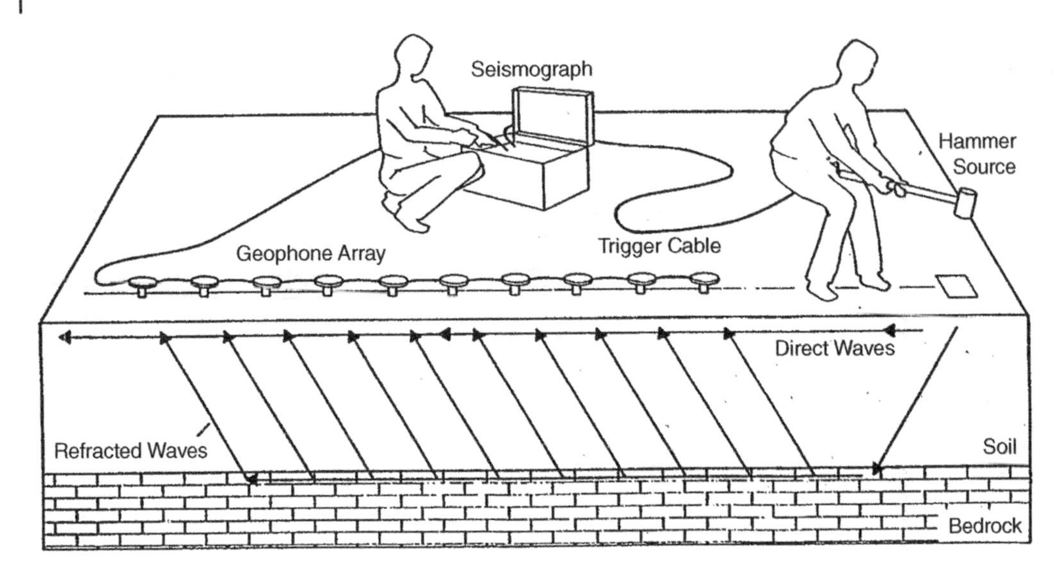

Figure 5.10 Schematic of seismic refraction layout of a 12-channel seismograph showing the path of direct waves and refracted waves in a two-layer soil/rock system (USEPA 1993a).

velocity can be used to determine the geologic layering in the area, such as depth to bedrock, slope of the bedrock, depth to groundwater table, and in some cases, the general lithology.

Both methods acquire data through a seismograph. Seismic reflection methods share the similar equipment with seismic refraction methods. The seismic reflection has been highly developed in the petroleum industry. In hydrogeological applications, this method is particularly useful in finding the thickness of unconsolidated materials overlying bedrock. The loose material in unconsolidated layer transmits seismic waves more slowly than consolidated bedrock. Seismic reflection can also identify groundwater table, because saturated zone has a greater seismic velocity than unsaturated equivalent soil unit.

5.3 Hydrogeologic Site Investigation

The determination of groundwater flow direction and flow rate requires hydrogeologic parameters such as hydraulic gradient, hydraulic conductivity, and effective porosity that we defined in Chapter 3. Effective porosity can be readily estimated in the lab. Monitoring water elevations (or generally hydraulic head) and hydraulic conductivity are most commonly measured in the wells. The discussions below start with well installation and development, followed by the determination of hydraulic head, flow direction, and hydraulic conductivity.

5.3.1 Well Installation, Development, and Purging

During the hydrogeologic investigation, a well is installed into an aquifer at a certain depth where the screen part of the vertical casing is exposed to surrounding groundwater. Well construction is quite a complex task that typically requires some specialized skills. Depending on the purposes of monitoring well (i.e. for elevation, water, NAPL, soil gas), well designs vary. Important considerations should be given to the dimension and materials of casings and well screen, selected materials for filter pack (mostly medium to coarse sand) and annular seal (commonly bentonite). For example, the well screen must be long enough to intersect the water table over the range of annual fluctuation. In most applications the minimum screen length for water table monitoring well is 10 ft with 5 ft above and 5 ft below the water table. Incorrect placement of a multipurpose monitoring well will not be able to measure the position of water table and the LNAPLs (light non-aqueous phase liquids) floating on the water surface.

A **piezometer** is a small diameter, nonpumping well used to measure the elevation of groundwater table or potentiometric surface. Here, a **potentiometric surface** is a hypothetical surface representing the level to which groundwater would rise if not trapped in a confined aquifer. In an unconfined aquifer, the potentiometric surface is equivalent to the water table. A piezometer should have a relatively short screen length, 2–5 ft, so the recorded pressure is representative of only a small vertical section of the aquifer. A piezometer can also be used to collect groundwater samples.

Well development involves removing fine particles (sand, silt, and clay) from the drilling process after a well has been installed. A nicely developed well allows groundwater to flow freely through the well screen into the well without the interference of fine particles. It also reduces the turbidity during subsequent groundwater sampling. The most common well development methods are: surging, jetting, overpumping, and bailing. **Surging** involves raising and lowering a surge block inside the well. The resulting surging motion forces water into the formation and loosens sediment, pulled from the formation into the well. **Jetting** involves lowering a small diameter pipe into the well and injecting a high velocity horizontal stream of water or air through the pipe into the screen openings. This method is especially effective at breaking down filter cakes (mud) developed during rotary drilling. Simultaneous air-lift pumping is usually used to remove fines. **Overpumping** involves pumping at a rate rapid enough to draw the water level in the well as low as possible, and then allowing the well to recharge to the original level. This process is repeated until sediment-free water is produced. **Bailing** includes the use of a simple manually operated check-valve bailer to remove water from the well. The bailing method, like other methods, should be repeated until sediment-free water is produced. Bailing may be the method of choice in a shallow well or well that recharges slowly.

Well purging is employed to remove stagnant water in the well borehole and adjacent sandpack so that the groundwater to be sampled will be stabilized prior to sampling and representative groundwater can be obtained. Various methods for determining the necessary extent of well purging have been recommended. For example, the USGS recommended pumping the well until water quality parameters (pH, temperature, and conductivity) are stabilized. A "stabilized" condition can be determined by *in situ* measurement of the following parameters: DO: ±0.3 mg/L; turbidity: ±10% (for samples greater than 10 Nephelometric Turbidity Unit or NTUs); specific conductivity: ±3%; oxidation–reduction potential (ORP): ±10 mV; pH: ±0.1 unit; temperature: ±0.1 °C. This approach is also specified in the USEPA's Groundwater Guidelines for Superfund and RCRA Project Managers (USEPA 2002). The USEPA also recommends the removal of three well-casing volumes prior to sampling. A handy formula in English measurement unit system is as follows (Bodger 2003).

$$V = 7.48 \times \pi r^2 h \qquad \text{Eq. (5.1)}$$

where V = well volume (gallons), r = radius of monitoring well in ft; h = the difference between depth of well and depth to water (i.e. the height of the water column in ft); 7.48 is the conversion factor (the gallons of water per ft^3 of water).

5.3.2 Hydraulic Head and Flow Direction

We have defined the concept of hydraulic head in Chapter 3. Briefly described below are the measurement of hydraulic head and the use of hydraulic head to determine groundwater flow direction.

5.3.2.1 Methods to Measure Hydraulic Head

The concept of hydraulic head and its relation to a well's water level is discussed in Chapter 3. The measurement of a well's water level is typically conducted prior to sampling. The easiest and most precise is a steel tape. Before the tape is lowered down the well, the bottom 1–2 ft is coated with carpenter's chalk. The tape is then lowered into the well until the lower part is submerged. Measurements are made to 1/100 of a foot. This method is now mostly replaced by electrical tape water level indicators. When an electrical probe is lowered into the water, an electrical circuit is completed causing a buzzer or light to be activated. Using a pressure transducer is another method for testing an aquifer where the water level changes rapidly. With a pressure transducer, water level data can be automatically transferred to a data logger for a continuous measurement.

5.3.2.2 Groundwater Flow Direction

Groundwater flow direction can be readily determined after the hydraulic heads have been measured in wells. Knowing the direction of groundwater flow is extremely important. Imagine, for example, if our measurements result in the groundwater flowing north when it is actually flowing south, sampling expenses will be wasted and the whole remediation effort will be in vain. Unlike surface water, groundwater flows very slowly (several feet to several hundred feet per year) and the flow direction cannot be visualized but has to be determined by other means, such as using wells.

Groundwater flows in the direction from the higher hydraulic head to the lower hydraulic head. The flow direction can be estimated based on the hydraulic head of three adequately spaced but connected wells as shown in Figure 5.11. The general procedures and the accompanied Example 5.2 are as follows.

- Step 1: Draw a line (i.e. line AC) between two wells with the highest (h_1) and the lowest hydraulic head (h_3).
- Step 2: Draw a line (i.e. BD) between the intermediate-head well (h_2) and the spot D on the above line (AC) that corresponds to the head at the intermediate well. In other words, the location of point D is selected such that the hydraulic head at point D along line AC is the same as the hydraulic head at the intermediate-head well B. Line BD is the **equipotential line,** meaning the head anywhere along the line is constant.
- Step 3: From the lowest-head well C, draw a line (i.e. line CE) perpendicular to the equipotential line. This is a "flow line," the groundwater flow is in a direction parallel to this line. The flow direction should be pointed from the higher hydraulic head to the lower hydraulic head (i.e. arrow pointing from E to C rather than from C to E).
- Step 4: Determine the hydraulic gradient $dh/dL = (h_2 - h_1)/L$, where L is the distance from well C to the line BD (i.e. L = CE).

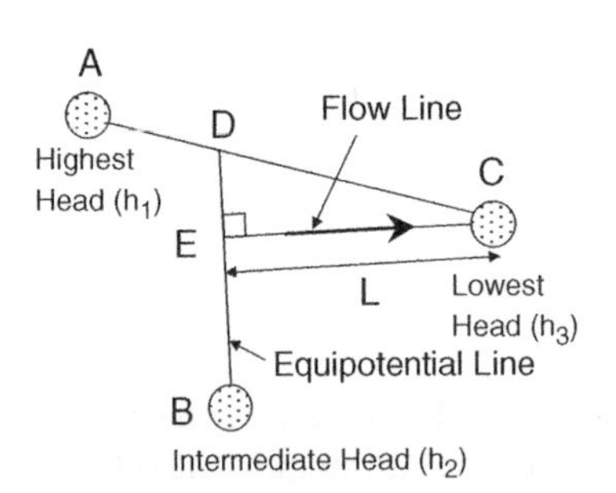

Figure 5.11 Determination of groundwater flow direction.

Example 5.2 Determine groundwater flow direction

Two wells are drilled 300 m apart along an east-west direction. The east well and west well have hydraulic heads of 20.0 and 20.4 m, respectively. The third well is located 200 m due south of the east well and has a head of 20.2 m. Schematically determine groundwater flow direction and estimate hydraulic gradient.

Solution

General Step (refer to Figure 5.12):

1) Draw a line (AB) between two wells with the highest (A) and lowest (B) heads (i.e. head: A > C > B)
2) Locate the point (D) on line A–B that corresponds to the well with intermediate head (C). Draw the line CD. This line is an equipotential line.
3) Draw a line BE perpendicular to line CD. Line EB is the flow line, meaning the groundwater flows in a direction parallel to this line. Since the water flows from the higher to the lower hydraulic head along line EB, the groundwater flows from E to B rather than from B to E.
4) Determine the hydraulic gradient:

$$\text{Hydraulic gradient} = \frac{\text{Head of well C} - \text{Head of well B}}{\text{Distance from B to the equipotential line }(CD)}$$

$$= \frac{20.2 - 20.0}{120} = 0.00167\ (\text{Dimensionless})$$

Note that point D happens to be in the middle of line AB, because the head of south well (20.2 m) is at the midpoint of head in east well (20.0 m) and west well (20.4 m). The geometric calculation of distance BE would be more complicated if this is not the case.

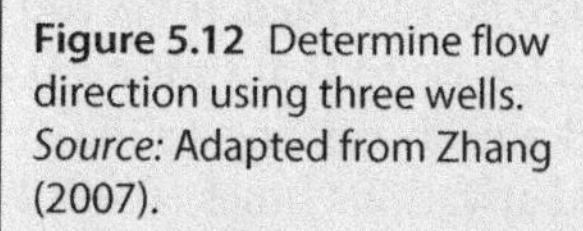

Figure 5.12 Determine flow direction using three wells. *Source:* Adapted from Zhang (2007).

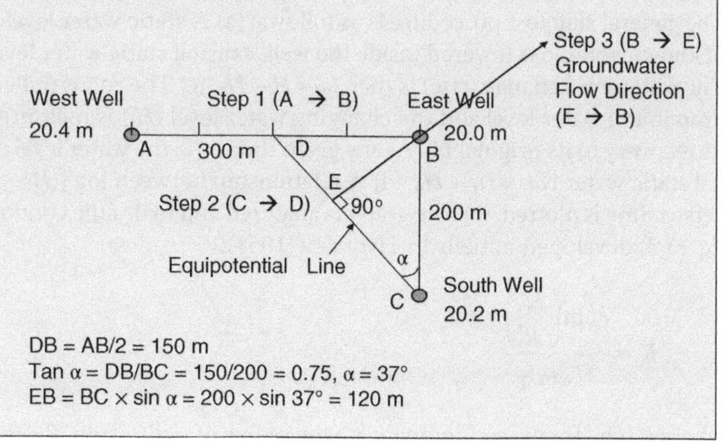

5.3.3 Aquifer Tests to Estimate Hydraulic Conductivity

Aquifer properties (hydraulic conductivity, transmissivity, storage coefficient) are not measured directly, but instead they are calculated through model equations or graphical methods using aquifer test data acquired in the field. Sometimes, using more than one method to determine aquifer properties is recommended.

Table 5.3 is a summary of these methods, including a laboratory device known as the **permeameter**, and *in situ* method using a single well (**slug test**) or multiple wells (**pumping test**). A simple empirical method based on soil properties like pore size, grain size distributions and soil texture is also used for approximate estimation. The hydraulic conductivity values measured in the laboratory using reconstituted disturbed aquifer materials are approximate and applicable only to small-scale situations. Hence, an *in situ* test under the controlled condition is always recommended. The discussions below focus on two common field aquifer tests, i.e. slug test and pumping test.

Table 5.3 Summary of methods to measure saturated hydraulic conductivity values in the laboratory and field.

Method	Application
Constant-head permeameter	Laboratory method to determine hydraulic conductivity of typically granular aquifer materials within a range from 1.0 to 10^3 cm/s.
Falling-head permeameter	Laboratory method to determine hydraulic conductivity of aquifer materials such as fine-grained to coarse-grained soils.
Slug test	Field aquifer test for confined aquifers with fully penetrating wells screened along the entire aquifer thickness. Single-well test to obtain hydraulic conductivity. Can be quickly done by one person without the need of power.
Pumping test	Field aquifer test for complex multiple-wells for confined aquifers with fully or partially penetrating wells. Used for a wide range of aquifer hydraulic conductivities. Test wells can be used for sampling. Test a relatively large volume of the aquifer. Need several days of pumping and the disposal of water.

5.3.3.1 Slug Test: Hvorslev Method

A commonly used slug test is to quickly remove a rod (slug) from groundwater in a well, and the recovery of the water level to the original water level is continuously measured over time. The general slug test procedure is as follows: (a) A static water level (H_0) is recorded, (b) A rod of known volume is lowered inside the well, causing static water level to rise to H at time $t = 0$. The water level change (rise) is then $h_0 = H - H_0$; (c) The rod is pulled quickly causing a sudden drop in the water level and the changing water level (H_t) is measured at various times (t) until its recovery to its original H_0. At any given time $t > 0$, the water level change relative to the original static water is $h = H_t - H_0$. (d) A relationship between log $[(H_t - H_0)/(H - H_0)]$, or log h/h_0, versus time is plotted. (e) The graph is analyzed and hydraulic conductivity is determined using Eq. (5.2) developed initially by Hvorslev (1951):

$$K = \frac{r^2 \ln\left(\frac{L_e}{R}\right)}{2 L_e t_{0.37}} \qquad \text{Eq. (5.2)}$$

where K = hydraulic conductivity, r = the radius of well casing, R = the radius of well screen plus sand pack (bore hole), L_e = effective length of well screen, $t_{0.37}$ = time required for the water level to fall to 37% of the initial change.

The value of $t_{0.37}$ can be obtained from a semi-logarithmic plot of h/h_0 vs. t, where $h/h_0 = 0.37$ (Figure 5.13). Note that H_0 = water level at $t = 0$, H = initial water level prior to the removal of slug, H_t = recorded water level at $t > 0$. After $t_{0.37}$ is obtained, Eq. (5.2) can be used to calculate K. Equation (5.2) is applicable when the ratio of effective lengths of the well screen to the effective radius of the well screen $L/R > 8$, which is a common design for monitoring wells.

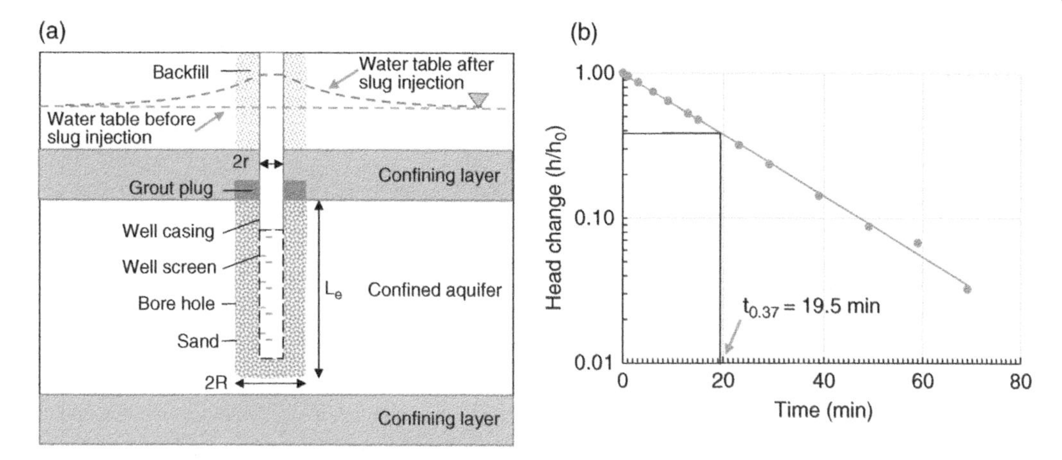

Figure 5.13 (a) Well construction parameters needed for slug test in a confined aquifer, (b) Analysis of slug test data by the Hvorslev method.

5.3.3.2 Slug Test: Bouwer and Rice Method

The most common slug test for hydraulic conductivity determination is through the method designed by Bouwer and Rice (1976), which was initially used for unconfined aquifer, but it can be used for confined or stratified aquifers if the top of the screen is some distance below the upper confining layer. The governing equation is:

$$K = \frac{r \ln\left(\frac{R_e}{R}\right)}{2L_e} \frac{1}{t} \ln \frac{y_0}{y_i} \qquad \text{Eq. (5.3)}$$

where r = the radius of the well casing, R = the radius of sand pack (bore hole), R_e = the effective radial distance over which y is dissipated (which varies with well geometry), L_e = the length of the screen or open section through which water can enter, y_o = drawdown (vertical difference between water level inside well and water level outside) at time $t = 0$, y_t = the drawdown at time t. Note that values of r, R, L_e, and y_0 in Eq. (5.3) can be obtained from the well construction information. The effective radial distance R_e was based on well construction geometry and empirical values (Bouwer and Rice 1976). Using any two points of the linear portion of the plot between ln y_t and time (t) (Figure 5.14), the value of $(1/t)\ln(y_0/y_i)$ can be calculated. The K thus estimated by Bouwer and Rice method is more time consuming than the Hvorslev method.

Since the rise in hydraulic head reflects the near-well aquifer characteristics, cautions should be exercised when using hydraulic conductivity values measured from slug tests to represent the variability in hydraulic conductivity for the aquifer. While slug test results provide information on the portion of the aquifer in the immediate vicinity of the test location, they are not recommended for the design of remediation systems.

5.3.3.3 Pumping Test: Theis Type-Curve Method

A typical pumping test procedure is as follows: (a) A well is pumped at a constant rate for a fixed period (e.g. 24 hours for confined aquifer to several days for an unconfined aquifer). (b) Water level drawdown is observed in the pumping well and one or more observation wells. (c) The pump is turned off and the water level is recovered. (d) The drawdown is then plotted versus time and aquifer hydraulic parameters (K, T, S) are calculated as described below.

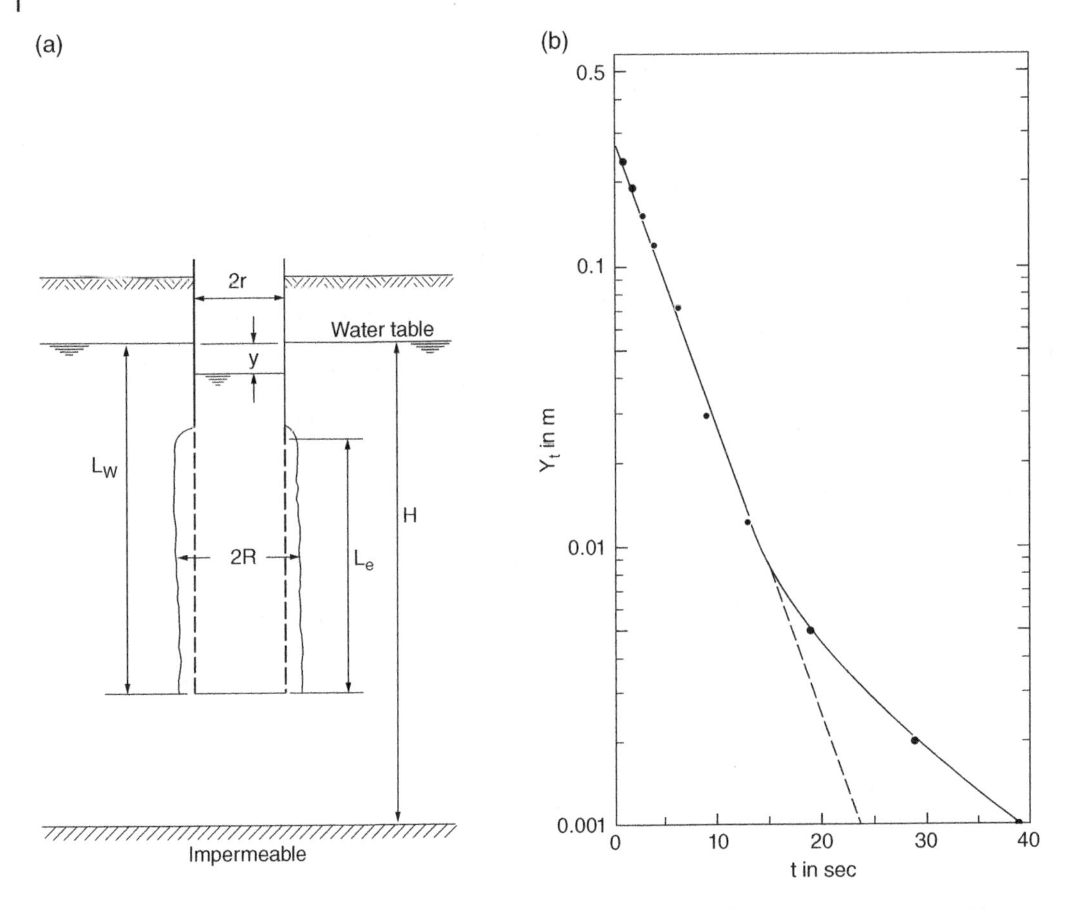

Figure 5.14 Slug test by Bouwer and Rice method: (a) geometry and symbols; (b) Plot of $\ln y_t$ and time (t). *Source:* Bouwer and Rice (1976), ©John Wiley & Sons.

In Section 3.3.3 of Chapter 3, we described a **steady-state** radial flow to a pumping well and derived Eq. (3.18) as commonly referred to as the **Thiem solution**. They are derived from the application of Darcy's law to cylindrical shell control volumes. Under an **unsteady-state** condition, however, the **Theis equation** derived by Charles Vernon Theis of the US Geological Survey in 1935 can be used for two-dimensional radial flow to a point source in an infinite, homogeneous aquifer:

$$h_0 - h = \frac{Q}{4\pi T}\int_u^{\infty}\frac{e^{-a}}{a}da = \frac{Q}{4\pi T}W(u) \qquad \text{Eq. (5.4)}$$

where the integral of Eq. (5.4), $W(u)$ can be approximated by an infinite series; so the Theis equation becomes:

$$h_0 - h = \frac{Q}{4\pi T}\left[-0.5772 - \ln u + u - \frac{u^2}{2\times 2!} + \frac{u^3}{3\times 3!} - \frac{u^4}{4\times 4!} + \cdots\right] \qquad \text{Eq. (5.5)}$$

$$u = \frac{r^2 S}{4Tt} \qquad \text{Eq. (5.6)}$$

where $h_0 - h$ is the drawdown (change in hydraulic head at a point since the beginning of the test), u is a dimensionless time parameter, Q is the discharge (pumping) rate of the well (volume per unit time, or m^3/s), T and S are the transmissivity and storativity of the aquifer around the well (m^2/s and unitless), r is the distance from the pumping well to the point where the drawdown was observed (m or ft), t is the time since pumping began (minutes or seconds), and $W(u)$ is the **well function** commonly referred to as the exponential integral in nonhydrogeology literature.

The drawdown data ($h_0 - h$) is measured in the well, r, t, and Q are observed. The solutions to Theis equations (Eqs. 5.4 and 5.6) require the graphical method to find the average T and S values near a pumping well. This graphical method, termed **type-curve method**, employed two superimposed curves, $W(u)$ vs. $1/u$ and measured drawdown ($h_0 - h$) vs. t. The $W(u)$ vs. $1/u$ plot (the Theis type curve) represents the theoretical response of a confined aquifer to a constant pumping stress.

The graphical approach is illustrated in Figure 5.15. Once the two curves are in a proper matching position, any arbitrary match point can be chosen on the overlapping graphs. The calculations would be easier when the match point is selected at intersections of major axes of $W(u)$ vs. $1/u$ plot where $W(u)$ and $1/u$ are equal to even powers of 10 (i.e. it is not necessary to choose the matching point on the Theis type curve). Two of the matching point values, $W(u)$ and drawdown $h_0 - h$ (s in Figure 5.15), are substituted into Eq. (5.4) to calculate T. After T is calculated, its value is substituted into Eq. (5.5) to solve the storage coefficient (S, not the drawdown s) using the match point values of t and u (Example 5.3).

Compared to the slug test, the pumping test provides more accurate estimates of aquifer transmissivity, hydraulic conductivity, and storage coefficient under conditions over a large portion of the aquifer. It is suitable for the detailed design of groundwater remediation systems, but more expensive and time-consuming than slug tests. Slug tests are often used at hazardous waste sites where large volumes of contaminated water do not have to be used, as in the case of pump test.

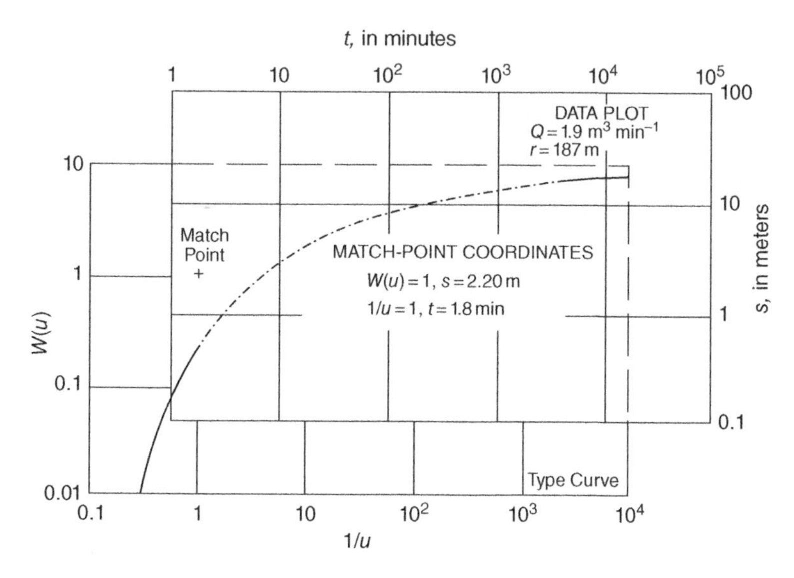

Figure 5.15 Plot overlay with $W(u)$ vs. $1/u$ and drawdown (s) vs. time (t) for Theis type curve method and corresponding match point values (Heath 1983).

Example 5.3 Use of Theis equation

A well in a confined aquifer (thickness = 25 m) is pumped at a flow rate of 2500 m^3/day. This aquifer has a hydraulic conductivity of 10 m/d and a storativity of 0.005. Determine the drawdown at a distance 10 m from the pumping well after two days of pumping?

Solution

First we calculate the transmisivity according to:

$$T = K\,b = 10\frac{\text{m}}{day} \times 25\,\text{m} = 250\frac{\text{m}^2}{day}$$

Using Eq. (5.6), we can calculate the dimensionless time parameter:

$$u = \frac{r^2 S}{4Tt} = \frac{(10\,\text{m})^2 \times 0.005}{4 \times 250\frac{\text{m}^2}{day} \times 2\,\text{d}} = 0.00025(\text{dimensionless})$$

Now we use the infinite series term of Eq. (5.5) to approximate *W(u)*:

$$W(u) = -0.5772 - \ln u + u - \frac{u^2}{2 \times 2!} + \frac{u^3}{3 \times 3!} - \frac{u^4}{4 \times 4!} + \cdots$$
$$= -0.5772 + 8.294 + 0.00025 + 1.562 \times 10^{-8} + 8.68 \times 10^{-13} + 4.069 \times 10^{-17} + \cdots \approx 7.72$$

Note that for $u < 0.09$, the contribution of the fourth and subsequent terms in *W(u)* can be neglected. By plugging the value of *W(u)* into Eq. (5.4), we can estimate the drawdown at a distance of 10 m away after two days of pumping at 2500 m^3/day:

$$h_0 - h = \frac{Q}{4\pi T} W(u) = \frac{2500\frac{\text{m}^3}{day}}{4 \times \pi \times 250\frac{\text{m}^2}{day}} \times 7.72 = 6.15\,\text{m}$$

5.4 Environmental Sampling and Analysis

Sampling and analysis for contaminant characterization in various environmental matrices (air, water, soil, waste, etc.) can be found in a number of standard methods and books (Zhang 2007; Zhang et al. 2013; USEPA 2017). Below is a synopsis of procedures pertaining to sampling of soil, groundwater, soil gas, and water in the vadose zone, and the types of laboratory instruments for the analysis of soil, groundwater, and soil gas samples.

5.4.1 Common Soil Samplers

Typical tools used in soil sampling are a scoop (soft soil depths of 1–10 ft), hand auger (depth of 3 in. to 10 ft), tube sampler, split spoon sampler, as well as drilling rig and geoprobes depending on site-specific conditions. Some representative soil samplers are schematically shown in Figure 5.16.

Hand-driven/machine-driven auger: This sampler typically consists of a short cylinder with a cutting edge attached to a rod and handle. It is advanced by a combination of rotation and downward force (Figure 5.16a and b). Samples taken this way are disturbed because the

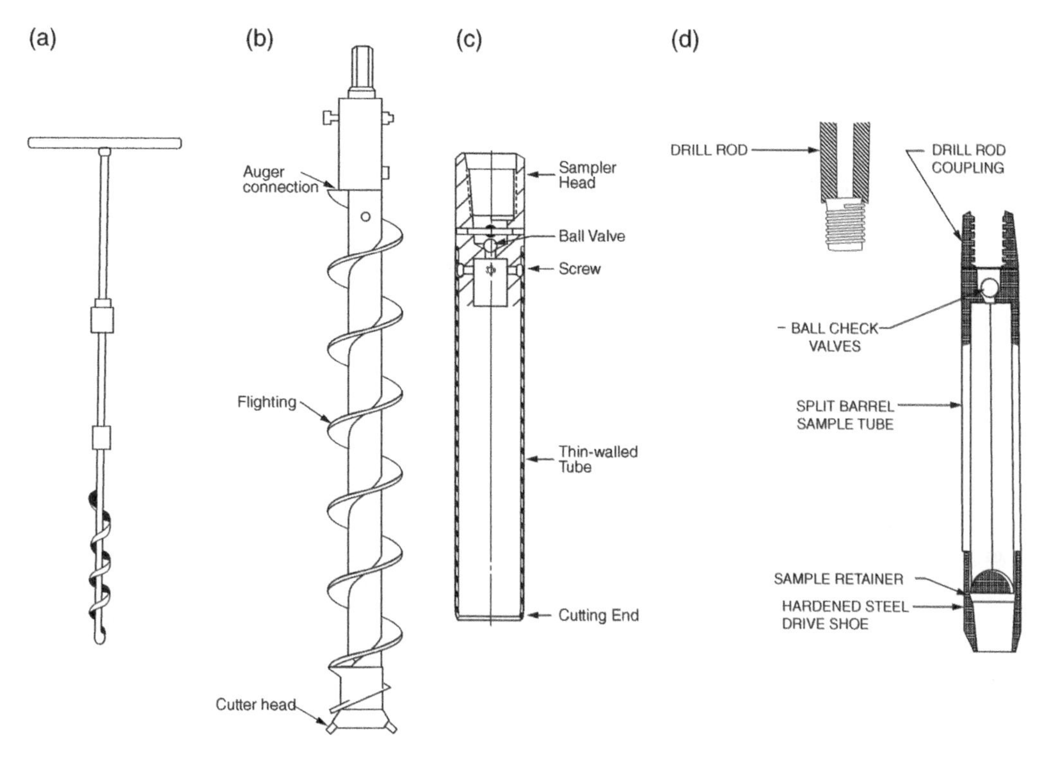

Figure 5.16 Common soil samplers: (a) hand auger, (b) solid-flight auger, (c) standard thin-walled (Shelby) tube sampler, (d) standard split spoon sampler (USEPA 1993a; US Army Corps of Engineers 2001)

in situ properties (e.g. soil texture) of the soil are not retained. Hand augers work best with depth less than 10 ft, and are particularly useful for areas where a truck cannot access.

Shelby tube sampler: This ASTM standard sampler (ASTM D 1587), called **Shelby tube sampler**, consists of a thin-walled tube with a cutting edge at the toe (Figure 5.16c). A sampler head attaches the tube to the drill rod, and contains a check valve and pressure vents. Generally used in cohesive soils, this sampler is advanced into the soil layer, generally 6 in. less than the length of the tube. The vacuum created by the check valve and cohesion of the sample in the tube cause the sample to be retained when the tube is withdrawn. A standard ASTM dimension of 3 in. OD and 3 ft long, 16 gauge thickness (1.59 mm) is used for most environmental drilling. Soil sampled in this manner is considered undisturbed, and therefore is good for further testing on strength, permeability, compressibility, and density.

Split-spoon/split-barrel sampler: This sampler is listed in the "Standard Test Method for Standard Penetration Test (SPT) and Split-Barrel Sampling of Soils" (ASTM D 1586). **Split-spoon/split barrel sampler** is typically a 18–30 in long, 2.0 in outside diameter (OD) hollow tube split in half lengthwise. A hardened metal drive shoe with a 1.375 in opening is attached to the bottom end, and a one-way valve and drill rod adapter at the sampler head (Figure 5.16d). It is driven into the ground with a 140-pound hammer falling 30 in Generally used for noncohesive soils, samples taken this way are considered disturbed.

Direct push sampler: These samplers can be mounted on a pick-up truck rather than large conventional drilling equipment. **Direct push sampler** is advanced through hydraulic push, augmented by a rapid percussion hammer to penetrate pavements and hard formations. It generally consists of steel tubes or split barrels fitted with acrylic or brass liners, which may be cut open or extruded in the field, or capped and sent directly to a lab for analysis. Some direct push samplers have a plug at the opening which is held in place by a piston or other mechanisms. When the piston is released at a pre-determined depth, undisturbed soil column can be collected. Direct push can be used for the installation of monitoring wells.

5.4.2 Groundwater Sampling

Groundwater sampling is probably the most complex and inherently the most difficult of all compared to other environmental matrices. We have thus far described well construction, well development and well purging as part of the good practice in groundwater sampling. Below are some additional descriptions of groundwater sampling tools and cross-contamination issues pertaining to groundwater sampling.

5.4.2.1 Groundwater Sampling Tools

Groundwater samples are collected from a well by a **bailer** or pumps of various types (Figure 5.17). A bailer is a pipe with an open top and a check valve at the bottom (3 ft long with a 1.5 in internal diameter and 1 L capacity). A line is used to mechanically lower the bailer into the well to retrieve a volume of water. A bailer is easy to use and transport but it may generate turbulence when the bailer is dropped down to the well and it is exposed to atmospheric O_2 when the sampled water is poured into a container. **Bottom-filled bailers**, which are more commonly used, are suitable provided that care is taken to preserve VOCs.

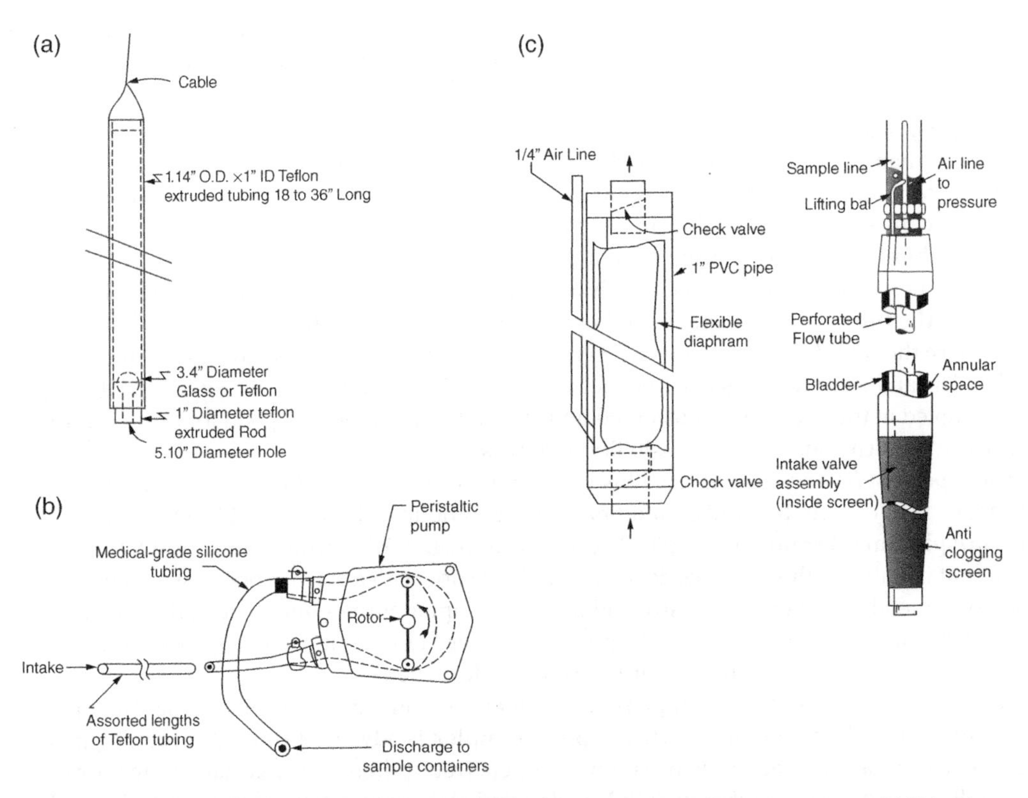

Figure 5.17 Common sampling tools used for groundwater: (a) bailer, (b) peristaltic pump, and (c) bladder pump (US Army Corps of Engineers 2001).

A **peristaltic pump** consists of a rotor with ball-bearing rollers. Dedicated tubings are attached to both ends of the rotor. One end is inserted into the well and the other end is a discharge tube. A peristaltic pump is suitable for sampling wells of small diameter (e.g. 2 in) and has a depth limitation of 25 ft. Cross-contamination is not of concern because dedicated tubing is used and the sample does not come in contact with the pump or other equipment. It can result in a potential loss of VOCs due to sample aeration.

A **bladder pump** consists of a stainless steel or Teflon housing that encloses a Teflon bladder. It is operated using a compressed gas source (bottled gas or an air compressor). Groundwater enters the bladder through a lower check valve, compressed gas moves the water through an upper check valve and into a discharge line. A bladder pump can be used to sample a depth of approximately 100 ft. It is recommended for VOC sampling because it causes a minimal alteration of sample integrity. The pump is somewhat difficult to decontaminate and should be dedicated to a well.

For well development (Section 5.3.1) and other purposes, various other pumps are available. A **submersible pump** can be used for high flow rates (gals/min), but it is hard to decontaminate. A **suction pump** is good for well development, when the proposed well area has a lot of sediment. The disadvantage is that it is gasoline-driven and the pump needs to be primed, meaning that water needs to be added to make the pump fittings air-free. It is effective only to a depth of 20 ft. An **air-lifter pump** operates by releasing compressed air. The air mixes with water in the well to reduce the specific gravity of the water column and lift the water to the surface. It is therefore used in well development rather than sampling.

5.4.2.2 Cross-Contamination in Groundwater Sampling

The prevention of cross-contamination is an important part of the quality assurance/quality control (QA/QC) in groundwater sampling. There are many potential contamination sources, including contamination during well construction (well casings), well purging (pump and tubing), and sampling-related equipment (sampling tools, water level indicator, monitoring probes). Table 5.4 is a list of materials used in well construction, pumping and tubing, and the potential contaminants (artifacts) contributed to groundwater samples by such materials. All sampling devices or well casings should be thoroughly cleaned prior to groundwater sampling.

Table 5.4 Contaminants potentially contributed to groundwater samples by materials used in sampling devices, well casings, and other items.

Material	Contaminants contributed
PVC-threaded joints	Chloroform
Stainless steel pump and casing	Cr, Fe, Ni, and Mo
Polypropylene or polyethylene tubing	Phthalates and other plasticizers
Polytetrafluoroethylene (Teflon) tubing	None detected
Soldered pipes	Sn and Pb
PVC-cemented joints	Methyl ethyl ketone, toluene, acetone, methylene chloride, benzene, ethyl acetate, tetrahydrofuran, cyclohexanone, three organic Sn compounds, and vinyl chloride
Glass container	B and Si
Marker used in labeling	Volatile chemicals
Duct tape for all purposes	Solvent

5.4.3 Vadose Zone Soil Gas and Water Sampling

Soil gas and water samples in the vadose zone are commonly collected to monitor the performance of a regulated facility such as a landfill before it contaminates groundwater. Compared to the saturated zone, the vadose zone has distinct geochemical characteristics that present unique challenge for representative sampling. It has the presence of soil air/gas in the pore spaces, extensive gas phase and gas exchange, abundant organic matter, and considerable heterogeneity in hydraulic conductivity.

Soil gas samples arc collected by driving a probe into the soil, and applying a vacuum that brings the soil gas into fluorocarbon bags or syringes (Figure 5.18). These samples can be analyzed on-site in a mobile lab or in a laboratory. Some mobile environmental laboratories can analyze soil gas, soil, and water for VOCs such as halocarbons, chlorinated hydrocarbon solvents, and fuel constituents (e.g. BTEX). The presence of soil gas can provide a valuable screening tool for locating VOCs.

However, water in the vadose zone is held under tension between the soil and water. Since water cannot flow freely by gravity in the vadose zone, a suction or vacuum **lysimeters** must be applied to extract liquid water from soil. Suction or vacuum lysimeters consist of a hollow porous cup typically located at the end of a hollow tube (Figure 5.19). When in operation, a suction or vacuum draws soil water into the cup. After a sufficient volume of soil water has entered the lysimeter, either a suction or a positive pressure is applied to bring the water sample to the surface through the tubing. In theory, an applied suction should be able to lift water up to 32 ft (10 m); however, in practice anywhere from 15 to 25 ft (4.5–8 m) is the upper limit of a suction lysimeter's effectiveness.

5.4.4 Instruments for Chemical Analysis

The site characterization, as we have discussed thus far in this chapter, generally falls within the work scope of geological professionals. Following soil and sample collection, subsequent chemical analyses are entirely the routine of chemists. Even though many monitoring parameters can be done *in situ*, an astonishing number of laboratory instrumental methods are available

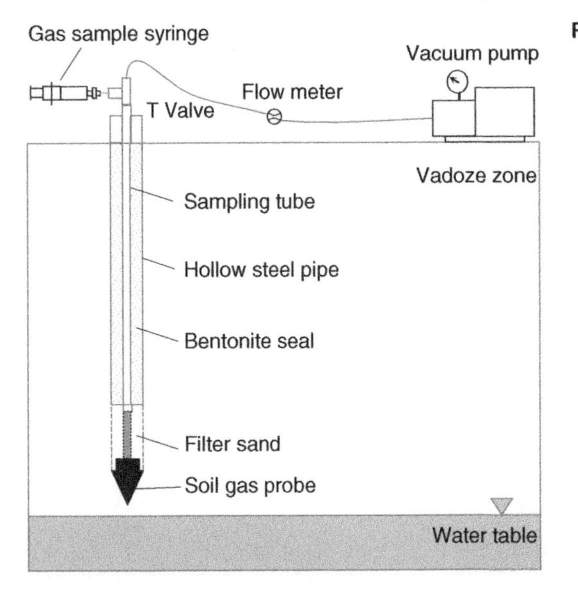

Figure 5.18 Vadose zone soil gas sampling.

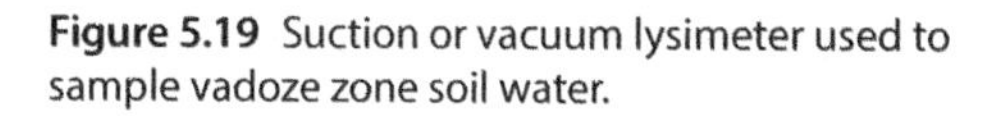

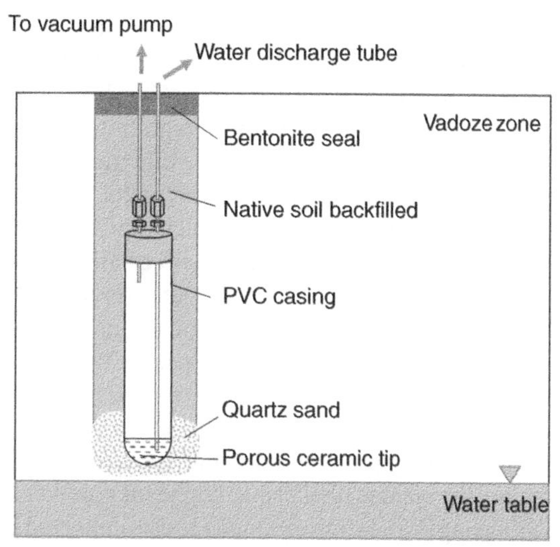

Figure 5.19 Suction or vacuum lysimeter used to sample vadoze zone soil water.

for more accurate and specific analysis of chemical constituents in soil and groundwater. Selected analytical techniques for specific constituents of geochemical interest are given in Table 5.5. The instrument of choice is often determined by the detection limit, the availability of instrument, sample throughput, analytical costs, and other factors.

The USEPA, USGS, and OSHA publish a large number of analytical methods in their websites. For example, USEPA's **SW-846 methods** are a detailed set of methods directly relevant to the analysis of solid and hazardous materials (USEPA 2018). Apart from these regulatory methods, many other professional organizations such as APHA, ASTM, and NIOSH publish their methods in the form of consensus methods, many of which have been recommended by the USEPA and other regulatory agencies. The guidelines for Phase I, II, and III described in this chapter are a good example of ASTM standards used for site characterization, and APHA method is a set of gold standard methods for the analysis of chemical and biological constituents in

Table 5.5 Common instrumental methods grouped according to types of analytes.

Type of chemicals	Instrument
Most metals	AA, GFAA, ICP-OES, ICP-MS
Mercury	CVAA
Ionic species such as NO_3^-, NO_2^-, NH_4^+, SO_4^{2-}, $Cr_2O_4^{2-}$	IC-CD, IC-UV
Halogenated compounds, chlorinated pesticides, PCBs	GC-ECD
Hydrocarbons, BTEX, PAHs	GC-FID
Most VOCs and SVOCs, PAHs, PCBs, pesticides	GC-MS
SVOCs, pesticides, explosives	HPLC

AA = atomic absorption spectroscopy, GFAA = graphite furnace atomic absorption spectroscopy, ICP-OES = inductively coupled plasma – optical emission spectroscopy (ICP-OES), ICP-MS = inductively coupled plasma – mass spectrometry, CVAA = cold vapor atomic absorption spectroscopy, IC-CD = ion chromatography – conductivity detector, IC-UV = ion chromatography – UV detector, GC-ECD = gas chromatography – electron capture detector, GC-FID = gas chromatography – flame ionization detector, GC-MS = gas chromatography – mass spectrometry, HPLC = high-performance liquid chromatography.

water and wastewater. The detail is out the scope of this text; interested readers can find a guide to these standard methodologies in Zhang (2007).

Last, but not least, is the QA/QC in both sampling and analysis of soil and groundwater. Quality assurance (QA) is the integrated management program designed to generate quality data including viewing QC data, evaluating parameters, taking corrective action, planning for process and personnel involved. Quality control (QC) are specific technical checks that are performed to yield data with expected precision, accuracy, and method detection limit (MDL). Prior to site characterization, a **data quality objective** (DQO) should be defined to determine what quality of data is required. Data should be of sufficient quality to withstand scientific and legal challenges and relevant to the intended use.

A **QA/QC program** should be part of the site characterization workplan. It will lessen errors made throughout sampling, analysis, and calculations in measurements. QA/QC samples are needed when doing field work where errors in collection, preservation, and transportation can occur. For example, the following QC samples should be collected to detect/minimize errors: equipment blanks, field blank, trip blanks, and field duplicates (Zhang et al. 2013). QA/QC samples during the laboratory stage are also collected to include various blanks and spikes to account for errors in sample preparation, cross-contamination, interferences, and other sources.

Bibliography

Alter, B. (2012). *Environmental Consulting Fundamentals: Investigation and Remediation*. Boca Raton, FL: CRC Press.

APHA (2017). *Standard Methods for the Examination of Water and Wastewater*, 23e. APHA.

ASTM (1996). *ASTM standards E1528-96 (Transaction Screen Process), American Society of Testing and Materials*. West Conshohocken, PA: ASTM.

ASTM (2011), *Standard Practice for Environmental Site Assessments: Phase II Environmental Sites Assessment Process*, Standard E1903-11, ASTM West Conshohocken, PA.

ASTM (2013). *Standard Practice for Environmental Site Assessments: Phase I Environmental Sites, Assessment Process, ASTM standards E1527-13*. West Conshohocken, PA: ASTM.

Bair, E.S. and Lahm, T.D. (2006). *Practical Problems in Groundwater Hydrology: Problem-Based Learning Using Excel Worksheets*. Upper Saddle River, NJ: Pearson Prentice Hall.

Barrows, L. and Rocchio, J.E. (1990). Magnetic survey for buried metallic objects. *Groundwater Monitoring Remediation* 10 (3): 204–211.

Bedient, P.B., Rifai, H.S., and Newell, C.J. (1999). *Ground Water Contamination: Transport and Remediation*, 2e. Upper Saddle River, NJ: Prentice Hall.

Bodger, K. (2003). *Fundamentals of Environmental Sampling*. Rockville, MD: Government Institutes.

Bouwer, H. and Rice, R.C. (1976). A slug test for determining hydraulic conductivity of unconfined aquifers with completely or partially penetrating wells. *Water Resources Research* 12 (3): 423–428.

Connecticut Department of Environmental Protection (2010). Site Characterization Guidance Document, 48pp.

Evans, R.B. and Schweitzer, G.E. (1984). Assessing hazardous waste problems: geophysical techniques are becoming more useful for locating hazardous wastes and estimating groundwater contamination. *Environ. Sci. Technol.* 18 (1): 330–339A.

Fetter, C.W. (2001). *Applied Hydrogeology*, 4e. New Jersey: Prentice Hall, Upper Saddle River.

Heath, R. C. (1983). Basic Ground-Water Hydrology: U. S. Geological Survey Water-Supply Paper 2220, 86pp.

Hvorslev, M.J. (1951). *Time Lag and Soil Permeability in Ground Water Observations*, Bulletin No. 36, 50. US Army Corps of Engineers.

Keith, L.H. (1996). *Principles of Environmental Sampling*, 2e. Washington, DC: American Chemical Society.

LaBrecque, D.J., Ramirez, A.L., Daily, W.D. et al. (1996). ERT monitoring of environmental remediation processes. *Meas. Sci. Technol.* 7: 375–383.

Ohio Environmental Protection Agency (2017). Chapter 6: drilling and subsurface sampling. In: *Technical Guidance Manual for Hydrogeologic Investigations and Ground Water Monitoring*. Ohio Environmental Protection Agency.

Pichtel, J. (2007). Chapter 5: environmental Site Assessment. In: *Fundamentals of Site Remediation: For Metal- and Hydrocarbon-Contaminated Soils*. Government Institute.

Splajt, T., Ferrier, G., and Frostick, L.E. (2003). Monitoring of landfill leachate dispersion using reflectance spectroscopy and ground-penetrating radar. *Environ. Sci. Technol.* 37 (18): 4293–4298.

The Hazardous Materials Training and Research Institute (HMTRI) (2002). Chapter 1: site investigation. In: *Site Characterization: Sampling and Analysis*. John Wiley &Sons.

US Army Corps of Engineers (2001). *Requirements for the Preparation of Sampling and Analysis Plans: Engineer Manual*, EM 200-1-3. US Army Corps of Engineers.

USEPA (1984). *Geophysical Techniques for Sensing Buried Wastes and Waste Migration*, EPA/600/S7-84/064. USEPA.

USEPA (1991). *Site Characterization for Subsurface Remediation*, EPA/625/4-91/026. USEPA.

USEPA (1993a). *Subsurface Characterization and Monitoring Techniques: A Desk Reference Guide, vol. 1, Solids and Ground Water, Appendices A and B*, EPA/625/R-93/003a. USEPA.

USEPA (1993b). *Suggested Operating Procedures for Aquifer Pumping Tests*, EPA/540/S-93/503. USEPA.

USEPA (1996). *A Guideline for Dynamic Workplans and Field Analytics: The Keys to Cost-Effective Site Characterization and Cleanup*, EPA 542-B-96-002. USEPA.

USEPA (1997). *Federal Facilities Forum Issue: Field Sampling and Selecting On-Site Analytical Methods for Explosives in Soil*, EPA 540-R-97-501. USEPA.

USEPA (1998a). *Leak Detection for Landfill Liners: Overview of Tools for Vadose Zone Monitoring*, EPA 542-R-98-019. USEPA.

USEPA (1998b). *Innovations in Site Characterization Case Study: Hanscom Air Force Base, Operable Unit 1*, EPA 542-R-98-006. USEPA.

USEPA (1998c). *Quality Assurance Guidance for Conducting Brownfields Site Assessments*, EPA 540-R-98-038. USEPA.

USEPA (1999). *Environmental Technology Verification Site Characterization and Monitoring Technologies Pilot*, EPA 542-F-99-009. USEPA.

USEPA (2001), *Technical Approaches to Characterizing and Cleaning Up Brownfields Sites*. Report Number EPA/625/R-00/009.

USEPA (2002). Ground-Water Sampling Guidelines for Superfund and RCRA Project Managers, Ground Water Forum Issue Paper, 53 pp.

USEPA (2013). Field Sampling and Analysis Technologies Matrix, Version 1.0 http://www.frtr.gov/site/ (accessed December 2018).

USEPA (2017). *Site Characterization and Monitoring Technology Support Center FY16 Report*, EPA/600/R-17/409. USEPA.

USEPA (2018). Test Methods for Evaluating Solid Waste/SW-846. https://www.epa.gov/hw-sw846 (accessed December 2018).

van Biersel, T.P., Bristoll, B.C., Taylor, R.W., and Rose, J. (2002). Abandoned underground storage tank locatipn using fluxgate magnetic surveying: a case study. *Ground Water Monitoring Rev.* **Winter**, 116–120.

Zhang, C. (2007). *Fundamentals of Environmental Sampling and Analysis, Chapter 4, and Chapter 5*. John Wiley & Sons.

Zhang, C., Mueller, J.F., and Mortimer, M.R. (2013). *Quality Assurance and Quality Control of Environmental Field Sampling*. London, UK: Future Science.

Questions and Problems

1 You are sent out by your project manager for site reconnaissance in dealing with the extent of contamination of a former transformer facility in a redevelopment zone of the city. What types of information would you like to obtain, and what would you do?

2 What should one expect from Phase I and Phase II report for a potentially contaminated site or property?

3 Search the Internet for an example of Phase I and Phase II environmental site assessment reports and summarize their differences in purposes, contents each covered, data requirements, conclusions, etc.

4 Is it essential to have a Phase I in order to do Phase II ESA?

5 What does each of the following terms entail: (a) stratigraphy, (b) lithology, and (c) structural geology?

6 What are the common drilling methods used to construct monitoring wells?

7 What are the ASTM standard methods for environmental site assessment?

8 Describe several common soil-sampling tools and the types of soils applicable.

9 Why is dry drilling particularly suitable for environmental purpose?

10 Compare the pros and cons of auger drilling and rotary drilling soil and geological sampling.

11 Briefly describe the following terms: (a) Phase I Environmental Site Assessment, (b) Borehole logging, (c) Cone penetrometer testing (CPT), (d) Ground-penetrating radar (GPR), (e) Well development, (f) Pumping test.

12 What surface geophysical method(s) are able to detect metal drum and abandoned tanks buried in the subsurface?

13 Can electrical resistivity be used as a geophysical tool to delineate a plume of saline groundwater as a result of salt-water intrusion, or a leachate from a landfill? Explain why or why not.

14 What soil sampler(s) are suitable to obtain undisturbed samples?

15 The difference between the depth of well and depth to water is 5 ft, and well internal diameter is 5 in. Estimate the approximate volume of water need to be purged prior to sampling.

16 What are the four common aquifer tests used to determine hydraulic conductivity?

17 Discuss the pros and cons of the slug test and pumping test.

18 Why is vadose zone soil gas and water sampling unique?

19 Two wells 250 m apart are located in the south–north direction. The south well and north well have hydraulic heads of 18.6 and 18.0 m, respectively. The third well is located 200 m due west of the north well and has a head of 18.3 m. Schematically determine groundwater flow direction and estimate hydraulic gradient.

20 A well has a casing internal diameter of 10 cm and screen plus sand pack diameter of 15 cm. The length of well screen is 1.8 m. The static water level measured from the top of casing is 2.45 m. A slug test is conducted and pumping done to lower the water level to 3.05 m. The time-drawdown in the unconfined aquifer is recorded every three seconds. The time required for the water level to rise to 37% of the initial change (T_0) is calculated to be 15 s, obtained from a semi-logarithmic plot of $(h - H_0)/(H - H_0)$ vs. t, where $(h - H_0)/(H - H_0) = 0.37$. Determine the hydraulic conductivity of the aquifer by the Hvorslev method.

21 Calculate the value of well function $W(u)$ for an u value of (a) 1.5×10^{-4} and (b) 4×10^{-3}, respectively. For time-saving, use Excel to demonstrate that the fourth to sixth terms are indeed negligible for u values in the range of 1×10^{-10} to 1×10^{-1}. In Excel, the function for factorial is @FACT(), for example, FACT(3) = 3! = 3 × 2 × 1 = 6.

$$W(u) = -0.5772 - \ln u + u - \frac{u^2}{2 \times 2!} + \frac{u^3}{3 \times 3!} - \frac{u^4}{4 \times 4!} + \cdots$$

22 A well is located in a confined aquifer with a thickness of 35 m, hydraulic conductivity of 10 m/day, and storativity of 0.015. The well is pumping at a constant rate of 2150 m^3/day. Calculate u, $W(u)$, and predict the drawdown of groundwater for a point 10 m away from this pumping well after one day of pumping.

23 What are the common tools for groundwater sampling?

24 What are the common types of cross-contamination in groundwater sampling? Why are they important?

25 Describe the purposes of (a) well development and (b) well purging for environmental analysis.

26 Describe the installation and use of (a) lysimeter and (b) piezometer.

27 Describe some common instrumentations in conducting analysis of inorganic and organic contaminants in groundwater.

28 Provide a source of standard analytical methods for soil and groundwater analysis.

Chapter 6

Overview of Remediation Options

LEARNING OBJECTIVES

1. Develop a broad view and understanding of the various soil and groundwater remediation technologies currently available
2. Differentiate among various remediation technologies based on locations relative to sources, environmental matrices, and mechanisms
3. Recognize the pros and cons of the established and innovative remediation techniques employed at contaminated sites
4. Develop strategic approaches for the remediation of various groups of contaminants such as VOCs, SVOCs, fuels, explosives, inorganics, and radionuclides – both in soil and groundwater
5. Understand the remedial investigation/remedial feasibility processes, from bench to pilot study, to full-scale application; and understand what it takes to succeed
6. Use Treatment Technologies Screen Matrix to screen through various *in situ* and *ex situ* remediation technologies according to screening variables such as contaminants, development status, overall cost, cleanup time, etc.
7. Be aware of recent progress and trend in the remediation industry by applying the green and sustainability principle
8. Rank remediation technologies most commonly used in the cleanup of contaminated soil and groundwater
9. Identify available technology options for the *in situ* and *ex situ* (after excavation) remediation of contaminated soil
10. Identify available technology options for the *in situ* and *ex situ* (after pumping) remediation of contaminated groundwater
11. Suggest a treatment train from a given soil and groundwater remediation scenario

We have thus far covered the fundamental principles and concepts that are essential to the understanding of soil and groundwater remediation, including chemistry, geology, regulations, risk assessment, cost considerations, and site characterization techniques. The primary objective of this chapter is to provide readers an overview of various soil and groundwater remediation technologies that will be described in detail in the remainder of this text. We will briefly learn the types of remediation technologies, how are they developed, how can we select remediation technology through various screening criteria, and what are the general applications of these remediation technologies. Since no single technology can generally remediate an entire contaminated site, the concept of treatment train is briefly introduced at the end of this chapter for the combined use of several treatment technologies.

The details of these remediation technologies, including governing equations, engineering design, cost-effectiveness, and case studies will be provided in each of these subsequent

Soil and Groundwater Remediation: Fundamentals, Practices, and Sustainability, First Edition. Chunlong Zhang.

Companion website: www.wiley.com/go/Zhang/Remediation_1e

chapters. Readers should note that the term **remediation** throughout this text refers to soil (sometimes sediment) and groundwater remediation only. Oftentimes, environmental professionals deal with other situations that require the remediation, such as the case of indoor asbestos abatement, mold survey and remediation, and radon survey and remediation. These topics require special expertise from the conducting environmental engineers. They are beyond the scope of this text and interested readers can find some needed details in Alter (2012).

6.1 Types of Remediation Technologies

Two goals are served for the discussions of various types of remediation technologies. First, by introducing a full spectrum of remediation options currently available, stakeholders are benefited for a better and informed decision. Second, even though common remediation technologies will be detailed in Chapters 7–12, there are other less common but appropriate technologies for certain remediation scenarios. It is thus important to have a broad view and understanding of all potential remediation technologies for our consideration in decision-making.

6.1.1 Classifications of Remediation Technologies

In selecting a remedial alternative, a **remedial project manager** (RPM), who oftentimes has adequate training in environmental science or environmental engineering, will face the option of a number of remediation technologies. The RPM typically focuses on a few **presumptive remedies** established by the USEPA. These presumptive remedies are the preferred technologies for common categories of contaminated sites, based on historical patterns of remedy selection, and the EPA's scientific and engineering evaluation of performance data on technology implementation. By focusing on few presumptive remediation technologies, a significant amount of time and effort will be saved to reduce site characterization data needs (USDoD 1994). For the convenience of our discussions, remedial technologies can be grouped by many means, for instance:

- From a location standpoint, remediation technologies can be *in situ* or *ex situ*, sometimes on-site, relative to the contamination source.
- From the contaminant media standpoint, remediation technologies are mainly for soil and groundwater, but sometimes they also include air emission (off-gas) treatment. The remediation of contaminated soils can imply sediment, bedrock, sludge, and other solid matrices, whereas groundwater remediation often entails surface water and particularly leachate from landfills. (However, surface water treatment for drinking water purpose and sewage treatment are referred to as "treatment" rather than "remediation." Water and wastewater treatment are the well-established subjects that are commonly taught in the environmental engineering curriculum.)
- From a contaminant standpoint, remediation technologies can vary significantly depending on what the contaminants are, such as VOCs (volatile organic compounds), SVOCs (semivolatile compounds), metals, fuels, energetics, or mixed wastes.
- From the activity standpoint, there are passive and active treatments (e.g. pumping vs. collection systems).
- From the developmental standpoint, remediation technologies can be established, innovative, or emerging.

There are other means to classify these different remediation techniques. Altogether, these are elaborated below.

In Situ versus Ex Situ Remediation: As their names would suggest, *in situ* and *ex situ* remediation technologies differ regarding their applications and limitations. *In situ* treatment technologies treat or remove contaminants from source media without excavation or removal of the source media, or from groundwater without extracting, pumping, or otherwise removing the groundwater from the aquifer. *Ex situ* treatment technologies require the excavation or removal of the contaminated sources media or extraction of groundwater from an aquifer before treatment may occur above ground. Examples of some common *in situ* and *ex situ* technologies are given as follows:

***In situ* treatment technologies:**	***Ex situ* treatment technologies:**
• Monitored natural attenuation	• Bioslurry
• *In situ* bioremediation	• Composting
• Air sparging	• Incineration
• Soil vapor extraction	• Soil washing
• Soil flushing	• Biopiles
• Slurry wall	• Landfarming
• Reductive dehalogenation	• Separation

Several factors can make *in situ* treatment the ultimate choice for contaminated site remediation. They include: (a) sites with high or moderate hydraulic conductivity or sites with low hydraulic conductivity regions surrounded by regions of moderate or high horizontal conductivity; (b) sites with deep groundwater table and/or competent aquitard; (c) sites containing wastes that are difficult to treat by land disposal; (d) when remediation completion time is not critical; (e) sites containing wastes that are not accessible due to structures; (f) contaminated matrices that are difficult or too expensive to excavate; and (g) when a poor transportation infrastructure or large volume of wastes containing low concentrations of contaminants makes *ex situ* treatment impractical.

Soil versus Groundwater Remediation: The media-based classification (Table 6.1) groups remediation technologies into: (a) soil and other solid matrices (sediment, bedrock, and sludge), and (b) groundwater and other liquid matrices (surface water and leachate). This classification has its own practical value; however, soil and groundwater remediation are not always separable at most contaminated sites. The cleanup of groundwater requires the removal of contaminants in their bound/residual form associated with soil particles. Additionally, the cleanup of contaminated soil oftentimes relies on removing the contaminant of interest in groundwater in equilibrium with the contaminated soil.

Table 6.1 also refers to the corresponding chapters (Chapters 7–12) that will cover the major remediation techniques in the remainder of this text. Several other remediation techniques that are of major or minor importance (i.e. solidification/stabilization, electrokinetic remediation) are excluded for the detailed discussion, but are defined later in this chapter. Techniques of common physical/chemical nature (e.g. redox, ion exchange, precipitation) or techniques in specialized fields (e.g. constructed wetlands) are also omitted.

Physical/Chemical, Biological, and Thermal Remediation: As noted in Table 6.1, various remediation technologies can be further divided, according to the underlying mechanisms, into

Table 6.1 Types of soil and groundwater remediation technologies.

A. Soil, sediment, bedrock, and sludge	Chapter	B. Groundwater, surface water, and leachate	Chapter
A1. *In situ* biological treatment		B1. *In situ* biological treatment	
Bioventing	9	Enhanced bioremediation	9
Enhanced bioremediation	9	Monitored natural attenuation	9
Phytoremediation	9	Phytoremediation	9
A2. *In situ* physical/chemical treatment		B2. *In situ* physical/chemical treatment	
Chemical oxidation	6/Appx B	Air sparging	8
Electrokinetic separation	6/Appx B	Bioslurping	9
Fracturing	7	Chemical oxidation	6/Appx B
Soil flushing	11	Directional wells (enhancement)	7
Soil vapor extraction	8	Due phase extraction	7
Solidification/stabilization	6/Appx B	Thermal treatment	10
A3. *In situ* thermal treatment		Hydrofracturing enhancements	7
Thermal treatment	10	In-well air stripping	8
A4. *Ex situ* biological treatment (assuming excavation)		Passive/reactive treatment walls	12
Biopiles	9	B3. *Ex situ* biological treatment	
Composting	9	Bioreactors	9
Landfarming	9	Constructed wetlands	9
Slurry phase biological treatment	9	B4. *Ex situ* physical/chemical treatment (assuming pumping)	
A5. *Ex situ* physical/chemical treatment (assuming excavation)		Adsorption/absorption	7
Chemical extraction	6/Appx B	Advanced oxidation processes	6/Appx B
Chemical reduction/oxidation	6/Appx B	Air stripping	8
Dehalogenation	6/Appx B	Activated carbon adsorption	7
Separation	6/Appx B	Groundwater pumping/pump and treat	7
Soil washing	11	Ion exchange	6/Appx B
Solidification/stabilization	6/Appx B	Precipitation/coagulation/flocculation	6/Appx B
A6. *Ex Situ* thermal treatment (assuming excavation)		Separation	6/Appx B
Hot gas decontamination	10	Sprinkler irrigation	6/Appx B
Incineration	10	B5. Containment	
Open burn/open detonation	6/Appx B	Physical barriers	7
Pyrolysis	10	Deep well injection	7
Thermal desorption	10		
A7. Containment			
Landfill cap	9		
Landfill cap enhancements/alternatives	9		

6/Appx B = Topics not covered or scattered throughout this book, but a brief explanation can be found in Chapter 6 and Appendix B.

physical/chemical, biological, and thermal technologies. Technologies based on physical/chemical principles include, but are not limited to, soil vapor extraction (SVE), solidification/stabilization, physical separation/chemical extraction, soil washing, pump-and-treat with granular activated carbon (GAC), and air stripping. There are many varieties of biologically based remediation for both contaminated soil and groundwater, including bioremediation, bioventing, composting, slurry-phase bioremediation, and solid-phase bioremediation. Thermal treatments are unique in environmental remediation, from the well-known incineration to thermally enhanced SVE and thermal desorption using steam, as well as high-temperature pyrolysis and vitrification.

Remediation Based on Contaminant Immobilization/Containment, Extraction, or Destruction: Remedial technologies fall into three distinct groups based on whether the contaminant is "immobilized/contained," "extracted," or totally "destructed" (Figure 6.1). The "immobilization" treatments include solidification/stabilization, containment barriers, and capping. These immobilized/contained techniques "lock" contaminants such that their exposure risks are minimized or eliminated. The "extraction" technologies include soil excavation, activated carbon, vacuum extraction, air stripping, surfactant flushing, and *in situ* thermal treatment. For example, SVE only removes contaminants from unsaturated soil through its vapor form, whereas surfactant flushing removes contaminants from soil into a liquid waste. These extraction technologies just change the phase for the contaminants present, but require additional treatments for a total cleanup. The "destruction" technologies include bioremediation, chemical oxidation, oxidative

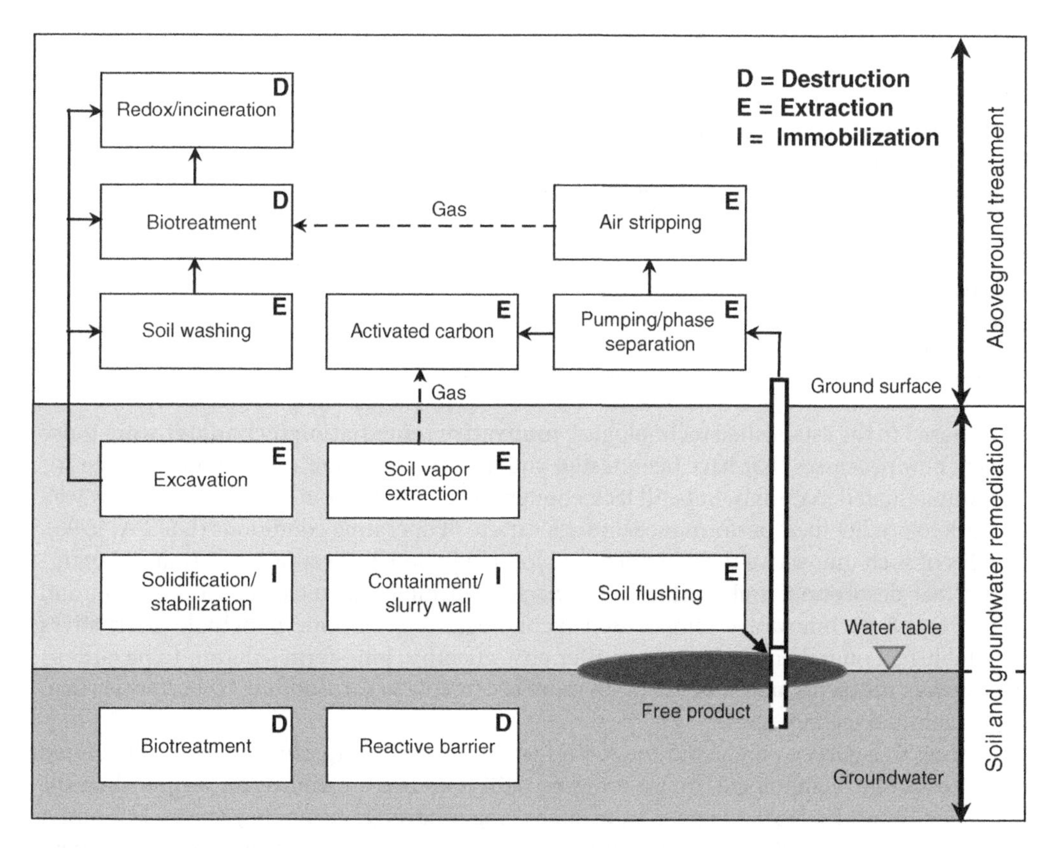

Figure 6.1 Classification of common soil and groundwater remedial technologies by function: D = Destruction; E = Extraction; I = Immobilization. Dashed lines denote gas phase transfer.

dechlorination, and incineration. Ideally, all remedial technologies involve contaminant "destruction" to some extent, i.e. contaminants are degraded/decomposed into a nontoxic form such as CO_2 and water, if all operated adequately.

Remediation Based on Treatment, Containment, or Institutional Control: Another interesting classification is to examine remediation technologies according to the source control remedy type: (a) **Source control treatment** includes any of the *in situ* or *ex situ* source control treatment technologies; (b) **Source control containment** is the containment of contaminants including the use of caps, liners, covers, and landfilling both on- and off-site; (c) Institutional controls and other nonengineered measures such as monitoring and population relocation (see Box 6.1).

6.1.2 Common and Frequently Used Remediation Technologies

Some common remediation technologies given in Table 6.1 are listed separately below according to the two major contaminated matrices (i.e. soil and groundwater). Of these common technologies, only incineration, landfarming, and solidification/stabilization are entitled to **established remediation technologies**. These were traditionally the most widely used at Superfund sites in the United States and contaminated sites worldwide.

Soil remediation:	Groundwater remediation:
• Bioremediation	• Air sparging
• Capping	• *In situ* bioremediation
• Dechlorination	• Natural attenuation
• Destruction	• Physical barrier
• Electrokinetic remediation	• Pump-and-treat
• Immobilization	• Reductive dehalogenation
• Incineration	
• Landfarming	
• Phytoremediation	
• Separation	
• Soil excavation	

As opposed to the established technologies, **innovative remediation technologies** are those relatively new processes that have been tested and used as treatment for hazardous wastes or other contaminated materials, but still lack enough information about their cost and how well they work to predict their performance under a variety of operating conditions (USEPA, 1996). Examples of such innovative technologies are bioremediation, soil vapor extraction, air sparging, thermal desorption, soil washing, *in situ* soil flushing, chemical dehalogenation, and solvent extraction. Innovative remediation technologies can become preferable alternatives over established ones, because they may offer cost-effective, long-term solution to hazardous waste cleanup problems, and they are often more acceptable to surrounding communities than some established technologies.

According to a survey conducted for the remediation of the Superfund sites in the United States, most of the technologies in use today are still classified as "innovative" rather than the "established" ones. By 1990, the percent of sites using innovative technologies was 15% and it was increased to 48% in 2004. Until recently, the percent of projects using innovative treatment has become nearly equal to those using more established treatment approaches.

Box 6.1 Institutional (nonengineered) controls for site remediation

Throughout the text, our focus is always on the engineered measures to clean up the contaminated soil and groundwater. However, nonengineered controls or institutional controls (ICs) can also become essential. In fact, ICs have received an increased acceptance among various stakeholders involved in remediation. **Institutional controls** (ICs) are nonengineered instruments, such as administrative and legal controls, that help minimize the potential for human exposure to contamination and/or protect the integrity of the remedy. ICs do not involve construction or physically changing the site. As such, fences that restrict access to sites are not considered by the EPA to be an institutional control.

ICs are meant to supplement engineering controls, therefore, they will rarely be the sole remedy at a site. Nevertheless, ICs play an important role in site remedies because they reduce exposure to contamination by limiting land or resource use, modifying behavior, and providing information to people. They are used when contamination is first discovered, when remedies are ongoing, and when residual contamination remains on-site at a level that does not allow for unrestricted use and unlimited exposure after cleanup.

There are four general types of ICs: (a) Government Controls: including local laws or permits such as county zoning, building permits, and groundwater use restriction; (b) Proprietary Controls: including property use restrictions based on private property law such as easements, covenants, and state use restriction; (c) Enforcement Tools: including documents that require individuals or companies to conduct or prohibit specific actions such as environmental cleanup consent decrees, administrative orders, or permits; (d) Informational Devices: including deed notices or public advisories that alert and educate people about a site.

A site manager's guide for selecting institutional controls is available from USEPA (2000), and details of various IC types, definitions, examples, benefits, and limitations can be found at the USEPA website: https://www.epa.gov/superfund/superfund-institutional-controls.

Figure 6.2 reveals additional details regarding the most common technologies for the Superfund site remediation in the United States. Many *ex situ* treatment projects have been completed but *in situ* treatment and pump-and-treat projects tend to have longer operation times. Approximately 93% of the groundwater remediation relies on pump-and-treat by 2004, which will be fully discussed in Chapter 7. Other common technologies of significant importance are soil vapor extraction/air parging (Chapter 8), bioremediation (Chapter 9), thermal technology such as incineration and thermal desorption (Chapter 10), soil washing and *in situ* flushing (Chapter 11), and reactive barrier (Chapter 12). Data shown in Figure 6.2 are not inclusive to all remediation technologies, such as institutional controls and containment that become more commonly used in recent years. Readers should check the most recent update from EPA websites. More recent summaries for Superfund remedy are available from USEPA (2013, 2017).

6.1.3 Technologies from Contaminant Perspectives

In Chapter 2, we introduced some frequent soil and groundwater contaminants and their properties. Here we further describe the applicable remediation technologies from the contaminant perspectives. In general, technologies differ fundamentally between organic compounds and inorganic metals.

For organic contamination in soil, the most commonly used technology is soil vapor extraction for VOCs, and in some cases for SVOCs through the enhanced soil permeability or

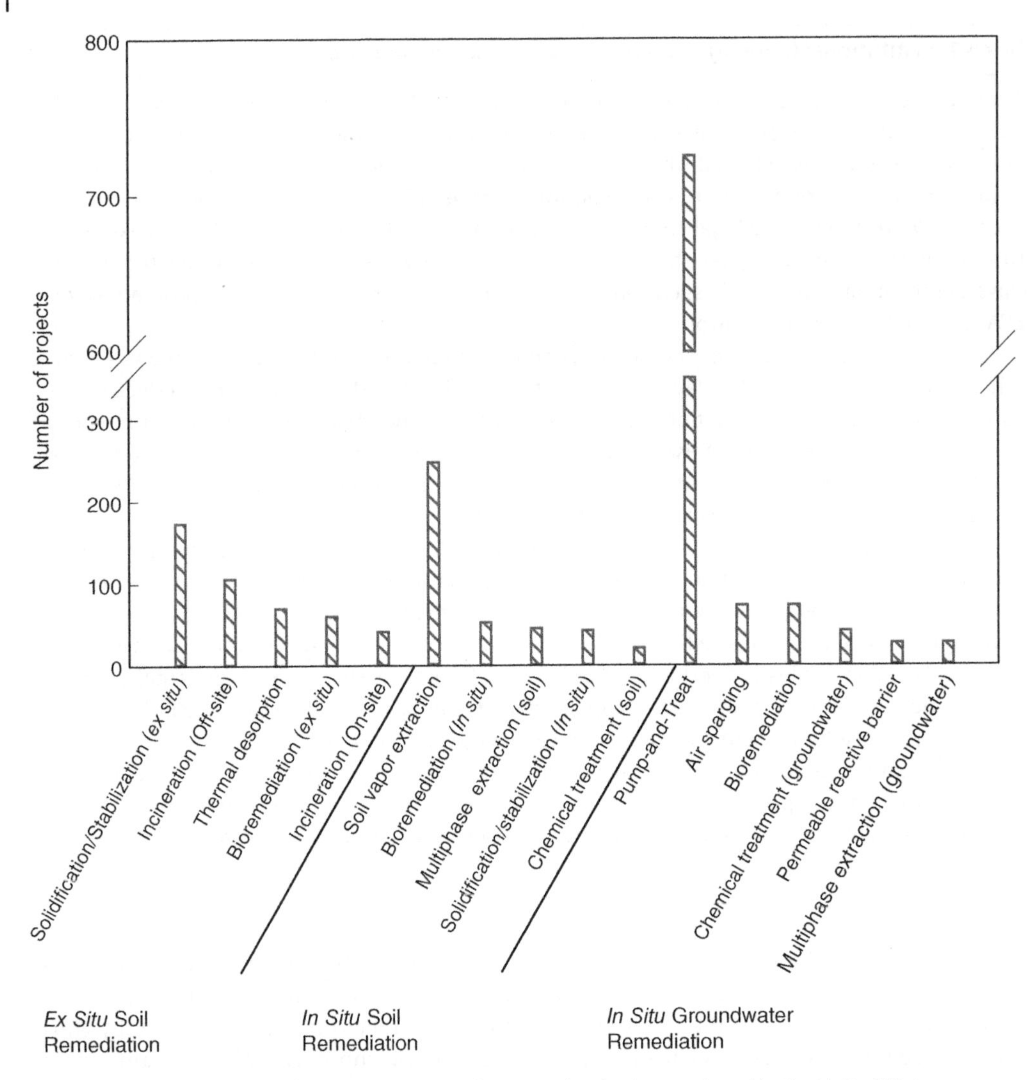

Figure 6.2 Common soil and groundwater remediation technologies employed in the United States. *Source:* USEPA (2007); based on 1982–2005 data.

contaminant volatility. Bioremediation is the second most commonly used for petroleum hydrocarbons and PAHs, and thermal desorption is the third most commonly used for VOCs and SVOCs.

For inorganic metals in soil, fixation with stabilization/solidification (S/S) is still the most commonly used. Technologies generally also differ between soil and groundwater. In the following discussion of some representative contaminant groups (USDoD 1994), we refer to "soil" as any solid matrices including soil, sediment, and sludge, and refer "groundwater" to common liquid waste including pumped groundwater, leachate, or surface water.

VOCs in Soil: For VOCs present in soil, sediment, and sludge, **soil vapor extraction, thermal desorption,** and **incineration** are the presumptive remedies for VOC-contaminated sites. When incineration is employed for halogenated VOCs, special off-gas treatment and scrubber water treatment should be implemented. **Bioventing**, an approach by injecting air to provide oxygen for aerobic degradation, is another commonly used technology. Generally, halogenated

VOCs are less amenable to biodegradation than nonhalogenated compounds. The general rule is that the more halogens attached to the compound, the more recalcitrant it is toward biodegradation.

VOCs in Groundwater: For VOCs in groundwater, leachate, pumped groundwater, or surface water, **air stripping** using a packed air stripper tower is the common method of choice. The air stripper can be installed on a concrete pad on-site as a permanent installation or on a skid/trailer as a temporary installation. Another method for VOCs in liquid is the liquid phase **carbon adsorption,** which is a well-established treatment for municipal and wastewater treatment. VOC-contaminated groundwater is pumped through a series of vessels containing activated carbon to remove dissolved VOCs at low concentration. Effluents with high concentrations of VOCs, however, are not cost-effective because of the frequent need for carbon regenerations.

VOCs in Air or Off-Gases: For the treatment of VOCs in emitted air or off-gases, **carbon adsorption, catalytic oxidation,** and **thermal oxidation** are the three commonly used methods. Similar to the aforementioned liquid phase carbon adsorption, granular activated carbon (GAC) packed beds in series or parallel can remove VOCs in the air streams. Catalytic oxidation often utilizes less expensive nonprecious metals (e.g. nickel, copper, or chromium, rather than noble metals such as platinum and palladium) in the temperature range of 600–1000 °F in remediation use. In thermal oxidation, oxidizers with refractory liners and burner using propane or natural gas are used.

SVOCs in Soil: Although grouped as SVOCs, this group of compounds includes a variety of structurally diverse chemicals from many different sources, such as chlorinated aromatics, phenols, ether, phthalate, PAHs, PCBs, and pesticides. Common treatment technologies for SVOCs in soil, sediment, and sludge include **bioremediation**, **incineration**, and **excavation with off-site disposal**. Lower-molecular-weight PAHs such as naphthalene, phenanthrene, and anthracene are generally biodegradable, whereas higher-molecular-weight PAHs such as pyrene, chrysene, and fluoranthene are less amenable to bacterial attack. Both *in situ* nutrient injections and *ex situ* bioreactors can be designed for the bioremediation of SVOCs. PCBs and chlorinated aromatics (e.g. PCP) are generally considered recalcitrant to biodegradation, which is considered not amenable to bioremediation for practical applications. Incineration can achieve destruction and removal efficiency (DRE) of >99.99% and even 99.9999% for PCBs and dioxins in a properly operated incinerator. An off-site disposal technique such as **landfill** for hazardous materials has become increasingly difficult and cost-prohibitive due to the increasingly stringent regulatory control.

SVOCs in Groundwater: If SVOCs are present in liquid streams such as groundwater, surface water, or leachate, the same liquid phase **carbon adsorption** used for VOCs is particularly applicable for PAHs, PCBs, and pesticides because their highly hydrophobic nature makes them particularly affinitive to activated carbon sorbent. Besides, UV-based **advanced oxidation processes** (AOPs) can also be used by employing UV light in combination with catalysts and oxidizers such as ozone (O_3) and hydrogen peroxides (H_2O_2).

Fuels in Soil: Fuels are mixtures of aliphatic hydrocarbons (e.g. alkane) and aromatic hydrocarbons (e.g. BTEX and a small amount of PAHs) from leaking underground storage tanks, petroleum refinery sites, degreasing solvents, and vehicle maintenance areas. Fuels may be present in the free phase NAPLs. Since they are mixtures of VOCs and SVOCs, the general applications of presumptive remedies for VOCs and SVOCs described previously are applicable for fuels. For fuels in contaminated soil and sediment, biodegradation of all types, both *in situ* and *ex situ*, can be used, including **bioventing, composting, slurry-phase bioreactors,** and **landfarming.** Bioremediation of fuels under anaerobic conditions may also be feasible, although sometimes biotransformation rather than mineralization is taking place (Box 2.2).

Incineration is desired if halogenated compounds are also present in fuel-based hydrocarbons. For fuels in surface soil and vadose zone, SVE works very well. **Low-temperature thermal desorption** in the temperature range of 90–315 °C can be used to separate volatile components from fuels.

Fuels in Groundwater: Fuel-contaminated groundwater is treated frequently by **air stripping, carbon adsorption,** and **free product recovery**. These are all *ex situ* treatment technologies requiring groundwater extraction. **Free products** (undissolved liquid phase hydrocarbons floating on the water table) are removed by pumping or a passive collection system.

Explosives in Soil: TNT, cyclo-1,3,5-trimethylene-2,4,6-trinitramine (RDX or cyclonite), high melting explosives (HMX), and tetryl (2,4,6-trinitrophenylmethylnitramine) are among the most prevalent explosives at military sites (see Figure 2.5 for the structures of TNT, RDX, and HMX). Since explosives are in the group of SVOCs, the aforementioned technologies described for SVOCs should generally apply. Safety precautions should be taken when remediating explosives-contaminated soils. For example, if contamination is above the limit of 10% explosives by mass in certain areas, contaminated materials can be blended by dilution. After blending is completed, contaminated soils can be treated by bioremediation or **incineration**. Other promising treatment technologies include reuse/recycle, **solvent extraction**, and **soil washing**.

Bioremediation is generally most effective for low-concentration explosives. TNT in the crystalline form is difficult to treat biologically. The biological pathways (see Chapter 9) for explosives vary depending on the site condition. TNT cannot be degraded completely under aerobic conditions, thus monoamine-, diamino-, hydroxylamine-DNT, and tetranitro-azoxynitrotoluenes can be accumulated. RDX and HMX can be degraded into carbon dioxide and water under anaerobic conditions. Under certain optimal aerobic conditions, both 2,4-DNT and 2.6-DNT can be mineralized into carbon dioxide, water, and nitrite (Zhang et al. 2000). DoD currently has been developing or implementing biological treatments for explosives-contaminated soils: **aqueous-phase bioreactor treatment, composting, land farming,** and **white rot fungus treatment**, which are solid phase treatments; and ***in situ* biological treatment**.

Incineration of materials containing less than 10% explosives by weight can be considered as a nonexplosive operation upon the approval of the DoD's Explosives Safety Board. The US Army primarily uses three types of incineration devices: **rotary kiln incinerator, deactivation furnace,** and **contaminated waste processor** (USDoD 1994).

Explosives in Groundwater: For explosives-contaminated groundwater, GAC and UV-based AOPs (advanced oxidation processes) are available technologies as we have described for SVOCs. **GAC** can be used to treat explosives-contaminated water, including process waters from the manufacture and demilitarization of munitions (pink water) and groundwater contaminated from the disposal of these waters. UV oxidation has not been used extensively for remediating water contaminated with explosives because of the widespread use of GAC treatment. Nevertheless, **UV-based AOPs** can be an effective treatment for explosives-contaminated water and, unlike carbon treatment, actually destroy target compounds rather than just transforming them to a more easily disposable medium.

Inorganics in Soil: Inorganic compounds of remedial importance include metals, radionuclides from nuclear wastes (discussed separately), cyanide, and asbestos. Arsenic and selenium are not true metals, but they are classified as two of the eight RCRA metals (arsenic, barium, cadmium, chromium, lead, mercury, selenium, and silver). **Solidification/stabilization (S/S)** is the most commonly used treatment for metals in soil, sediment, and sludge. It involves the addition of materials such as fly ash to reduce the solubility and mobility of waste constituents.

Unlike organic compounds, metals cannot be destroyed or biodegraded. Metals can only be converted into immobile (biologically unavailable) form and/or nontoxic species. However, they can be biotransformed from one species to another via chemical and biological pathways. For example, the carcinogenic Cr(VI) can be transformed into a nontoxic form of Cr(III) through chemically or biologically mediated reduction processes. Arsenate, As(VI), can be anaerobically reduced to a more toxic form of arsenite, As(III), which is also more subject to leaching owing to its higher solubility.

Metals are generally considered to be nonvolatile, therefore they are not subject to any volatilization-based remediation technologies such as soil vapor extraction and air sparging. The exceptions for their volatilization are lead (Pb) and mercury (Hg). Lead becomes volatile when it is biomethylated to form tetramethyl and tetraethyl lead. The metallic form of mercury (Hg^0) or dimethylmercury after methylation of mercury are both volatile. The methylated forms of Pb and Hg are more hazardous after evaporation into the atmosphere, and the formation pathways of both should be avoided.

Inorganics in Groundwater: For metals in liquid streams such as groundwater, surface water, and leachate, **precipitation/flocculation, filtration,** and **ion exchange** are widely used *ex situ* treatment technologies. These technologies are well-established in the traditional water and wastewater treatment plants.

Radionuclides: Unlike nonradioactive metals, the decay of radionuclides into nonradioactive chemicals could take an extremely long period of time (half-life: ^{239}Pu, 24 100 years; ^{235}U, 7×10^8 years). The storage, treatment, and disposal of radioactive wastes are, therefore, present unique challenges. In the United States, the nuclear weapons facilities are responsible for at least 10 times more wastes than the nuclear power industries, and the radioactive wastes from other sources (hospitals, research laboratories, some industries, and mining and processing of uranium) only account for a small quantity (Girard 2014). Since the cleanup of radioactive wastes is considered a specialized effort, a very brief background information is provided in Box 6.2.

Table 6.2 summarizes the treatability for various groups of contaminants in soils and groundwater. These contaminants are classified into eight groups: nonhalogenated VOCs, halogenated VOCs, nonhalogenated SVOCs, halogenated SVOCs, fuels, inorganics, radionuclides, and explosives. The treatability is defined as "above average" = effectiveness demonstrated at pilot or full scale; "average" = limited effectiveness demonstrated at pilot or full scale; "below average" = no demonstrated effectiveness at pilot or full scale. The use of this matrix to screen remediation technologies is briefly illustrated in Examples 6.1 and 6.2.

Example 6.1 Contaminants subjected to destruction at the wood preserving sites

The effectiveness of contaminant destruction as an option for the remediation of contaminated wood preserving sites is provided in the table below (USEPA 1992). Briefly explain the effectiveness of these remediation techniques toward different contaminants.

Contaminants	Incineration	Chemical dehalogenation	Chemical oxidation	Bioremediation
Dioxins/PCBs	●	●	○	○
PCP, cresols	●	◑	◑	◑
PAHs	●	○	◑	●
Polar organic compounds	●	○	◑	●
Metals	○	○	◑	○

● Demonstrated effectiveness; ◑ Potential effectiveness; ○ No expected effectiveness

Solution

Based on what we have described in Section 6.1.3, the effectiveness of various remedial options toward selected groups of contaminants become mostly self-explanatory. For example, one would expect that incineration can destroy all organic compounds except metals because metals cannot be changed. Chemical dehalogenation can occur for halogenated compounds given adequate dehalogenated reaction conditions. Dioxin, PCBs, and metals are recalcitrant to biodegradation, whereas low-molecular-weight PAHs and most polar organic compounds are amenable to fast biodegradation kinetics. The use of chemical oxidation is least likely feasible for compounds like dioxins and PCBs, but it is possible through the selection of adequate oxidizing agents, such as H_2O_2 as the oxidizing agent and Fe^{2+} as the catalyst (known as the **Fenton reagents**).

Box 6.2 Radioactive wastes: It is not just NIMBY, it is NOPE!

Radioactive wastes are divided into two major categories: low-level radioactive wastes (LLWs) and high-level radioactive wastes (HLWs). **High-level radioactive wastes (HLWs)** contain ^{239}Pu (plutonium), consisting of spent fuel rod assemblies from nuclear reactors in both commercial power plants and weapons plants and certain other highly radioactive wastes from nuclear weapons facilities. All other radioactive wastes are classified as low-level wastes, which typically contain dilute and shorter half-life radionuclides from hospitals, research labs, certain industries, as well as diluted transuranium (also known as transuranic) wastes (elements with atomic number greater than that of uranium, i.e. 92) from nuclear weapons plants.

US Congress passed the **Low-Level Radioactive Waste Policy Act**, which set 1986 (later extended to 1992) as the deadline for each state to dispose of the LLWs. The initial treatment of low-level radioactive radionuclides involves **separation, concentration/volume reduction,** and **immobilization**. After a significant decay of radioactivity has been achieved, LLWs can be disposed of in specially designed trenches, landfills, and pits, such as the one in Hanford, Washington and Waste Isolation Pilot Plant (WIPP) in Carlsbad, New Mexico. The WIPP is the world's third deep geological repository licensed to permanently dispose of transuranium radioactive wastes for as long as 10 000 years.

Special geologic repositories should be developed for high-level radioactive wastes such as spent nuclear fuel for permanent disposal. The **Nuclear Waste Policy Act** made the U.S. Department of Energy responsible for finding a site. After years' efforts, the DoE narrowed down to three sites (Hanford, Washington; Deaf Smith County, Texas; and Yucca Mountain, Nevada) and finally chosen the Yucca Mountain in Nye County, Nevada as the geologic repository designed to accept HLWs (Lehr et al. 2002). However, interim storage continued at multiple sites in the United States. Apparently, the opposition to radioactive waste disposal sites is not just a case of **NIMBY** (not in my backyard), but it is one of the NOPE (not on planet Earth)!

6.2 Development and Selection of Remediation Technologies

This section introduces how various remediation technologies are generally developed through various feasibility studies: bench-scale, pilot-scale, and field demonstration. Subsequently, a screening matrix and criteria are presented to help the selection of appropriate remediation technologies. Finally, some recent progress in soil and groundwater remediation that incorporates the concept of sustainability is introduced. This concept will be reinforced in future chapters where more details will be incorporated into various remediation technologies.

Table 6.2 Treatability screening matrix for various groups of soil and groundwater contaminants.

Common soil and groundwater remediation technologies	Nonhalogenated VOCs	Halogenated VOCs	Nonhalogenated SVOCs	Nonhalogenated SVOCs	Fuels	Inorganics	Radionuclides	Explosives
A. Soil, sediment, bedrock, and sludge								
A1. *In situ* biological treatment								
Bioventing	●	◈	●	○	●	○	◈	○
Enhanced Bioremediation	●	●	●	◈	●	◈	◈	●
Phytoremediation	◑	◑	◑	◈	◑	◑	○	○
A2. *In situ* physical/chemical treatment								
Chemical oxidation	◑	◑	○	◑	○	◈	○	◑
Soil flushing	●	●	◑	◑	◑	●	○	○
Soil vapor Extraction	●	●	○	○	●	○	○	○
Solidification/stabilization	○	○	◑	◑	○	●	●	○
A3. *In situ* thermal treatment: thermal treatment	●	●	●	●	●	○	○	○
A4. *Ex situ* biological treatment with excavation								
Biopiles	●	●	◑	◈	●	◈	○	○
Composting	◑	◑	◑	◈	●	○	○	●
Landfarming	◑	◑	●	◑	●	○	○	◈
Slurry phase biological treatment	◑	●	●	◈	●	◈	○	●
A5. *Ex situ* physical/chemical treatment with excavation								
Chemical extraction	◑	◑	●	●	◑	●	◑	○
Chemical reduction/oxidation	◑	◑	◑	◑	◑	●	○	◑
Dehalogenation	○	●	○	●	○	○	○	◑
Soil washing	◑	◑	◑	◑	◑	◑	○	○
Solidification/stabilization	○	○	◑	◑	○	●	●	○

(*Continued*)

Table 6.2 (Continued)

Common soil and groundwater remediation technologies	Nonhalogenated VOCs	Halogenated VOCs	Nonhalogenated SVOCs	Nonhalogenated SVOCs	Fuels	Inorganics	Radionuclides	Explosives
A6. *Ex situ* thermal treatment with excavation								
Incineration	●	●	●	●	●	○	○	●
Pyrolysis	◑	◑	●	●	◑	○	○	○
Thermal Desorption	●	●	●	●	●	○	○	●
A7. Containment: landfill cap and enhancements/alternatives	◑	◑	◑	◑	◑	◑	○	◑
A8. Other treatment: excavation, retrieval, off-site disposal	◑	◑	◑	◑	◑	◑	○	◑
B. Ground water, surface water, and leachate								
B1. *In situ* biological treatment								
Enhanced Bioremediation	●	◈	●	◈	●	◈	○	◑
Monitored Natural Attenuation	●	◑	◑	◑	●	○	○	○
Phytoremediation	◑	◑	◑	◑	◑	◈	○	○
B2. *In situ* physical/chemical treatment								
Air Sparging	●	◑	◑	◑	●	○	○	○
Bioslurping	◑	◑	●	●	●	◑	○	○
Chemical oxidation	◑	◑	○	◑	○	◈	○	◑
Directional wells (enhancement)	◑	◑	◑	◑	◑	◑	○	◑
Thermal treatment	◑	●	●	●	●	○	○	○
In-well air stripping	◑	◑	◑	○	◑	○	○	○
Passive/reactive treatment walls	●	●	●	●	◑	◈	○	●
B3. *Ex situ* biological treatment								
Bioreactors	●	●	●	◈	●	○	○	●
Constructed wetlands	◑	◑	◑	◈	◑	●	○	●

Common soil and groundwater remediation technologies	Nonhalogenated VOCs	Halogenated VOCs	Nonhalogenated SVOCs	Nonhalogenated SVOCs	Fuels	Inorganics	Radionuclides	Explosives
B4. *Ex situ* physical/chemical treatment with pumping								
Adsorption/absorption	◑	◑	◑	◑	○	●	◈	○
Advanced oxidation processes	●	●	●	●	●	◈	◈	●
Air stripping	●	●	○	○	○	○	○	○
Granulated activated carbon/liquid phase carbon adsorption	●	●	●	●	●	◈	○	◈
Groundwater pumping/pump and treat	◑	◑	◑	◈	◑	◑	○	◑
B5. Containment: physical barriers	●	●	●	●	●	●		●
B6. Air emissions/off-gas treatment								
Oxidation	●	●	●	●	●	○	I/D	◑
Vapor phase carbon adsorption	●	●	●	●	●	◑	I/D	●

Source: Adapted from the USEPA (2018).

● Above average; ◑ Average; ○ Below average; ◈ = Level of effectiveness highly dependant upon specific contaminant and its application; I/D = insufficient data.

6.2.1 Remedial Investigation/Remedial Feasibility Study

In remedial investigation/remedial feasibility (RI/RF) stage, the **treatability study** is the key component with its primary objectives focusing on the following: (i) To provide sufficient data to allow treatment alternatives to be fully developed and evaluated during the detailed analysis to support the remedial design of a selected alternative; (ii) To reduce cost and performance uncertainties for treatment alternatives to an acceptable level so that a remedy can be selected. A treatability study performed during an RI/RF is used to adequately evaluate a specific technology, including evaluating performance, determining process sizing, and estimating costs in sufficient detail to support the remedy selection process. It is not meant to be used solely to develop detailed design or operating parameters that are more appropriately addressed during the remedial design phase.

There are three levels of a treatability study (Figure 6.3). The level chosen depends on the information available about the site, the technology, and the nature of information that is needed. The quickest and least expensive treatability study is the laboratory screening and the most expensive one is the field-scale study by testing vendor supplied equipment. **Laboratory screening** is performed to learn more about the characteristics of the waste and to determine if it would be treatable by a particular technology. A laboratory screening test takes a matter of days to weeks and generally costs in an approximate range of \$10 000–\$50 000. A successful laboratory screening may lead to more sophisticated treatability studies.

The next level of a treatability study is the **bench-scale study** which collects more performance data (and in some cases, the cost data) of a technology by simulating the treatment process with a very small quantity of wastes or contaminated matrix. The objective of this bench-scale test is to determine whether the technology can meet the cleanup standards set for the site. The approximate costs of these tests may range between \$50 000 and \$250 000. Bench tests are generally used to determine if the "physics," "chemistry," or "microbiology" of the process works and are usually performed in batch (e.g. well-mixed "jar tests") or soil columns, with treatment parameters varied one at a time. Because small volumes and inexpensive reactors (e.g. bottles, beakers, columns) are used, bench tests can be used economically to test a relatively large number of both performance and waste composition variables.

At the next level, the **pilot-scale treatability study** is usually conducted in the field or the laboratory and requires installation of the treatment equipment. This study is used to provide performance, cost, and design data for the treatment technology. Due to the high cost (e.g. >\$250 000), it is used almost exclusively to fine-tune the design of the technology. Pilot studies are intended to simulate the physicochemical and/or biological parameters of a full-scale process; therefore, the treatment unit sizes and the volume of the wastes to be processed in pilot systems greatly increase over those of bench scale tests. As such, pilot tests are intended to

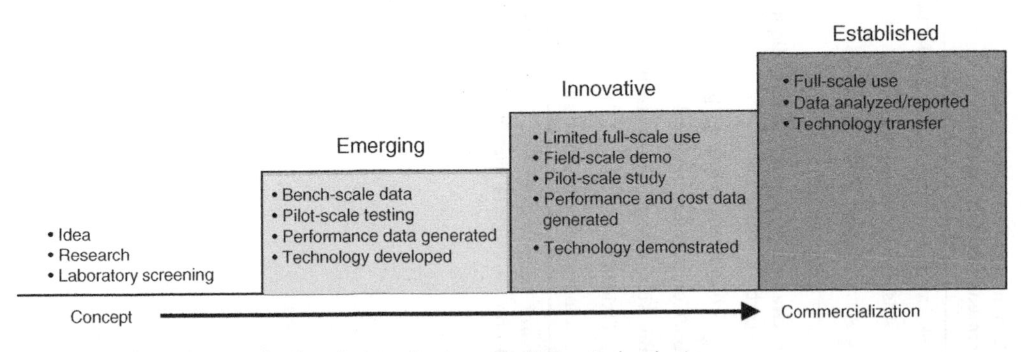

Figure 6.3 Development of soil and groundwater remediation technologies.

bridge the gap between bench level analyses and full-scale operation, and are intended to more accurately simulate the performance of the full-scale process.

Pilot units are usually sized to minimize the physical and geometric effects of test equipment on treatment performance to simulate full-scale performance. Examples of these effects include mixing, wall effects, accurate settling data, and generation of sufficient residues (sludge, off-gas, etc.) for additional testing (dewatering, fixation, etc.). Pilot units are operated in a manner as similar as possible to the operation of the full-scale system. In other words, if the full-scale system will be operated continuously, then the pilot system would usually be operated continuously. Table 6.3 provides a summary of comparison of bench and pilot tests.

Figure 6.4 shows an example of pilot-scale testing for the biodegradation of explosives in soils using slurry phase bioreactors. The bioreactor device uses 80-L commercially available slurry reactor in a batch-wise fill-and-draw mode of operation. The process starts with sieving of contaminated soils to remove gravels, followed by soil washing using tap water to remove the bulk of explosive compounds that can be dissolved in water. The resulting slurries are manually placed in a bioreactor system, which continuously monitors and adjusts the pH change for optimal growth of bacteria. The solid residue from the bioreactor is further treated in a holding tank to separate treated soil from the liquid effluent prior to discharge (Zhang et al. 2001).

It is important to note that not all treatments need to conduct treatability studies, including those well-established processes such as activated carbons and air stripping. Also, not all treatability studies on innovative technologies will lead to its successful utilization and commercialization in remediation. This is because some technologies need more research and development, some technologies need good field data, and some technologies need cost

Table 6.3 Bench and pilot study parameters.

Parameter	Bench study	Pilot study
Purpose	To determine the performance and optimal process parameters, including active mechanisms, chemical doses, effects of environmental factors, material compatibility, and process kinetics	To determine performance and cost in a large scale and fine-tune operating criteria and remedial design
Size	Laboratory or bench top (jars and soil column tests)	1–100% of full scale
Location	Laboratory	Onsite
Providers	Product vendors, universities, national labs, independent labs	Commercial R&D labs and environmental consulting firms
Number of Variables Tested	Can be many variables	Few limited site-specific variables
Materials and Wastes	Small to moderate amounts	Relatively large amounts
Time Required	Days to weeks	Weeks to months
Typical Cost	0.5–2% of capital costs of remedial action	2–5% of capital costs of remedial action depending on the total cost
Limitations	Wall, boundary and mixing effects; volume effects; difficulty in simulating solids processing; transportation of sufficient waste volume	Costly to test large number of variables; generate large volume of wastes; safety, health and other risks; required disposal of process wastes

Source: Adapted from USEPA (1988).

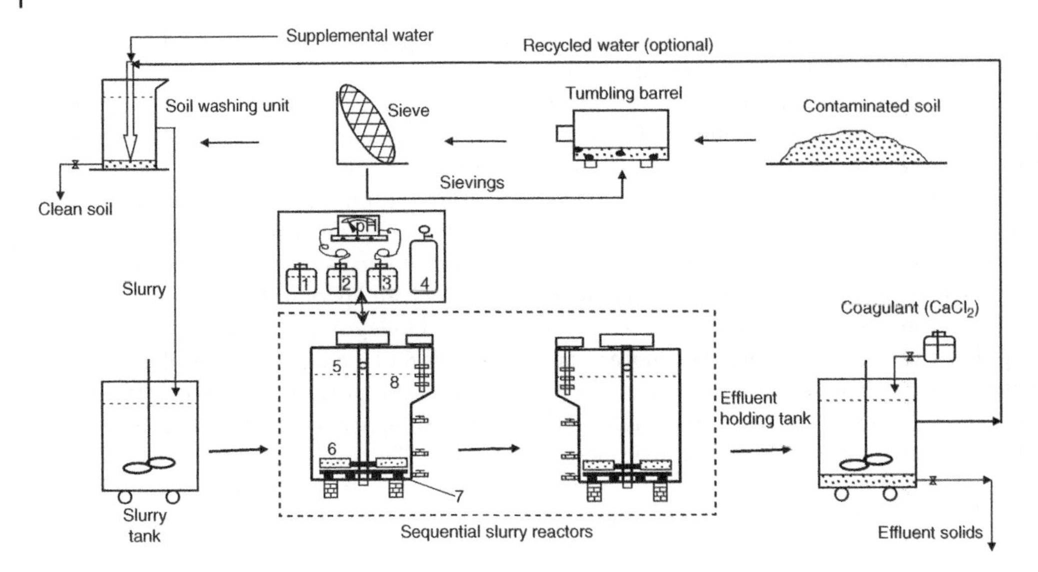

Figure 6.4 Schematic of a pilot-scale testing on slurry-phase biodegradation of explosives in soils: 1 = Nutrients; 2 = HCl; 3 = NaOH; 4 = Compressed air; 5 = Air-lifter; 6 = Air diffuser; 7 = bottom rakers; 8 = foam breakers. *Source:* Adapted from Zhang et al. (2001).

reduction. Certain technologies will also disappear for various reasons, such as high cost, failure, bad press, misapplication, poor system design, or inherent limits. Some technologies may work at one site, but they may not work at another site.

Technologies that are most likely to succeed include those that have simplicity and low cost, carry success stories, and have large organization interest (e.g. venting, sparging, bioremediation). Technologies that will facilitate their success include those with credible white paper for marketing or sales, detailed protocols to follow, user-friendly implementation tools, and large organization utilization. The USEPA encourages the use of innovative remediation technologies, although some major deterrents may be encountered. These deterrents to the use of new remediation technologies may include liability concerns, lack of usable data from technology developers, unrealistic cost and performance data, lack of accurate market data to estimate capital needs, high risk and poor return, overly long time needed for commercialization (regulations, permitting, etc.), lack of stakeholder partnerships, and lack of focused technology transfer contacts and information.

6.2.2 Remediation Technologies Screening and Selection Criteria

Under the auspices of the Federal Remediation Technologies Roundtable (FRTR), the U. S. Army Environmental Center (USAEC) led a multiagency effort to create the Remediation Technologies Screening Matrix and Reference Guide. Its purpose is to create a comprehensive "Remediation Technologies Yellow Pages" for use by those responsible for environmental cleanup (Teefy 1997). Table 6.4 summarizes pertinent information for screening each of the major treatment technologies according to the following factors:

- Development status (scale status of an available technology)
- Treatment train (Is the technology only effective as part of the treatment train?)
- Relative overall cost and performance (O&M, capital, system reliability and maintainability, relative costs, time required to cleanup a "standard" site)
- Availability (number of vendors that can design, construct, and maintain the technology)

Table 6.4 Soil and groundwater remediation technology screening matrix.

Common soil and groundwater remediation technologies	Development Status	Treatment Train	Relative overall cost and performance					
			O&M	Capital	System reliability and maintainability	Relative costs	Time	Availability
A. Soil, sediment, bedrock, and sludge								
A1. *In situ* biological treatment								
Bioventing	●	●	●	●	●	●	◑	●
Enhanced Bioremediation	●	●	○	◑	◑	●	◑	●
Phytoremediation	●	●	●	●	○	●	○	◑
A2. *In situ* physical/chemical treatment								
Chemical oxidation	●	●	○	◑	◑	◑	●	●
Soil flushing	●	●	○	◑	◑	◑	◑	●
Soil vapor extraction	●	○	○	◑	●	●	◑	●
Solidification/stabilization	●	●	◑	○	●	●	●	●
A3. *In situ* thermal treatment: thermal treatment	●	○	○	○	●	◑	●	●
A4. *Ex situ* biological treatment with excavation								
Biopiles	●	●	●	●	●	●	◑	●
Composting	●	●	●	●	●	●	◑	●
Landfarming	●	●	●	●	●	●	◑	●
Slurry phase biological treatment	●	○	○	○	◑	◑	◑	●
A5. *Ex situ* physical/chemical treatment with excavation								
Chemical extraction	●	○	○	○	◑	◑	◑	●
Chemical Reduction/oxidation	●	◑	◑	○	●	◑	●	●
Dehalogenation	●	◑	○	○	○	○	◑	◑
Soil washing	●	○	○	○	●	◑	●	●

(*Continued*)

Table 6.4 (Continued)

Common soil and groundwater remediation technologies	Development Status	Treatment Train	Relative overall cost and performance					
			O&M	Capital	System reliability and maintainability	Relative costs	Time	Availability
Solidification/stabilization	●	●	◑	○	●	●	●	●
A6. *Ex situ* thermal treatment with excavation								
Incineration	●	●	○	○	◑	○	●	●
Pyrolysis	●	●	○	○	○	○	●	●
Thermal desorption	●	●	○	○	◑	◑	●	●
A7. Containment: landfill cap and enhancements/alternatives	●	●	◑	○	●	●	○	●
A8. Other treatment: excavation, retrieval, off-site disposal	●	●	●	●	●	◈	●	●
B. Ground water, surface water, and leachate								
B1. *In situ* biological treatment								
Enhanced bioremediation	●	●	○	◑	◑	●	◈	●
Monitored natural attenuation	●	●	○	◑	◑	●	◈	●
Phytoremediation	●	●	●	●	○	●	○	◑
B2. *In situ* physical/chemical treatment								
Air sparging	●	●	●	●	●	●	●	●
Bioslurping	●	◑	●	●	◑	●	◑	●
Chemical oxidation	●	●	◑	○	◑	◑	◑	●
Directional wells (enhancement)	●	●	◑	○	◑	◑	◑	●
Thermal treatment	●	○	○	○	◑	◑	●	●
In-well air stripping	●	◑	◑	○	◑	◑	○	●
Passive/reactive treatment walls	●	●	◑	○	●	◑	○	●

Common soil and groundwater remediation technologies	Development Status	Treatment Train	Relative overall cost and performance					
			O&M	Capital	System reliability and maintainability	Relative costs	Time	Availability
B3. *Ex situ* biological treatment								
Bioreactors	●	●	◑	○	◑	●	◑	●
Constructed wetlands	●	●	◑	○	◈	◑	◈	○
B4. *Ex situ* physical/chemical treatment with pumping								
Adsorption/absorption	●	●	○	◑	◑	○	○	●
Advanced oxidation processes	●	◑	○	○	◑	◑	○	●
Air stripping	●	◑	○	◑	●	●	○	●
Granulated activated carbon/liquid phase carbon adsorption	●	◑	○	◑	●	◑	○	●
Groundwater pumping/pump and treat	●	◑	○	○	●	○	○	●
B5. Containment: physical barriers	●	●	●	●	●	○	○	●
B6. Air emissions/off-gas treatment								
Oxidation	●	N/A	●	●	●	●	I/D	●
Vapor phase carbon Adsorption	●	N/A	●	●	●	●	I/D	●

● Above average; ◑ Average; ○ Below average; ◈ = Level of effectiveness highly dependant upon specific contaminant and its application; I/D = insufficient data; N/A = not applicable. The definitions of symbols used in the treatment technologies screening matrix are detailed in the USEPA (2018).

Additional factors, not listed in Table 6.4, may include some of the following:

- Residuals produced such as solid, liquid, and vapor
- Minimum contaminant concentration achievable
- Consideration of toxicity, mobility, or volume
- Long-term effectiveness and performance
- Awareness of remediation consulting firms and communities
- Regulatory and permitting acceptability
- Community acceptability

The example below shows the use of this screening matrix based on the major criteria listed in Table 6.4. For detailed screening analysis, readers should be aware of all the considerations in regard to engineering, regulations, economics, public health, and community perception.

Example 6.2 Use of Treatment Technologies Screen Matrix

What would be your comments as an environmental professional if a stakeholder, for example, a representative from a local community or a regulator who is involved in a remedial project, approaches you regarding the remediation of (a) nonhalogenated SVOCs in soils from a coal gasification facility, (b) halogenated SVOCs surrounding an old burn pit? The two presumptive remedies in both cases are *in situ* bioremediation and incineration.

Solution

(a) The nonhalogenated SVOCs are most likely PAHs from a coal gasification facility. Both *in situ* bioremediation and incineration are technically feasible to remediate PAHs. The *in situ* bioremediation is advantageous in its lower capital requirement and remediation cost; however, the *ex situ* incineration is advantageous in its cleanup time. (b) The compounds of interests might be PCBs, chlorinated pesticides, or other halogenated organic compounds. While halogenated SVOCs can be completely destructed (99.9999%, if needed) by *ex situ* incineration, such halogenated SVOCs are typically not amenable to microbial degradation.

6.2.3 Green and Sustainable Remediation

The environmental remediation sector for the cleanup of contaminated soil and groundwater started in the late 1970s in the United States, Canada, and most of the European countries following the environmental movement. As we have learned in the previous section, the selection of remediation technologies has been traditionally based on the cost-effectiveness (capital and O&M costs, remedial efficiency, and time) of the specific technique to meet the cleanup requirement without much consideration of the impacts beyond the site. It is, therefore, not surprising that more invasive and energy-intensive remediation technologies such as pump-and-treat, soil excavation, and incineration (Table 6.5) were the primary remedial options to clean up Superfund sites since its inception of the nation's remediation programs. With the increasing recognition of the cost-effective issues, these established technologies have been gradually shifted to other so-called innovative remediation technologies around 1990, such as soil vapor extraction, *in situ* bioremediation, and natural attenuation.

Since around 2000, however, an increased attention has been paid to the sustainability of contaminant remediation. Particularly during the past decade, a significant progress has been made in the sustainable remediation field, through the publication of guidance, strategies, and policies by government agencies, remediation practitioners, and industry associations (e.g. USEPA 2008, 2009a, 2009b, 2010a, 2012; SURF 2009).

Table 6.5 Evolution of environmental remediation technologies (Zhang 2013).

Period (approximate)	1970-	1990-	2000-
Examples of favorable technologies	Soil excavation/ Incineration/ Pump-and-treat	Soil vapor extraction/ *In situ* bioremediation/ Natural attenuation	Holistic approach considering site characterization, design, construction, operation, and monitoring
Primary determinant for remedial selection	Effectiveness	Cost; Effectiveness	Cost; Effectiveness; Sustainability
Spatio-temporal boundary	Local; Short-term	Local; Long-term (generally)	Regional to global; Life-cycle
Pros and cons	More efficient (supposedly), invasive, but more costly	Less invasive and costly, but more time (generally)	A harmony balancing social, economic, and environmental benefits

Sustainable remediation is defined as a remedy or combination of remedies whose net benefit on human health and the environment is maximized through the judicious use of limited resources (SURF 2009). Organizations sometimes refer to **green remediation**, which can be defined as the practice of considering all environmental effects of remedy implementation and incorporating options to maximize the net environmental benefit of cleanup actions (USEPA 2008). Like the sustainability of other sectors, sustainable remediation is aimed to "create and maintain conditions, under which human and nature can exist in a productive harmony, that permit fulfilling the social, economic, and other requirements of present and future generations" in accordance with President Obama's Executive Order 13514 (The White House 2009).

Sustainable practices in contaminant remediation are not currently enforced by any regulatory provisions in the United States. The USEPA considers science-based green program a voluntary complement to its traditional "command-and-control" regulation of industry toward achieving environmental protection and sustainable development (Hjeresen et al., 2001). Therefore, the most notable challenges for green remediation practices lie in regulatory and economic drivers. Fortunately, there has been increased awareness of sustainability in the society due to increasing concerns about global warming, climate change, and other environmental effects. Corporate leaders also become more conscious about the regional and global impacts from their activities. The motivation of sustainability increasingly comes from within the industry, whose own desire is to improve its cost-saving (at least for some best management practices, BMPs) and improve the corporate's own image of sustainability.

It is anticipated that sustainability will become a deciding factor in the approval of future site Remedial Action Plans. There appears to be a good deal of sustainable practice in the United States and in several European countries. Specific applications of green and sustainable principles in soil and groundwater remediation will be incorporated into Chapters 7–12.

6.3 A Snapshot of Remediation Technologies

Many remediation technologies have been mentioned in this chapter. The following sections mainly provide a snapshot of most, if not all, common remediation technologies, since selected technologies will be detailed in the remainder of this text. In many contaminated sites, a single approach is rarely used to accomplish the cleanup goal. Two or more of these technologies are generally combined to form what is commonly referred to as the treatment train.

6.3.1 Description of Various Treatments

Appendix B can be used to find a brief definition of various remediation technologies. Readers should be cautioned that novel remediation technologies continue to evolve from the research community and remediation industries. It is also noteworthy that not all these remediation technologies could become equally important for a specific case. Rather, each innovative technology may present its own promising feature and suit to a particular remedial application scenario. In Appendix B, the remediation technologies are grouped according to the media to be treated: (a) soil, sediment, and sludge; (b) groundwater, surface water, and leachate; and (c) air emission/off-gas treatment (Federal Remediation Technologies Roundtable 2018). For (a) and (b), technologies are further grouped based on *in-situ* or *ex situ*.

6.3.2 Treatment Train

Two or more remediation technologies described in Appendix B are often used together in site remediation by what is referred to as the **treatment trains**, which are either integrated processes or a series of sequential treatments to achieve a specific treatment goal. Some treatment trains are employed when no single technology is capable of treating all the contaminants in a particular medium. For example, soil contaminated with organics and metals may be treated first by bioremediation to remove organics and then by S/S to reduce the leachability of metals. In other cases, a treatment train might be used to render a medium more easily treatable by another subsequent technology, reduce the amount of waste prior to a more expensive technology, or prevent the emission of volatile contaminants during excavation and mixing.

Figure 6.5 identifies some common treatment trains used in remedial actions in the US Superfund sites (USEPA 2007). As can be seen, innovative treatment technologies may be used with established technologies or with other innovative technologies. According to a Superfund survey, the most common treatment trains are air sparging used in conjunction with SVE and bioremediation followed by SVE or S/S. In the case of air sparging used with SVE, the air sparging is used to remove contaminants from groundwater *in situ*, while the SVE captures the contaminants removed from the groundwater and removes contaminants from the soil above the groundwater table (the vadose zone). When *in situ* technologies are used in a treatment train, a more aggressive technology may be applied to remediate areas with high contaminant concentrations or NAPLs (hot spots), followed by application of a less aggressive technology to remediate a larger area that includes the former hot spot area. Given two different scenarios, the concepts of treatment train are illustrated with Examples 6.3 and 6.4.

Example 6.3 Treatment train for a site with TCE-contaminated groundwater

A former military training ground was found to have caused a significant TCE migration into groundwater with the potential presence of a pooled DNPAL phase associated with TCE. Initial remediation with *in situ* chemical oxidation (ISCO) using Fenton reagents through the injection of hydrogen peroxide (H_2O_2) succeeded very well during several months of the operation. However, TCE concentrations in groundwater rebound several months after the cessation of H_2O_2 injection. Further site characterization and computer modeling revealed that the contaminated area can be grouped into target treatment zones. Zone 1 is the source zone with high concentrations of free-phase TCE and Zone 2 has aqueous phase TCE dissolved in groundwater between the source area and the property boundary. Based on the information provided, suggest a treatment train.

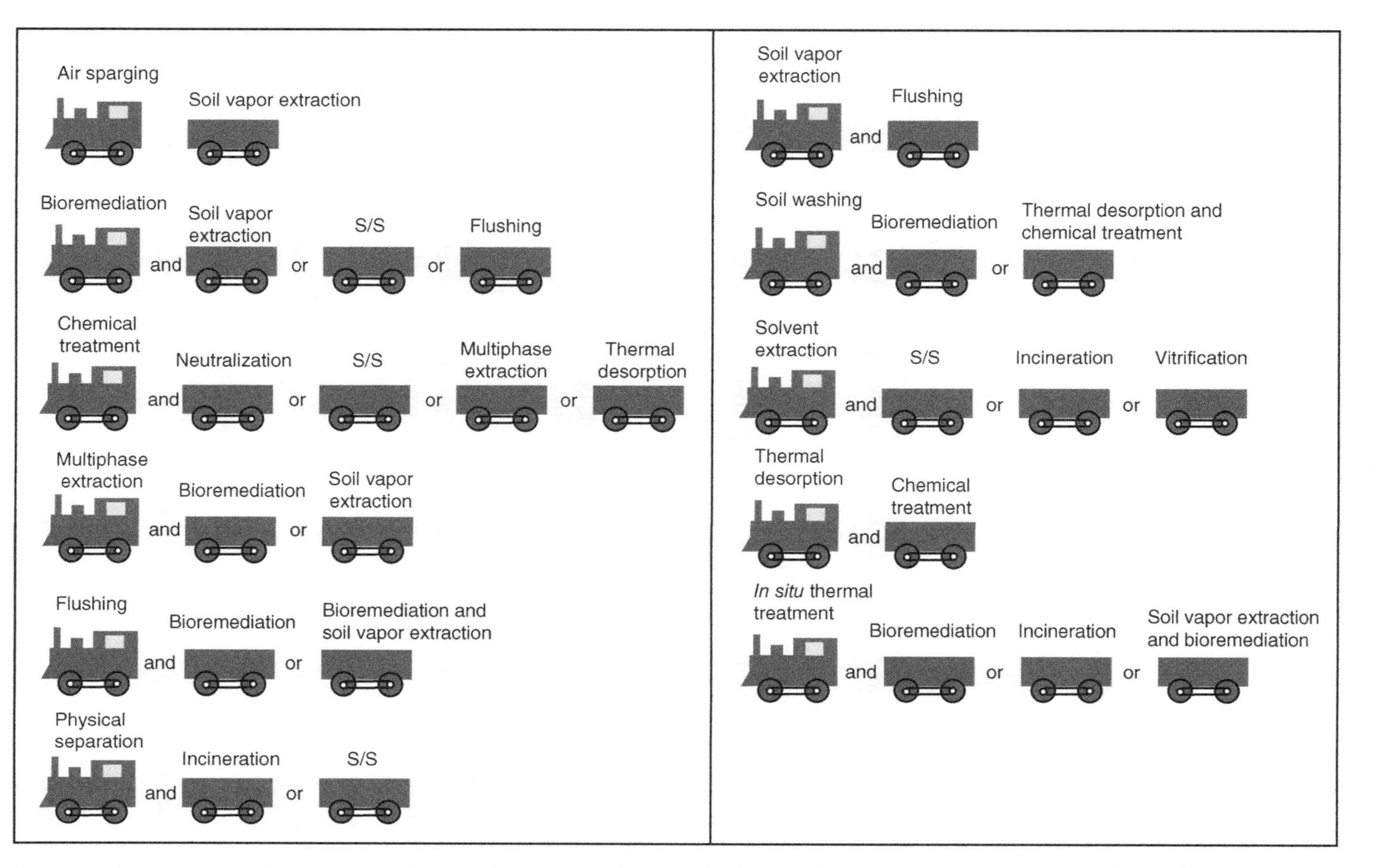

Figure 6.5 Treatment trains with innovative soil and groundwater remediation technologies in the United States Superfund sites. S/S = solidification/stabilization. *Source:* Adapted from USEPA (2007).

Solution

The answer to questions of this type may vary depending on the site-specific aquifer conditions, the remedial goal, and many other factors. Given the conditions, it is reasonable to first suggest two different treatment trains for these two zones.

For Zone 1 with potential presence of free-phase TCE, an invasive Free Product Recovery using active pumping can be employed. Alternatively, a chemically mediated reductive dechlorination can be employed by the injection of reducing agents. Following these invasive processes, the bulk of TCE should be removed. The residual TCE can be gradually cleaned up with a microbe-based polishing step. Certain dehalogenating bacteria such as *Dehalococcoides* can convert TCE into ethane as the harmless end-product given that enough organic electron donors are present in the contaminated groundwater. If not, injection of molasses and dechlorinating bacteria will accelerate dechlorination process in Zone 1.

For the lightly contaminated Zone 2, following the initial ISCO, a less aggressive monitored natural attenuation (MNA) can be recommended, assuming conditions for MNA are favorable, such as the presence of terminal electron accepting processes (TEAPs) and microbial activities as described for the polishing step in Zone 1. MNA should be monitored continuously to assure contaminant concentrations are declined over time in this target treatment zone.

Example 6.4 Treatment train for a mixed waste containing PAHs and heavy metals

High residual concentrations of high-molecular-weight PAHs and four metals (Cu, Pb, Zn, Cd) are present in a small area of surface soils in an abandoned former manufactured gas plant. Suggest a general treatment train for the contaminated soil.

Solution

The answer may vary depending on other site-specific details. The co-occurrence of PAHs and metals complicates the remediation because of the vastly different properties between PAHs and metals. Thermally enhanced desorption or incineration can do well in removing/destroying PAHs in soil, but would not change the metals in soil. If the amount of soil to be treated is manageable with an on-site soil washing unit, soil washing can be pursued using a mixed solution containing both a surfactant and a chelating agent (such as EDTA) to remove hydrophobic PAHs and metals, respectively. Cleaned soils can be back-filled on-site or hauled away to a landfill site. Liquid waste stream containing PAHs and metals can be treated by carbon adsorption or disposed of in a nearby wastewater treatment plant if available.

A bench-scale treatability study should be conducted, for example, to test the type and dosage of the washing solutions as well as the removal efficiency of the activated carbon process.

As a concluding remark of this chapter, readers are reminded that the scope of this chapter is to serve as an overview of available remediation technologies. This is a rapidly evolving field since more and more R&D (research and development) data will become available. Readers are encouraged to check some useful websites for the most recent update. Some useful online resources on hazardous waste remediation and characterization are the USEPA's Superfund Innovative Technologies and Brownfields Initiative, websites by the EPA Office of Solid Waste and Emergency Response (OSWER), Strategic Environmental Research & Development Program (SERDP), Defense Environmental Network & Information Exchange (DENIX), and DoE Office of Environmental Management (EM), to just name a few.

Bibliography

Alter, B. (2012). *Environmental Consulting Fundamentals: Investigating and Remediation*. Boca Raton, FL: CRC Press.

Federal Remediation Technologies Roundtable (FRTR) (2018). Remediation Technologies Screening Matrix and Reference Guide, Version 4. http://www.frtr.gov/matrix2/top_page.html (accessed December 2018).

Girard, J.E. (2014). *Principles of Environmental Chemistry*, 3e. Burlington, MA: Jones & Bartlett.

Hadley, P.W. and Ellis, D.E. (2009). Integrating sustainable principles, practices, and metrics into remediation projects. *Remediation J.* 19 (3): 5–114.

Hjerssen, H.L., Anastas, P., Ware, S., and Kirchhoff, M. (2001). Green chemistry progress and challenges. *Environ. Sci. Technol.* 115A–119A.

Lehr, J., Hyman, M., Gass, T.E., and Seevers, W.J. (2002). *Handbook of Complex Environmental Remediation Problems*. New York, NY: McGraw-Hill.

Sustainable Remediation Forum (SURF) (2009). Integrating sustainable principles, practices, and metrics into remediation projects. *Remediation J.* 19 (3): 5–114. editors P. Hadley and D. Ellis, Summer 2009.

Teefy, D.A. (1997). Remediation technologies screening matrix and reference guide: version III. *Remediation J.* 8 (1): 115–121.

The White House (2009). *Federal Leadership in Environmental, Energy, and Economic Performance*. Office of the Press Secretary, 5 October 2009.

USDoD (1994). *Remediation Technologies Screening Matrix and Reference Guide*, 2e, 611. DoD Environmental Technology Transfer Committee.

USEPA (1988). *Guidance for Conducting Remedial Investigations and Feasibility Studies Under CERCLA*, EPA/540/G-89/004. USEPA.

USEPA (1990). *The Superfund Innovative Evaluation Program: A Fourth Report to Congress*, EPA/540/5-91/004. USEPA.

USEPA (1992). *Contaminants and Remedial Option at the Wood Preserving Sites*, EPA 600/R-92/182. USEPA.

USEPA (1996). *A citizen's guide to innovative treatment technologies for contaminated soils, sludge, sediments, and debris*, EPA 542-F-96-001. USEPA.

USEPA (2000). *Institutional Controls: A Site Manager's Guide to Identifying, Evaluating and Selecting Institutional Controls at Superfund and RCRA Corrective Action Cleanups*, EPA 540-F-00-005, OSWER 9355.0-74FS-P. USEPA.

USEPA (2007). *Treatment Technologies for Site Cleanup: Annual Status Report*, 12e, EPA-542-R-07-012. USEPA.

USEPA (2008). *Green Remediation: Incorporating Sustainable Environmental Practices into Remediation of Contaminated Sites*, EPA 542-R-08-002. USEPA.

USEPA (2009a). *Principles for Greener Cleanups*, U.S. Environmental Protection Agency, Office of Solid Waste and Emergency Response.

USEPA (2009b)). *Green Remediation Best Management Practices: Pump and Treat Technologies*, Office of Solid Waste and Emergency Response.

USEPA (2010a). *Superfund Green Remediation Strategy*, U.S. Environmental Protection Agency, Office of Solid Waste and Emergency Response, Office of Superfund Remediation and Technology Innovation.

USEPA (2010b). *Superfund Remedy Report*, 13e, EPA-542-R-10-004. USEPA.

USEPA (2012). *Methodology for Understanding and Reducing a Project's Environmental Footprint*, EPA 542-R-12-002, U.S. Environmental Protection Agency.

USEPA (2013). *Superfund Remedy Report*, 14e, EPA-542-R-13-016. USEPA.

USEPA (2017). *Superfund Remedy Report*, 15e, EPA-542-R-17-001. USEPA.
USEPA (2018), *Contaminated Site Clean-Up Information (CLU-IN)*. USEPA http://www.clu-in.org/ (accessed December 2018).
Zhang, C., Hughes, J.B., Daprato, R.C. et al. (2001). Remediation of dinitrotoluene contaminated soils from former ammunition plants: soil washing efficiency and effective process monitoring in bioslurry reactors. *J. Haz. Materials* 2676: 1–16.
Zhang, C. (2013). Incorporation of green remediation into soil and groundwater cleanups. *Int. J. Sustainable Human Dev.* 1 (3): 128–137.
Zhang, C., Hughes, J.B., Nishino, S.F., and Spain, J. (2000). Slurry-phase biological treatment of 2,4-dinitrotoluene and 2,6-dinitrotoluene: role of bioaugmentation and effects of high dinitrotoluene concentrations. *Environ. Sci. Tech.* 34 (13): 2810–2816.

Questions and Problems

1 Why are *in situ* remediation technologies preferred, and what factor(s) make *ex situ* remediation an essential option?

2 What are the presumptive soil remedial technologies generally for (a) halogenated VOCs, (b) halogenated SVOCs, (c) nonhalogenated VOCs, and (d) nonhalogenated SVOCs?

3 What are the presumptive remedial technologies for metals and radionuclides in soils?

4 What are the respective presumptive remedies for the remediation of VOCs in soils and VOCs in groundwater?

5 What remediation approaches can be considered for (a) the bioremediation of explosives in contaminated soils, (b) remediation of explosives in contaminated groundwater extracted by pumping?

6 What special precaution must be taken when remediating explosives-contaminated solids and hazardous materials?

7 Give a list of destructive remediation technologies for organic contaminants.

8 What are the respective presumptive remedies for the remediation of SVOCs in soils and VOCs in groundwater?

9 In the remediation of Superfund sites in the United States, on average which one is the mostly used: established remediation technologies or innovative remediation technologies?

10 Under what circumstances, institutional controls can become critical in the overall remedial strategy? Illustrate your point by using an example of a closed hazardous landfill site or a delisted Superfund site.

11 Provide a list of common institutional controls (ICs). Why fences used to restrict access to a contaminated site are not considered as ICs by the U.S. EPA definition?

12 Describe whether *in situ* bioremediation is generally applicable for (a) halogenated VOCs, (b) halogenated SVOCs, (c) nonhalogenated VOCs, and (d) nonhalogenated SVOCs.

13 Indicate whether bioremediation (*in situ* or *ex situ*) can be generally employed for the remediation of soil and groundwater contaminated with (a) PAHs, (b) PCBs, (c) fuels, and (d) radionuclides.

14 Indicate whether bioremediation (*in situ* or *ex situ*) can be generally employed for the remediation of soil and groundwater contaminated with (a) BTEX, (b) phenols, (c) metals, and (d) explosives.

15 What is unique about the remediation of metals in comparison with organic contaminants in general?

16 Using the remediation screening matrix, compare the pros and cons for the treatment of fuel hydrocarbon-contaminated soils using (a) landfarming and (b) incineration.

17 What are some of the most commonly used innovative remediation technologies currently used in Superfund sites in the United States?

18 The potential applicability of a contaminant can usually be gauged by considering the basic mechanisms by which the technology works and the chemical characteristics of the contaminant. Based on the information provided in this chapter, which technologies should effectively remove PCE and PCBs in contaminated aquifer on a former solvent recovery facility site (see the table below).

	Contaminant	
Remediation Options	**PCE**	**PCBs**
Excavation of contaminated soils then incineration	?	?
Pump + above ground air stripping	?	?
Pump + above ground carbon adsorption	?	?
In situ bioremediation	?	?

Give your answer to each of the question mark in the table explicitly (yes/no) and provide rationales as to why or why not. Note that PCBs are recalcitrant to bacterial degradation, while certain bacteria in soil are known to degrade PCE anaerobically in a great extent.

19 Soil washing, solvent extraction, and thermal desorption can be considered as "separation" technologies applicable for the separation of contaminants from soils. Based on the underlying principles, indicate in the table whether they are likely to be "demonstrated effectiveness" (●), "potential effectiveness" (◑), or "no expected effectiveness" (○). Note that exceptions are always likely, if so, please justify.

Contaminants	Soil washing	Solvent extraction	Thermal desorption
Dioxins/PCBs			
PCP			
PAHs			
Polar organic compounds			
Metals			

20 Describe the following remediation technologies: (a) soil washing (b) *in situ* soil flushing (c) vitrification (d) solidification/stabilization.

21 Describe the following remediation technologies: (a) air sparging (b) soil vapor extraction (c) natural attenuation (d) reactive barrier.

22 Can the following remediation techniques be used in both *in situ* and *ex situ*: (a) soil vapor extraction (b) solidification/stabilization (c) vitrification?

23 What are the common oxidizing agents used for *ex situ* chemical redox for the remediation of contaminated soils?

24 What are the temperature ranges used for the following *ex situ* thermal treatment of excavated soils: (a) high-temperature thermal desorption (b) incineration (c) low-temperature thermal desorption?

25 What differentiate pyrolysis from incineration?

26 What mechanisms is natural attenuation based on?

27 For *in situ* biological treatment of groundwater, what are the respective purposes of adding methane, nitrate, oxygen, air, and hydrogen peroxide?

28 Describe the different uses of air sparging and air stripping. Do both employ the use of air?

29 Among the various physical/chemical treatments of pumped groundwater, specify what can be used for the treatment of inorganics, and what can be used for the treatment of organics?

30 What are the remedial options for air emission or off-gas treatment?

31 If someone representative of a local township approaches you to conduct a pilot-scale treatability testing on the bioremediation of a former military training ground with explosives-contaminated surface soils, what would you propose in regard to the timeframe, cost, testing objectives, and variables of the treatability test. The city would like to invest a total of approximately $2 million to have this site cleaned up. The site is 5 mi away from the residential and commercial area, so the remediation completion time is not essential.

32 Suggest a treatment train for (a) a site with shallow soils contaminated with both metals (lead, cadmium) and PCBs, and (b) a site with some recent BTEX spill as well as some historically contaminated soils containing explosives.

33 Suggest a treatment train for (a) a site with halogenated VOCs in both vadose zone and saturated zone, and (b) a site with recent BTEX-containing gasoline spills in both surface soil and shallow aquifer.

34 Describe the evolution of remediation technologies from the 1970s in respect to the realization of sustainability. Toward this goal, what have been changed as the primary determinants for the selection of remediation technologies?

Chapter 7

Pump-and-Treat Systems

LEARNING OBJECTIVES

1. Delineate various configurations of extraction wells, injection wells, and physical and hydraulic barriers used to either extract (remove) or contain contaminant plumes
2. Understand the design parameters (well number, location, and pumping rate) to achieve the optimized performance of pump-and-treat systems
3. Calculate capture zone and optimum well spacing using the governing equations under simplified conditions of an ideal aquifer
4. Know the general options available for the aboveground treatment of common inorganic and organic contaminants
5. Understand the parameters (including Henry's law constant) that control the performance of air stripping and the governing equations for the design calculation of an air stripping tower
6. Design an activated carbon process for the estimation of daily carbon usage and vessel size by employing sorption isotherm data
7. Understand the pros and cons of pump-and-treat and know when it works and when it does not work regarding the contaminant removal and hydraulic containment
8. Delineate the factors that limit pump-and-treat through quantitative analysis of dissolved phase and residual saturation of NAPLs
9. Identify various causes of tailing and rebound such as slow dissolution, desorption, diffusion, and variable groundwater flow velocities
10. Propose various mitigation approaches to improve the conventional pump-and-treat, including, but are not limited to, chemical enhancement, horizontal wells, adjusted pumping, and induced fractures
11. Apply the concepts and principles of green and sustainable remediation by proposing best management practices for conventional pump-and-treat systems

Pump-and-treat (P&T) systems have been frequently used in the cleanup of most soil and groundwater contamination sites. In the United States, about three-quarters of the Superfund sites have been using conventional P&T or one of its altered forms. The purpose of this chapter is to introduce the appropriate uses of conventional P&T, the general design considerations in both pumping and subsequent aboveground treatments of extracted contaminated groundwater, the limitations of pumping extraction, and briefly some of the variations of the conventional P&T. After reading through this chapter, readers are expected to understand what an appropriate P&T is designed for, when P&T works or does not work, and what are the innovative ways to improve P&T systems. The descriptions and discussions are geared more toward organic contaminants, but the basic consideration of principles and operations should apply to the remediation of inorganic contaminants in soil and groundwater.

Soil and Groundwater Remediation: Fundamentals, Practices, and Sustainability, First Edition. Chunlong Zhang.

Companion website: www.wiley.com/go/Zhang/Remediation_1e

7.1 General Applications of Conventional Pump-and-Treat

Pump-and-treat involves pumping contaminated groundwater to the surface, subsequently treating the contaminated groundwater aboveground, and in many cases returning the treated water to the subsurface by reinjections. In a broad sense, pump-and-treat refers to any system where the withdrawal from or injection into groundwater is part of a remediation system. The pump-and-treat described in this chapter refers to the conventional pump-and-treat systems. These conventional P&T systems were started in the late 1970s to the early 1980s in the United States for Superfund site remediation. Such P&T systems were considered to be the only remedy for the cleanup of contaminated groundwater. It was only after about a decade of its operations that remediation engineers started to realize its applications and limitations (Travis and Doty 1990). Nevertheless, P&T remains an essential component of aquifer restoration. Based on decades of experience, we now know that the general applications of conventional P&T fall into the following two categories:

- *Contaminant Removal via Extraction:* Pumping can be efficient during the initial stage of remediation for the removal of "dissolved phase" contaminants through a recovery well or series of recovery wells. Pumping can also remove free-phase LNAPLs when a pool of LNAPLs can be identified accurately through site characterization.
- *Hydraulic Containment:* Pumping (including injection in a broad sense) can be used successfully for the "containment" of contaminants (hydraulic barrier) to control the plume migration through the use of extraction and injection wells.

In the following discussions of this section, we will expand the above concepts regarding the appropriate use of pump-and-treat for contaminant removal and hydraulic containment. Although contaminant removal via extraction and hydraulic containment represent separate goals, oftentimes aquifer restoration efforts are undertaken to achieve a combination of both. To this end, the layout of extraction wells and injections wells designed to optimize pump-and-treat will be presented prior to our quantitative analysis of groundwater capture zone in Section 7.2.

7.1.1 Contaminant Removal versus Hydraulic Containment

Hydraulic Containment is a process involving hydraulically controlling the movement of contaminated groundwater, preventing the continued expansion of the contaminated zone. It can be done via extraction (pumping) well alone, injection well alone, and the combination of both extraction and injection wells (Figure 7.1). Hydraulic containment can also function when it is in conjunction with the physical barrier made of impermeable materials in various configurations. The term **hydraulic barrier** is used interchangeably with hydraulic containment.

Figure 7.1a is a schematic of extraction wells used in pump-and-treat remediation. These **extraction wells** are the primary forms of wells used to "remove" contaminants in a "dissolved" phase, but they serve as a hydraulic containment by controlling the direction of groundwater flow with horizontal or vertical capture zones. The control is accomplished by a *decreased* hydraulic head to which nearby groundwater flows.

The **injection wells** used in hydraulic control (Figure 7.1b) are the pressure ridge systems where treated groundwater or uncontaminated water is injected into the subsurface through a line of injection wells located upgradient or downgradient of a contamination plume. The primary purpose of a pressure ridge is to obtain an *increased* hydraulic gradient and hence the velocity of clean groundwater moving into the plume, thereby increasing the flow to recovery wells for the cleanup of contaminated aquifer. In other words, the pressure ridge serves to

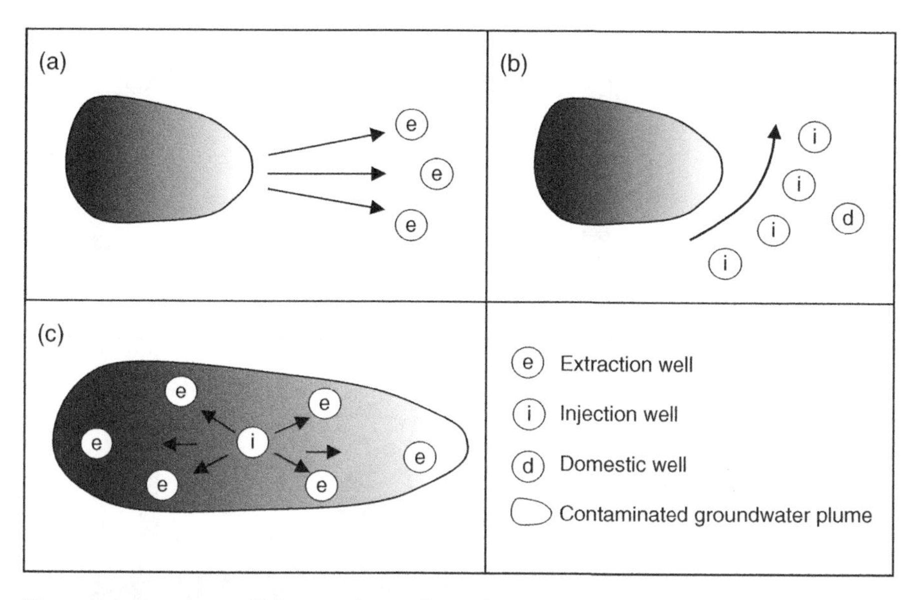

Figure 7.1 Plan views of (a) extraction wells used to remove contaminated groundwater plume in a dissolved phase as well as to direct (contain) contaminated groundwater, (b) injection wells used to divert contaminated groundwater, and (c) a combined extraction well and injection well system used to divert and remove contaminants.

increase pore-volume exchange rates rather than functioning as a barrier. Barrier pressure ridge systems are created by placing injection wells along the perimeter of a contaminant plume. Upgradient pressure ridges also serve to divert the flow of uncontaminated groundwater around the plume, whereas downgradient pressure ridges prevent further expansion of the contaminant plume. Typically, treated groundwater from extraction wells within a contaminant plume supplies the water for injection wells to create a pressure ridge.

Extraction wells and/or injection wells can also be combined with physical barriers. The bird's eye view on two configurations of slurry walls are schematically shown in Figure 7.2a and b. The **physical barriers** are constructed of low-permeability material and serve to keep fresh groundwater from entering a contaminated aquifer zone. They also help prevent existing contaminated areas from moving into an area of clean groundwater or releasing additional contaminants to a dissolved contaminant plume. Most systems involving physical barriers require groundwater extraction to ensure containment by maintaining a hydraulic gradient toward the contained area. The advantage of physical barriers is to reduce the amount of groundwater that must be extracted compared to the amount when using hydraulic controls.

The major types of physical barriers include caps (or covers), grout curtains, sheet piling, and slurry walls. **Caps** are made of low-permeability material at the ground surface. They can be constructed of native soils, clays, synthetic membranes, soil cement, bituminous concrete, or asphalt. Capping prevents or reduces infiltration of rainfall through the contaminated soil. It can be ineffective if water table fluctuates within the zone of contamination or when NAPL vapor is present. **Grout curtains** are created by injecting stabilizing materials (e.g. Portland cement, sodium silicate) under pressure into the subsurface to fill voids, cracks, fissures, or other openings in the subsurface. Grout can also be mixed with soil using large augers. **Sheet piling** cutoff walls are constructed by driving sheet materials, usually steel, through unconsolidated materials with a pile driver or more specialized vibratory drivers. For **slurry trench walls**, soils are first excavated at the proper location and to the desired depth. The resulting

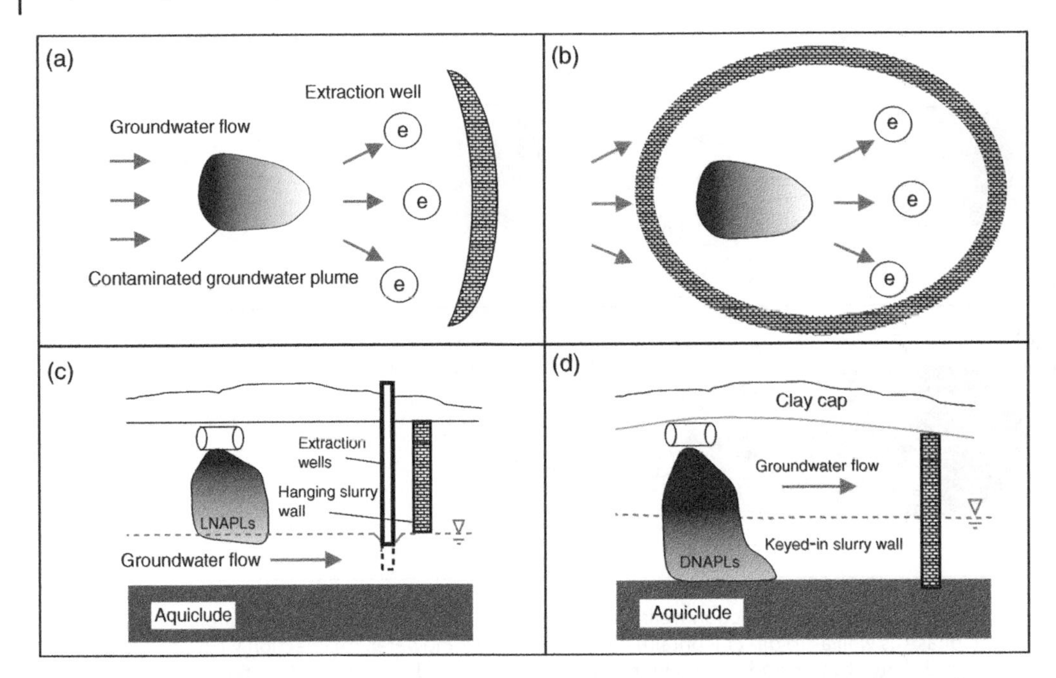

Figure 7.2 A bird's eye view of (a) a slurry wall positioned downgradient of a contaminant plume and extraction wells and (b) a circumferential slurry wall along with extraction wells. A vertical view of two types of slurry walls: (c) a hanging slurry wall for floating LNAPLs and (d) a keyed-in slurry wall for DNAPLs.

trench is filled with a clay slurry, and the trench sidewalls are backfilled with soil bentonite, cement bentonite, or concrete mixture to keep them from collapsing.

Figure 7.2c and d schematically shows the two types of slurry walls: the hanging slurry walls and the keyed-in slurry walls. Depending on whether the contaminant plume is the result of LNAPLs or DNAPLs, slurry walls can be hung above a confined wall, or keyed into the confined layer. The purpose of these various configurations is to block or surround the contaminant source from flowing into the protected usage area of groundwater such as domestic wells.

The above-mentioned containment and isolation methods for contaminated groundwater plumes have the following advantages: (a) The containment, either hydraulic or physical barriers, is a simple and robust technology; (b) Containment typically is inexpensive compared to treatment, especially for large source areas; (c) A well-constructed containment system almost completely eliminates contaminant transport to other areas and thus prevents both direct and indirect exposures; (d) In unconsolidated soils, containment systems substantially reduce the mass flux and the potential for the migration from source; and (e) Containment systems can be combined with an *in situ* treatment to control the migration of either contaminants or reagents.

The primary goal of containment is to prevent further spread of the contaminant plume. Containment is an essential initial step in remediation when contaminant removal is not feasible, as is often the case with DNAPLs residing below the water table. Several inherent limitations associated with containment are as follows: (a) Containment does not reduce source zone mass, concentration, or toxicity unless it is used in combination with treatment technologies; (b) Containment systems such as slurry walls are not strictly impermeable and hence provide containment over a finite period; (c) Field data are limited concerning the long-term integrity of the different types of physical containment systems; and (d) Long-term monitoring is essential to assure that contaminants will not migrate from the containment system.

Example 7.1 is used to show that pumping with extraction wells will reduce the contaminant concentration or mass and its effectiveness depends on the dissolved phase concentration and the pumping rate (*C* and *Q*). On the contrary, containment will not reduce the contaminant mass.

Example 7.1 Dissolved phase contaminant removal: Effects of pumping rates in extraction wells

A single contamination source depicted in Figure 7.1a releases TCE at a constant rate of 15 g/day. The pumping rate of each of the three downgradient extraction wells is 450 L/min. Assume that the contaminated plume is fully captured in the extraction wells, and there are no loss mechanisms (volatilization, abiotic and biotic transformation) from the source zone to the extraction wells. (a) Estimate the TCE concentration in the extraction wells, (b) If the pumping rate is doubled in each well, what would be the TCE concentration to be detected?

Solution

a) The concentration (mass/volume) can be calculated by the release rate in mass/volume divided by the volumetric pumping rate in volume/time as follows:

$$C=\frac{M}{Q}=\frac{15\frac{\mathrm{g}}{\mathrm{day}}}{450\frac{\mathrm{L}}{\mathrm{min}}\times 3}\times\frac{1\,\mathrm{day}}{24\times 60\,\mathrm{min}}=7.72\times 10^{-6}\frac{\mathrm{g}}{\mathrm{L}}=7.76\frac{\mu\mathrm{g}}{\mathrm{L}}$$

b) If the pumping rate is doubled, then the concentration to be detected in the wells is reduced in half. That is:

$$C=\frac{M}{Q}=\frac{15\frac{\mathrm{g}}{\mathrm{day}}}{900\frac{\mathrm{L}}{\mathrm{min}}\times 3}\times\frac{1\,\mathrm{day}}{24\times 60\,\mathrm{min}}=3.86\times 10^{-6}\frac{\mathrm{g}}{\mathrm{L}}=3.86\frac{\mu\mathrm{g}}{\mathrm{L}}$$

This example illustrates that the contaminant concentration in downgradient wells will be increased for wells with low pumping rates. This is because wells with lower pumping rate have less dilution for the contaminants. The practical implication of this demonstrated calculation is that small private water supply wells may more often be at greater risk than large municipal systems pumping hundreds to thousands of liters per minute (Einarson and Mackay 2001). Caution should be exercised in drawing this conclusion, because we have assumed a constant mass release rate. This example also implies that for a site with a slow-releasing contaminant source, merely increasing the pumping rate will not necessarily improve the total removal of contaminant mass.

7.1.2 Schemes of Injection/Extraction Well Placement

The deciding factors for the effectiveness of extraction and containment of groundwater contaminants are the number of injection and extraction wells, their placement locations, screen/open interval depth, and pumping rates. Computer models have been available to optimize the well patterns. Although the optimum injection/extraction well schemes depend on site-specific conditions, objectives, and constraints, some generalizations can be made from some of the modeling studies.

The first modeling analysis simulated three alternative pumping strategies for an idealized site with a uniform medium, linear equilibrium sorption, a single nondegrading contaminant, and a continuing release (Figure 7.3). The three plume management strategies included: (a) downgradient pumping, (b) source control with downgradient pumping, and (c) source control with a mid-plume and downgradient pumping (Faust et al. 1993; USEPA 1997). As shown, the downgradient pumping by itself allows and increases the movement of highly contaminated groundwater throughout the flow path between the release area and the downgradient recovery well. This alternative results in an expansion of the highly contaminated plume and makes it more difficult to achieve cleanup. The importance of source control is clearly demonstrated by comparing the management alternatives. **Source control** through pumping prevents a continued off-site migration, thereby facilitating the cleanup of downgradient contaminated groundwater. The combined source control, mid-plume, and downgradient pumping alternative reduces the flow path and travel time of contaminants to extraction wells and diminishes the impact of processes which cause tailing. As such, with a more aggressive P&T, cleanup is achieved more quickly and the volume of groundwater that must be pumped for cleanup is less than that for the other alternatives.

The second modeling study compared the effectiveness of seven injection/extraction well schemes at removing a contaminant plume (Satkin and Bedient 1988). At first, it appears difficult to tell the relative effectiveness, but the contaminant transport model made it possible to assess the performance of each scheme under eight different hydrogeologic conditions by varying maximum drawdown (high > 10 ft, low < 5 ft), longitudinal dispersivity (high = 30 ft, low = 10 ft), and regional hydraulic gradient (high 0.008, low 0.0008) (Figure 7.4). The effectiveness was evaluated on the basis of simulated cleanup, flushing rate, and the volume of water requiring treatment. Findings of this study demonstrated the following: (i) Multiple extraction wells located along the plume axis (the center line scheme) reduce the cleanup time by shortening contaminant travel paths and allowing higher pumping rates; (ii) The doublet, three-spot, and double-cell schemes were effective under low hydraulic gradient conditions, but require on-site treatment and reinjection; (iii) The three-spot pattern outperformed the other schemes for simulations incorporating a high regional hydraulic gradient; (iv) The centerline well pattern was the most effective in achieving up to 99% contaminant reduction under both low- and high-gradient conditions, but it may present a water disposal problem. (v) The five-spot pattern with an extraction well in the center was the least effective scheme for the cleanup.

The hydraulic head and groundwater flow velocity of a five-spot pumping scheme (in this case, an injection well in the center and four adjacent extraction wells) were delineated by model simulation (Figure 7.5) (USEPA 1994, 1996). The stagnation zones in this pumping scheme are clearly visible, which should be minimized during pump-and-treat operation.

7.2 Design of Pump-and-Treat Systems

From the examples illustrated in the preceding section, we know that the number and location of wells are critical in P&T systems. In this section, we introduce the governing equations for the engineering design of pumping wells under a very simplified aquifer condition. Subsequently, we introduce the design of two most commonly used aboveground treatments: air stripping and activated carbon processes.

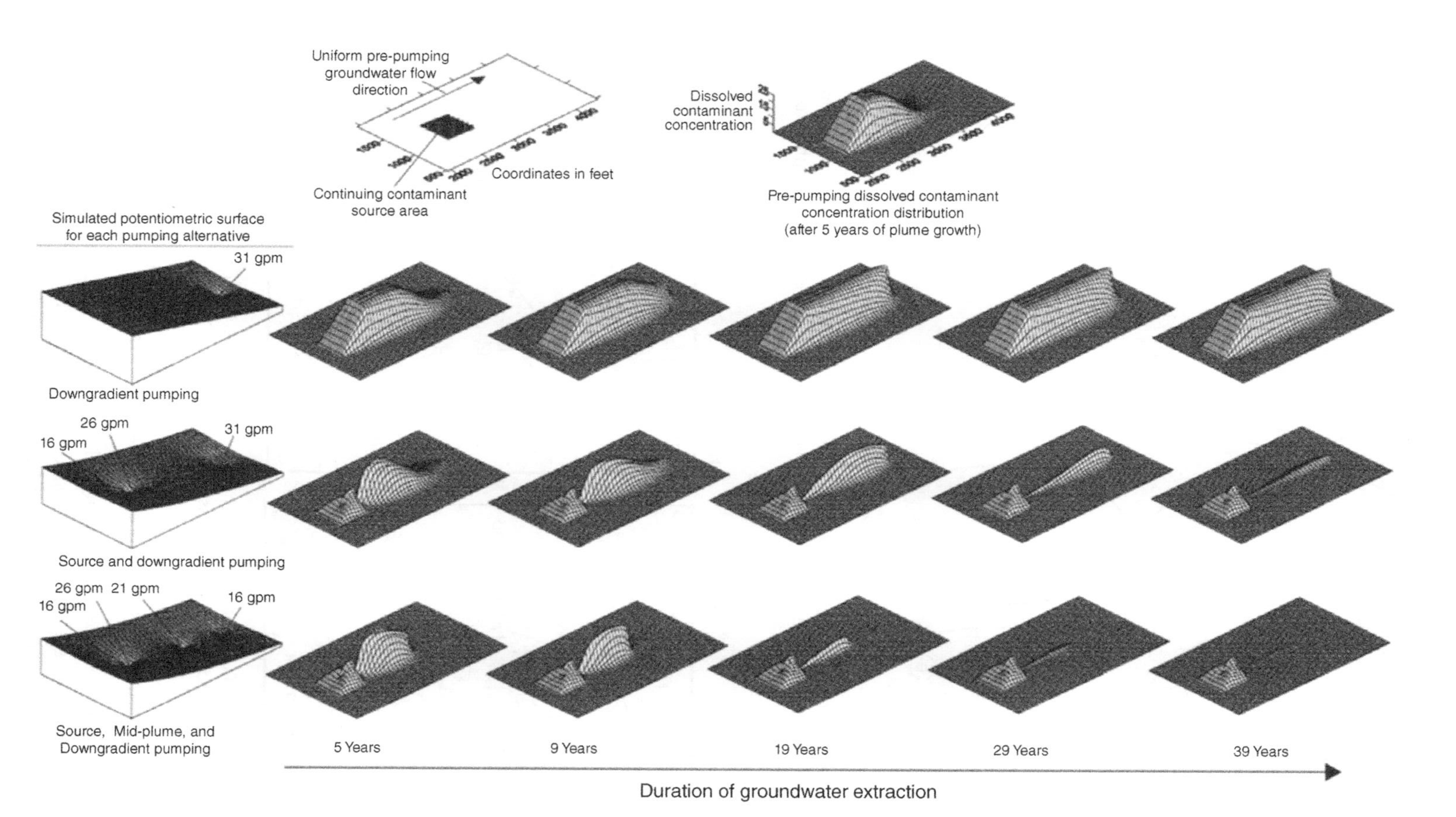

Figure 7.3 Simulation results of three P&T alternatives for an idealized site showing dissolved contaminant concentrations with time of pumping. Assume uniform media, linear equilibrium sorption, and a single nondegrading contaminant (USEPA 1997).

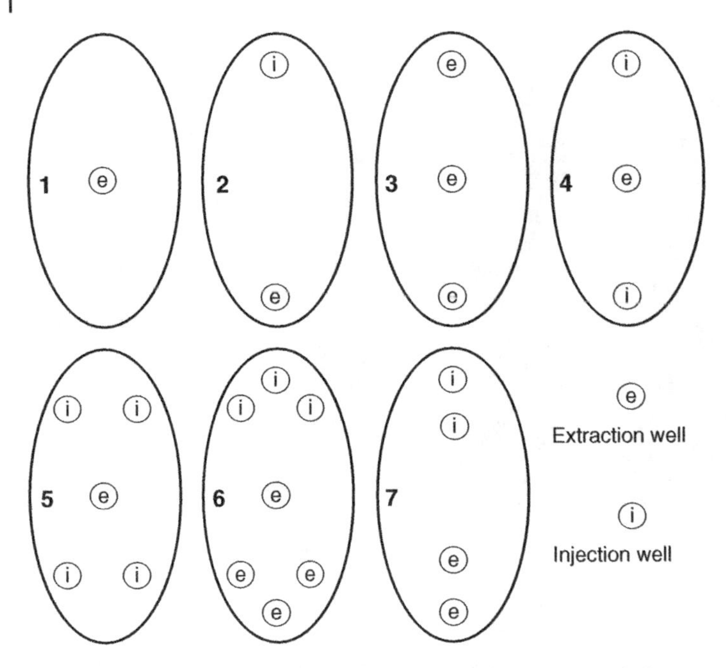

Figure 7.4 Seven well schemes for a comparison of contaminated plume removal: 1 = single, 2 = doublet, 3 = centerline, 4 = three-spot, 5 = five-spot, 6 = double triangle, 7 = double cell. *Source:* adapted from Satkin and Bedient (1988).

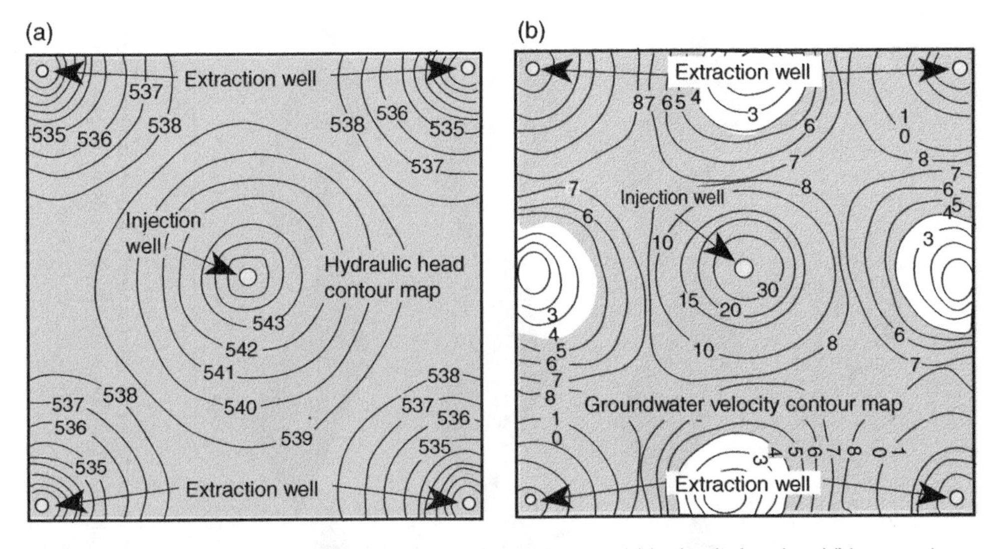

Figure 7.5 Flow pattern of a five-spot pumping scheme showing (a) hydraulic head and (b) stagnation zones (white area) where the groundwater velocity is less than 4 L/T (USEPA 1996).

7.2.1 Capture Zone Analysis of Pump-and-Treat Optimization

The concept of capture zone was introduced in Chapter 3 (Section 3.3.4). It is essential to continue our discussion here because the size of capture zone is central to the design and optimization of contaminant extraction and containment using P&T. Numeric models for groundwater flow, such as MODFLOW, MODPATH, MOC, MT3D, are readily available for remedial engineers in the real-world design of well network under complex hydrogeological conditions. However, it is more intuitive to illustrate the concept with some simple examples. The capture-zone type curves developed by Javandel and Tsang (1986) is a simple graphical method used to determine minimum pumping rates and well spacing needed to capture contaminated groundwater. This method assumed one, two, or three pumping wells along a line perpendicular to the regional groundwater flow direction in a confined aquifer. In referring to Figure 7.6, the general equation defining the **capture zone** under simplified conditions of an ideal aquifer (i.e. homogeneous, isotropic, uniform in cross section, and infinite in width) is as follows (Javandel and Tsang 1986):

$$y = \pm\frac{Q}{2bv} - \frac{Q}{2\pi bv}\tan^{-1}\frac{y}{x} \qquad \text{Eq. (7.1)}$$

where x is the coordinate in the flow direction (the distance from the well), y is the coordinate in the direction along the width of the aquifer that is perpendicular to the groundwater flow, Q is the pumping rate from the well (m^3/day), b is the aquifer thickness (m), v is the Darcy velocity (m/day), and $\tan^{-1}$ is arctangent function (i.e. the inverse of tangent).

The above equation can be rewritten in terms of an angle ϕ (radians) drawn from the original to the x, y coordinates on the capture-zone curve (Figure 7.6).

$$\tan\phi = \frac{y}{x} \qquad \text{Eq. (7.2)}$$

Equation (7.1) can then be rearranged to the following for $0 \le \phi \le 2\pi$:

$$y = \frac{Q}{2bv}\left(1 - \frac{\phi}{\pi}\right) \qquad \text{Eq. (7.3)}$$

We now can use Eq. (7.3) to deduce some important features of the capture zone. First, the **width of capture zone** ($2Y_{max}$) is the y-value when x approaches infinity, i.e. the angle $\phi = 0$:

$$Y_{max} = \frac{Q}{2bv} \qquad \text{Eq. (7.4)}$$

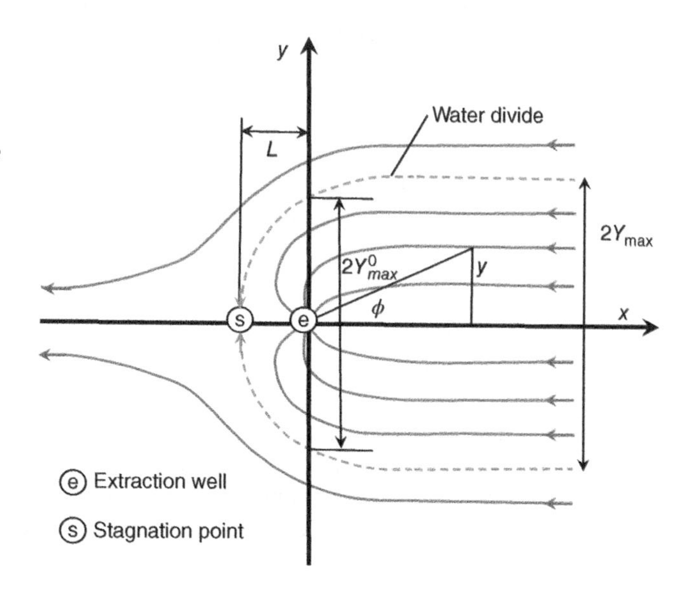

Figure 7.6 Width of capture zone ($2Y_{max} = Q/bv$) and stagnation point ($L = -Q/2\pi bv$). The dashed flow lines that separate the water flowing to the well and the water flowing around the well are called separating streamlines or water divide lines. The area in between the water divide is the capture zone of the well.

Referring to Figure 7.6, the capture zone width ($2Y_{max}$) will be *twice* the y-value in Eq. (7.3) (i.e. $2Y_{max}$ vs. Y_{max}). The above equation indicates that the capture zone width increases with increasing pumping rates, but decreases with the increasing product of Darcy's velocity and aquifer thickness. In other words, it requires higher pumping rates to capture the same area of contaminant plume when Darcy's velocity is higher. Since the amount of water that can be pumped is limited by the drawdown, the pumping rate Q will reach a maximum limit. This also clearly implies that the width of the capture zone will be limited by this maximum pumping rate. In order to increase the capture zone width, multiple wells must be lined up.

Second, one can obtain the width of capture zone along the y-axis at the line of pumping wells when $x = 0$ (i.e. y-axis at the well location, Y^0_{max}) using Eq. (7.3). When $x = 0$, the angle $\phi = \pi/2$, then:

$$Y^0_{max} = \frac{Q}{2bv}\left(1 - \frac{\frac{\pi}{2}}{\pi}\right) = \frac{Q}{4bv} \qquad \text{Eq. (7.5)}$$

Thus the width of the capture zone at the well location along the y-axis ($2Y^0_{max}$) is only half as broad as it is far from the well ($2Y_{max}$). Again, the y-value must be multiplied by 2 to calculate the width of the capture zone at $x = 0$ (i.e. when $x = 0$, the width of capture zone $= 2 \times Q/4bv = Q/2bv$).

Third, a point of special interest is the **stagnation point** located downgradient from the well at a distance (L) on the x-axis ($y = 0$). If we use the minus sign to denote the x-axis downgradient from the well, the stagnation point can be derived as the following:

$$L = -\frac{Q}{2\pi bv} \qquad \text{Eq. (7.6)}$$

It is important to note that multiple extraction wells are needed if the contaminant plume is wider than the capture zone developed by the maximum pumping rate. One concern with multiple extraction wells is that their capture zones must overlap, otherwise water can bypass between the wells without being captured. This minimum required distance between two adjacent wells is the **optimum well spacing** (d), which can be calculated as:

$$d = \frac{Q}{\pi bv} \qquad \text{Eq. (7.7)}$$

Analogous to Eq. (7.3) for a single well, a general equation for the positive half of the capture-zone type curve for n optimally spaced wells arranged symmetrically along the y-axis is as follows:

$$y = \frac{Q}{2bv}\left(n - \frac{\sum_{i=1}^{i=n}\phi_i}{\pi}\right) \qquad \text{Eq. (7.8)}$$

The widths of capture zone for $n = 1$, 2, and 3 and the optimum well spacing for $n = 2$ and 3 are summarized in Table 7.1. Finding the optimum well spacing between two adjacent wells when n is large becomes quite cumbersome, and this is not listed in Table 7.1. For $n = 4$, this value is approximately equal to $1.2Q/\pi bv$. Example 7.2 illustrates the use of some of the equations described above.

Table 7.1 Characteristic features of capture-zone curves.

Number of wells	$2Y_{max}$	$2Y^0_{max}$	D
$n=1$	Q/bv	$Q/2bv$	—
$n=2$	$2Q/bv$	Q/bv	$Q/\pi bv$
$n=3$	$3Q/bv$	$3Q/2bv$	$\sqrt[2]{2}Q/\pi bv$
$n=4$	$4Q/bv$	$2Q/bv$	$1.2Q/\pi bv$
$n=n$	nQ/bv	$nQ/2bv$	—

$2Y_{max}$ = Width of capture zone (x → ∞): Distance between streamlines far upstream from the wells; $2Y^0_{max}$ = Width of capture zone (x → 0): Distance between streamlines at the line of wells; and D = Optimum well spacing between each pair of pumping wells.

Example 7.2 Calculation of capture zone and well spacing

A family-owned food processing and package plant is located in a fruit farm of a small town. It has been an influential fruit jelly producer in the area, but only until recently it has received some bad press for its use of potentially contaminated groundwater. The owner suspects the leaking of two nearby gasoline stations and wants to file a law suit against the owners of these two gasoline stations. The jelly plant owner has contacted you as an environmental geologist/hydrogeologist for some preliminary survey. The data available to you are as follows: The jelly plant pumps groundwater at 50 gallons per minute (gpm). The farm area is situated in a homogeneous and isotropic aquifer in an infinite flow domain. The aquifer has a thickness of 10 m, hydraulic conductivity of 3.5×10^{-4} m/s, and hydraulic gradient of 0.003 in the flow direction of east to west. Gasoline station A is 1500 m east and 50 m north, whereas gasoline station B is 200 m east and 200 m south. What would be your professional findings?

Solution

We first calculate the Darcy's velocity using Darcy's law:

$$v=-K\frac{dh}{dl}=-\left(3.5\times10^{-4}\,\frac{\text{m}}{\text{s}}\right)\times(-0.003)=1.05\times10^{-6}\,\frac{\text{m}}{\text{s}}=9.07\,\frac{\text{cm}}{\text{day}}$$

Now we can use the formulas listed in Table 7.1 for a single well ($n=1$) to calculate the y-value in the y-axis at points $x=0$ and $x=$ infinity, respectively:

$$2Y^0_{max}=\frac{Q}{2bv}=\frac{50\,\frac{\text{gal}}{\text{min}}\times\frac{1\,\text{min}}{60\,\text{s}}\times\frac{1\,\text{m}^3}{264\,\text{gal}}}{2\times10\,\text{m}\times1.05\times10^{-6}\,\frac{\text{m}}{\text{s}}}=\frac{0.003157\,\frac{\text{m}^3}{\text{s}}}{0.000021\,\frac{\text{m}^2}{\text{s}}}=150\,\text{m}$$

$$2Y_{max}=\frac{Q}{bv}=300\,\text{m}$$

These values are labeled in Figure 7.7, in relation to the locations of the well of this fruit farm and two gasoline stations.

Since the y-value is half of the capture width, the dividing line would pass $y=\pm75$ m at $x=0$ and $y=\pm150$ m at $x=\infty$. If one sketches the capture-zone curves and the locations of the two stations

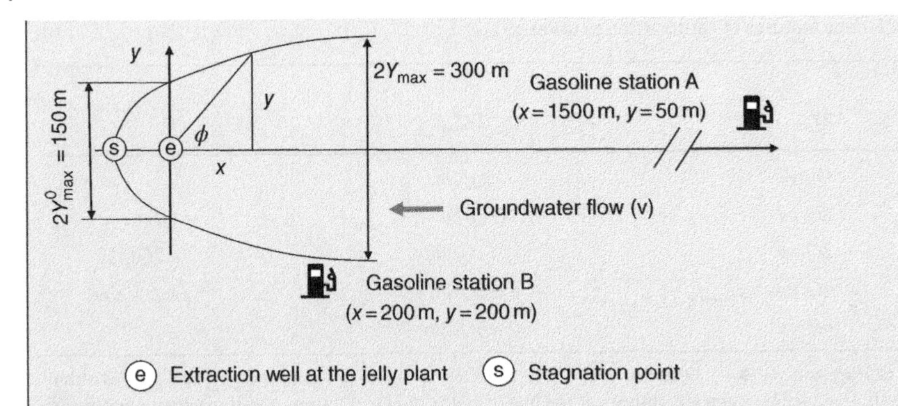

Figure 7.7 Schematic of capture zone in relation to the locations of a local jelly plant and two gasoline stations (not to scale).

relative to the well (Figure 7.7), i.e. Station A ($x = 1500$ m, $y = 50$ m), Station B ($x = 200$ m, $y = -200$ m), and the well ($x = 0$, $y = 0$), then it becomes clear that only Station A is within the dividing line of the capture zone. Station A is much further away from the farm (1.5 km east) than Station B (200 m east), but it is closer to the center line of the well and flow direction than Station B. It is very likely that only Station A is potentially responsible for the contaminated groundwater in the farm.

7.2.2 Aboveground Treatment of Contaminated Groundwater

After contaminated groundwater has been extracted by extraction wells, it is brought to the surface for aboveground treatment. The aboveground treatment is relatively straightforward, since the designs of such processes are well-established in the water and wastewater treatment industries (e.g. AWWA 2011; Tchobanoglous et al. 2014). Here we briefly present a list of aboveground treatment options applicable to the treatment of various contaminants in groundwater, and then focus on the design equations and calculations of two commonly used treatment technologies. These two common technologies are air stripping and activated carbon processes.

7.2.2.1 General Treatment Technologies Available

Table 7.2 summarizes the applicability of various treatment technologies to groundwater contaminated by a broad category of inorganic and organic contaminants. For example, most heavy metals (e.g. Cu, Zn, Pb, Cd, and Ni) in groundwater can be treated by coagulation/precipitation, ion exchange, electrochemical, and filtration. Since hexavalent chromium, arsenic, and mercury have their unique physicochemical properties, the applicable technologies might be different. VOCs and SVOCs have different applicable treatment technologies. For example, while VOCs can be treated by air/steam stripping, activated carbon, and chemical oxidation, SVOCs can be treated by activated carbon, steam stripping, UV/ozone, or chemical oxidation. Certain organic compounds such as ketones, pesticides, and PCBs can be removed through various physical, chemical, or biological mechanisms.

Since air stripping and activated carbon are the two predominant processes employed in treating organic compounds, we will further illustrate these processes by describing their governing equations in the engineering design. Readers should, however, note that packed air stripper and activated carbon towers are commercially available. An important

Table 7.2 Applicability of treatment technologies to contaminated groundwater.

Contaminants	Precipitation	Coprecipitation/ coagulation	UV / Ozone	Chemical oxidation	Reduction	Air stripping	Steam stripping	Activated carbon	Gravity separation	Flotation	Membrane separation[a]	Ion exchange	Filtration	Biological	Electrochemical
Heavy metals	●	●	×	×	○	×	×	○	●	×	●	●	●	×	●
Cr(VI)	●	×	×	×	●	×	×	○	×	×	○	●	×	×	●
Arsenic	○	●	○	○	×	×	×	○	○	×	●	●	●	×	×
Mercury	●	●	×	×	●	×	×	●	○	×	○	●	●	×	×
Cyanide	×	×	●	●	×	×	×	×	×	×	●	●	×	○	○
VOCs	×	×	○	●	×	●	●	●	×	×	○	○	×	○	×
SVOCs	○	○	●	●	×	×	●	●	○	○	●	●	×	●	×
Ketones	×	×	○	●	×	●	●	×	×	×	×	×	×	●	×
Pesticides	○	○	●	●	×	×	○	●	○	○	●	●	●	○	×
PCBs	●	●	●	●	×	×	×	●	●	●	●	●	●	○	×
Oil & Grease	●	●	×	×	×	×	×	×	●	●	●	●	○	○	×

Source: Adapted from the USEPA (1996).

[a] Technology includes several processes such as reverse osmosis and ultrafiltration.

● Applicable; ○ Probably applicable; × Not applicable.

consideration will be the adequacy of treatment capacity of the selected technologies for groundwater applications.

7.2.2.2 Design Considerations for Air Stripping

Air strippers are widely used to remove most VOCs, and, to a limited extent, some SVOCs from extracted groundwater. VOCs typically have high vapor pressure and Henry's law constant; however, certain VOCs have high vapor pressure but low Henry's constant (see Section 2.3.1). These VOCs have low removal efficiencies by air stripping, including acetone, *t*-butyl alcohol (TBA), MTBE, naphthalene, 1,2-dichloroethane, tetrachloroethane (1,1,1,2 or 1,1,2,2), 2-butanone, methyl isobutyl ketone, and 1,1,2-trichloroethane. Air stripping systems have been in use and reliably treating groundwater for decades with two common configurations: air stripper packed towers and tray strippers. Because air stripping transfers contaminants from water to air, the air stripper off-gas may need further treatment such as catalytic oxidation or vapor-phase carbon adsorption.

In a typical air stripping process, groundwater containing VOCs is countercurrently contacted with air in a packed tower (Figure 7.8). The packed tower is so named because it is filled with packing materials for an increased contact area for mass transfer. The flow of water and VOCs stream is countercurrent, because clean water enters at the top, flows down by gravity, and VOCs-contaminated water is fed upward along with the airflow from the blower at the bottom. The VOCs are transferred to the gas phase during the intimate gas–liquid contact. The equilibrium distribution between gas and water phases can be described by Henry's law (Eqs. 2.3 and 2.4, Chapter 2). The VOCs from the exhaust must be treated by passing the

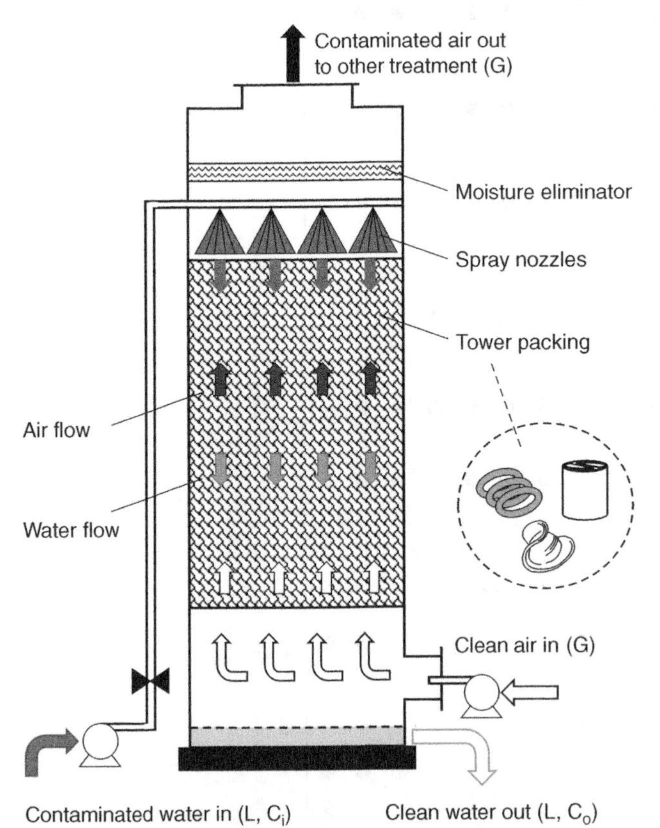

Figure 7.8 Design of an air stripping tower for the treatment of contaminated groundwater.

VOC-laden air in an activated carbon column. The major design parameters affecting an air stripper performance are the Henry's law constant (H), the liquid loading rate (L), and the gas-to-liquid ratio (G/L).

The major design equation for the required height of air stripper tower is (AWWA 2011):

$$z = \frac{L}{K_{La}} \frac{R}{R-1} \ln \left[\frac{\frac{C_i}{C_o}(R-1)+1}{R} \right] \quad \text{Eq. (7.9)}$$

where z = packing height (m), L = liquid loading (m^3/m^2-h), K_{La} = overall mass transfer coefficient (h^{-1}) which is a function of Henry's constant and the individual mass transfer coefficient of the VOC in both liquid and gas phases, C_i = concentration of VOC in the influent, C_o = concentration of VOC in the effluent, and R = stripping factor (unitless), which is defined as:

$$R = \frac{HG}{L} \quad \text{Eq. (7.10)}$$

where H = dimensionless Henry's law constant for the VOC and G = airflow (m^3/m^2-h). The stripping factor is crucial in determining the ability of an air stripper to remove a specific contaminant. Since the operating G/L ratio is based upon the volumetric gas and liquid phase loadings in the stripper, Eq. (7.10) indicates that contaminants with lower Henry's constants will require higher air-to-water ratios to achieve successful removal rates. In theory (Eq. 7.9), a stripping factor R = 1 requires an infinite tower height (z) to attain 100% VOC removal. Thus, in practices the stripping factor should be significantly greater than unity to approach a complete VOC removal with a reasonable tower height. The value of Henry's constant of a VOC can then determine whether a VOC can be stripped or not with a practical G/L ratio. It is generally considered that air strippers are effective for volatile chemicals with unitless Henry's law constant greater than 0.01. The removal efficiency of VOCs can be 95–99% or higher in a well-designed air stripper. Air stripping is particularly useful when treating a large quantity of low concentration of VOCs in the waste streams. For a gas containing the lower H or higher concentration (>100 mg/L), steam stripping may be employed (Davis and Cornwell 2013). Example 7.3 illustrates the use of the design equations described previously.

Example 7.3 Design of a packed air stripping tower

A well water is contaminated with 500 µg/L perchloroethylene. The water must be treated to the maximum contaminant level (MCL) of 5 µg/L using a packed air stripper. What is the required height for the following conditions? L=80 m^3 water/m^2 tower cross section-hour, G=2000 m^3 air/m^2 tower cross section-hour, unitless H=0.34 (20 °C), and K_{La} based on a laboratory study = 50 h^{-1}.

Solution

We first apply Eq. (7.10) to calculate the stripping factor:

$$R = \frac{HG}{L} = \frac{0.34 \times 2000}{80} = 8.5(\text{unitless})$$

Now we use Eq. (7.9) to estimate the height of air stripping tower:

$$z = \frac{L}{K_{La}} \frac{R}{R-1} \ln \left[\frac{\frac{C_i}{C_o}\left(R-1\right)+1}{R} \right] = \frac{80}{50} \times \frac{8.5}{8.5-1} \ln \left[\frac{\frac{500}{5}\left(8.5-1\right)+1}{8.5} \right] = 8.1\,\text{m}$$

When using Eq. (7.9), consistent units must be used as illustrated in this example.

7.2.2.3 Design Considerations for Activated Carbon

Carbon adsorption is very effective for the removal of SVOCs and some VOCs of hydrophobic nature in both liquid and gaseous streams. SVOCs that can be effectively removed include chlorinated solvents, total petroleum hydrocarbons (TPH), PAHs, and PCBs. Activated carbon process is not appropriate for some hydrophilic VOCs with low molecular weights such as methanol, ethanol, and acetone. Activated carbon is capable of adsorbing some metals (Cu, Zn, Pb, Cr, and Ni) to a limited extent (typically with a removal efficiency of <90%). There are two types of activated carbons in use: **granular activated carbon (GAC)** having a diameter of 0.2–2.4 mm and **pulverized activated carbon (PAC)** having a diameter less than 0.44–0.074 mm (200 mesh). GAC is usually packed in fixed bed columns, and PAC is usually mixed into water and then recycled or separated. PAC is less commonly used in groundwater treatment. Spent carbon may be landfilled or regenerated by thermal treatment to remove and destroy organic contaminants.

The design variables for adsorption include how much sorbing materials are needed (lb/day) and the sizing of the adsorption unit. The starting equation is one of the adsorption isotherms (linear, Langmuir, and Freundlich) describing the relationship between sorbed phase concentration associated with the activated carbon and the liquid (or air) phase concentrations at equilibrium. Common steps for the design of activated carbon process are as follows:

First, calculate the amount of chemical to be removed (X) in lb/min, which is dependent on the amount of water treated (Q, gpm) and the concentration (mg/L) of contaminant in the groundwater.

$$X = f \times Q \times \left(C_i - C_o\right) \qquad \text{Eq. (7.11)}$$

where C_i and C_o are, respectively, the influent and effluent concentrations in mg/L, f is the unit conversion factor ($f = 8.34 \times 10^{-6}$) when the units of Q and C are, respectively, gpm and mg/L and X is reported as lb/min $\left(\frac{\text{gal}}{\text{min}} \times \frac{\text{mg}}{\text{L}} \times \frac{3.785\,\text{L}}{1\,\text{gal}} \times \frac{1\,\text{lb}}{454\,000\,\text{mg}} = 8.34 \times 10^{-6}\,\frac{\text{lb}}{\text{min}}\right)$. This unit conversion factor is the same as the commonly used unit conversion factor of 8.34 when Q is expressed in million gallons per day (MGD) in wastewater treatment plants. For a **linear isotherm**:

$$\frac{X}{M} = K C_o \qquad \text{Eq. (7.12)}$$

where X/M = the mass of contaminant adsorbed per unit mass of carbon (mg/g), K = partitioning coefficient (L/kg) that can be measured in a lab test. Equation (7.12) is virtually the same as Eq. (2.9) in Chapter 2 used to describe the sorption to soil, where K_d is the sorption coefficient, and C_s is used to denote the adsorbed phase concentration in soil ($C_s = X/M$). If we use the non-linear **Freundlich isotherm**:

$$\frac{X}{M} = K C_o^{1/n} \qquad \text{Eq. (7.13)}$$

Table 7.3 The *K* and 1/*n* parameters in the Freundlich isotherm.

Contaminant	Toluene	Chlorobenzene	Lindane	PCE	TCE	Methylene chloride
$K\,(\text{mg/g})(\text{L/mg})^{1/n}$	100	100	285	51	28	1.3
$1/n$	0.45	0.35	0.43	0.56	0.62	1.16

Parameters from USEPA (1980), EPA-600/8-80-023.

where the partitioning coefficient *K* has a unit of $(\text{mg/g})\times(\text{L/mg})^{1/n}$, $1/n$ = experimentally measured constant to fit the Freundlich isotherm (see Table 7.3 for the reported values for several common contaminants). Note that C_o is used because it is the equilibrium concentration of the adsorbent (activated carbon). The **daily carbon usage** (*M*, lb/day) based on isotherm data can be calculated as follows:

$$M\left(\frac{\text{lb}}{\text{day}}\right)=\frac{X\left(\dfrac{\text{lb}}{\text{day}}\right)}{\left(\dfrac{X}{M}\right)}=\frac{X\left(\dfrac{\text{lb}}{\text{day}}\right)}{K\times C_o^{1/n}} \quad \text{Eq. (7.14)}$$

As will be shown in Example 7.3, consistent units must be used to calculate the daily carbon usage.

The next design variable is the sizing of GAC unit, which should account for the **empty bed contact time** (EBCT) between the process water and the GAC as well as convenient GAC changeout schedule. EBCT is a measure of the time during which water to be treated is in contact with the treatment medium in a contact vessel, assuming that all liquid passes through the vessel at the same velocity. EBCT is equal to the volume of the empty bed (L^3) divided by the flow rate (Q in L^3/T). The EBCT should generally be between 15 and 30 minutes per vessel. The approximate **vessel size** expressed in the mass of GAC can then be calculated as follows:

$$\text{Vessel size}=\text{EBCT}(\text{min})\times Q\left(\frac{\text{Gal}}{\text{min}}\right)\times\frac{1\,\text{ft}^3}{7.48\,\text{gal}}\times\frac{30\,\text{lb of GAC}}{1\,\text{ft}^3} \quad \text{Eq. (7.15)}$$

Finally, the breakthrough time (i.e. the time when carbon should be replaced) should be considered. Ideally, the vessel should be sized to allow changeouts to occur on a quarterly or semi-annual basis when other site activities are conducted. Example 7.4 illustrates the design calculations using Eqs. (7.11) through (7.15).

Example 7.4 Design of an activated carbon system

Design the GAC unit in treating 15 gpm groundwater contaminated with 10 mg/L of toluene. The maximum concentration limit (MCL) for toluene in drinking water is 1 mg/L.

Solution

We use MCL as the effluent concentration and substitute it into Eq. (7.11) to calculate the amount of toluene to be removed:

$$X=f\times Q\times(C_i-C_o)=8.34\times10^{-6}\times15\,\text{gpm}\times(10-1)\frac{\text{mg}}{\text{L}}=1.13\times10^{-3}\frac{\text{lb}}{\text{min}}$$
$$=1.62\frac{\text{lb}}{\text{day}}$$

Next, we use the K and $1/n$ values in Table 7.3 and input their values in Eq. (7.14) to estimate the daily usage of activated carbon:

$$M\left(\frac{\text{lb}}{\text{day}}\right)=\frac{X\left(\frac{\text{lb}}{\text{day}}\right)}{\left(\frac{X}{M}\right)}=\frac{X\left(\frac{\text{lb}}{\text{day}}\right)}{K\times C_0^{1/n}}=\frac{1.62\frac{\text{Lb}}{\text{day}}\times\frac{454\,000\text{ mg}}{1\text{ lb}}}{100\left(\frac{\text{mg}}{\text{g}}\right)\times\left(\frac{\text{L}}{\text{mg}}\right)^{0.45}\times\left(1\frac{\text{mg}}{\text{L}}\right)^{0.45}}=7360\frac{\text{g}}{\text{day}}$$

$$=7360\frac{\text{g}}{\text{day}}\times\frac{1\text{ lb}}{454\text{ g}}=16.2\frac{\text{lb}}{\text{day}}$$

By assuming a 20 minutes of empty bed contact time (EBCT in the range of 15–30 minutes), the vessel size according to Eq. (7.15) can be estimated as:

$$\text{Vessel size}=\text{EBCT}(\text{min})\times Q\left(\frac{\text{Gal}}{\text{min}}\right)\times\frac{1\text{ ft}^3}{7.48\text{ gal}}\times\frac{30\text{ lb of GAC}}{1\text{ ft}^3}$$

$$=20\text{ min}\times 15\frac{\text{Gal}}{\text{min}}\times\frac{1\text{ ft}^3}{7.48\text{ gal}}\times\frac{30\text{ lb of GAC}}{1\text{ ft}^3}=1200\text{ lb GAC}$$

7.3 Pump-and-Treat Limitations and Alterations

We have thus far realized how pump-and-treat is essential to achieve the hydraulic containment of contaminated plumes and/or extraction of contaminated groundwater. In the following discussions, we will examine the limitations of pump-and-treat. That is, what processes are likely to hamper the effectiveness of pump-and-treat, what makes it a long and enduring effort to achieve the cleanup goal in many of the contaminated sites, and what improvement and modification can be made on the conventional pump-and-treat systems?

7.3.1 Residual Saturations of Nonaqueous Phase Liquid

7.3.1.1 Dissolved Contaminant from NAPLs

In Section 7.1, we were cautioned that the conventional pump-and-treat systems extract the contaminant only in its "dissolved" phase. This results in one of the fundamental limitations for the P&T if free phase of NAPLs and residual saturations of NPALs are present in a contaminated aquifer. We will use Example 7.5 to quantitatively illustrate how small the percentage of "dissolved" phase can be. We will subsequently introduce the concept of "residual saturation" at the soil microscopic scale to examine quantitatively how this residual saturation can further limit the success of P&T.

Example 7.5 Dissolved phase versus NAPL phase from a gasoline spill

A leaking underground storage tank released 1000 gallons of gasoline (density ~0.9 g/mL and 1% benzene) to the subsurface. After one year, the resulting dissolved benzene plume is 100 ft long, 50 ft wide, and 10 ft deep. The average dissolved benzene concentration of the plume is 0.10 mg/L, and the porosity of the aquifer is 0.30. Assuming no hydrocarbon is lost through volatilization or biodegradation, (a) estimate the total benzene spilled in kg; (b) estimate the total volume of the contaminated groundwater plume; (c) estimate how much the spilled benzene is in the dissolved phase, and how much is in the NAPL phase; and (d) what does this imply to the groundwater remediation? (adapted from Bedient et al. 1999)

Solution

a) Mass of benzene released in gasoline = volume of benzene × density of benzene = (volume of gasoline × 1%) × density of benzene

$$\text{Mass of benzene} = 1000\,\text{gal} \times 1\% \times 0.9\frac{\text{g}}{\text{mL}} \times \frac{3.78\,\text{L}}{1\,\text{gal}} \times \frac{1000\,\text{mL}}{1\,\text{L}} \times \frac{1\,\text{kg}}{1000\,\text{g}} = 34.02\,\text{kg}$$

b) Volume of contaminated groundwater plume = volume of plume × porosity

$$\text{Volume of contaminated groundwater} = (100 \times 50 \times 10)\,\text{ft}^3 \times \frac{28.3\,\text{L}}{1\,\text{ft}^3} \times 0.3$$
$$= 424\,500\,\text{L}$$

c) Mass of dissolved benzene = volume × concentration

$$\text{Mass of dissolved benzene} = 424\,500\,\text{L} \times 0.1\frac{\text{mg}}{\text{L}} \times \frac{1\,\text{kg}}{10^6\,\text{mg}} = 0.04245\,\text{kg} = 42.45\,\text{g}$$

d) Based on the total mass of benzene in NAPLs and in dissolved phase (NAPL mass = 34.02 kg, dissolved = 0.042 45 kg), the % of benzene in the dissolved phase is only 0.042 45/34.02 = 0.12%. The dissolved benzene concentration of 0.1 mg/L is much lower than the reported benzene solubility of 1780 mg/L.

This example clearly demonstrates that benzene is not in its equilibrium state, and NAPL can act as a long-term source of dissolved contaminants to groundwater. It will take a significant volume of groundwater and hence prolong time to extract benzene from groundwater using a network of extraction wells.

7.3.1.2 Residual Saturation

The presence of contaminants in the form of "residual saturation" is another important factor limiting the effectiveness of pump-and-treat. **Saturation** (S) is the fraction of total pore space containing a particular fluid (air, water, NAPL) in a porous medium. For NAPLs, the saturation is defined as follows:

$$S = \frac{V_{\text{NAPLs}}}{V_{\text{pore space}}} \qquad \text{Eq. (7.16)}$$

Thus, a saturation (S) value of 20% implies that a 20% of the pore volume is filled with NAPLs. When NAPL content is high in an aquifer, it can form a continuous phase (free phase) within the soil pores. This continuous phase is able to freely flow by gravity; therefore, it is pumpable (recoverable by pumping). However, when the content of NAPLs continues to reduce, it reaches to a point that NPALs will form a discontinued phase (Figure 7.9). This is the point at which the residual saturation (S_r) is defined. **Residual saturation** is the fraction of total pore volume occupied by residual NAPLs under ambient groundwater flow conditions. Residual saturation is where a continuous NAPL becomes *discontinuous* and is immobilized by capillary forces and therefore it cannot be pumped (extracted) from the subsurface. Simply put, residual saturation defines the fraction of NAPLs that cannot be recovered from soil by pumping. For instance, if a residual saturation value (S_r) of 10% occurs in a contaminated aquifer, this implies 10% of pore is filled with residual NAPLs and is not amenable to pumping by extraction wells. Typical values of residual saturation are 5–20% in the vadose zone and 15–50% in the saturated zone.

As shown in Figure 7.9, NAPLs tend to form entrapped ganglia in a saturated soil, which are discontinuous blobs of an organic liquid within the center of the larger pores. In vadose zone,

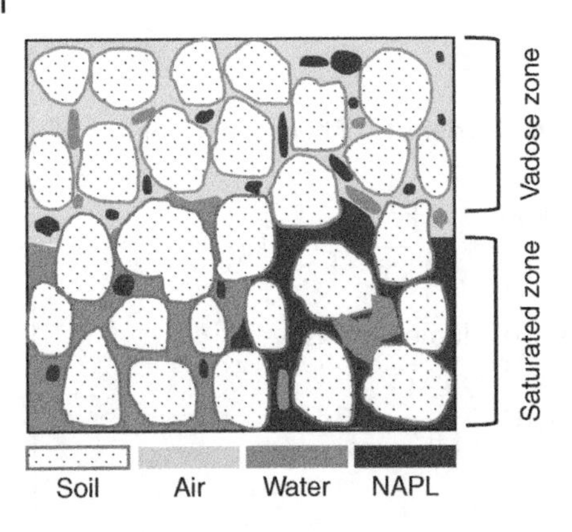

Figure 7.9 A microscopic examination of the distribution of residual NAPL in a saturated zone and NAPL in an unsaturated zone. Note that the discontinuous blobs of an organic liquid are residual saturation, whereas the continuous NAPL phase is a free phase.

water preferentially wets the solid surface and tends to occupy the entire volume of the smallest pores. The center of the largest pores, however, is now occupied by the soil gas in the unsaturated zone. The entrapped organic liquid tends to spread along the water–gas interfaces, forming continuous films, lenses, or wedges. Studies have shown that water is almost always the wetting fluid when mixed with air or NAPLs in the subsurface. NAPLs serve as the wetting fluid when mixed with air, but act as the nonwetting fluid when combined with water in the subsurface (Domenico and Schwartz 1997).

The method for the determination of saturation (S) and residual saturation (S_r) values originated from the petroleum industries, including the simple but less accurate retort method and the more time-consuming yet more accurate Dean–Stark extraction method. For the **retort method**, soil sample is placed in a sealed aluminum cell and heated in a distillation device (called retort) to 100 °C to collect water and then 650 °C to measure the amount of NAPLs. The volume of water and NAPLs in conjunction with total soil pore volume are used to estimate the water saturation, NAPL saturation, and air saturation (see Example 7.6). The **Dean–Stark method**, a standardized ASTM procedure (ASTM D95), uses the vapor of a solvent (toluene) from a heated flask to displace water and NAPLs from soil pores.

Example 7.6 Use of retort method to measure saturation of soil fluids

Following the retort method, the volumes of NAPL (oil) and water recovered were 3.50 and 3.00 mL, respectively. Prior to this experiment, the bulk volume of the test soil sample was measured to be 35.50 mL and the soil grain volume was 26.25 mL. Determine the saturations of water, soil, and DNPALs in this sample.

Solution

The following stepwise procedure is presented.

a) The pore volume of the soil sample is: $V_p = V_b - V_g = 35.50 - 26.25 = 9.25$ mL
b) The porosity of the sample is: $n = 9.25/35.50 = 0.261$ (26.1%)
c) Applying Eq. (7.16), the NAPL (oil) saturation is: $S_{NAPL} = 3.50/9.25 = 0.379$ (37.9%)
d) Similarly, we can calculate the water saturation: $S_{water} = 3.00/9.25 = 0.324$ (32.4%)
e) Since all soil pores must be occupied by fluids in the form of NAPLs, water, or air, the air saturation is: $S_{air} = 1 - S_{NAPL} - S_{water} = 1 - 0.379 - 0.324 = 0.297$ (29.7%)

For practical purpose in environmental remediation, the readily measurable **total petroleum hydrocarbons (TPHs)** in soil can be used to estimate the NAPL saturation. TPHs are extracted by a solvent (fluorocarbon -113) and measured by infrared analysis according to the EPA Method 418.1. The conversion from milligrams of hydrocarbon per kilogram of soil (mg/kg TPH in soil) to liters of oil per liter of soil pore space (% saturation) is as follows:

$$S = \frac{\rho_b \times \text{TPH}}{\rho_n \times n \times 10^6} \qquad \text{Eq. (7.17)}$$

where S = NAPL saturation (unitless), ρ_b = soil bulk density (g/cm^3), ρ_n = NAPL density (g/cm^3), TPHs = total petroleum hydrocarbons (mg hydrocarbon/kg dry soil). In using the above equation with a unit conversion factor of 10^6, the units of bulk density of soil (ρ_b) and the density of NAPLs (ρ_n) are both g/cm^3, and the unit of TPHs as commonly reported in mg/kg can be used directly without an additional unit conversion. Equation (7.17) is applicable when TPH is greater than 5000 mg/kg, otherwise a substantial error may arise if the TPH method used for testing does not fully cover the range of NAPL constituents present (biased low). Example 7.7 shows the use of Eq. (7.17) to calculate saturation from the experimentally measured TPH, and Example 7.8 further elaborates the difference in saturation, residual saturation, and free-phase NAPLs with important remediation implications.

Example 7.7 Estimate the NAPL saturation based on TPHs in soil

A contaminated soil has a measured TPH value of 30 000 mg/kg. It has a bulk density of 1.85 g/cm^3 and a porosity of 0.35. The contamination of the TPHs was organic solvents in a storage tank, mostly BTEX compounds having a density of 0.80 g/cm^3. Estimate NAPL saturation. Neglect a small portion of NAPLs present in a dissolved phase and a sorbed phase.

Solution

By applying Eq. (7.17), we obtain:

$$S = \frac{\rho_b \times \text{TPH}}{\rho_n \times n \times 10^6} = \frac{1.85 \times 30000}{0.80 \times 0.35 \times 10^6} = 0.198(19.8\%)$$

A saturation value of 19.8% implies that 19.8% of the soil pore volume (not the soil volume) is occupied by the spilled BTEX. If a much lower TPH content is detected for the same site, for example, 5000 mg/kg TPHs, then the NAPL saturation can be calculated as $(1.85 \times 5000)/(0.80 \times 0.35 \times 10^6)$ = 3.3%. With this low NAPL saturation, it is most likely that NAPL is present in the residual state and this fraction of NAPL is immobile.

Example 7.8 Calculation of saturation, residual saturation, and free-phase NAPLs following an LNAPL oil spill

In the early fall of 2018, a farm tank truck overturned, resulting in an estimated amount of 550 gallons (4114 ft^3) of diesel (density = 0.85 g/mL) spill in a remote rural location in a small town in western Texas, USA. The accidental spill occurred rapidly within hours onto a sandy loam soil with an estimated surface contaminated area of 20 ft by 20 ft (400 ft^2), and it missed the window of opportunity for the emergency response crew to clean up the free-flowing diesel on the ground surface. Assume all spilled diesel was infiltrated through the vadose soil which sits 7 ft above the groundwater table, and subsequently formed a free-phase NAPL around the groundwater table. For the simplicity of calculation, ignore any evaporation loss of diesel into the air. In addition, we assume a negligible amount of vapor phase diesel in the vadose zone and the dissolved phase diesel in the soil pore water and groundwater plume. Other available data for this spill site include the parameters for vadose zone (porosity, 0.38; air saturation, 0.15; water saturation, 0.20; NAPL residual saturation, 0.10), and the parameters for the saturated zone (porosity, 0.35; residual saturation, 0.32).

a) Estimate the volume of NAPL saturation in the vadose zone.
b) Estimate the volume of NAPL residual saturation in the vadose zone.
c) Estimate the volume of the free-phase NAPL pool sitting on the groundwater table.

Solution

a) The vadose soil pores constitute three phases, i.e. air, water, and NAPLs. We first estimate the volumes of contaminated soil, soil pore, residual water, and residual air in the vadose zone.

V_s (soil) $= 400\,\text{ft}^2 \times 7\,\text{ft} = 2800\,\text{ft}^3$
V_v (void) $= V_s \times \text{porosity} = 2800\,\text{ft}^3 \times 0.38 = 1064\,\text{ft}^3$
V_w (water) $= V_v \times \text{residual water} = 1064\,\text{ft}^3 \times 0.20 = 212.8\,\text{ft}^3$
V_a (air) $= V_v \times \text{residual air} = 1064\,\text{ft}^3 \times 0.15 = 159.6\,\text{ft}^3$

The above calculations provide the volumes of residual water ($212.8\,\text{ft}^3$) and residual air ($159.6\,\text{ft}^3$) contributing to a total of $1064\,\text{ft}^3$ vadose pore volume. The remaining pore volume is occupied by the NAPL phase:

$$V_{NAPL} = 1064 - 212.8 - 159.6 = 691.6\,\text{ft}^3$$

b) The volume of $691.6\,\text{ft}^3$ can be considered as the initial saturation immediately following the spill. If all of this NAPL becomes immobile in the vadose zone, the residual saturation in the vadose zone of 0.65 ($=691.6\,\text{ft}^3/1064\,\text{ft}^3$) is considered to be very high. This high residual saturation is very unlikely considering the downward NAPL flow due to gravity in relation to the capillary forces by soil particles. Since we were given the residual saturation of 0.1 in the vadose zone, the equivalent volume of residual NAPL is: $1064\,\text{ft}^3 \times 0.1 = 106.4\,\text{ft}^3$. This is the volume of residual NAPL or discontinuous NAPL phase in the vadose zone that cannot be recovered by pumping. Since $106.4\,\text{ft}^3$ of NAPL will reside in the vadose zone, a total of $691.6 - 106.4 = 585.2\,\text{ft}^3$ of NAPL will be subsequently dispersed into soil vapor phase or contribute to the free-phase pool above the groundwater table. The fate and transport of this portion of the NAPL in the vadose zone depend on the time frame, the hydraulic conductivity, and other physicochemical parameters (e.g. temperature in relation to vapor pressure).

c) If we further assume no loss mechanism of NAPL in the vadose zone other than the gravity flow in the short time frame, the volume of free-phase NAPL floating on the groundwater table can be estimated by the difference between the total amount of spill and the NAPL retained in the vadose zone, i.e. $4114 - 691.6 = 3422.4\,\text{ft}^3$ (458 gallons).

This calculation reveals that for the above spill scenario, the emergency response crew can still take further action to recover most of the free-phase NAPL before it spreads away from the source downgradient. If we assume a higher residual saturation of 0.32 in the saturated zone, the volume of free-phase diesel that can be removed by pumping at the spill site would be: $3422\,\text{ft}^3 \times (1 — 0.3\ 2) = 2327\,\text{ft}^3$ (311 gallons), which is 57% of the initially spilled diesel. Note that the above example is used to illustrate the concept of saturation and residual saturation. Over-interpretations of the data should be cautioned because we have made numerous assumptions to simplify our calculations.

7.3.2 Tailing and Rebound Problems

The widespread use of pump-and-treat systems during the late 1970s to early 1990s was supposed to clean up many contaminated sites in the United States and around the world. Unfortunately, the inherent complexities of aquifers and chemicals (NAPLs) made what appear to be possible in theory impossible in practice in many of these contaminated sites. Monitoring results of groundwater in P&T sites often revealed "tailing" and "rebound" phenomena. **Tailing** refers to the progressively slower rate of dissolved contaminant concentration decline observed with continued operation of a P&T system (Figure 7.10). The concentration of tailing contaminant may exceed cleanup standards. **Rebound** is the fairly rapid increase in contaminant concentration that can occur after pumping has been discontinued due to a temporary attainment of a cleanup standard. This increase may be followed by stabilization of the contaminant concentration at a somewhat lower level.

The implications of tailing and rebound on remediation efforts are significant because of the longer treatment times and the exceedance of the cleanup standard. The tailing effect significantly increases the time pump-and-treat systems must be operated to achieve groundwater restoration goals. Indeed, monitoring data suggest that pumping may need to be conducted for hundreds of years rather than tens of years for many sites with NAPLs and complex geologic conditions. The rapid decline in the rate of contaminant concentrations corresponds to the initial pumping. When a tailing occurs, the decline in concentration is more gradual, which eventually stabilizes at an apparent residual concentration level above the cleanup standard.

Several chemical and hydrodynamic factors contribute to tailing and rebound, including: (a) the slow *dissolution* of NAPLs in their residual saturation or free phase; (b) the slow *desorption* of sorbed organic contaminants or dissolution of inorganic precipitates from porous media to the groundwater; (c) the slow *diffusion* of contaminants trapped in low-permeability matrix regions inaccessible to the flowing groundwater; and (d) variable groundwater flow velocities and different flow paths taken by contaminants to extraction wells. Because of their importance in determining the ultimate effectiveness of pump-and-treat systems, these four factors are further elaborated as follows.

7.3.2.1 Slow NAPL Dissolution

Although NAPLs are relatively insoluble in water, they are often sufficiently soluble to cause concentrations in groundwater to exceed MCLs by the dissolution process from NAPLs' residual saturation or pooled free phase (Figure 7.11a). In particular, the free phase present in source

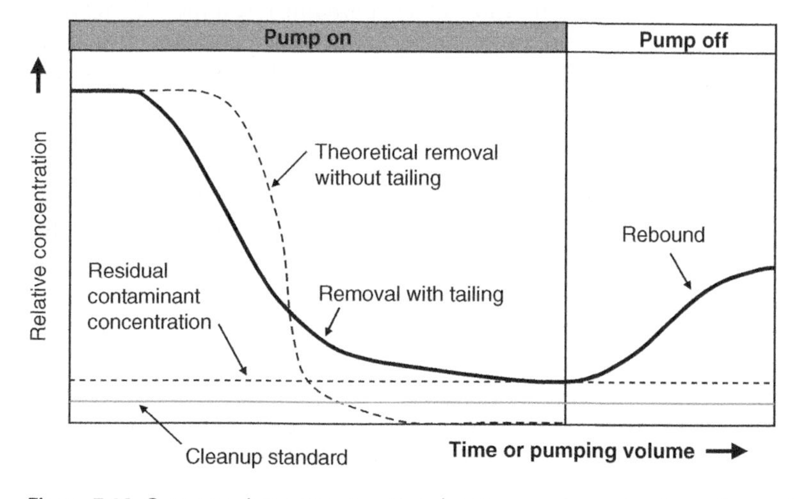

Figure 7.10 Concentration versus pumping duration or volume showing tailing and rebound effects.

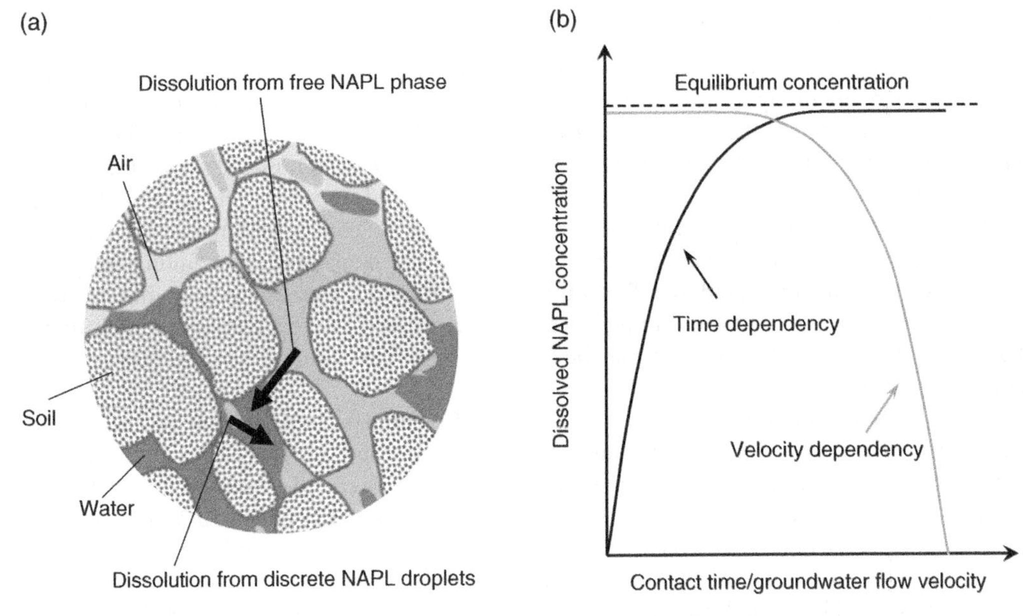

Figure 7.11 (a) NAPL dissolution from discrete NAPL droplets (residual saturation) or free phase (liquid–liquid partitioning). (b) Dependency of dissolved NAPL concentration through dissolution on contact time and groundwater flow velocity.

zone often acts as long-term reservoirs supplying the dissolving phase with groundwater contaminants for periods of time. The dissolution process depends on many site-specific factors. For example, when groundwater moves very slowly or the contact time is long, time is sufficient for the equilibrium to establish between the pure NAPLs' liquid phase and dissolved phase. In other words, NAPLs could approach the solubility limit (Figure 7.11b). Although pump-and-treat systems increase groundwater velocity, which causes an initial decrease in concentration, the decline in concentration will later tail off until the NAPLs' rate of dissolution is in equilibrium with the velocity of the pumped groundwater. If pumping stops, the groundwater velocity slows and concentrations can rebound. The rebound may be rapid at first and then gradually reach the equilibrium concentration, unless pumping is resumed. Compared to LNAPLs, DNAPL pools are especially more problematic because the contaminant will be dissolved even more slowly than residual DNAPL. It may take decades to remove a small amount of contaminant from a DNAPL pool (NRC 1994).

7.3.2.2 Slow Contaminant Desorption/Precipitate Dissolution

Studies have shown that sorbed organic contaminants are often sequestered, meaning that contaminants are held tightly by organic matters or mineral components in soil. Consequently, the desorption of these sorbed contaminants can be a very slow process. As dissolved contaminant concentrations are reduced by pump-and-treat system operation, contaminants sorbed to subsurface media only slowly desorb from the soil matrix into the groundwater. Similar to the slow dissolution from residual and free-phase NAPLs (Figure 7.11), slow desorption also causes a significant tailing if pumping continues. Moreover, contaminant concentrations resulting from sorption and desorption show a relationship to groundwater velocity and contact time similar to that of NAPLs, causing the tailing during pumping and rebound after pumping has stopped.

For inorganic contaminants, precipitation–dissolution equilibrium is the determining factor in controlling the dissolved phase concentration. For example, the dissolved phase concentration of chromium can be controlled by a large quantity of inorganic contaminant deposit such as chromate in $BaCrO_4$ found in its crystalline or amorphous precipitates in the subsurface (Palmer and Wittbrodt 1991). Figure 7.12 illustrates a tailing curve where the contaminant concentration is controlled by the solubility, which relates to the K_{sp} value of the precipitate ($K_{sp} = 2.3 \times 10^{-10}$ for $BaCrO_4$; see also Section 2.2.3 and Example 2.3 for the calculation using K_{sp}). In this situation, if pumping stops before the depletion of a solid phase, rebound can occur.

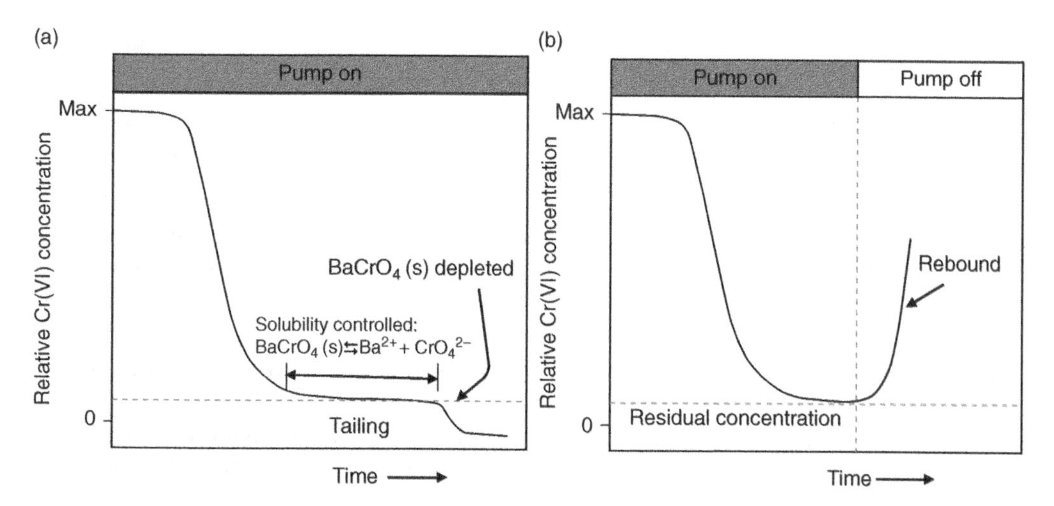

Figure 7.12 Schematic of dissolved Cr(VI) concentration in groundwater pumped from a recovery well versus time in a formation that contains a solid phase $BaCrO_4$ precipitate: (a) tailing and (b) rebound after cessation of pumping. *Source:* adapted from Palmer and Wittbrodt (1991).

7.3.2.3 Slow Matrix Diffusion

Matrix diffusion is the contaminant movement from low-permeability zone to high-permeability zone (e.g. from silt, clay, rock, or fracture to clay) in a heterogeneous aquifer due to concentration gradient (Figure 7.13a). Matrix diffusion occurs more likely when dissolved contaminants are not strongly sorbed, such as inorganic anions and some less hydrophobic organic chemicals. During a pump-and-treat operation, dissolved contaminant concentrations in the relatively higher-permeable zones are reduced by advective flushing, whereas concentrations in the less permeable regions are increased. This slow diffusion process is depicted in Figure 7.13b, based on model calculations of a hypothetical release of chlorinated solvents (Seyedabbasi et al. 2012). For example, it takes approximately 200 years to deplete the source zone PCE and immediately following that, the matrix diffusion takes approximately additional 40 years. Matrix diffusion accounts for 17 and 69% of the total mass for PCE and TCE, respectively. The significance of matrix diffusion increases as the length of time between contamination and cleanup increases. In heterogeneous aquifers, the contributions of matrix diffusion to tailing and rebound can be expected, as long as contaminants have been diffused into less permeable materials. This tailing and potential rebound can lead to an extended remediation time.

7.3.2.4 Groundwater Velocity Variation

Tailing and rebound also result from the variable times for contaminants to travel through different flow paths toward an extraction well. For example, groundwater at the edge of a capture zone created by a pumping well travels a greater distance under a lower hydraulic gradient than

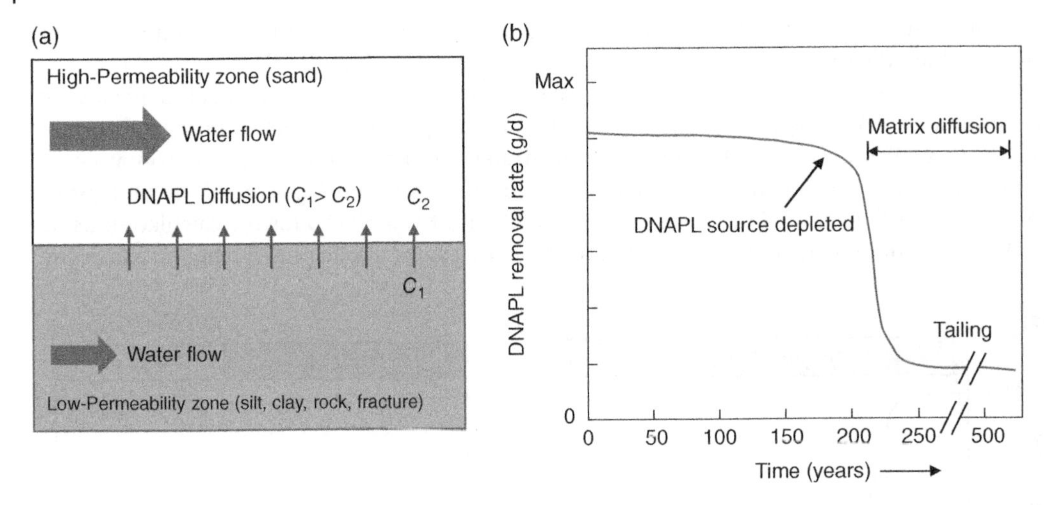

Figure 7.13 (a) Schematic of matrix diffusion of DNAPL in a heterogeneous aquifer and (b) DNAPL removal rate as a function of time illustrating the slow process of DNAPL source depletion followed by the tailing due to matrix diffusion.

groundwater closer to the center of the capture zone (Figure 7.14a). Additionally, contaminant-to-well travel time varies as a function of the hydraulic conductivity in heterogeneous aquifers. As shown in Figure 7.14b, tailing occurs at time t_2 when clean water from the upper gravel strata mixes with still-contaminated groundwater in the lower sand strata (USEPA 1994).

7.3.3 Alterations of Conventional Pump-and-Treat

There are numerous variations, enhancements, and alterations of the conventional pump-and-treat. The improved efficiency of pump-and-treat can be achieved by (i) chemical enhancement to increase the mobility and dissolved phase concentrations such as the use of cosolvents and surfactants; (ii) use of wells other than conventional vertical wells, such as horizontal wells, inclined wells, interceptor trenches, and drains; (iii) change of pumping operation such as the use of phased extraction wells, adaptive pumping, and pulsed pumping to avoid unnecessary

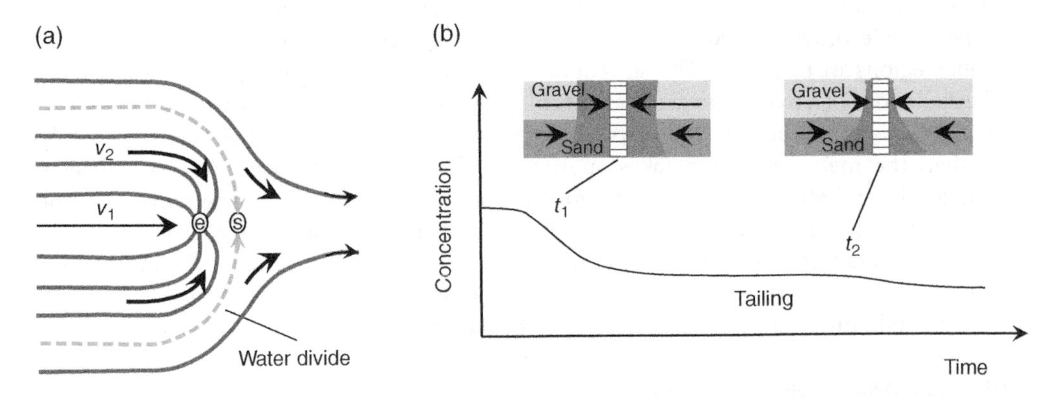

Figure 7.14 Tailing resulting from groundwater velocity variations: (a) horizontal variations in the velocity of groundwater moving toward a pumping well ($v_1 > v_2$) and (b) in a stratified sand and gravel aquifer. *Source:* adapted from USEPA (1994).

pumping of a large volume of groundwater during the tailing period prior to rebound; (iv) change of an aquifer hydraulic conductivity such as induced fractures, and (v) pumping/injection wells in conjunction with physical barriers including nonreactive slurry walls and reactive barrier used in funnel-and-gate treatment systems. Some of these topics will be detailed in future chapters, such as cosolvent and surfactant use (Chapter 11) and reactive barriers (Chapter 12). A brief summary of the variations of conventional pump-and-treat systems is provided as follows.

7.3.3.1 Chemical Enhancement to Increase Contaminant Mobility and Solubility

With the use of cosolvents such as ethanol and surfactants, aqueous solubilities of organic compounds can be greatly enhanced. Some of these compounds, particularly surfactants, will also mobilize NAPLs by reducing the interfacial tension thereby increasing the mobility of NAPLs (detailed in Chapter 11). When NAPLs become mobile, its removal by pumping can be improved. For inorganic metals, solubility and mobility can be increased by the addition of chelating compounds. Some chemical enhancements transform contaminants into less toxic or nontoxic forms in the subsurface.

7.3.3.2 Horizontal Wells, Inclined Wells, Interceptor Trenches, and Drains

Conventional P&T systems usually involve extraction wells and injection wells placed vertically in an aquifer. Recent developments in directional drilling technology in the petroleum industries also make use of horizontal wells or inclined wells a feasible and attractive approach to improve the recovery of contaminated groundwater. Directional drilling methods can create wellbores with almost any trajectory. Figure 7.15a and b shows the flow regimes in typical unconfined aquifers, particularly thin unconfined aquifers (Kawecki 2000). The flow regimes developed in such a horizontal orientation are especially suited to the remediation applications, such as the elongated plume, beneath building structure, and intersecting vertical fractures (Figure 7.15c). Horizontal wells can also be more sustainable than the traditional vertical wells (see Box 7.1, Tables 7.4, and 7.5).

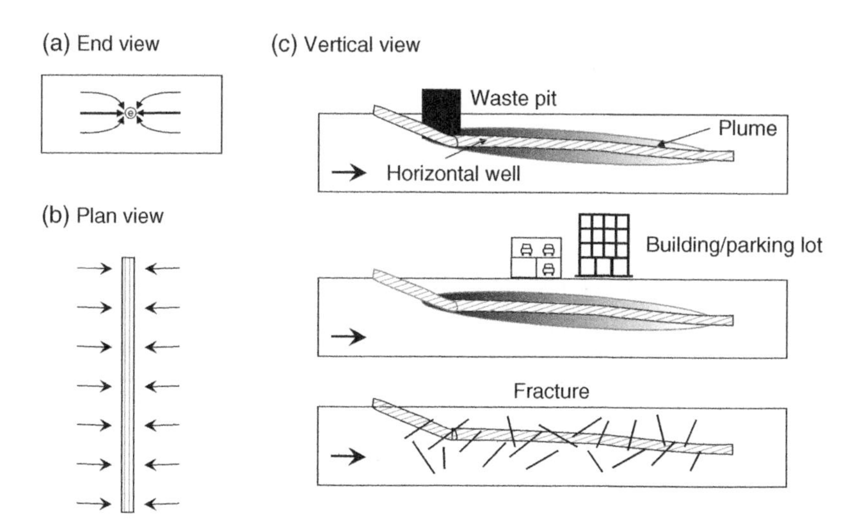

Figure 7.15 (a) End view of horizontal well flow; (b) Plan view of horizontal well flow; (c) Example applications of horizontal wells (from top to bottom): intercepting plume elongated by regional gradient, access beneath building and parking lot, intersecting vertical fractures.

Box 7.1 Comparison of sustainability: Horizontal versus vertical drilling

Current pump-and-treat systems rely on age-old conventional vertical drilling and wells. Little effort has been made to improve the sustainability of this energy-intensive technique. Although not universally applicable or desirable, **horizontal directional drilling** (HDD) was suggested to be a better choice over vertical auger drilling to achieve sustainability in terms of construction and operation. HDD can substantially reduce the time spent in constructing well network (a week for a single horizontal well compared to several weeks required to install multiple vertical wells). This implies a significant amount of energy saving associated with the reduced equipment usage, fewer man-hours on site, and fewer trips to the site during construction. An HDD system typically requires only a single pump to extract contaminated groundwater over hundreds of feet long. Compared to the vertical well systems, a combination of fewer/smaller pumps or blowers, and reduced size or number of transfer pumps for conveyance lines saves energy and hence reduces emission during operation (Lubrecht 2012).

For a hypothetical plume of 1000 ft in length with a groundwater table at 50 ft, a comparison of horizontal directional drilling (HDD) versus vertical wells for 310 m plume capture is given in Table 7.4. All wells have a configuration of 4-in diameter and 8-in bore diameter.

Table 7.5 gives further comparison of air emission between vertical auger drilling and horizontal directional drilling. The data from Lubrecht (2012) suggest a moderate reduction of environmental footprints using HDD in comparison with the conventional vertical drilling using hollow stem augers. HDD installations use drilling mud which reduces airborne emissions of VOCs or dust. The use of mud recyclers, which remove drill cuttings from the mud, enables a relatively small amount of mud to be reused continuously during operation.

Table 7.4 Comparison of HDD versus vertical wells for 1000 ft (310 m) plume capture.

Description	Vertical well	Horizontal well
Number of wells	12	1
Well spacing	83 ft	NA
Depth (length) of wells	70 ft	1480 ft
Screen length per well	20 ft	1000 ft
Total screen length	240 ft	1000 ft
Combined riser length	600 ft	480 ft
Combined well length	840 ft	1480 ft
Well material	PVC	HDPE

Table 7.5 Comparison of various air emissions between auger drilling and HDD.

Air emission (metric tons)	Vertical auger drilling	Horizontal directional drilling
Greenhouse gas emission	23	21
NO_x emission	0.137	0.130
SO_x emission	1.95×10^{-2}	1.35×10^{-2}
PM_{10} emission	2.05×10^{-2}	1.90×10^{-2}

Interceptor trenches are also widely used for controlling subsurface fluids and recovering contaminants. They function similarly to horizontal wells, but also can have a significant vertical component, which cuts across and allows access to the permeable layers in interbedded sediments. For shallower applications, trenches can be installed at a relatively low cost using conventional equipment. Recent innovations allow trench excavation and well screen installation to be complete in a single step for depths up to 20 ft (USEPA 1994). Where depth is not a constraint, interceptor trenches are generally superior to vertical wells. In such situations, they are especially effective in low-permeability materials and heterogeneous aquifers.

7.3.3.3 Phased Extraction Wells, Adaptive Pumping, and Pulsed Pumping

An efficient P&T system considers the change of groundwater quality and incorporates the most recent monitoring data to refine and optimize the initial design of P&T. For example, monitoring data allow **phased extraction wells** to be employed by appropriately siting subsequent wells for an improved plume capture. The second related approach is to use the **adaptive pumping**, which involves designing the well field such that extraction and injection can be varied to reduce zones of stagnation. Extraction wells can be periodically shut off, others turned on, and pumping rates varied to ensure that contaminant plumes are remediated at the fastest rate possible. Computer modeling revealed that **adaptive pumping** can significantly reduce the time required for site cleanup from about 100 to 50 years at the Lawrence Livermore National Laboratory Superfund site (USEPA 1994).

The third approach shown in Figure 7.16 illustrates the concept of **pulsed pumping**. During the resting period without pumping, contaminant concentrations increase due to diffusion, desorption, and dissolution in slowly moving groundwater. Once pumping is resumed, groundwater with a higher concentration of contaminants is removed, thus increasing mass removal during pumping. Pulsed pumping has the potential to increase the ratio of contaminant mass removed to the groundwater volume where mass transfer limitations restrict dissolved contaminant concentrations. Under a rapidly pulsed pumping regime, the improved mass transfer between the poorly connected pores and the well-connected pores was attributed to two mechanisms: vortex ejection and deep sweep (Kahler and Kabala 2016). The vortex ejection advects the newly ejected contaminant away from the poorly connected pores with the sudden

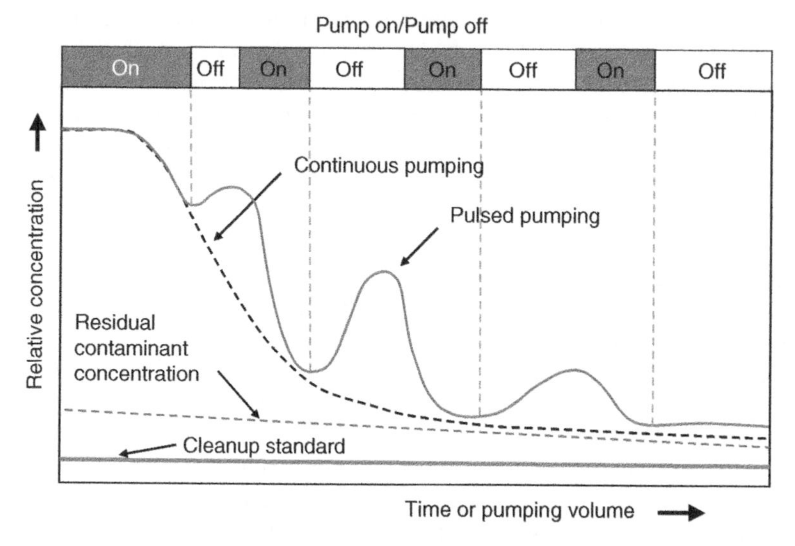

Figure 7.16 Schematics of pulse pumping versus conventional continuous pumping.

decrease and subsequent increase in flow, whereas the deep sweep plunges water into the poorly connected pores with the sudden increase in the flow.

7.3.3.4 Induced Fractures

Low-permeability geologic formation can severely limit the circulation of water, hence the effectiveness of the intended remedy based on the delivery of nutrients, air, or other carrier fluids such as those used in bioremediation and soil vapor extractions. Fractures can be induced in low-permeability sites, clays, silty sands, and bedrock materials such as shale, limestone, and sandstone to enhance their permeability. A network of induced fractures helps improve the advective transport and shorten the diffusive transport pathways. Induced fractures in the subsurface will also improve the yield of wells and therefore the total contaminant removal. Although widely used by the petroleum industry, the use of induced fractures is considered an emerging technology in groundwater remediation with applications limited to contaminated groundwater in low-permeability materials (see Box 7.2).

Box 7.2 Hydraulic fracturing: The good and bad

Hydraulic fracturing, more commonly known as "fracking," is the process of injecting a large amount of water, sand, and chemicals into geologic formations at high pressure. The high-pressure mixture causes the geological formations to crack, and these cracks are held open by the injected sand particles. Hydraulic fracturing has become an increasingly popular technology in the oil and gas industries for the improved production of hydrocarbons by significantly stimulating the movement of oil and gas from impermeable geological formations. Just over the past few years, advances in fracking technology have made tremendous reserves of natural gas in the United States economically recoverable for the first time. In the United States, for example, natural gas supply from shale beds increased from 1% in 2000 to nearly 25% primarily because of the emerging fracking technology.

However, there are growing concerns about the fracking technologies due to the potential environmental impact, mainly the use of a large amount of water as a fracking fluid as it flows back to the surface for storage and treatment. The recovered water was reported to contain high concentration of salts and naturally occurring radioactive chemicals. It was reported that the water may contain up to 750 chemicals (many of these proprietary slurries can include hydrochloric acid, ethylene glycol, aluminum phosphate, and 2-butoxyethanol), and 29 of which are either likely or known carcinogens (U.S. News, 29 November 2011). The fracking process also produces uncontrolled airborne pollutants such as methane (stronger greenhouse gases than CO_2), benzene, and sulfur oxide. Consequently, fracking is considered as a very controversial practice. Poland has embraced it, whereas France has banned it. The United States and other countries are still debating the topic.

Fracking used in soil and groundwater remediation industries holds the same principle as the fracking used in petroleum industries. The main difference lies in the motivation that in soil and groundwater remediation, it is the contaminants that are now targeted for their removal by the significantly improved hydraulic conductivity. Enhanced removal of subsurface contaminants by fracking is an emerging technology. As such, more data are needed. However, it does appear to hold promise because of the differences in the situation of fracking used in remediation versus oil/gas recovery. Compared to fracking used for the improved production of oil and gas, there are several factors that favor hydraulic fracturing used for soil and groundwater remediation, including: (i) much reduced (1 000–10 000 times smaller) amount of water usage in remedial fracture, (ii) much shallow groundwater depth (~100 ft vs. ~1000 ft), (iii) much lower pressure (several psi vs. thousands of psi), and (iv) safer use of chemicals.

7.3.3.5 Pumping in Conjunction with Permeable and Impermeable Barriers

In Section 7.1.1, we introduced the use of physical barriers such as slurry walls and sheet piling to complement the extraction wells for hydraulic control and contaminant removal. While slurry walls and sheet piling are impermeable barriers, permeable and reactive walls can also be used with or without pumping (i.e. under the natural hydraulic gradient). The funnel-and-gate system has received the most attention because numerous possible configurations can be developed to address different types of contaminant plumes and geologic settings. It typically combines impermeable barriers to contain and divert the flow of the contaminant plume toward a reactive barrier. Depending on the contaminants present in the plume, the reactive zone can use a combination of physical, chemical, and biological processes (e.g. barrier packed with activated carbon, zero-valent iron). The great promise of *in situ* reactive barriers is that they will require little or no energy input once installed, yet provide more active control and treatment of the contaminant plume than intrinsic bioremediation. The main engineering challenges involve provision of suitable amounts of reactive materials in a permeable medium and proper placement to avoid short-circuiting the contact between the gate and the cutoff wall. Chapter 12 will provide more details regarding various reactive materials and the barrier design.

Case Study The Marine Corp Air Station, Camp Lejeune, NC, USA

The Marine Corps Air Station is colocated with the Marine Corps Base in Camp Lejeune, North Carolina. A ROD for this Superfund site was signed for two operable units (OU) at the installation. OU1 (described below) consists of three sites (Sites 21, 24, 78) where pesticides, PCBs, waste oils, and other industrial wastes had been disposed. OU2 consists of three sites where solvents, oils, spent ammunition, and other wastes had been disposed. An "operable unit," as defined by the National Oil and Hazardous Substances Pollution Contingency Plan (more commonly called National Contingency Plan or NCP), is a discrete action that comprises an incremental step toward comprehensively addressing site problems.

In OU1, site 21 has portion of the area historically used as a pesticide mixing and cleaning for pesticide application equipment from 1958 to 1977, and a pit (20–30 ft long by 6 ft wide and 8 ft deep) for transformer oil disposal during a one-year period between 1950 and 1951. Site 24 was used for the disposal of fly ash, cinders, solvents, used paint stripping compounds, sewage sludge, and water treatment sludge from the late 1940s to 1980. Site 78 had leaks and spills of mostly petroleum-related products and solvents from buildings and facilities including maintenance shops, gas stations, administrative offices, commissaries, warehouses, and storage yards.

The RODs specified pump-and-treat to remediate groundwater at OU1. The pumping employed three extraction wells (6-in diameter stainless steel casing and wire-wrap screen; 35 ft deep), 18 shallow monitoring wells (screened 5–25 ft deep), two intermediate monitoring wells (screened 55–75 ft deep), and two deep monitoring wells (screened 130–150 ft deep). The aboveground treatment in OU1 system employed oil/water separation, flocculation/filtration, air stripping, and granular activated carbon adsorption. The design flow rate was 80 gpm, and treated water was discharged to a sanitary sewer.

After 2.5 years of operation, the cumulative mass removed was estimated to be 12 pounds of total VOCs (6 pounds removed during first 3 months of operation, and 6 pounds removed during past 27 months or at 0.22 pounds per month). The monthly total VOC influent concentrations was relatively low (<400 μg/L). The effluent from treatment plant had consistently met discharge limits; however, low hydraulic conductivity of the shallow aquifer resulted in influent treatment plant flow rates of <9% of the design capacity. The average cost per pound of contaminant removed was $28 277, and monthly O&M costs for the treatment plant were $12 300 during 1999.

The case study cited above is typical for the dilemma we are facing in the remediation of many contaminated sites using pump-and-treat. It can be generally concluded that the cleanup time with P&T depends not only on pumping rate but also on many other factors. Typical situations show that predicted cleanup times range from a few years to tens (for simple homogeneous aquifer), hundreds, and even thousands of years (for DNAPLs in heterogeneous aquifers). The cost of pump-and-treat operations was reported to be \$50 000–\$5 million, but for most cases the costs can be much greater. It is also generally agreed that returning the groundwater to drinking water standards may not be possible at many sites. Pump-and-treat groundwater remediation, while successful in containing contaminated groundwater plumes and reducing the concentration of groundwater contaminants, cannot be relied on to bring contaminant levels down to environmentally accepted standards.

We conclude this chapter by citing the following statements: "Groundwater scientists and engineers generally agree that complete aquifer restoration is an unrealistic goal for many, if not most, contaminated sites" (USEPA 1996), and "Conventional pump-and-treat systems are therefore an inherently inefficient method for removing contaminants, even if they are effective in some cases" (National Research Council 1994). Regardless of many criticisms, one should recognize that pump-and-treat will continue to be employed in many contaminated sites and will be an indispensable means for site remediation around the world. With recent progress in green remediation, the costly pump-and-treat systems can be made more sustainable by implementing some best management practices (BMPs) (see Box 7.3).

Box 7.3 Green remediation and BMPs in Pump-and-Treat

Opportunities for sustainable practices existed long before the concept of green remediation. Elements of green remediation technologies vary depending on the stage of remediation and the specific remedial option chosen. Sustainable practices exist particularly for those resource-intensive remediation techniques including pump-and-treat. The USEPA has published a series of fact sheets regarding the green remediation BMPs for various activities, beginning with site assessment/investigation and ending through remedy operation and closure during the whole life cycling of remediation (USEPA 2009a, 2009b, 2010).

According to the USEPA (2008), five energy-intensive technologies used for Superfund sites, in their decreasing order of the estimated annual average energy consumptions are: pump-and-treat (79.2%), thermal desorption (15.0%), multiphase extraction (3%), air sparging (1.6%), and soil vapor extraction (1.1%). The total annual energy of these five cleanup technologies is estimated to be 6.18×10^8 kWh. By assuming 1.37 lb CO_2 emitted into the air for each kWh of electricity generated in the United States, the use of these five technologies at NPL sites in 2008 through 2030 is predicted to have CO_2 emission totaling 9.2 million metric tons. This is equivalent to operating two coal-fired power plants for one year. A similar claim was also made for the remediation projects in New Jersey where the difference between two proposed remedies could be as high as 2% of the annual greenhouse gas emissions for the entire state (Ellis and Hadley 2009).

For pump-and-treat, sustainable design can be incorporated into processes and parameters such as better well placement, extraction rates, pumping duration, and more efficient above-ground treatments (activated carbon/air striping). For example, wells can be better placed by considering land reuse plans, local zoning, maintenance, and monitoring of any engineering and institutional controls. Pumps, blowers, and heaters should not be oversized, and pulsed pumping schemes should be adopted when necessary.

The BMPs introduced during the construction of pump-and-treat can continue during remedy operation. Sustainable practices can encompass the use of clean fuels and renewable energy sources for vehicles and equipment, retrofitting diesel machinery and vehicles for improved emission controls, reusing construction and routine operational materials, reclaiming demolition or processing wastes, and installing maximum controls for storm water runoff. Sustainable constructions for pump-and-treat also include well placement compatible with reuse and zoning, green chemicals and materials, storm water discharge control, green structure and housing for aboveground treatment train (USEPA 2009b).

A typical pump-and-treat system during the operation stage constitutes 39% labor, 23.5% utilities, 16% materials, 13% chemical analysis, and 8.5% disposal cost, which are all subjective to the deployment of sustainable practices such as renewable energy, green acquisition, recycled or bio-based materials (e.g. surfactants). Renewable energy for pumps, blowers, and heaters have been used in various contaminated sites in the United States, including solar energy through photovoltaics (PV) direct or indirect heating and lighting systems, or concentrating solar power; wind energy as an alternative in coastal areas or at high altitudes common to many mining sites. A wind speed of 9 mph (>13 mph for a wind farm) is sufficient for groundwater pumping. Renewable energy systems can operate independently or tie to an existing utility power grid.

Bibliography

American Water Works Association (AWWA) (2011). *Water Quality and Treatment: A Handbook on Drinking Water*, 6e. McGraw-Hill.

Bedient, P.B., Rifai, J.S., and Newell, C.J. (1999). Chapter 11: nonaqueous phase liquids. In: *Groundwater Contamination: Transport and Remediation*. New Jersey: Prentice-Hall, Inc.

Davis, M.L. and Cornwell, D.A. (2013). *Introduction to Environmental Engineering*, 5e. McGraw-Hill.

Domenico, P.A. and Schwartz, F.W. (1997). *Physical and Chemical Hydrogeology*, 2e. New York, NY: John Wiley & Sons.

Einarson, M.D. and Mackay, D.M. (2001). Predicting impacts of groundwater contamination. *Environ. Sci. Technol.* 35 (3): 66A–73A.

Ellis, D.E. and Hadley, P.W. (2009). Sustainable remediation white paper – integrating sustainable principles, practices, and metrics into remediation projects. *Remediation J.* 19 (3): 5–114.

Faust, C.R., Sims, P.N., Spalding, C.P. et al. (1993). *FTWORK: Groundwater Flow and Solute Transport in Three Dimensions. Version 2.8*. Sterling, VA: GeoTrans, Inc.

Javandel, I. and Tsang, C. (1986). Capture-zone type curves: A tool for aquifer cleanup. *Ground Water* 24 (5): 616–625.

Kahler, D.M. and Kabala, Z.J. (2016). Acceleration of groundwater remediation by deep sweeps and vortex ejections induced by rapidly pulsed pumping. *Water Resour. Res.* 52: 3930–3940.

Kawecki, M.W. (2000). Transient flow to a horizontal water well. *Ground Water* 38 (6): 842–850.

Lubrecht, M.D. (2012). Horizontal directional drilling: a green and sustainable technology for site remediation. *Environ. Sci. Technol.* 46 (5): 2484–2489.

MacDonald, J.A. and Kavanaugh, M.C. (1994). Restoring contaminated groundwater: an achievable goal? *Environ. Sci. Technol.* 28 (8): 362A–368A.

Mackay, D.M. and Cherry, J.A. (1989). Groundwater contamination: pump-and-treat remediation. *Environ. Sci. Technol.* 23 (6): 630–636.

Masters, G.M. and Ela, W.P. (2007). *Introduction to Environmental Engineering and Science*, 3e. New Jersey: Prentice Hall, Upper-Saddle River.

National Research Council (1994). *Alternatives for Groundwater Cleanup*. Washington, DC: National Academy Press.

Nyer, E.K. (1992). *Practical Techniques for Groundwater and Soil Remediation*. London: Lewis Publishers.

Nyer, E.K. (2000). Chapter 1: limitations of pump and treat remediation methods. In: *In Situ Treatment Technology*, 2e. Lewis Publishers.

Palmer, C.D. and Wittbrodt, P.R. (1991). Processes affecting the remediation of chromium-contaminated sites. *Environ. Health Perspect.* 92: 25–40.

Palmer, C.D. and Fish, W. (1992). Chemical enhancements to pump-and-treat remediation, U.S. EPA groundwater issue paper, EPA/540/S-92/001, U.S. EPA.

Satkin, R.L. and Bedient, P.B. (1988). Effectiveness of various aquifer restoration schemes under various hydrogeologic conditions. *Ground Water* 26 (4): 488–498.

Seyedabbasi, M.A., Newell, C.J., Adamson, D.T., and Sale, T.C. (2012). Relative contribution of DNAPL dissolution and matrix diffusion to the long-term persistence of chlorinated solvent source zones. *J. Contam. Hydrol.* 134–135: 69–81.

Tchobanoglous, G., Stensel, H.D., Tsuchihashi, R., and Burton, F. (2014). *Wastewater Engineering: Treatment and Resource Recovery*, 5e. McGraw-Hill.

The White House (2009). Federal Leadership in Environmental, Energy, and Economic Performance. Office of the Press Secretary, 5 October 2009.

Travis, C.C. and Doty, C.B. (1990). Can contaminated aquifers at Superfund sites be remediated? *Environ. Sci. Technol.* 24 (10): 1464–1468.

USEPA (1980). *Carbon Adsorption Isotherm for Toxic Organics*, EPA-600/8-80-023. USEPA.

USEPA (1989). *Performance Evaluations of Pump-and-Treat Remediation*, EPA/540/4-89/00. USEPA.

USEPA (1990). *Air/Superfund National Technical Guidance Study Series: Air Stripper Design Manual*, EPA-450/1-90-003. USEPA.

USEPA (1994). *Methods for Monitoring Pump-and-Treat Performance*, EPA/600/R-94/123. USEPA.

USEPA (1996). *Pump and Treat Ground-Water Remediation: A Guide for Decision Makers and Practitioners*, EPA/625/R-95/005. USEPA.

USEPA (1997). *Design Guidelines for Conventional Pump-and-Treat Systems*, EPA/540/S-97/504. USEPA.

USEPA (2001). *Groundwater Pump and Treat Systems: Summary of Selected Cost and Performance Information at Superfund-financed Sites*, EPA 542-R-01-021b. USEPA.

USEPA (2005). *Cost Effective Design of Pump and Treat System: One of Series on Optimization*, EPA 542-R-05-008. USEPA.

USEPA (2008). *Green Remediation: Incorporating Sustainable Environmental Practices into Remediation of Contaminated Sites*, EPA 542-R-08-002. USEPA.

USEPA (2009a). *Principles for Greener Cleanups*, U.S. Environmental Protection Agency, Office of Solid Waste and Emergency Response.

USEPA (2009b). *Green Remediation Best Management Practices: Pump and Treat Technologies*, EPA 542-F-09-005. USEPA.

USEPA (2010). *Superfund Green Remediation Strategy*, U.S. Environmental Protection Agency, Office of Solid Waste and Emergency Response, Office of Superfund Remediation and Technology Innovation.

USEPA (2012). *Methodology for Understanding and Reducing a Project's Environmental Footprint*, EPA 542-R-12-002, USEPA.

Zhang, C. (2013). Incorporation of green remediation into soil and groundwater cleanups. *Int. J. Sustainable Human Develop.* 1 (3): 128–137.

Questions and Problems

1 Give examples of permeable and impermeable barriers used in groundwater remediation.

2 What are the underlying mechanisms for the containment of plume using hydraulic versus physical barriers? Can extraction wells and injection wells both be employed in hydraulic containment?

3 Why hydraulic containment is particularly important in the initial stage of the contamination and plume formation?

4 Discuss the general design considerations (factors) for pump-and-treat.

5 What conclusions (regarding source control, plume containment, and extraction) can be drawn from the study on three pumping scenarios by Faust et al. (1993)?

6 What general conclusions can be drawn from the modeling study by Satkin and Bedient (1988) in regard to the optimal placement of extraction wells and injections wells?

7 Is this statement correct: The more the extraction wells and/or the injections wells, the better they are for the cleanup of the contaminated sites?

8 Using the governing equations for the calculation of capture zone, what one can do to increase the capture zone?

9 Why the capture zones must be overlapped between neighboring wells if multiple extraction wells are used? What is the maximum spacing between wells if the total number of wells (n) is 2 or 3?

10 Schematically draw the plume growth patterns of (a) LNAPLs versus (b) DNPLAs from a leaking tank in the following aquifer. The bed rock and clay lens are the confining layers, and the graph shows the initial spill and the initial groundwater table.

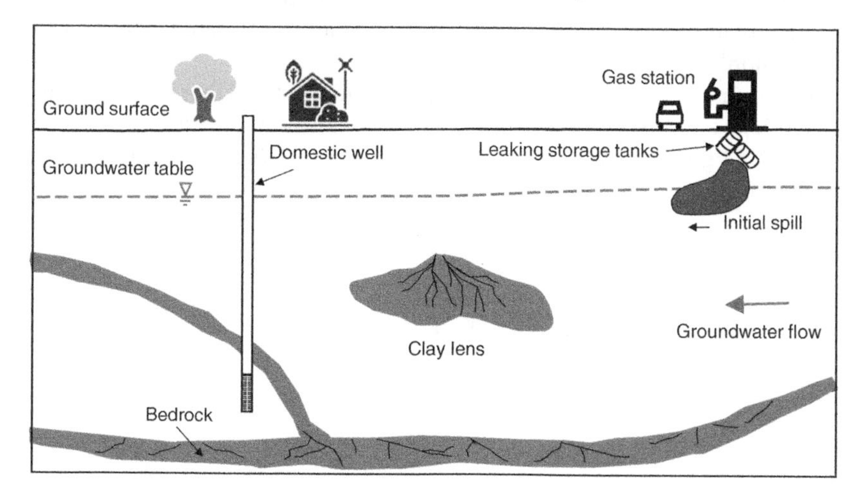

11 A contaminated site has been surveyed and a contaminated region 150 ft × 100 ft × 15 ft ($L \times W \times H$) was delineated. The average concentration of total petroleum hydrocarbons (TPHs) in the soil is 10 000 mg/kg.

a. What is the total *mass* of contaminants in kilograms at the site? Assume the bulk density of the soil is about 128 lb per cubic foot (specific gravity = 2).

b. Using the results from (a), estimate the total *volume* of petroleum hydrocarbons initially released, assuming 50% of the hydrocarbons have been lost to volatilization, biodegradation, and dissolution (report the answer in gallons). Assume that the hydrocarbon released at the site was gasoline with a specific gravity of 0.8 (i.e. 0.8 kg/L), and that the density of the gasoline did not change over time in the soil.

c. Based on the TPHs, bulk density of soil, density of TPHs, and soil porosity (n), estimate the *residual saturation* (in %) of the hydrocarbon–soil system. Assume a soil porosity (n) of 0.35.

12 An aquifer with a unit volume of 1 m^3 contains 20 L of pure trichloroethylene (TCE) released from a leaking storage tank. TCE has its solubility of 1100 mg/L and a specific gravity of 1.47 g/mL. The aquifer has a porosity of 0.35, and groundwater moves through it with an actual velocity of 0.025 m/day. The dissolved TCE concentration is 20% of its solubility. (a) Estimate the total mass of TCE, mass of dissolved TCE, and the percentage of dissolved TCE (b) Estimate the time needed to remove all the TCE (assume that the unit cross-sectional area perpendicular to the flow is 1 m^2) (after Masters and Ela 2007).

13 A sampling program at a contaminated site indicated the following DNAPL zones:

A "pool" of free-phase DNAPL in a stratigraphic depression in an unfractured clay. The pool is 200 ft^2 in area and averages 5 ft in thickness.

A zone of residual DNAPL extending directly underneath an old pit 100 ft^2 in area. The residual zone extends through the 5 ft thick unsaturated zone and 15 ft through the saturated zone until it reaches the DNAPL pool.

Other data: Laboratory tests of soil hydrocarbons and engineering judgment provided the following saturation data: Residual saturation in the unsaturated zone = 0.10; Residual saturation in the saturated zone = 0.35; Saturation in the free-phase zone = 0.70. Assume a typical porosity value of 0.3, which should be considered an upper range estimate of total recoverable volume. The actual volume is usually much lower.

a. What is the total volume of contaminated soils (i.e. unsaturated zone + saturated zone + depression)?

b. What is the estimated total volume of DNAPL at the site (i.e. DNAPL in unsaturated zone + DNAPL in saturated zone + DNAPL in depression as a free phase)?

c. How much DNAPL is pumpable from a theoretical basis (Note that residual NAPL is not pumpable)?

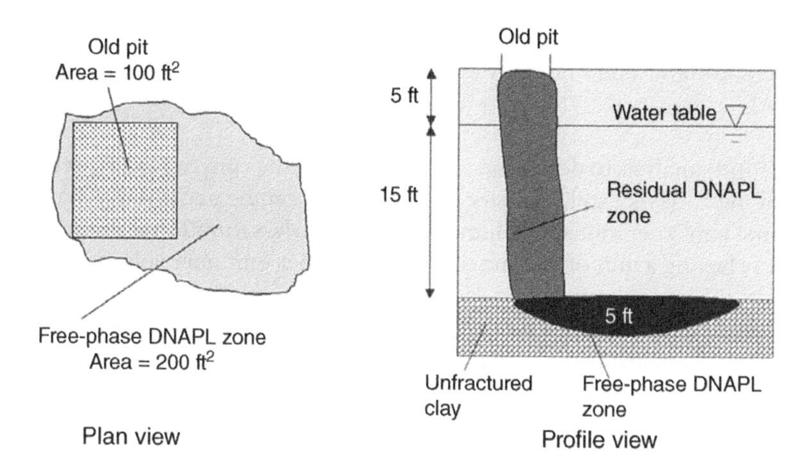

14 A well water is contaminated with 200 μg/L carbon tetrachloride. The groundwater must be treated to the maximum contaminant level (MCL) of 5 μg/L using a packed air stripper. What is the required height for the following conditions? $L = 50\,\text{m}^3$ water/m^2 tower cross section-hour, $G = 2500\,\text{m}^3$ air/m^2 tower cross section-hour, unitless $H = 0.97$ (20 °C), and K_{La} based on a laboratory study = $25\,\text{h}^{-1}$.

15 A confined aquifer has a thickness of 30 m, hydraulic conductivity of 1.75×10^{-3} m/s, and hydraulic gradient of 0.0006. The maximum pumping rate was determined to be $3 \times 10^{-3}\,\text{m}^3/\text{s}$. A recent contaminated plume was found to be 75 m in width. Determine the location of a single extraction well relative to the plume location such that the plume can be removed from this single extraction well (after Masters and Ela 2007).

16 Consider the same situation as the problem above, but instead two pumping wells are used to extract the contaminated plume. (a) Determine the optimal distance of the two wells. (b) If these two optimally located wells are aligned along the leading edge of the plume, what minimum Q would assure a complete plume capture for a plume of 75 m in width? (c) If the plume is 800 m long and the aquifer porosity is 0.35, how long would it take to pump an amount of water equal to the volume of water contained in the plume?

17 The retort method was used to measure the saturation of a soil core sample collected from an oil contaminated field. The volumes of oil and water recovered from the retort method were 4.43 and 1.95 mL, respectively. Prior to this experiment, the bulk volume was measured to be 35.00 mL and the grain volume was 26.50 mL. Determine the oil saturations of this soil sample.

18 The water in a domestic well contains 375 μg/L chloroform. The water must be treated to achieve a 95% removal rate using a packed air stripper tower. What is the required height of the stripping tower for the following conditions? Given: $L = 65\,\text{m}^3$ water/m^2 tower cross section-hour, $G = 2300\,\text{m}^3$ air/m^2 tower cross section-hour, unitless $H = 0.21$ (20 °C), and K_{La} based on a laboratory study = $35\,\text{h}^{-1}$.

19 Design the GAC unit in treating 40 gpm groundwater contaminated with 1 mg/L of TCE. The maximum concentration limit (MCL) for TCE in drinking water is 0.005 mg/L.

20 Estimate the NAPLs' saturation for a contaminated aquifer if the content of TPHs was measured to be 40 000 mg/kg and a porosity to be 0.35? The bulk density of the soil is 1.8 g/cm^3 and the DNAPLs' density is 1.75 g/cm^3.

21 Use dimensional (unit) analysis to derive Eq. (7.17), including the conversion factor of 10^6. TPH has a unit of mg TPHs/kg soil. Porosity is unitless, but can be perceived as having a unit of pore volume (cm^3) per unit soil volume (cm^3). Similarly, saturation is unitless, but can be perceived as having a unit of volume of TPHs (cm^3) per unit pore volume (cm^3).

22 Air stripping is designed to effectively remove VOCs, but why certain VOCs such as acetone and MTBE are not amenable to air stripping treatment?

23 Which of the following compound is the least amenable to the efficient removal by activated carbon: BTEX, PAHs, PCBs, phenols, and metals?

24 Smearing is the up-and-down movement of LNAPLs corresponding to the rise and fall of groundwater table in the aquifer. Explain how smearing of LNPAL can lead to the contamination of an aquifer.

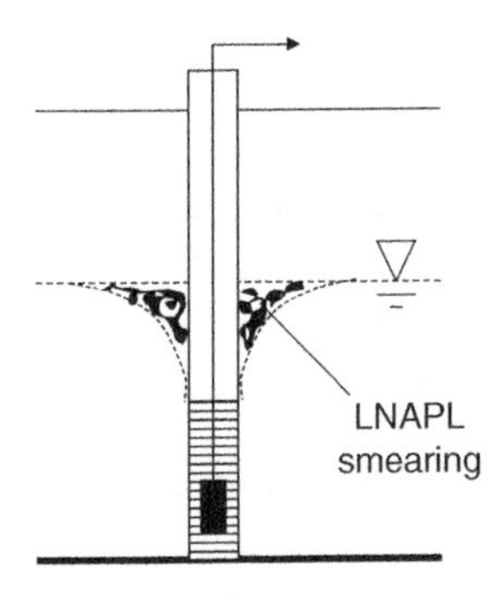

25 Explain how the following partitioning equilibrium affects the tailing and rebound of contaminants in groundwater: (a) NAPL dissolution, (b) sorption–desorption, (c) precipitation–dissolution. What are the major differences for the causes of tailing between organic and inorganic compounds?

26 Why less hydrophobic compounds may have a larger impact on tailing due to the matrix diffusion?

27 What are the potential applications of horizontal wells in groundwater remediation? Why horizontal directional drilling (HDD) can be advantageous over the conventional vertical rotary drilling from the sustainable standpoint?

28 Give examples of best management practices applicable to pump-and-treat systems.

Chapter 8

Soil Vapor Extraction and Air Sparging

LEARNING OBJECTIVES

1. Delineate the difference and similarity between two vapor-based technologies for the remediation of contaminated vadose zone soil (soil vapor extraction) and groundwater (air sparging)
2. Recognize the applicability and limitations of soil vapor extraction (SVE) as well as *in situ* air sparging (ISA), and describe how volatility and permeability affect their performance
3. Correlate airflow patterns with aquifer properties (e.g. grain size) in soil vapor extraction and air sparging
4. Use Henry's law constant and soil–water partitioning coefficient to calculate the concentrations in vapor phase in relation to the dissolved and sorbed phases
5. Use Raoult's law to calculate vapor pressure and vapor concentration in equilibrium with NAPLs in the vadose zone
6. Understand the kinetic aspect of mass transfer limits due to vapor sorption, dissolution, volatilization, and diffusion in porous media
7. Apply Darcy's law to the advective flow of vapor under steady-state conditions and estimate radius of influence of venting wells
8. Evaluate quantitatively whether soil venting is appropriate using a screening model developed by Johnson et al. (1990a)
9. Comprehend the limitations of SVE/IAS and the strategy to improve vapor extraction such as heating, surface seal, and ozone sparging
10. Estimate well numbers, flow rates, and well locations during the design of soil venting
11. Know various vapor treatment options following soil vapor extraction and air sparging
12. Be aware of the sustainable practices in soil vapor extraction and air sparging

Two vapor-based technologies, soil vapor extraction (SVE) and *in situ* air sparging (IAS), are introduced in this chapter. These two processes are often employed together at contaminated sites and leaking underground storage tank locations for the removal of organic contaminants in their gas (vapor) phase. SVE and IAS are therefore complementary to the conventional pump-and-treat systems which directly remove contaminants in their dissolved phase. SVE process relies on the vacuum extraction (negative pressure) of VOCs from the vadose zone (i.e. unsaturated zone), whereas IAS removes VOCs from the saturated zone with the injection of air (positive pressure). The purpose of this chapter is to introduce the general applications and limitations of these two processes, the fundamental chemical and geologic parameters, principles that govern the vapor behavior and their removal, and lastly the mathematical equations as a tool to determine the appropriateness of soil vapor extraction and its design, including airflow rate and the number of wells. When reading through various mathematical equations, readers should make an effort in understanding the underlying physical and chemical mechanisms, the model

Soil and Groundwater Remediation: Fundamentals, Practices, and Sustainability, First Edition. Chunlong Zhang.

Companion website: www.wiley.com/go/Zhang/Remediation_1e

framework, and the logical design process unique to the vapor transport rather than the mathematical derivations. For the design of soil venting and particularly the air sparging process, since no fixed "recipe" exists, one should always have to make assumptions and an intelligent decision by incorporating site-specific conditions and parameters.

8.1 General Applications and Limitations of Vapor Extraction

Both SVE and IAS are designed for *in situ* vapor extraction. However, they are in a significant contrast regarding system approaches, applications, and limitations. These are elaborated in the following sections.

8.1.1 Process Description and System Components

Soil vapor extraction (SVE) extracts contaminants from soils in the vapor form by applying a vacuum through an extraction well. It removes volatile organic compounds (VOCs) and some semivolatile organic compounds (SVOCs) from soil in the unsaturated zone. The extracted vapors are then treated, as necessary, and discharged to the atmosphere or reinjected to the subsurface (where permissible). Often times, in addition to vacuum extraction wells, air injection wells are installed to increase the airflow and contaminant removal rate. The injection of ambient air can be achieved primarily by active injection wells through a blower, or passively through leaky boundaries and/or wells (Figure 8.1).

Air sparging (IAS) is used along with SVE for contaminated groundwater below the groundwater table (saturated zone). It involves the injection of contaminant-free air into the subsurface saturated zone, enabling a phase transfer of contaminants from a dissolved state to a vapor phase. The air is then vented through the unsaturated zone. Air sparging is most often used together with soil vapor extraction (SVE), but it can also be used with other remedial technologies such as bioventing for an enhanced biodegration of contaminnats. When air

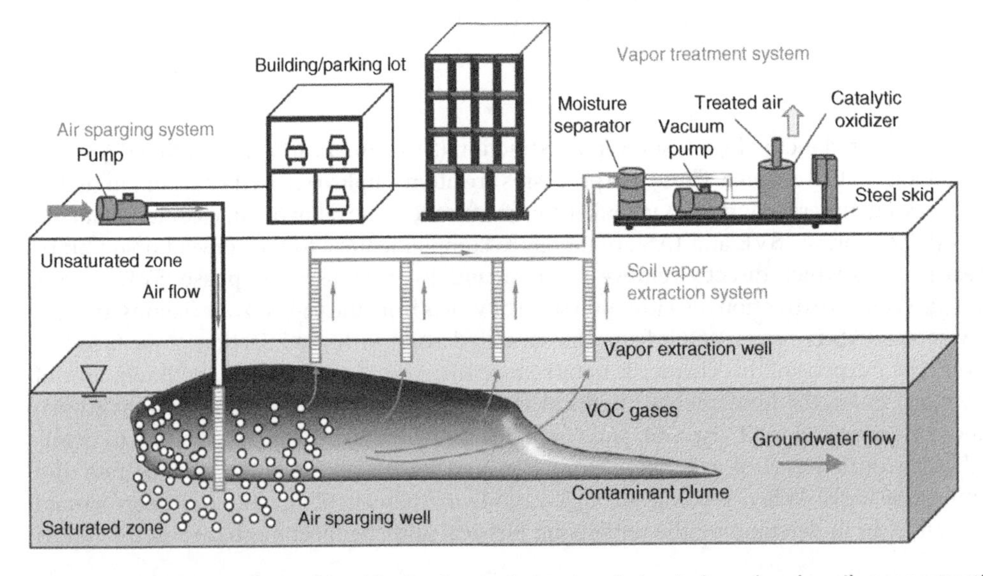

Figure 8.1 Schematics of a combined *in situ* air sparging system (saturated zone) and a soil vapor extraction system (unsaturated zone) followed by a vapor treatment system.

sparging is combined with SVE, the SVE system creates a negative pressure in the unsaturated zone through a series of extraction wells to control the vapor plume migration. This combined system is called **SVE/IAS**. Air sparging wells are commonly proposed for the removal of residual contaminants from an aquifer, the remediation of VOCs in equilibrium with dissolved phase in a contaminant plume, and/or use as migration barriers for VOCs.

As depicted in Figure 8.1, the SVE and IAS systems typically consist of a network of sparging (injection) and extraction wells, an off-gas treatment, and sometime the treatment for a small amount of the liquid condensed from the vapor. The vapor extraction system usually consists of a partially screened pipe placed in a permeable packing (coarse sand or gravel) in the vadose soil, a vapor–liquid separator, vacuum pumps and/or blowers, and monitoring wells for vapor concentration and pressure measurement. Vacuum pumps and blowers are used to reduce or increase gas pressure in the extraction wells and induce contaminated vapor flow to the wells. The vapor–liquid separator is optional but essential for the protection of pumps and blowers and if thermal gas treatment units are used. Other optional components include barrier wells and impermeable caps on the ground surface. If impermeable caps (plastic membranes, buildings, parking lots, soil layers of low permeability) are in place, the purpose is to minimize infiltration of water from the surface, increase the system's radius of influence by preventing short circulation, and control the horizontal movement of inlet air which can bypass contaminants.

Both SVE and IAS are nondestructive, hence they require the off-gas treatment to further remove contaminants from the vapor stream before it can be discharged to the atmosphere. These include a variety of methods such as thermal destruction (vapor combustion, catalytic oxidation; see Chapter 10), vapor condensation, wet scrubber, bioreactor, and activated carbon units. Recovered condensates should be treated in accordance with pretreatment requirement, such as air stripping and activated carbons (Chapter 7), prior to their discharge to a publicly owned wastewater treatment plant.

In the literature, SVE has other names, including *in situ* volatilization/vaporization, soil venting, soil vacuum extraction, and enhanced volatilization. Air sparging is also known as *in situ* air stripping and *in situ* volatilization. In this chapter, we use soil vapor extraction and air sparging to distinguish these two processes. At some point, we use vapor extraction or venting to loosely denote either processes or the combined use of these two processes for the convenience of discussions.

8.1.2 Chemical and Geologic Parameters Affecting Vapor Extraction

From the above-mentioned process description, it is clear that both SVE and IAS are for the remediation of volatile chemicals from the subsurface. SVE removes contaminants from the permeable vadose zone, whereas IAS removes contaminants from the permeable saturated zone. As elaborated below, the chemical and geologic parameters affecting these two processes bear a great deal of similarities.

First, the appropriateness of SVE and IAS is a function of a compound's volatility, which in turn can be determined by the vapor pressure, boiling point, or Henry's law constant. As we have discussed in Chapter 2, there is no clear dividing line between a volatile and a semivolatile compound. It is, however, generally agreed that compounds with a vapor pressure of at least 0.5 mmHg (equivalent to 6.6×10^{-4} atm, 66 Pa, 0.66 millibar), a boiling point lower than 250–300 °C, or a dimensionless Henry's constant of at least 0.01 (DePaoli 1996) are considered to be amenable to SVE. This rule applies to air sparging as well. For example, benzene as a very volatile compound (vapor pressure: 0.125 atm; boiling point: 80.1 °C; Henry's constant: 0.24) is amenable to SVE and IAS. The less volatile PAHs such as naphthalene (vapor pressure: 3×10^{-4} atm; boiling point: 218 °C; Henry's constant: 0.049) can still be treated by SVE and IAS, but the higher-molecular-weight PAHs such as pyrene (vapor pressure: 3.29×10^{-9} atm; boiling point: 404 °C; Henry's constant: 4.66×10^{-4}) are not amenable to these vapor-based techniques.

Low-molecular-weight/volatile compounds such as TCE, PCE, and most lighter fractions of gasoline constituents (butane, pentane, hexane, the aromatic benzene including BTEX, and other alkylbenzene) are well removed by SVE. Chemicals that are less applicable for SVE include trichlorobenzene, acetone, and heavier petroleum fuels (diesel fuel, crude petroleum, heating oils, used oil, and fuel oil). Compounds that cannot be removed by SVE include heavy oils, heavy metals, PCBs, and dioxins. SVE and IAS can remove nearly all gasoline constituents and a portion of kerosene and diesel fuel constituents can be removed from the saturated zone by SVE/IAS. Heating and lubricating oils cannot be removed by SVE/IAS (Table 8.1). However, SVE/IAS can promote biodegradation of semivolatile and nonvolatile constituents.

Secondly, soil property is another important factor for vapor extraction. Conditions favorable to SVE include permeable and homogeneous soils, and soils with low organic (f_{oc}) and moisture content. Heterogeneities influence the movement of air and contaminants in a more unpredictable way than that in a homogeneous aquifer. Thus, the presence of heterogeneities makes it more difficult to position extraction and injection wells. Soil with a low moisture content is more favorable for SVE because it is easier to draw air through a dried soil. This also makes SVE one of the few techniques applicable for an aquifer with a deep groundwater table where contamination extends to the groundwater table as deep as more than 40 ft (12 m). Studies indicated that a complete removal can be achieved for gasoline within 100 days of release from porous soils. However, other contaminants in petroleum products with higher boiling points can still remain at high levels after 120 days of vapor extraction.

Air permeability (i.e. soil permeability to air) is probably the single most important factor for the success of vapor extraction. As we defined in Chapter 3, **permeability** is a measure of how well a porous media transmits a fluid (in our case here, it is the air). Since it is independent of the fluid property, it is termed **intrinsic permeability**. Permeability has a unit of $(\text{length})^2$, such as cm^2 or m^2. Hydraulic conductivity, a term related to permeability, is defined as a measure of how easily water moves through the porous media. It depends on the permeability of the matrix, as well as the property of the fluid. It has a unit of length/time, such as cm/day or m/day. When air is the fluid of concern, the permeability determines the rate at which soil vapor can be extracted from the vadose zone (SVE) or the air can be injected into the saturated zone (IAS). The permeability to air is also the determining factor for the mass transfer rate of the contaminants from the dissolved phase to the vapor phase.

Table 8.1 Petroleum product boiling point ranges.

Products	Approximate number of carbon atoms	Boiling point range (°C)
Petroleum ether	4–10	35–80
Gasoline	4–13	40–225
Kerosene	10–16	180–300
Diesel fuel	12–20	200–338
Heating oil	14–24	250–350
Lubricating oil	20–50	<350
Asphalt	80% Carbon Viscous Liquid	>300
Coke	>90% Carbon Solid	Not relevant

Source: modified from USEPA (2017).

Table 8.2 Intrinsic permeability in relation to the effectiveness of soil vapor extraction and air sparging (USEPA 2017).

	Intrinsic permeability (*k*)	
Effectiveness of SVE or IAS	**Soil vapor extraction**	**Air sparging**
Generally effective	$k \geq 10^{-8}$ cm^2	$k \geq 10^{-9}$ cm^2
May be effective, needs further evaluation	$10^{-8} \geq k \geq 10^{-10}$ cm^2	$10^{-9} \geq k \geq 10^{-10}$ cm^2
Marginally effective to ineffective	$k < 10^{-10}$ cm^2	$k < 10^{-10}$ cm^2

Coarse-grained soils (e.g. sands) have a greater intrinsic permeability than fine-grained soils (e.g. clays or silts). For the wide range of earth materials, intrinsic permeability can vary over 13 orders of magnitude (from 10^{-16} to 10^{-3} cm^2). For most soil types, air permeability is in a more limited range of 10^{-13}–10^{-5} cm^2. The ability of a soil to transmit air is reduced by the presence of soil water, which can block the soil pores and reduce airflow. This is especially important in fine-grained soil which tends to retain water. Table 8.2 lists the range of intrinsic permeability for the effectiveness of SVE and IAS.

The combined effect of chemical property (vapor pressure), soil air permeability, and the time after the release of contaminants on the applicability of SVE can be visualized semi-quantitatively as shown in Figure 8.2. This **nomogram** (also called nomography) allows for the approximate graphical computation of the likelihood of success for SVE. To use the nomogram, start at the time of contaminant release and draw a horizontal line to the appropriate soil air permeability. From the intersection point, draw a straight line to the vapor pressure. The point where this line intersects the continuum indicates how likely SVE would succeed at a contaminated site with that set of conditions.

Note that SVE is employed to remove VOCs from vadose zone. If VOCs are present at the same site, air sparging complements the use of SVE to remove VOCs dissolved in groundwater. SVE can be employed in vertical wells ranging 5–300 ft (1.5–90 m). When the groundwater is near the surface (e.g. <5 ft or 1.5 m), SVE is not appropriate. In addition, SVE cannot be used when groundwater levels fluctuate over short time periods, for example, during a seasonal fluctuation of precipitation and tidal movement (Cole 1994).

8.1.3 Pros and Cons of Vapor Extraction and Air Sparging

Soil Vapor Extraction: Several apparent advantages make SVE the most widely used innovative remediation technology at contaminated sites including many leaking underground storage tank locations in the United States. SVE requires few mechanical parts, and has low operation and maintenance (O&M) costs. It employs conventional off-the-shelf equipment and materials, and can be installed rapidly. It does not require reagents, and provides permanent remediation. This technology is not limited by the depth to groundwater, and requires no excavation of soil. SVE is not disruptive (no interruption of business operation or residential living habits), but it shows immediate results. It can be used to remediate soil underneath buildings or other structures. Another added advantage of SVE is that it may enhance biodegradation of compounds by aerobic bacteria (bioventing).

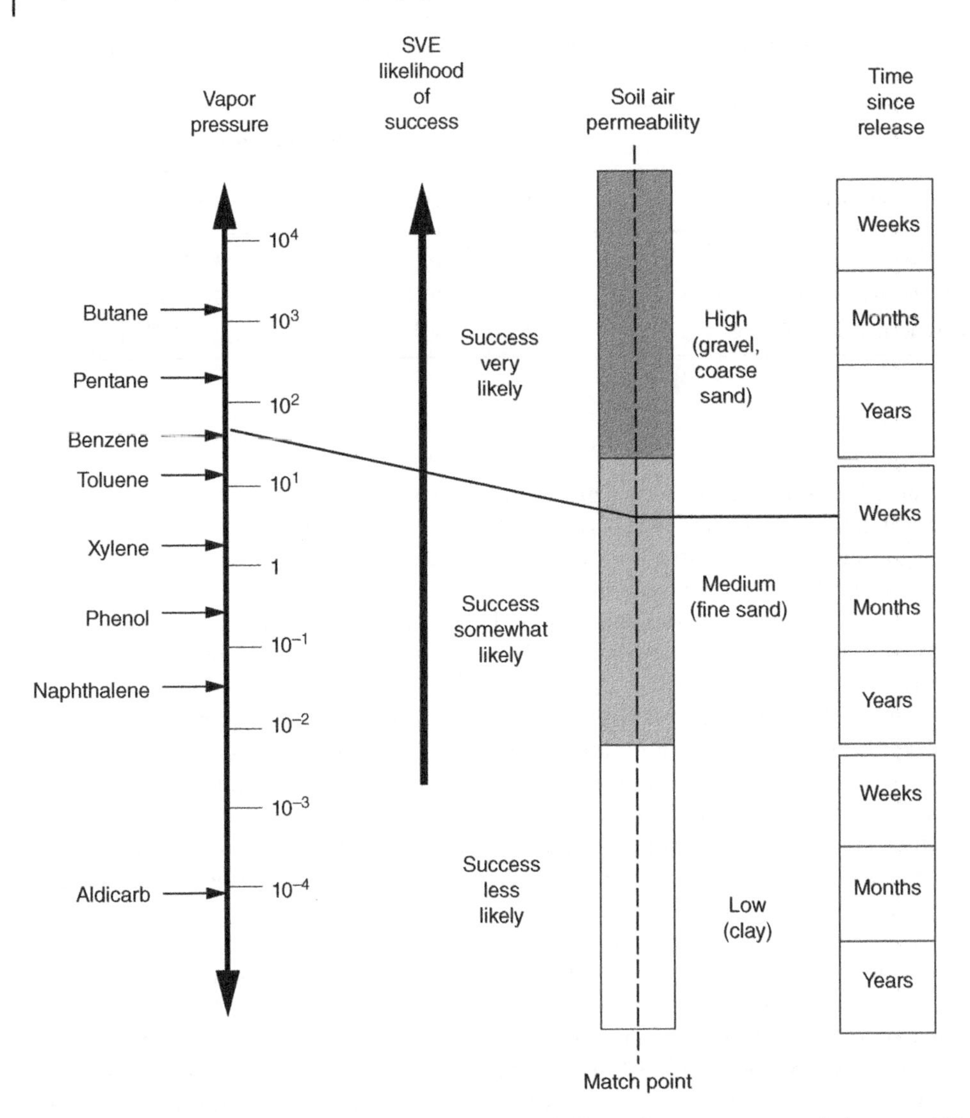

Figure 8.2 Soil vapor extraction nomograph showing the effects of vapor pressure, soil air permeability, and time after contaminant release on the likelihood of SVE success (USEPA 1991a).

Limitations of SVE are as follows: SVE has poor removal efficiencies for chemicals with low vapor pressure. The use of SVE is also limited if vapors are impeded by fine-grained soils, or if resulting vapors become an air quality concern and require further aggressive treatment. Since SVE does not alter chemical structure, the toxicity of contaminants is not changed and additional treatment technologies may be needed. The effectiveness and time of cleanup may be unpredictable for sites with complex strata and permeability. Another possible negative effect is that there may be an **upwelling** in the water table due to vacuum pumping. The upwelling could result in the spreading of the contaminant into the upwelled water, thereby requiring remediation of the groundwater as well. This issue can be mitigated, however, by adding a groundwater pumping well near the plume in order to stabilize the water table level.

Air Sparging: Air sparging removes gas phase contaminants by advective flow (i.e. advection) in the saturated zone. This advective flow is advantageous over the diffusion-controlled flow in conventional pump-and-treat systems. Another advantage of air sparging over pump-and-treat is the fact that contaminants desorb more readily into the gas phase than into groundwater. Air sparging can be used to treat contamination in the capillary fringe and/or below the groundwater table (in contrast to soil vapor extraction techniques). It is possible that existing monitoring wells could be used for air sparging. Because of the low O&M costs, air sparging can be particularly cost-effective when large quantities of groundwater must be treated.

Air sparging can become unfavorable when contaminants form complexes with the soil matrix, thereby decreasing volatilization rates. As indicated earlier, the effectiveness of air sparging is a function of vapor pressure, solubility, and Henry's constant. Constituents with relatively high solubilities and low Henry's law constants, such as MTBE and ethylene dibromide, could be dissolved in water and not removed effectively by air sparging. Similar to the limitations of SVE, IAS does not work for fine-grained and low-permeability soils that would decrease airflow through groundwater and in the vadose zone. Heterogeneous soils may cause preferential movement (channeling) of air through high-conductivity layers, or possibly bypass area of contamination. The complex airflow conditions associated with heterogeneity will also make it difficult to predict and/or control. If the depth of contaminated groundwater is less than 5 ft (1.5 m) below ground surface (BGS) or where the saturated thickness is small, it may require a prohibitive number of wells to ensure a full coverage of the area of concern. When a free product is present, air sparging can create groundwater mounding which could potentially cause the free product to migrate and contamination to spread. Thus, any free product must be removed prior to initiating air sparging. IAS cannot be used when nearby basements, sewers, or other subsurface confined spaces are present at the site. Potentially dangerous constituent concentrations could accumulate in basements unless a vapor extraction system is used to control vapor migration. IAS cannot be used when contaminated groundwater is located in a confined aquifer system, because the injected air would be trapped by the saturated confining layer and would not be able to escape to the unsaturated zone.

8.2 Soil Vapor Behavior and Gas Flow in Subsurface

The behavior of air and contaminant vapor in subsurface deserves some special attention in this chapter because air is injected through IAS and contaminant vapor is removed through SVE. In a broad sense, the behavior of gas (vapor) in subsurface shares some similar principle with that of water (dissolved contaminant), such as Darcy's law to describe the advection flow. Vapor phase contaminants are related to dissolved phase contaminants through various mass transfer processes to be discussed in this section.

8.2.1 Airflow Patterns in Subsurface

The pattern of air/vapor flow around air sparging or vacuum extraction wells is a crucial determinant of system effectiveness, and an important factor influencing system design and cost. Knowing the air/vapor flow pattern is important to evaluate site suitability for air sparging/vacuum and to optimize SVE/IAS system design and operation. For example, an overly optimistic estimate of radius of influence will lead to inefficient vapor removal. Lundegard and LaBrecque (1998) used electrical resistance tomography (ERT) to map the airflow patterns around sparging wells for two sites. As shown in Figure 8.3 (top) with a site consisting of relatively homogeneous dune sand, the principal region of airflow was approximately symmetric about the sparge well and only 2.4 m in radius. In another site (Figure 8.3, bottom) with a highly heterogeneous glacial till (grain sizes ranges from silt to cobbles), the pattern of airflow was more complex but a major horizontal component was exhibited.

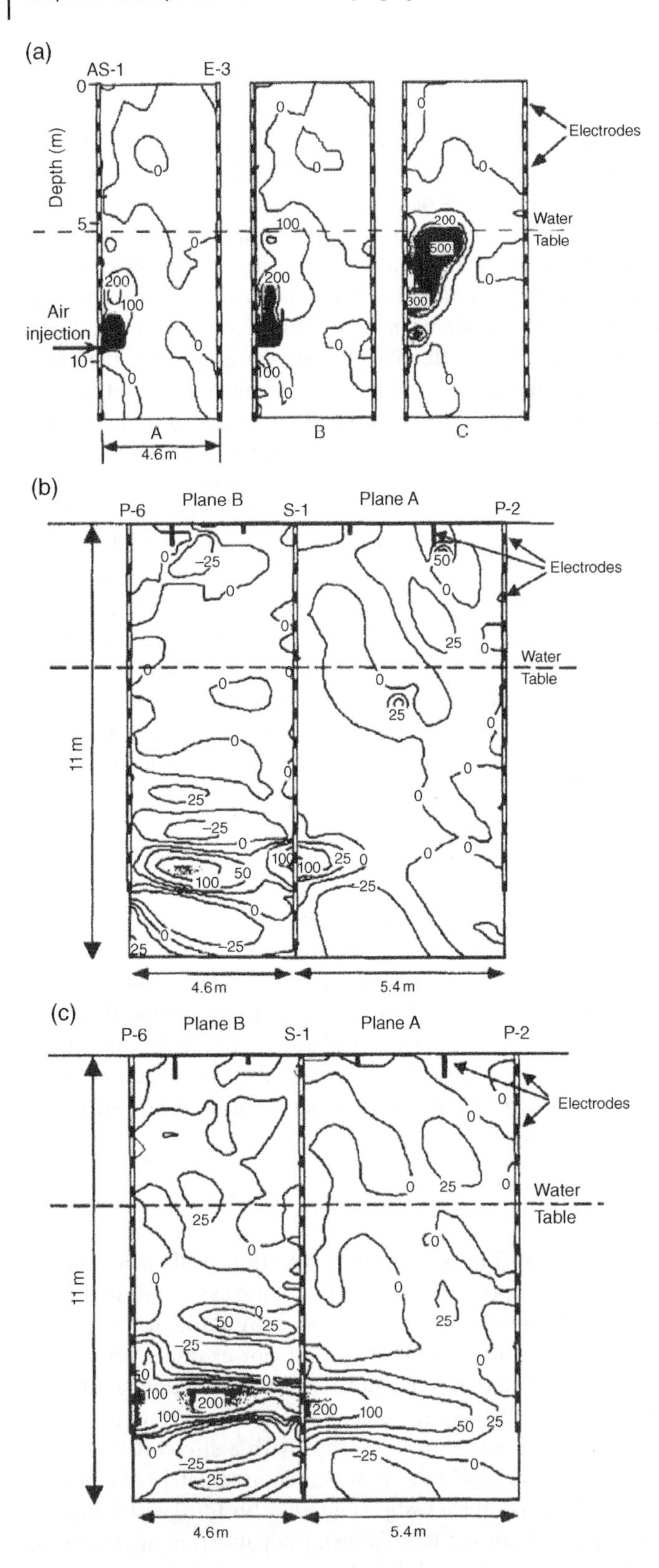

Figure 8.3 Electrical resistance tomography (ERT) showing two different airflow patterns around injection wells (AS1 and S-1) in saturated zones on two sites with very different subsurface conditions: Site A consists of relatively homogeneous dune sand (a), and Site B consists of a highly heterogeneous glacial till (b and c). The contours depict percent change in resistivity relative to pre-sparging background data. Site A was continuously sparged with air for 20 minutes, 2, and 48 hours for A, B, and C, respectively. Site B was continuously sparged with air for 2 hours (b) and 28 hours (c). E3 is the monitoring well with ERT electrode, and P-2 and P-6 are deep piezometers with ERT electrodes. *Source:* Lundegard and LaBrecque (1998), Reprinted with permission from John Wiley & Sons.

The vertical airflow in homogeneous sandy aquifers suggests that the radius of influence of sparge wells may in fact be overestimated. In contrast, the horizontal (lateral) airflow in low-permeability aquifer implies very little injected air could reach the vadose zone, and thus VOCs may be trapped in the saturated zone in the immediate vicinity of the sparge well.

8.2.2 Vapor Equilibrium and Thermodynamics

Both SVE and IAS directly remove contaminant "vapor," but contaminants in soil and groundwater are present in several other forms. In vadose zone, four contaminant phases can be potentially present: vapor phase, dissolved phase, adsorbed phase, and free phase. In saturated zone where soil air is absent, three contaminant phases are possible: dissolved phase, adsorbed phase, and free phase. These various contaminant phases are interrelated through various partitioning equilibrium as shown in Figure 8.4. For example, VOCs adsorbed on soil particles must be desorbed first, then partitioned into vapor phase, and finally transported as a free gas/vapor via advection, diffusion, and dispersion. Whichever the rate-limiting step will be the controlling factor for the vapor extraction. It is, therefore, important to understand how "vapor" behaves in soil–water system and how vapor eventually escapes from contaminated soil and groundwater.

Vapor in Porous Media: Henry's law (Eqs. 2.3 and 2.4 described in Section 2.3) can be used to relate the vapor concentration in soil pores to the concentration of contaminants dissolved in

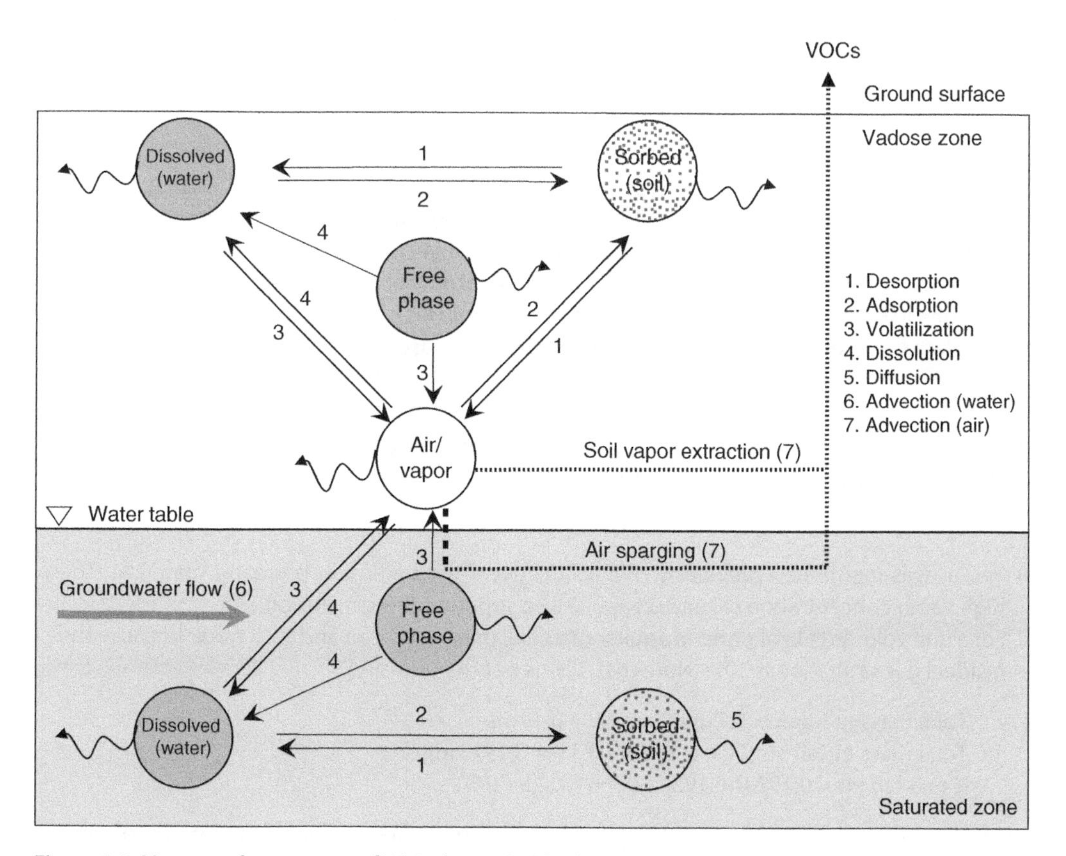

Figure 8.4 Mass transfer processes of VOCs during SVE (vadose zone) and IAS (saturated zone). The numbers in the figure denote: 1 = desorption, 2 = adsorption, 3 = volatilization, 4 = dissolution, 5 = diffusion, 6 = advection due to water flow, and 7 = advection due to forced airflow in SVE/IAS.

soil moisture. The soil–water partition coefficient (K_d), defined in Eq. (2.9) (Section 2.3) as the ratio of equilibrium concentrations in soil and that in water, can also be used in a similar manner to relate equilibrium concentrations in soil to concentrations of vapor in soil void. The volatilization and sorption of contaminant vapor can be assumed at equilibrium in cases such as diffusion-controlled or weakly advective groundwater environment. Hence, **Henry's law** constant (H) and sorption coefficient (k_d) are useful parameters to determine vapor partitioning among phases, i.e. soil moisture, soil air, and soil particles. They can be used to quantify the dissolved, vapor, and sorbed fractions at the assumed equilibrium state. Examples 8.1 and 8.2 illustrate their uses in a two-phase and a three-phase system, and how the results are relevant to vapor-based remediation technologies.

Note that Henry's law applies to ideal solutions with low concentrations of solute(s) in water. In the case of NAPLs dissolved in groundwater, for example, Henry's law relates the vapor pressure of the NAPLs in the soil air to the equilibrium concentration of NAPLs dissolved in groundwater. Since aqueous solubilities of these NAPLs are typically very low, Henry's law is certainly applicable to these dilute solutions. If there are multiple components dissolved in the groundwater such as the case of BTEX, the vapor pressure of these VOCs can be individually calculated using the Henry's law for such ideal solutions.

Two examples provided below illustrate the use of Henry's law and adsorption coefficient to conduct equilibrium calculation when dealing with air–water (two-phase) system or air–water–soil (three-phase) system. Analogous to the calculation of dissolve phase concentration for pump-and-treat in Chapter 7, Examples 8.1 and 8.2 are relevant for vapor-based remediation using SVE and IAS.

Example 8.1 Calculation of two-phase air-water equilibrium in saturated zone

An aquifer contaminated with vinyl chloride (VC) contains a residual gas saturation of 30% (i.e. 30% of porosity contains gas). The aqueous phase vinyl chloride concentration is 0.7 mg/L. Henry's law constant of vinyl chloride at 25 °C = 22.4 L-atm/mol or 0.916 (unitless). (a) Calculate the gas phase concentration in mg/L; (b) Is it feasible for this chemical to be removed by SVE? (c) Calculate the % of total vinyl chloride mass in the gas phase of this two-phase system.

Solution

a) The gas phase concentration can be calculated from the Henry's law. The unitless H is used below for the convenience of calculation.

$$C_g = C_w \times H = 0.7\,\text{mg/L} \times 0.916 = 0.641\,\text{mg/L of soil air.}$$

b) Yes, as was mentioned previously, SVE is effective if the unitless H is greater than 0.01. The high vapor concentration calculated above also supports this conclusion.
c) For a unit volume (1 L) of pores in aquifer material, there is 0.3 L air and 0.7 L water because the residual gas saturation is 30% (Note that 30% is not the porosity).

$$\text{Total mass in water} = 0.7\,\text{mg/L} \times 0.7\,\text{L} = 0.49\,\text{mg}$$
$$\text{Total mass in soil air} = 0.641\,\text{mg/L} \times 0.3\,\text{L} = 0.1923\,\text{mg}$$
$$\%\ \text{mass in air} = 0.1923/(0.1923 + 0.49) = 0.28\ (28\%)$$

Note that the calculated 28% of VC vapor does not imply SVE can only remove 28% of vinyl chloride. As VC volatilizes and continues to be removed, new equilibrium will be established for further removal by volatilization from soil moisture.

Example 8.2 Calculation of three-phase air-water-soil equilibrium in unsaturated zone

A benzene leak soaked into a patch of soil. Soil samples were taken and the average concentration of benzene in soil was determined to be 1000 mg/kg. Assume the K_d value for benzene is 0.88 L/kg, and the unitless Henry's constant is 0.24. (a) If rainwater percolates and saturates the surface soil, what concentration of benzene might initially leach from the soil and be found dissolved in the water? Assume that the water and soil are in equilibrium with respect to benzene. Also assume that the soil/water ratio is 4:1 (4 kg soil to 1 L water). (b) If portion of the water is quickly drained from this surface soil (which allows atmospheric air to make up the drained water), what concentration of benzene might be found dissolved in the water and vapor present in soil pore? Assume that the air/water/soil ratio is 1 : 1 : 8.

Solution

a) The assumption of soil/water ratio is important, because benzene concentration depends on the relative amount of water. If an infinite amount of water is present, the concentration of benzene would be very small. On the other hand, if the water content is relatively small, then one would expect a higher concentration of benzene dissolved in water.

Note also that the 1000 mg/kg is the initial concentration, not the equilibrium concentration, because the leakage through rainwater is a subsequent event after the spill. Otherwise this concentration would be equal to the sorbed concentration (C_s). When 1 L of water is in equilibrium with 4 kg of soil, we have the mass balance:

$$C_s\left(\frac{\text{mg}}{\text{kg}}\right)\times 4\,\text{kg} + C_w\left(\frac{\text{mg}}{\text{L}}\right)\times 1\,\text{L} = 1000\frac{\text{mg}}{\text{kg}}\times 4\,\text{kg soil}$$

This would become: $4\,C_s + C_w = 4000$

The second equation is to apply the linear isotherm (Eq. 2.9, Chapter 2) and the given K_d value of 0.88 L/kg:

$$\frac{C_s}{C_w} = 0.88,\ \text{i.e.}\ C_s = 0.88C_w$$

Substitute C_s into the first equation:

$$4\times(0.88C_w) + C_w = 4000$$

Thus $C_w = 885$ mg/L, and $C_s = 779$ mg/kg. If by mistake, $C_s = 1000$ mg/kg was assumed, then $C_w = C_s/K_d = 1000/0.88 = 1136$ mg/L.

With the calculated C_w and C_s, we can further determine the % of benzene in the water phase, and in the soil phase when benzene in the two phases reaches the sorption equilibrium.

$$\%\,\text{in water phase} = \frac{885\frac{\text{mg}}{\text{L}}\times 1\,\text{L water}}{779\frac{\text{mg}}{\text{kg}}\times 4\,\text{kg soil} + 885\frac{\text{mg}}{\text{L}}\times 1\,\text{L water}} = \frac{885\,\text{mg}}{3115\,\text{mg} + 885\,\text{mg}} = 0.22(22\%)$$

This number indicates that a substantial portion (22%) of the sorbed benzene can partition into the water, implying that benzene has a sufficiently high solubility to be mobile in the environment. This is predicable from its small log K_{ow} value of 2.13. The small value of log K_{ow} corresponds to a high aqueous solubility and mobility (see Chapter 2).

b) Different from (a), we now have three phases with three unknown concentrations, i.e. C_s (soil), C_W (water), and C_a (soil air). This three-phase system is more relevant to the unsaturated zone, assuming no additional free-phase NAPL is present. Note again that the 1000 mg/kg is the initial concentration in soil, not the equilibrium concentration. We now need three equations to solve contaminant concentrations in three phases (C_s, C_W, and C_a). When 1 L of water and 1 L of soil air are in equilibrium with 8 kg of soil, then we have the mass balance:

$$C_s\left(\frac{\text{mg}}{\text{kg}}\right)\times 8\,\text{kg} + C_w\left(\frac{\text{mg}}{\text{L}}\right)\times 1\,\text{L} + C_a\left(\frac{\text{mg}}{\text{L}}\right)\times 1\,\text{L} = 1000\frac{\text{mg}}{\text{kg}}\times 8\text{ kg soil}$$

This would become: $8\ C_s + C_W + C_a = 8000$

The second equation is to apply the linear isotherm (Eq. 2.9, Chapter 2):

$$\frac{C_s}{C_w} = 0.88,\ \text{i.e. } C_s = 0.88\ C_w$$

The third equation is to apply the Henry's law between air, water, and soil using $H = 0.24$:

$$\frac{C_a}{C_w} = 0.24,\ \text{i.e. } C_a = 0.24 C_w$$

Substitute C_s and C_a into the first equation to solve for C_W:

$$8\times(0.88C_w) + C_w + 0.24C_w = 8000$$

Thus $C_W = 966$ mg/L, and we can solve for C_a and C_s as: $C_a = 232$ mg/L and $C_s = 850$ mg/kg.

With the calculated C_a, C_W, and C_s, we can further determine the % of benzene in the air phase, in the water phase, and in the soil phase when benzene in the three phases reaches the equilibrium.

$$\%\text{ in water phase} = \frac{966\frac{\text{mg}}{\text{L}}\times 1\,\text{L water}}{850\frac{\text{mg}}{\text{kg}}\times 8\,\text{kg soil} + 966\frac{\text{mg}}{\text{L}}\times 1\,\text{L water} + 232\frac{\text{mg}}{\text{L}}\times 1\,\text{L air}}$$
$$= \frac{966\,\text{mg}}{6800\,\text{mg} + 966\,\text{mg} + 232\,\text{mg}} = 0.12(12\%)$$

$$\%\text{ in air phase} = \frac{232\frac{\text{mg}}{\text{L}}\times 1\,\text{L air}}{850\frac{\text{mg}}{\text{kg}}\times 8\,\text{kg soil} + 966\frac{\text{mg}}{\text{L}}\times 1\,\text{L water} + 232\frac{\text{mg}}{\text{L}}\times 1\,\text{L air}}$$
$$= \frac{232\,\text{mg}}{6800\,\text{mg} + 966\,\text{mg} + 232\,\text{mg}} = 0.029(2.9\%)$$

This result indicates that benzene is now partitioned approximately 2.9, 12, and 75.1% in air, water, and soil, respectively.

Vapor in Equilibrium with Free-Phase NAPLs: We now change our situation to the vapor of multicomponents in equilibrium with their corresponding pure phase liquids (e.g. a BTEX mixture rather than each individual BTEX dissolved in water), such as the case of multicomponent vapor from a gasoline spill. The governing equation in relating vapor pressure of each component in the mixture to the vapor pressure of each pure compound is the **Raoult's law,** which states that the vapor pressure of component i (p_i) is proportional to the vapor pressure of its pure component (p_i^0) and the mole fraction of the component i (x_i) in the **mixture:**

$$p_i = x_i\, p_i^0 \quad \text{Eq. (8.1)}$$

where mole fraction (x_i) can be obtained from the moles of chemical i divided by the total moles of all components ($i = 1, 2,\ldots, n$) in the mixture ($\sum_{i=1}^{n} x_i = 1$).

Equation (8.1) indicates that for a single compound in its pure liquid, i.e. $x_i = 1$, its vapor pressure (p_i) will be equal to the vapor pressure of its pure liquid. If there are multiple components present in the mixture, then the vapor pressure of each individual component will be proportionally reduced according to the corresponding mole fractions of the components. Take a binary mixture of benzene and toluene, for example, if the mole fractions for benzene and toluene are $x_1 = 0.6$ and $x_2 = 0.4$, respectively, then the vapor pressure of benzene will be 60% of its pure liquid vapor pressure and the vapor pressure of toluene will be 40% of its pure liquid vapor pressure. Here the vapor pressures of the pure liquids (p_i^0) are the tabulated values at a given temperature available from literature. The p_i^0 values at temperatures other than tabulated temperature conditions can be estimated by the **Clausius–Clapeyron equation** that describes the temperature dependency of vapor pressure.

$$\ln\frac{p_{i1}}{p_{i2}} = \frac{\lambda_i}{R}\left(\frac{1}{T_2} - \frac{1}{T_1}\right) \quad \text{Eq. (8.2)}$$

where p_{i1} = vapor pressure at temperature T_1 in Kelvin, p_{i2} = vapor pressure at temperature T_2 in Kelvin, λ_i = molar heat of vaporization (kJ/mol) assumed to be a constant at the temperature range of T_1 to T_2, and R = ideal gas law constant (0.0821 atm-L/mol-K).

Note that the partial pressure (p_i) of a chemical in the unit of atm in air can be easily converted into parts per million on the volume basis (ppm$_v$), since the total atmospheric pressure is equal to 1 atm at the sea level. For example, if the vapor pressure of benzene in air is 0.0015 atm, the benzene concentration is equal to $0.0015 \times 10^6 = 1500$ ppm$_v$. The vapor pressure can also be directly converted into the mass per unit volume concentration unit such as mg/L, or more commonly mg/m^3 for atmospheric pollutants, using the ideal gas law ($p\,V = n\,RT$, or $n/V = p/RT$) in the form of:

$$C_i = \frac{n}{V} \times \text{MW}_i = 10^6 \times \frac{p_{i} \times \text{MW}_i}{RT} \quad \text{Eq. (8.3)}$$

where C_i = vapor concentration of component i reported in mass (mg) per unit volume (m^3) converted from the product of mole concentration (n/V in moles per litter) and the molecular weight (MW$_i$). Equation 8.3 is the same as Eq. (2.2) in Chapter 2 for a single-component system. Examples 8.3 and 8.4 illustrate the use of Eqs. (8.1) and (8.3).

Example 8.3 Calculate vapor concentrations using ideal gas law

a) In a dry season, a tank of pure TCE is released into a vadose zone sandy loam soil, what would be its saturated vapor concentration in soil? Report the concentration in mg/m^3 and ppm$_v$.
b) SVE requires a minimal vapor pressure be greater than 0.5 mmHg (0.000 658 atm), what would be its concentration in soil vapor in mg$_{TCE}$/L? TCE has a molecular weight of 131, vapor pressure of 0.08 atm, and dimensionless Henry's law constant of 0.42 at 20 °C.

Solution

a) Since TCE is the only chemical, the mole fraction (x_i) in Eq. (8.1) would be 1.0. For a pure liquid, we can convert the vapor pressure ($p^\circ = 0.08\,\text{atm}$) directly into ppm_v:

TCE in ppm_v = 0.08 atm TCE per 1 atm air = 0.08 volume TCE per unit volume of air = 0.08×10^6 volume TCE per million volume of air = 80 000 ppm_v

The ideal gas law in the form of Eq. (8.3) is used to directly estimate its saturated vapor phase concentration:

$$C\left(\frac{\text{mg}}{\text{m}^3}\right) = 10^6 \times \frac{p_i \times MW_i}{RT} = 10^6 \times \frac{0.08\,\text{atm} \times \frac{131\,\text{g}}{1\,\text{mol}_{\text{TCE}}}}{0.0821\frac{\text{atm-L}}{\text{mol-K}} \times (273+20)\text{K}}$$

$$= 436\,000\frac{\text{mg}}{\text{m}^3_{\text{air}}}\left(= 436\frac{\text{g}}{\text{m}^3_{\text{air}}}\right)$$

b) Using the same equation as in (a) and substitute the vapor pressure of 0.5 mmHg minimally required for SVE. Note that mmHg should be converted into atm (1 atm = 760 mmHg).

$$\text{mole concentration} = \frac{n}{V} = \frac{p}{RT} = \frac{0.5\,\text{mmHg} \times \frac{1\,\text{atm}}{760\,\text{mmHg}}}{0.0821\frac{\text{atm-L}}{\text{mol-K}} \times (273+20)\text{K}} = 2.73 \times 10^{-5}\frac{\text{mol}_{\text{TCE}}}{\text{L}}$$

We can then convert the molar concentration into mass/volume concentration using the molecular weight:

$$C = 2.73 \times 10^{-5}\frac{\text{mol}_{\text{TCE}}}{\text{L}} \times \frac{131\,000\,\text{mg}}{1\,\text{mol}_{\text{TCE}}} = 3.58\frac{\text{mg}_{\text{TCE}}}{\text{L}_{\text{air}}} = 3\,580\frac{\text{mg}_{\text{TCE}}}{\text{m}^3_{\text{air}}}$$

In this example, the vapor pressure of TCE (0.08 atm) is much higher than the minimally required by the SVE (0.000 658 atm). If TCE is saturated in the vadose zone soil, the removal of its vapor by SVE would reach a concentration as high as 436 g/m^3 in air!

Example 8.4 Use of Raoult's law to estimate vapor phase concentrations in an NAPL mixture

A dry cleaning site is contaminated with a 10-kg pool (free phase) of solvents consisting of 50% PCE and 50% TCE by weight on the soil surface. This site is to be cleaned up by soil vapor extraction. Estimate the saturated vapor concentrations of PCE and TCE in the vadose zone soil at 20 °C. The vapor pressures of PCE and TCE at 20 °C are 0.02 and 0.08 atm, respectively. MW of PCE = 166 g/mol and MW of TCE = 131 g/mol.

Solution

This question regards the use of Raoult's law for a binary DNAPL system. For a total mass of 10 kg (10 000 g), PCE and TCE would have 5 000 g each. We now can calculate the moles (n) and mole fractions (x) for PCE and TCE.

$$n_1\,(\text{PCE}) = 5000\,\text{g}/(166\,\text{g/mol}) = 30.1\,\text{mol}$$
$$n_2\,(\text{TCE}) = 5000\,\text{g}/(131\,\text{g/mol}) = 38.2\,\text{mol}$$

x_1 (PCE) = 30.1/(30.1 + 38.2) = 0.441
x_2 (TCE) = 38.2/(30.1 + 38.2)=0.559

Note that the sum of mole fractions is $x_1 + x_2 = 0.441 + 0.559 = 1.0$ for this binary system. Using Raoult's law:

p_1 (PCE) = $x_1 \times p_1^0$ = 0.441 × 0.02 = 0.008 82 atm (=8 822 ppm$_v$)
p_2 (TCE) = $x_2 \times p_2^0$ = 0.559 × 0.08 = 0.04471 atm (=44710 ppm$_v$)

The volumetric (molar) composition of the extracted vapor = 0.00882/(0.00882+0.04471) = 0.165 (16.5%) for PCE, and 1 − 0.165 = 0.835 (83.5%) for TCE.

Applying Eq. (8.3) to calculate the mass concentration from their respective vapor pressures:

$$C_1\left(\frac{\text{mg}}{\text{m}^3}\right)(\text{PCE}) = 10^6 \times \frac{p_1 \times \text{MW}_1}{RT} = 10^6 \times \frac{0.008\,82\,\text{atm} \times \dfrac{166\,\text{g}}{1\,\text{mol}_{\text{PCE}}}}{0.0821\dfrac{\text{atm-L}}{\text{mol-K}} \times (273+20)\text{K}} = 60\,860\frac{\text{mg}}{\text{m}^3_{\text{air}}}$$

$$C_2\left(\frac{\text{mg}}{\text{m}^3}\right)(TCE) = 10^6 \times \frac{p_2 \times \text{MW}_2}{RT} = 10^6 \times \frac{0.04471F \times \dfrac{131\text{g}}{1\text{mol}_{\text{TCE}}}}{0.0821\dfrac{\text{atm-L}}{\text{mol-K}} \times (273+20)\text{K}} = 243\,480\frac{\text{mg}}{\text{m}^3_{\text{air}}}$$

The weight composition (w) of the extracted vapor can be calculated as follows:

w_1 (PCE) = 60.86/(60.86 + 243.48) = 0.20 (20.0%).
w_2 (TCE) = 243.48/(60.86 + 243.48) = 0.80 (80.0%)

The above results indicate that the extracted vapor has more TCE than PCE based on both volume and weight. If the mass percentages are compared between the extracted vapor and the original spilled solvents (i.e. 50% each), the extracted vapor has more TCE than PCE. This is typical for any SVE system, because the extracted vapor would contain more compounds with higher volatilities (here TCE is more volatile than PCE). As SVE operation proceeds, less volatile compounds will become increasingly predominant in the extracted vapor, resulting in lower removal efficiency by SVE over time (Kuo 1999).

As illustrated in Example 8.4, Raoult's law can be employed to determine the saturated vapor pressure in equilibrium with free-phase NAPLs. Raoult's law is also applicable to ideal aqueous solutions in predicting how solutes will impact the vapor pressure of the solvent (i.e. water). If a nonvolatile chemical (e.g. glucose) is dissolved in water, the vapor pressure of water will be reduced in the proportion of water's mole fraction. If a volatile chemical (e.g. benzene) is dissolved into the water, the vapor pressure of water will be reduced by its reduced mole fraction. Since benzene is volatile, the total vapor pressure of this solution will be the vapor pressure of the water vapor plus that of benzene vapor. The change of water vapor pressure might be minor, because of the low aqueous solubility of benzene as a solute.

8.2.3 Kinetics of Volatilization, Vapor Diffusion, and NAPL Dissolution

Volatilization: In Chapter 2, the ideal gas law was applied to relate the vapor phase concentration to the vapor pressure of its pure liquid at equilibrium (Eqs. 2.1 and 2.2). This calculated vapor phase concentration at equilibrium will be overestimated if volatilization is rate-limited kinetically. The volatilization rate (J, mol/m^2-h) is typically assumed to be directly proportional to the concentration gradient between the dissolved phase concentration (C_{aq}) (i.e. the actual concentration that can be measured) and the theoretical (calculated) aqueous phase concentration in equilibrium with the vapor phase concentration (C_i/H):

$$J = -K_L\left(C_{aq} - \frac{C_i}{H}\right) \quad \text{Eq. (8.4)}$$

where K_L = mass transfer coefficient (m/h), C_{aq} = aqueous phase concentration (g/m^3); C_i = gas phase concentration; H = Henry's law constant. From the **Conservation of Mass** ($dC/dt = -dJ/dL$), Eq. (8.4) can be written in the format of the change of aqueous phase concentration as a function of time:

$$\frac{dC_{aq}}{dt} = \frac{K_L}{L}\left(C_{aq} - \frac{C_i}{H}\right) \quad \text{Eq. (8.5)}$$

where L = depth in meter (i.e. unit volume interfacial area). Integration of Eq. (8.5) will result in the dissolved phase concentration of contaminant i at a given time t:

$$C_{aq} = \frac{C_i}{H} + \left(C_{aq}^0 - \frac{C_i}{H}\right)\exp\left[\left(-\frac{K_L}{L}\right)t\right] \quad \text{Eq. (8.6)}$$

When the gas phase concentration (C_i) is very low compared to the liquid phase concentration, which is usually true in actual remediation systems, the above equation can be simplified:

$$C_{aq} = \left(C_{aq}^0\right)\exp\left[\left(-\frac{K_L}{L}\right)t\right] \quad \text{Eq. (8.7)}$$

Equation (8.7) reveals that the volatilization can be described as the first-order kinetics. A laboratory column study simulating the volatilization of VOCs from the stagnant water in a monitoring well (2-in diameter, 4-ft long piezometer) revealed the first-order kinetics (McAlary and Barker 1987). Volatilization losses from stagnant wells reached 10% within 12 hours and 50% in 4 days. Their results imply that, without the pressure-induced airflow through vacuum or sparging, volatilization could be a rate-limited slow process if the groundwater recovery is slow, such as the case for low-permeability aquifer.

Vapor Diffusion: The diffusion of vapor is driven by the concentration gradient within a phase (intraphase chemical movement). The diffusion kinetics for compound i can be described by **Fick's second law**:

$$\frac{\partial C_{aq}}{\partial t} = D\frac{\partial^2 C_{aq}}{\partial^2 x} \quad \text{Eq. (8.8)}$$

where D = molecular diffusion coefficient (cm^2/s) and x = distance of diffusion along x-axis. The analytical solution to Fick's second law for C_{aq} at any time t and distance x, after defining one initial condition and two boundary conditions, can be obtained as follows (Grasso 1993):

$$\frac{C_{aq} - C^*}{C^0 - C^*} = \frac{2}{\sqrt{\pi}}\int_0^{x/\sqrt{4Dt}} e^{-n^2}\,dn = \text{erf}\left(\frac{x}{2\sqrt{Dt}}\right) \quad \text{Eq. (8.9)}$$

where C^0 = initial concentration at t = 0, C^* = saturation concentration, and erf is the **error function**. Using Eq. (8.9), one can determine the concentration at a specific time (t) and location (x) given the diffusivity constant and the concentration gradient. The slow diffusion process can become important in slow-moving groundwater or contaminant movement in engineering systems, for example, when contaminants in leachate move across landfill liners toward underlying groundwater.

NAPL Dissolution: The dissolution of NAPLs in the form of discrete blobs or pools of free phase will contribute greatly to the quality of subsurface soil and groundwater. The dissolution can be described as follows (Clement et al. 2004):

$$\frac{\rho}{\theta}\frac{dC_s}{dt} = -K_{La}\left(C^* - C\right) \qquad \text{Eq. (8.10)}$$

where ρ is the bulk density of the aquifer (M/L^3), θ is the aquifer porosity, C_s is concentration of NAPL trapped in soil (M/M, mg/kg), C^* is the aqueous phase solubility of NAPL (M/L^3), C is the aqueous phase concentration of NAPL (M/L^3), and K_{La} is lumped mass transfer rate coefficient (T^{-1}). The mass transfer rate coefficient K_{La} is a function of porous media properties, NAPL saturation, and geometry including interfacial area and flow condition and therefore, may exhibit spatial and temporal variations (Mobile et al. 2012).

At this point, one should note the "kinetic" aspect of the above processes (volatilization, diffusion, dissolution, etc.) because the **rate-limiting process** will be the controlling step for contaminant removal from soil-groundwater system. In other words, for a NAPL contaminant in the adsorbed phase (on soil particles), it must be desorbed from soil, dissolved in water, and then volatized to be removed (desorption + dissolution + volatilization). Any of these processes can become the rate-limiting step in the mass transfer. A modeling study by Garges and Baehr (1998) suggested the dominant transport mechanism of diffusion for PCE transport, advection for MTBE transport, and advection moving toward diffusion for benzene. The hydrocarbons at the Bemidji site, Minnesota were found to be transported mainly by diffusion (with some advection) in the unsaturated zone, and by advection and mechanical dispersion in the saturated zone (Essaid et al. 2011).

8.2.4 Darcy's Law for Advective Vapor Flow

The flow and transport of vapor in porous media in response to concentration gradient at the interfaces of air–water and NAPL free-phase water were discussed in the preceding section. Both diffusion (subsequent to dissolution and desorption) and advection are known to potentially cause the mass transfer limiting step. Under pressure-induced conditions such as the cases of vacuum extraction (SVE) or air injection (IAS), the pressure gradient will be the dominant driving force in controlling the advective flow of vapor in the soil–groundwater system. **Darcy's law** used to describe the advective *flow of water* (Chapters 2 and 3) can also be used to describe the *flow of vapor* by advection:

$$v = -K\frac{d\left(z + \dfrac{P}{\rho g}\right)}{dx} \qquad \text{Eq. (8.11)}$$

Note that the hydraulic head h is substituted by $z + P/\rho g$ in the vadose zone, where v is the darcian vapor velocity (cm/s), z is the potential head (the elevation above mean sea level, MSL), and $P/\rho g$ is the pressure head.

From Eq. (3.5) in Chapter 3, we recall the relationship between hydraulic conductivity (K) and the intrinsic permeability (k):

$$K = k\frac{\rho g}{\mu}$$

where μ is dynamic vapor viscosity (g/cm-s). By substituting the above equation into Eq. (8.11) and neglecting the potential head change ($dz/dx = 0$), we obtain:

$$v = -k\frac{\rho g}{\mu}\left(\frac{dz}{dx} + \frac{1}{\rho g}\frac{dP}{dx}\right) = -\frac{k}{\mu}\frac{dP}{dx} \qquad \text{Eq. (8.12)}$$

The above Darcy's law indicates that the Darcy's velocity for the vapor flow (v) is directly proportional to the pressure gradient in the soil vapor extraction with a proportionality constant related to the **intrinsic permeability** and the **dynamic viscosity**.

To derive a vapor flow equation, we need a **Continuity Equation** (conservation of vapor):

$$\frac{\partial n\rho_b}{\partial t} = -\nabla \rho b v \qquad \text{Eq. (8.13)}$$

where n = air-filled porosity, ∇ = gradient operator (cm^{-1}), b = aquifer thickness (cm), v = vapor velocity (cm/s), and ρ = vapor density (g/cm^3) which varies with the vapor pressure change according to the ideal gas law:

$$\rho = \rho_{atm}\frac{P}{P_{atm}} \qquad \text{Eq. (8.14)}$$

where ρ_{atm} = vapor density at the reference pressure P_{atm}. By substituting v (Eq. 8.12) and ρ (Eq. 8.14) into the continuity equation (Eq. 8.13) (Grasso 1993; USEPA 1995), we obtain:

$$\left(\frac{2n\mu}{k}\right)\frac{\partial P}{\partial t} = -\nabla^2 P^2 \qquad \text{Eq. (8.15)}$$

In the case of steady-state ($\partial P/\partial t = 0$) and radial flow subject to the boundary conditions of: $P = P_w$ at $r = R_w$; $P = P_{atm}$ at $r = R_I$ (Figure 8.5), the solution to Eq. (8.15) for a homogeneous or layered soil system is (Johnson et al. 1990a; USEPA 1995):

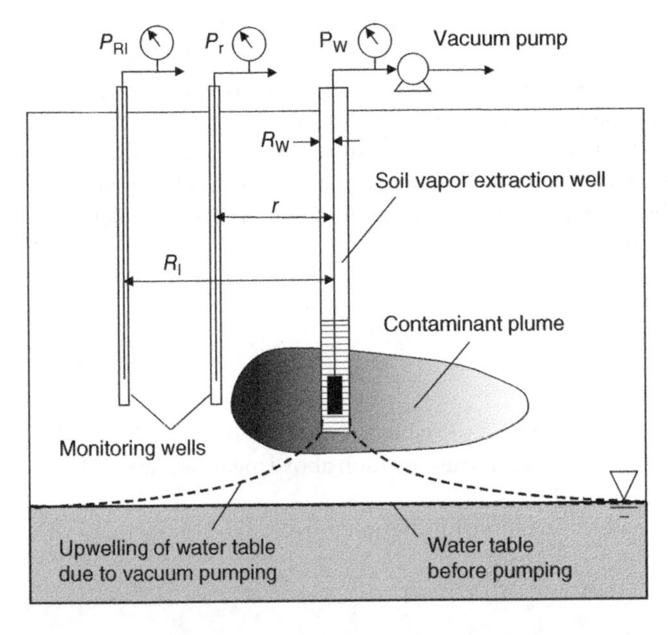

Figure 8.5 Schematic of soil vapor extraction well and the monitoring wells showing the well locations and vacuum pressures.

$$P_r^2 - P_w^2 = \left(P_{RI}^2 - P_w^2\right)\frac{\ln\left(\frac{r}{R_w}\right)}{\ln\left(\frac{R_I}{R_w}\right)} \qquad \text{Eq. (8.16)}$$

where P_r = pressure at a radial distance r from the vapor extraction well, P_w = pressure at the vapor extraction well, P_{RI} = pressure at the radius of influence typically set at atmospheric pressure (1 atm) or another preset value, r = radial distance from the vapor extraction well, R_I = radius of influence where pressure is equal to a preset value, and R_w = radius of the vapor extraction well. The use of Eq. (8.16) is illustrated in Example 8.5.

Example 8.5 Calculate radius of influence for a vapor extraction well

For a soil vapor extraction system, the pressure at an extraction well = 0.9 atm, and the pressure at a monitoring well 30 ft away from the extraction well = 0.98 atm. The diameter of the extraction well = 4 in. (0.33 ft). Determine the radius of influence of this soil vapor extraction well (adapted from Kuo 1999).

Solution

To calculate R_I by Eq. (8.16), we have the values of all parameters except P_{RI}, which can be assumed equal to the atmospheric pressure (1 atm). Thus,

$$0.98^2 - 0.9^2 = \left(1.0^2 - 0.9^2\right)\frac{\ln\left(\frac{30}{0.165}\right)}{\ln\left(\frac{R_I}{0.165}\right)}$$

$$R_I = 118\ \text{ft}$$

Note that Eq. (8.16) is derived from Eq. (8.15) for a steady-state radial flow of vapor in homogeneous porous media. Interested readers can also find the solution to Eq. (8.15) for vapor flow under unsteady-state condition. The governing equation developed by Johnson et al. (1990a) can be used to determine the intrinsic permeability of air in soil by conducting a field air permeability test analogous to pumping test for the acquisition of hydraulic conductivity (see Chapter 3). Intrinsic permeability (k) is best determined in the field by conducting permeability tests or SVE pilot studies, but it can also be estimated within one or two orders of magnitude accuracy in the laboratory using soil core samples from the site. At sites where soils in the saturated zone are similar to those within the unsaturated zone, hydraulic conductivity of soils is related to the air permeability of the soils (Eq. 3.5, Chapter 3). Since air permeability can be readily measured, hydraulic conductivity can also be estimated empirically by the regression. Box 8.1 provides some additional information regarding hydraulic conductivity, intrinsic permeability, and air permeability.

Box 8.1 Hydraulic conductivity, intrinsic permeability, and air permeability

a) Hydraulic conductivity (K_w) defines how easily "water" flows in a porous medium. It depends on the properties of both water (density and viscosity) and soil (porosity, soil texture, etc.). Hydraulic conductivity in the saturated zone is referred to as the **saturated hydraulic conductivity**, in comparison with the **unsaturated hydraulic conductivity** in the unsaturated zone. Since water is a variable in the unsaturated zone, unsaturated hydraulic conductivity depends on water-filled porosity (i.e. soil water content).
b) Intrinsic permeability of "water" depends solely on the property of porous media rather than the property of water (or in general, the fluid). This is why it is termed "intrinsic" permeability. Intrinsic permeability of water (k_w) is related to the saturated hydraulic conductivity according to $K = k(\rho g/\mu)$. Hence, hydraulic conductivity has a unit of length/time (cm/s), whereas intrinsic permeability has a unit of length^2 (cm^2 or darcy). For example, water has a dynamic viscosity (μ) of 0.01 g/cm-s and density (ρ) of 1 g/cm^3 at 20 °C, hence:

$$\frac{\rho g}{\mu} = \frac{1\frac{\text{g}}{\text{cm}^3} \times 981\frac{\text{cm}}{\text{s}^2}}{0.01\frac{\text{g}}{\text{cm} \times \text{s}}} = 9.81 \times 10^4 \frac{1}{\text{cm} \times \text{s}} \cong 10^5 \frac{1}{\text{cm} \times \text{s}}$$

It is clear that hydraulic conductivity (K, cm/s) can be calculated by multiplying intrinsic permeability (k, cm^2) by 10^5. Note that a common non-SI unit of intrinsic permeability is darcy, which is referenced to a mixture of unit systems. A medium with a permeability of 1 **darcy** permits a flow of 1 cm^3/s of a fluid with viscosity 1 centipoise (1 cP = 1 mPa·s = 0.01 g/cm/s) under a pressure gradient of 1 atm/cm acting across an area of 1 cm^2. From Eq. (8.11), we have:

$$v = -\frac{k}{\mu}\frac{dP}{dx}$$

By rearranging intrinsic permeability (k) as a function of other determining factors, we obtain:

$$k(\text{darcy}) = -\frac{v \times \mu}{dP/dx} = -\frac{\frac{1\text{cm}^3/\text{s}}{1\text{cm}^2} \times 1\text{mPa-s}}{1\frac{\text{atm}}{\text{cm}} \times \frac{1.01 \times 10^8\,\text{mP}}{1\text{atm}}} \cong 10^{-8}\,\text{cm}^2$$

Therefore, to convert k from cm^2 to darcy, one needs to multiply k value in cm^2 by 10^8. For example, a soil with an intrinsic permeability of 5×10^{-8} cm^2 will have 5 darcy. If we are further given the density and viscosity of the fluid (i.e. water) at another temperature, say a density of water at 0.999 703 g/cm^3 and viscosity of water at 0.013 07 g/s-cm at 10 °C, the hydraulic conductivity of this soil for the water flow at 10 °C is (Eq. 3.5, Chapter 3):

$$K = k\frac{\rho g}{\mu} = 5 \times 10^{-8}\,\text{cm}^2 \times \frac{0.999\,703\frac{\text{g}}{\text{cm}^3} \times 980\frac{\text{cm}}{\text{s}^2}}{0.013\,07\frac{\text{g}}{\text{s-cm}}} = \left(5 \times 10^{-8}\,\text{cm}^2\right) \times \left(7.5 \times 10^4\right) = 3.75 \times 10^{-3}\frac{\text{cm}}{\text{s}}$$

Air permeability (k_a) can be defined as a measure of the ability of soil to conduct air (not water) by convective flow, which is the movement of soil air in response to a gradient of total gas pressure.

Since the pores in the soil are generally occupied by air and water, air permeability is inversely related to soil water content, and can be expressed as a function of **air-filled porosity** (or volumetric air content). Also, similar to water conductivity (which depends on **water-filled porosity**, i.e. soil water content), air permeability is related to the total air-filled porosity, the pore size distribution, and the connectivity of the open pores in the soil (which is related to the type of soil). For a completely dry soil, air permeability is considered to be equal to the intrinsic permeability (k) of the soil (both have a unit of cm^2 or darcy). Generally, a saturated hydraulic conductivity (K_w) and air permeability (k_a) cannot be converted from one to the other, although studies have shown the log–log correlation between the two. Note that in this text, the subscript "w" for K_w and "a" for k_a are omitted, therefore K and k are used to denote hydraulic conductivity to water and permeability to air, respectively.

8.3 Design for Vapor Extraction and Air Sparging Systems

The quantitative analysis and design equations in this section are excerpted mostly from Johnson et al. (1990a, 1990b) for their pioneering work on soil vapor extraction. These mathematical equations are the simplifications of the real-world scenario. Regardless, they are very useful in screening level analysis to decide whether soil venting is an appropriate cleanup technology, and determine the primary design parameters for the successful implementation of SVE/IAS. Numerous models have been available for vadose zone analysis (Bedient et al. 1999), including screening models (e.g. BioSVE, HyperVentilate, see Box 8.2), airflow models (e.g. MODAIR, P3DAIR), multiphase models (e.g. T2VOC, Bioventing), and flow and transport models in an unsaturated zone (e.g. Chemflo, AIRFLOW/SVE, 3DFEMFAT). In the discussion below, we focus on the design basics for soil vapor extraction.

8.3.1 Quantitative Analysis for the Appropriateness of Soil Venting

In the screening model developed by Johnson et al. (1990b), a set of six questions are answered to decide if venting is appropriate. (1) What contaminant vapor concentrations are likely to be obtained? (2) Under ideal vapor flow conditions (i.e. 100–1000 scfm vapor flow rates), is this concentration great enough to yield acceptable removal rates? (3) Can the range of vapor flow rates be realistically achieved? (4) Will the contaminant concentrations and realistic vapor flow

Box 8.2 Computer models to assist SVE design: Hyperventilate

Computer models are available to predict the flow of vapor in porous media, such as Airtest, AIRFLOW, and AIR3D. These models, however, do not directly assist design engineers in designing SVE systems. Two other models, VENTING and HyperVentilate, are more complete SVE screening models based on the work by Johnson et al. (1990b) of Shell Oil Company. The basic equations have been introduced in Sections 8.3.1 and 8.3.2.

HyperVentilate can evaluate the soil permeability, estimate the airflow rate, contaminant residual concentrations, contaminant removal rates, and various SVE system design parameters, such as number of extraction wells needed, cleanup time, and system operating conditions. HyperVentilate can guide the users through a structured decision-making process to determine the appropriateness and effectiveness of a SVE system to remediate the site with required site-specific data.

rates produce acceptable removal rates? (5) Will a significant residual remain in the soil? Will vapor concentrations exceed the regulatory requirements? (6) Are there likely to be negative effects of soil venting? Negative answer to Question 2 or 4 will rule out the *in situ* use of soil venting as a practical treatment method.

The estimated vapor concentrations (C_{est}) in Question 1 can be calculated according to Raoult's law and ideal gas law (Eqs. 8.1, 2.1, 2.2) for mixed components such as gasoline. Note that C_{est} is the maximal concentration at the start of venting, or the best-case scenario regarding the contaminant removal. The actual vapor concentrations will decline over time due to the changes in composition (see Example 8.4), residual levels, or increased diffusional resistances.

The "acceptable" removal rate ($R_{acceptable}$) in Question 2 is calculated from the actual spill mass (M_{spill}) divided by the maximum acceptable cleanup time (t):

$$R_{acceptable} = \frac{M_{spill}}{t} \qquad \text{Eq. (8.17)}$$

The vapor concentration (C_{est}) can be used to estimate contaminant vapor removal (R_{est}) according to:

$$R_{est} = C_{est} \times Q \qquad \text{Eq. (8.18)}$$

Under an ideal condition, the vapor flow rate Q is in the range of 100–1000 scfm (standard cubic feet per minute) or 140–2800 L/min (1 ft^3 = 28.3 L). Johnson et al. (1990a) indicated a removal rate of greater than 1 kg/day will be needed for a continued use of soil venting. This minimally required mass removal rate is equivalent to a minimal vapor concentration of 0.3 mg/L (calculated from R = 1 kg/day, Q = 140 L/min), or an equivalent pure component vapor pressure of 0.0001 atm at the subsurface temperature. The contaminant concentration is sometimes reported in the form of methane equivalency (ppm_v CH_4). This unit is used because field analytical tools always report ppm_v values which are often calibrated with methane.

For example, if an average 6.25 kg/day (i.e. to clean up an approximately 500 gal or 1500 kg gasoline spill within eight months) is desired using a flow rate of 2800 L/min (100 cfm) in a venting process, Eq. (8.18) can be used to directly calculate the minimal vapor concentration required as follows:

$$C_{est} = \frac{R_{est}}{Q} = \frac{6.25\dfrac{\text{kg}}{\text{day}}}{2800\dfrac{\text{L}}{\text{min}}} \times \frac{1\,\text{day}}{24 \times 60\,\text{min}} \times \frac{10^6\,\text{mg}}{1\,\text{kg}} = 1.55\frac{\text{mg}}{\text{L}}$$

The calculated concentration of 1.55 mg/L (i.e. 1550 mg/m^3) is equal to the methane (CH_4) equivalent 2373 ppm_v as follows:

$$ppm_v\,CH_4 = 1550\frac{\text{mg}}{\text{m}^3} \times \frac{24.5}{16} = 2373$$

The above calculation is the conversion of mg/m^3 into ppm_v assuming a constant ideal gas volume of 24.5 L/mol at 25 °C, and molecular weight of 16 for CH_4.

Question 3 estimates the realistic vapor flow rates (Q) for the site-specific conditions according to the following equation:

$$\frac{Q}{H} = \frac{\pi k}{\mu} P_w \frac{\left[1 - \left(\frac{P_{atm}}{P_w}\right)^2\right]}{\ln \frac{R_w}{R_I}} \qquad \text{Eq. (8.19)}$$

where Q/H = flow rate (cm^3/s) per unit thickness of the well screen (cm). H is the length of the well screen, which is typically the same as the thickness of the contaminated aquifer (cm), k = soil permeability to airflow (cm^2, or darcy), μ = viscosity of air = 1.8×10^{-4} g/cm-s, P_w = absolute pressure at the extraction well (g/cm-s^2) or atm, P_{atm} = absolute ambient pressure (1.01×10^6 g/cm-s^2 or 1 atm), R_w = radius of a vapor extraction well (cm), and R_I = radius of influence of the vapor extraction well (cm).

Equation (8.19) indicates that the required flow rate increased linearly with the increased air permeability and it is inversely proportional to the viscosity among several other factors as shown in Figure 8.6. The strong dependence of vapor flow rate on air permeability (k) is evident, since k has several orders of magnitude difference depending on soil conditions. In Eq. (8.19), the extraction well vacuum pressure (P_w) is in the range of 0.40–0.95 atm and the radius of extraction well (R_w) by default is 5.1 cm (4 in). It was shown that Q/H is insensitive to the radius of influence (R_I) which is in the range of 9–30 m (30–100 ft). However, a default R_I value of 12 m can be assumed (see the top legend in Figure 8.6). If these default values can be used, Figure 8.6 can be employed as an alternative to Eq. (8.19) to predict the steady-state airflow rate per unit well screen thickness (i.e. the y-axis: Q/H) for a range of soil permeability (k) and applied vacuum (P_w). Corrections under other R_w and R_I conditions are detailed in Johnson et al. (1990b). Example 8.6 will illustrate these two methods (Eq. 8.19 or Figure 8.6) to estimate Q/H.

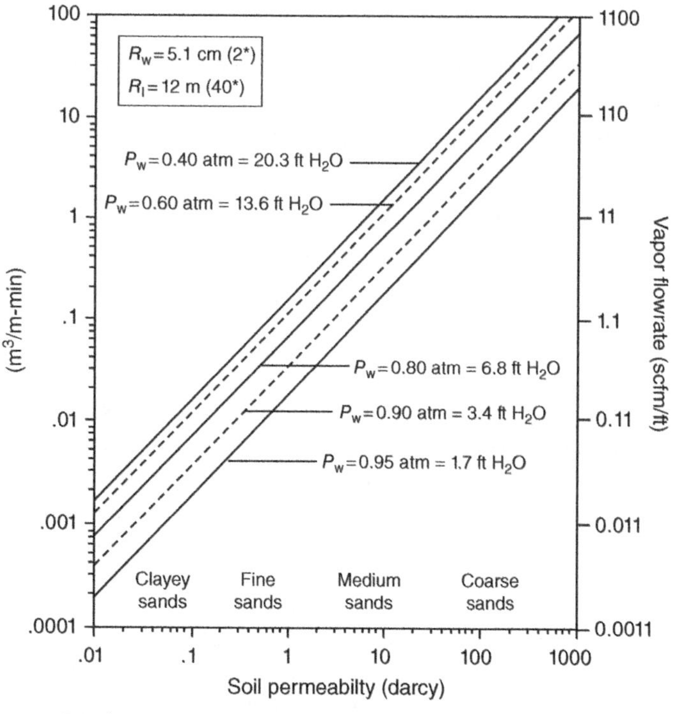

Figure 8.6 Predicted steady-state flow rates per unit well screen thickness (m^3/m-min) for a range of soil permeability (darcy) and applied vacuum (P_w in atm). Assumed values: R_w = 5.1 cm (2 in.); R_I = 12 m (40 ft).

Example 8.6 Design of required airflow rate in a vapor extraction well

Given the following conditions: air permeability (k) of 10^{-8} cm^2, radius of extraction well (R_w) = 5.1 cm, radius of vacuum influence (R_I) = 12 m, vacuum pressure (P_w) = 0.90 atm (i.e. 9.09×10^5 g/cm-s^2), and air viscosity of 1.8×10^{-4} g/cm-s. (a) Find the airflow rate per unit thickness of well screen Q/H by using Figure 8.6 and applying Eq. (8.19) to verify, (b) If the well is screened to the entire contaminated zone of 5 m thick, what is the airflow rate in cfm?

Solution

a) By multiplying the k value of 10^{-8} cm^2 by 10^8, we convert air permeability from 10^{-8} cm^2 to 1.0 darcy. From Figure 8.6, we locate 1.0 darcy on the x-axis, draw a vertical line to intersect the line of P_w = 0.90 atm. From this intersecting point, we draw a horizontal line toward the y-axis to get the value of Q/H, which is 0.04 m^3/m-min for k = 1.0 darcy, P_w = 0.90 atm, and default R_w = 5.1 cm. Alternatively, if we apply Eq. (8.19), we can also calculate the Q/H:

$$\frac{Q}{H}=\frac{\pi k}{\mu}P_w\frac{\left[1-\left(\frac{P_{atm}}{P_w}\right)^2\right]}{\ln\frac{R_w}{R_I}}=\frac{\pi\times10^{-8}\,\text{cm}^2}{1.8\times10^{-4}\,\frac{\text{g}}{\text{cm}\times\text{s}}}\times9.09\times10^5\,\frac{\text{g}}{\text{cm}\times\text{s}^2}\times\frac{\left[1-\left(\frac{1\,\text{atm}}{0.90\,\text{atm}}\right)^2\right]}{\ln\frac{5.1\,\text{cm}}{1200\,\text{cm}}}=6.81\,\frac{\text{cm}^3}{\text{cm}\times\text{s}}$$

Convert this unit into m^3/m-min:

$$\frac{Q}{H}=6.81\,\frac{\text{cm}^3}{\text{cm}\times\text{s}}\times\frac{1\,\text{m}^3}{10^6\,\text{cm}^3}\times\frac{100\,\text{cm}}{1\,\text{m}}\times\frac{60\,\text{s}}{1\,\text{min}}=0.041\,\frac{\text{m}^3}{\text{m}\times\text{min}}$$

This is in good agreement with the Q/H obtained schematically using Figure 8.6 (0.041 vs. 0.04 m^3/(m × min)).

b) If the well screen H = 5 m, then:

$$Q=0.041\,\frac{\text{m}^3}{\text{m}\times\text{s}}\times5\,\text{m}=0.205\,\frac{\text{m}^3}{\text{s}}=0.205\,\frac{\text{m}^3}{\text{s}}\times\frac{60\,\text{s}}{1\,\text{min}}\times\frac{35.3\,\text{ft}^3}{1\,\text{m}^3}=423.6\,\text{cfm}$$

This airflow rate (Q) falls in an ideal range of 100–1000 cfm for soil vapor extraction.

Question 4 concerns whether the estimated removal rate (R_{est}) meets the requirement of $R_{est} > R_{acceptable}$? The maximum removal rate is achieved only through the zone of contaminated soil and the mass transfer is limited. In other words, all vapors flow through contaminated soils and become saturated with contaminant vapors. This is the best-case scenario as estimated by Eq. (8.18). Other less optimal conditions are accounted for by the modified equations as detailed by Johnson et al. (1990b). These situations include: (i) If a fraction (ϕ) of vapor flows through an uncontaminated soil, then $R_{est} = (1-\phi)\,C_{est}\,Q$; (ii) If vapor flows parallel to, but not through, the zone of contamination, then $R_{est} = \eta\,C_{est}\,Q$, where η is the efficiency relative to the maximum removal rate (Johnson et al. 1990a for details). This is the case when a significant mass transfer resistance occurs due to vapor diffusion, for example, a layer of NAPL rests on top of an impermeable strata or the water table; (iii) If vapor flows primarily past, rather than through the contaminated soil zone, such as the case for a contaminated clay lens surrounded by sandy soils. In this case, vapor diffuses through the clay, resulting in a limited removal rate.

Question 5 concerns the residual contaminant left in soil in regard to the vapor compositions and concentrations compared to the regulatory requirements. As venting proceeds, a small percentage of residual soil contamination level decreases and the mixture compositions become richer in less volatile compounds. More vapor flow will be needed in order to remove this residual contamination. The maximum efficiency of a venting operation is limited by the equilibrium partitioning of contaminants between the soil matrix and vapor phases affected by the sorption and dissolution processes (Figure 8.4, and Section 8.2.2). This can be predicted by mathematical models, although a complete discussion is not appropriate here. Because of such limitations, one should consider other polishing techniques such as *in situ* bioremediation.

The last question is given to the consideration of any negative effects of soil venting, including the migration of off-site contaminants from other responsible parties in the neighboring property, or the **upwelling** (refer to water table rise in Figure 8.5) at the center of vacuum extraction well. The solution to the off-site gas migration is to establish a vapor barrier by allowing vapor flow into any perimeter groundwater monitoring wells, which then act as passive air supply wells. The solution for the water table rise is to install water pumping well(s) near the vacuum extraction wells.

8.3.2 Well Number, Flow Rate, and Well Location

If the screening model reveals that venting is a feasible option, design engineers will use the chemical and hydrogeological data along with other site information to come up with the number of extraction wells, well spacing, well construction, etc.

There are several methods to choose the number of vapor extraction wells. The greatest number of well from these methods will be selected for soil venting. The *first method* of estimating the well number (N_{wells}) is based on the ratio of acceptable removal rate (kg/day) to the estimated removal rate (kg/day) per well:

$$N_{wells} = \frac{R_{acceptable}}{R_{est}} \quad \text{Eq. (8.20)}$$

Equations (8.17) and (8.18) provide the formula to calculate $R_{acceptable}$ and R_{est} used in the above equation. The *second method* is to use the ratio of contaminated area to the area of influence of a single venting well:

$$N_{wells} = \frac{A_{contamination}}{\pi R_I^2} \quad \text{Eq. (8.21)}$$

Reported R_I values for permeable soils such as sandy soils at depth greater than 20 ft below ground surface or shallower soils beneath good surface seals are usually 10–40 m. R_I can be obtained by fitting radial pressure distribution data, $P(r)$, from the air permeability test (Johnson et al. 1990a) using Eq. (8.16) described previously.

The *third method* is based on the pore volume of soil. The number of wells is calculated by dividing the total extraction flow rates by the flow rate achievable with a single well. The total extraction flow rate is the rate at which the soil pore volume within the treatment area is exchanged in a reasonable amount of time (8–24 hours).

$$N_{wells} = \frac{\frac{nV}{t}}{v} \quad \text{Eq. (8.22)}$$

where n = soil porosity (m^3 vapor/m^3 soil), V = volume of soil in the treatment area (m^3 soil), v = vapor extraction rate from a single extraction well (m^3 vapor/h), and t = pore volume exchange time (hours).

Depending on the information available, all these three methods can be used to estimate the required number of vacuum wells for SVE to safeguard the removal of contaminant vapor. Example 8.7 is an exercise for this design calculation regarding the number of wells in SVE.

Example 8.7 Estimate the required number of vapor extraction wells

As a continuation of Example 8.6, if a spill of 500 gallon TCE occurs in an area of $200 \times 200\,ft^2$ of a medium-sized sandy soil, what would be the number of extraction wells required for SVE? TCE has a density of 1.46 g/mL.

Solution

Use Eq. (8.21) to calculate the number of wells, assuming the same radius of vacuum influence (R_I) = 12 m or 12×3.28 = 39.36 ft:

$$N_{wells} = \frac{A_{contamination}}{\pi R_I^2} = \frac{200 \times 200\ ft^2}{\pi \times (39.36\ ft)^2} = 8.3 \approx 8\ wells$$

If the site is required to be cleaned up within three months, then we can also use Eq. (8.20) to estimate the well number. First, we calculate the acceptable removal rate using Eq. (8.17):

$$R_{acceptable} = \frac{M_{spill}}{t} = \frac{500\ gal \times 1.46 \frac{g}{mL} \times \frac{3.79\ L}{1\ gal} \times \frac{1000\ mL}{1\ L} \times \frac{1\ kg}{1000\ g}}{90\ days} = 30.7 \frac{kg}{day}$$

Instead of using the equilibrium vapor phase concentration of 436 mg/L (see Example 8.3. This equilibrium concentration is typically not achievable under vacuum-induced advective flow condition), we assume C_{est} = 1 mg/L, and vapor flow Q of 423.6 ft^3/min (Example 8.6) to calculate the estimated removal rate (Eq. 8.18):

$$R_{est} = C_{est} \times Q = 1 \frac{mg}{L} \times 423.6 \frac{ft^3}{min} \times \frac{28.3\ L}{1\ ft^3} \times \frac{24 \times 60\ min}{1\ day} \times \frac{1\ kg}{10^6\ mg} = 17.3 \frac{kg}{day}$$

The required number of wells (Eq. 8.20):

$$N_{wells} = \frac{R_{acceptable}}{R_{est}} = \frac{30.7 \frac{kg}{day}}{17.3 \frac{kg}{day}} = 1.8 \approx 2\ wells$$

Alternatively, we can also estimate the number of wells based on the pore volume according to Eq. (8.22). Assuming 10 hours of pore volume exchange time, porosity of 0.4, and 200 cfm per well, the well number thus estimated is:

$$N_{\text{wells}} = \frac{\frac{nV}{t}}{v} = \frac{\frac{0.4 \times (200 \times 200 \times 39.36)\text{ft}^3}{10\,\text{h}}}{200\frac{\text{ft}^3}{\text{min}} \times \frac{60\,\text{min}}{1\,\text{h}}} = 2.2 \approx 2 \text{ wells}$$

To be conservative, we choose the greatest number of wells estimated from these three methods, so a total of eight wells is required for soil vapor extraction.

The next step of design is to choose the well locations for soil vapor extraction and air sparging. The basic strategy will be the same as we have discussed for the P&T systems in Chapter 7. The goal of proper well numbers, well spacing, and locations (configuration) is to ensure the radius of influence (ROI) and pore-gas velocity in the entire zone of contamination are both maximized. The radius of influence is based on the pressure-drop of the SVE system around the extraction wells. The volume contained within this radius is referred to as the **zone of influence** (ZOI). Within the ZOI, vapor will be contained, but this does not necessarily capture vapor. Another approach is based on the **zone of capture** (ZOC). ZOC uses a critical pore-gas velocity, which incorporates the effects of mass transfer limitations into the SVE design. ZOC-based design will ensure the meaningful airflow toward the extraction well. Typically, a cutoff vacuum of 2.54 cm of water (0.0025 atm) can be selected to define the boundary of ZOI, whereas a critical pore-gas velocity of 0.01 cm/s can be used to define ZOC (Dixon and Nichols 2006).

Similar to the well placement in P&T, well placement is also important in SVE/IAS. For example, if one well is sufficient, it should almost always be placed in the geometric center of the contaminated soil zone, unless vapor/water flow channeling is expected along a preferred direction. When the radii of influence of two wells overlap, dead zones are created. It was suggested to have 1.5 times of the radius of vacuum influence for SVE design. This is different from the well spacing consideration in P&T, because overlapping of wells is expected to avoid stagnant area.

The dead zone in SVE is caused by the vacuum forces of the two wells canceling each other out. In the dead zone, there is no flow and the contaminant does not move. The dead zone must be avoided by correctly placing air injection wells or passive air wells. This allows airflow to both competing wells. If three extraction wells are required at a given site, and they are installed in the triplate design shown in Figure 8.7, this would result in a large stagnant region in the middle of the three wells. The solution is to place a passive injection well or a forced air injection well in the center. The **passive well** is essentially a well open to the atmosphere to allow airflow toward three extraction wells. In practice, this passive injection well can be an existing monitoring well.

8.3.3 Other Design Considerations

Well Installations: A typical well is constructed of very basic materials such as PVC pipe with slots located in the contaminant zone, coarse sand or gravel filter packing, and bentonite pellets and cement grout over the filter packing. Vapor extraction wells are similar to groundwater monitoring wells in construction. Since vapor will flow in without the need of filtration before they enter the well, the filter packing should be as coarse as possible. Extraction wells should be screened only through the zone of contamination, unless the permeability to vapor flow is so low that removal rates would be greater if flows are induced in an adjacent soil layer. **Trenches** can also be used for vapor extraction from shallow aquifers (<4 m below ground surface).

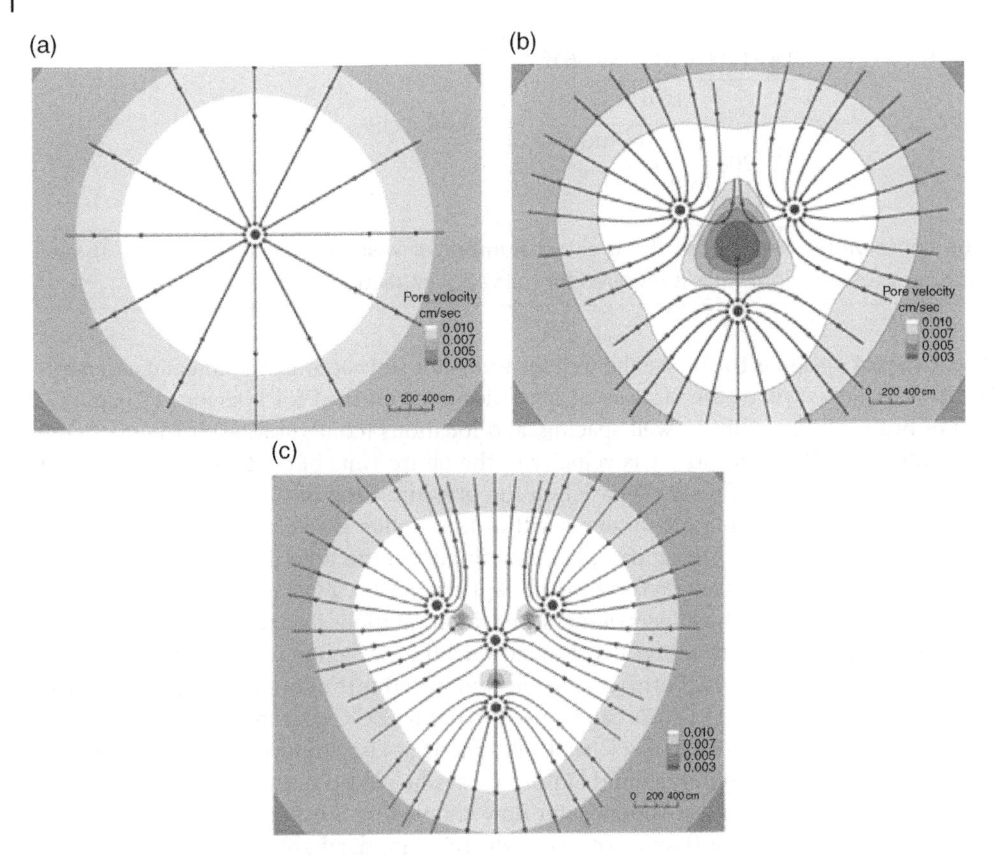

Figure 8.7 Venting well configuration: (a) one-well scenario, (b) three-well scenario, and (c) four-well scenario. The markers on flow lines represent 0.5-day intervals starting 100 cm from the well. The dark grey areas show the dead zone with a very small pore-gas velocity (cm/s). *Source:* Dixon and Nichols (2006); Reprinted with permission from John Wiley & Sons.

Monitoring for SVE Performance: To ensure SVE is working properly, operational parameters and characteristics of the soil and groundwater system need to be monitored. These include vapor flow rates and pressure, vapor concentrations and compositions, moisture contents, temperature of the air and soil, groundwater table, and contamination depth. As SVE progresses, vapor composition will change, i.e. more volatile compounds will decrease, and less volatile compounds will increase. If the total vapor concentration decreases without a change in composition, it is probably due to mass transfer resistance, water table upwelling, pore blockage, or leaks. If a decrease in vapor concentration is accompanied by a shift in composition toward less volatile compounds, it is most likely due to a change in the residual contaminant concentrations.

Field data in Figure 8.8 illustrates the efficiency of soil vapor extraction of PCE in a $9 \times 9 \times 3.3$ m block of surficial sand aquifer located at the Canadian Forces Base in Borden, Ontario. SVE was very effective for the initial stage when advection dominated the vapor removal. The concentrations (note the $\log_{10}$ scale in the figure) at both the 1 and 2 m depths were significantly decreased. The reduction in concentrations is particularly evident for areas with the highest PCE concentration, i.e. along the southwest corner to the northwest corner (1 m) and near the southwest corner and the north edge of the cell (2 m). In these areas, concentrations were reduced by 2–3 orders of magnitude from the background

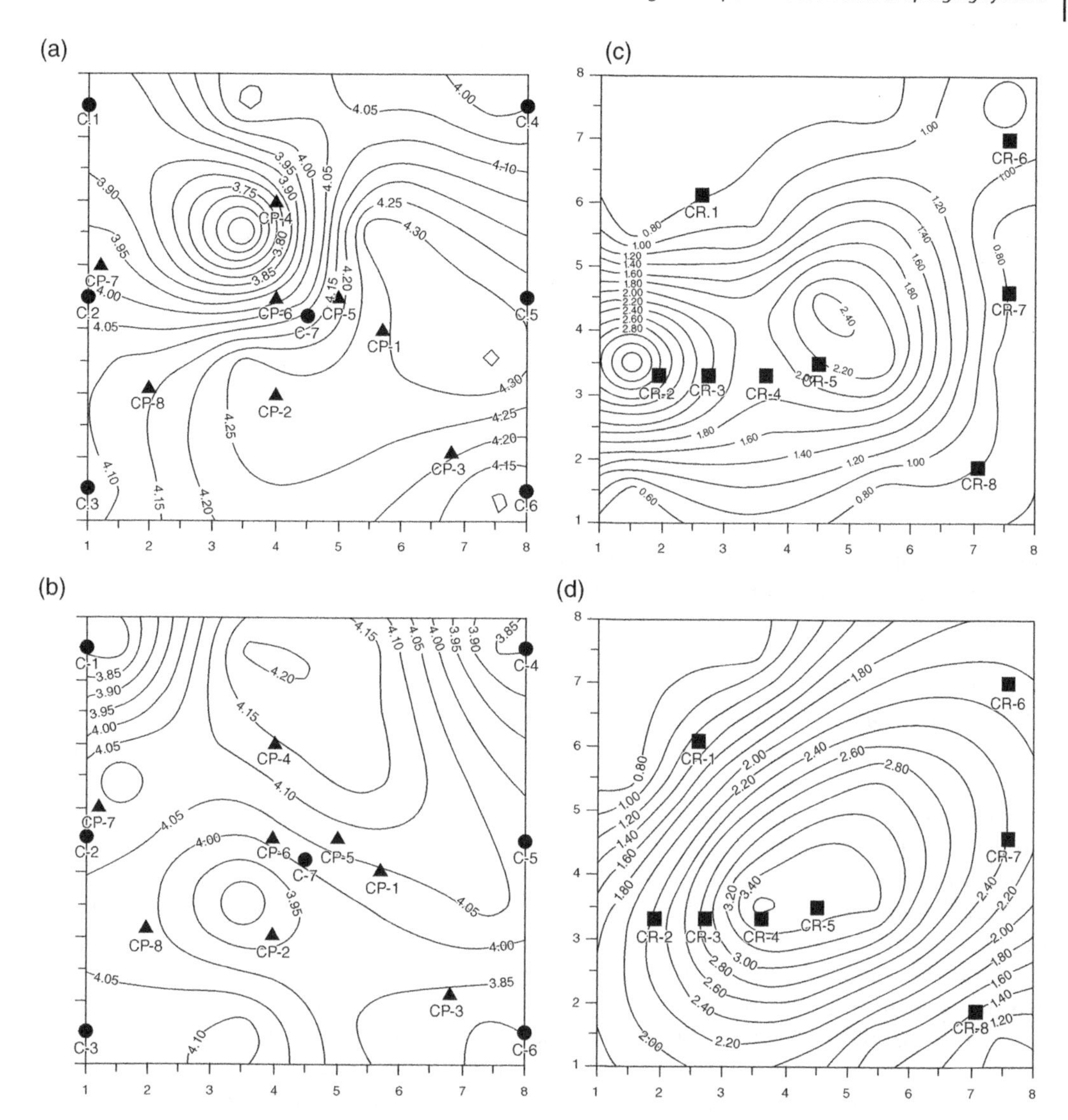

Figure 8.8 Spatial distribution of soil gas in $\log_{10}$ (ppm) for (a) background conditions at the 1 m depth (contour interval of 0.05); (b) background conditions at the 2 m depth (contour interval of 0.05); (c) at 170 days at the 1 m depth (contour interval of 0.2); and (d) at 170 days at the 2 m depth (contour interval of 0.2). *Source:* Thomson and Flynn (2000); Reprinted with permission from John Wiley & Sons.

conditions to 170 days of soil venting. Further examination of the data in Figure 8.8 reveals the elevated PCE concentrations near the center of the cell after 170 days. This elevated PCE saturation in the region was found to couple with the relatively high moisture content. The moisture creates a reduced gas phase permeability, resulting in the channeling of gas flow around this region (Thomson and Flynn 2000). This field data indicate the importance of comprehensive monitoring of the SVE system, including VOCs, site heterogeneity, and the operational parameters.

When to Stop SVE/IAS: If the spill amount is accurately known, then the cumulative mass removal is a good indicator to halt the SVE/IAS process. Vapor concentrations in extraction wells indicate how effective the venting process is. However, this decrease does not necessarily mean the decrease in actual contaminant concentration in soil. The decreases could be due to the water table increase, increased mass transfer resistance due to drying, or leaks in the extraction system. Since SVE will remove more volatile compounds first, the change of vapor composition toward less volatile compounds indicates the expected performance of vapor extraction.

Vapor Treatment Options: The various options of vapor treatment are only briefly described here. (a) Vapor combustion: A vast majority of the compounds pumped out of the soil via SVE can be combusted. However, their concentrations may be too low such that a fuel may have to be supplemented for a full combustion. This process will not be cost-effective if the contaminant concentration is less than about 10 000 ppm of the vapor. (b) Catalytic oxidation: The extracted vapors can be heated and passed through a catalyst bed to reach up to 95% of the contaminant removal. This procedure is useful when contaminant concentrations are lower than 8000 ppm. (c) Granular activated carbon (GAC): The vapor phase contaminants are sorbed onto the GAC. This method is very efficient for vapor flow rates less than about 100 g/day. (d) Biofiltration: Many organic compounds can be degraded ultimately to carbon dioxide via microorganisms. (e) Diffuser stacks: It is a nontreatment dilution option. Air transport models are used to assess ambient concentration when this option is used. (f) Liquid/vapor condensation: It is used when contaminant concentration is high, and the flow rate is low. It is usually accomplished by refrigeration and hence is expensive. For off-gas treatment using GAC and air stripping, readers are referred to Chapter 7 for details.

SVE Enhancements: For various reasons, SVE alone can prove to be inadequate or slow. However, with the help of one or more modifications, SVE may still be the best option for the remediation of a given site. SVE can be enhanced in conjunction with biodegradation through the increased air supply to aerobic microorganisms, or in conjunction with air sparging through the improved vapor transport from the saturated zone. Other enhancements include: (a) *In situ* heating: By pumping hot air into the subsurface, the vapor pressure is increased (Eq. 8.2), thereby increasing volatilization of contaminants with low vapor pressures. (b) Passive air injection: The addition of wells open to the atmosphere can increase airflow through the subsurface, thereby increasing the efficiency. (c) ***In situ* chemical oxidation**: The ozone sparging system was shown to be effective for the treatment of chlorinated solvents in a former disposal pit in southern Michigan (Dickson and Stenson 2011). (d) Surface seals and capping: By applying a surface seal in the vicinity of the well, airflow to the contaminant plume can be altered, which may improve the efficiency. **Surface seals** are used along with the extraction wells, which can promote a greater lateral flow of soil vapors, thereby enhancing SVE. It is effective for a shallow vadose zone (<5 m), and less effective or ineffective when it is below 8 m. The materials for the surface seals include a flexible membrane liner with synthetic materials such as high-density polyethylene (HDPE) (most common), clay or bentonite, and concrete or asphalt cap (works best at paved sites such as an industrial park or a gas station).

As an example, the effectiveness of surface seals on the performance of a SVE system was tested in a pilot scale setting consisting of an aluminum cylindrical tank (56 cm diameter), a height of 20 cm of sand packing, and 3 cm of overlying impermeable clay layer and a vacuum extraction well (3 cm diameter) in the center of the tank. Modeling data included pressure differential (mbar) and pore-gas velocity (cm/s). Surface seals increased the vacuum from 0.25 to 7 mbar, thus increasing the radius of influence of vapor extraction. Higher pore velocity was also attained in the covered surface compared with the uncovered surface, thus enhancing the volatilization of VOCs (Boudouch et al. 2009, Figure 8.9). The higher pore velocity is essential to ensure meaningful vapor movement in SVE.

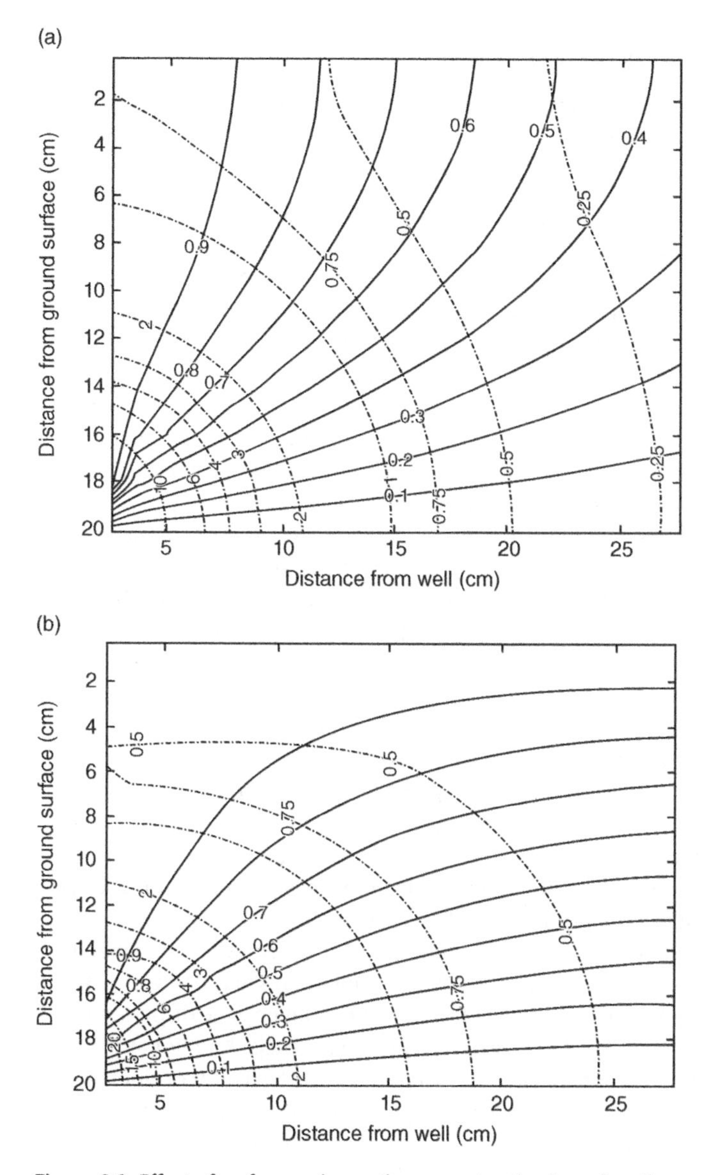

Figure 8.9 Effect of surface seal on soil vapor extraction based on the pore-gas velocity (cm/s dashed lines) and streamlines (solid lines) in (a) uncovered soil (open to atmosphere), and (b) covered soil under gas flow rate of 0.73 g/s. *Source:* Boudouch et al. (2009); Reprinted with permission from John Wiley & Sons.

SVE Economics: The short treatment duration makes SVE/IAS a significant cost-saving in contrast with P&T. P&T may take many years or even several hundred years at \$250 000–\$500 000 per project, whereas SVE takes 0.5–5 years duration. The cost of SVE is in the range of \$10–\$50/ton depending on whether an off-gas or water treatment is required. Wastewater treatment may add approximately 25% to the total cost. Small sites are typically more expensive, as much as \$150/ton. Economic advantages of SVE are most noticeable when the depth of the unsaturated zone is at least 10 ft. For both SVE and IAS, further improvement in cost-effectiveness and sustainability can be achieved through various green remediation and best management practices during the stages of design, well installation, operation, and monitoring (Box 8.3).

Box 8.3 Green remediation BMPs for SVE and air sparging

Many opportunities exist for reducing the footprints of SVE and IAS during their design, construction, and operation/monitoring stages. These include, but are not limited to, the following (USEPA 2010):

1) Designing an SVE or IAS System: During the design stage, soil-vapor flow models coupled with a thorough delineation of source areas and vapor-phase plumes help optimize well locations and screen depths. Energy consumption and air emissions can be reduced through BMPs such as the selection of vacuum pumps and blowers that accommodate changes in operating requirements as treatment progresses. Using piping of sufficient diameter can minimize pressure drops and resulting need for additional energy to operate blowers. Variable frequency drive motors can be used to automatically adjust energy use. Use of pulsed rather than continuous air exchange processes can also facilitate extraction of higher concentrations of contaminants. Barometric pumping can use barometric pressure differences thereby enhancing air throughput if an adequate response lag exists between the subsurface and the atmosphere.

   During the design state, an on-site pilot test is recommended (i) To assure a suitable sizing of equipment to be used in adding or withdrawing air to or from the subsurface, which will optimize energy use; (ii) To determine the minimum airflow rate that can meet the cleanup objectives and schedule while minimizing energy consumption; (iii) To evaluate the efficacy of air/vapor treatment for any opportunity in reducing material use or waste generation.
2) Constructing an SVE or IAS System: A significant portion of the environmental footprint is the construction of an SVE system involving well installation. Direct-push technology (DPT), for example, can be used to install standard 2-in diameter vacuum extraction wells, air injection wells, groundwater depression wells, and monitoring points. Use of DPT equipment rather than conventional drilling rigs can eliminate drill cuttings and associated waste disposal, avoid consumption or disposal of drilling fluids, and reduce drilling duration by as much as 50–60%. Another example is the use of small-diameter injection wells that can lead to large pressure drops and increase energy consumption of the system.
3) Operating and Monitoring an SVE or IAS system: SVE and IAS system operations can generate high levels of noise. Acoustic barriers with recycled or recyclable components may be constructed on-site or obtained commercially. Use of centrifugal blowers rather than positive displacement blowers and installation of air-line mufflers will also decrease noise levels. Additional reductions in land or ecosystem disturbance and efficiencies can be gained by an early consideration of anticipated reuse of the site. For example, an SVE or IAS pipe network could be constructed in a way to allow for future integration into the utility infrastructure of the site.

Case Study The Fairchild Superfund site, Mountain View, CA, USA

The Fairchild Superfund Site is located in south San Jose, CA, a former semiconductor manufacturing facility that was operated from 1977 until 1983. An estimated 60 000 gallons of solvents were released from an UST. The primary contaminants were 1,1,1-TCA and its degradation product 1,1-DCE, along with other contaminants such as xylene, acetone, Freon-13, and PCE.

Remediation activities started in 1982 by the removal of a damaged UST and associated piping, excavation of 3400 yd^3 of contaminated soils, and the removal of additional facilities such as acid waste tanks, and concrete slab beneath the former waste solvent tank. Groundwater pump-and-treat system was installed in 1982 to control the plume migration. In 1986, a bentonite slurry wall was constructed around the site perimeter to contain contaminated groundwater and to facilitate remediation of VOC "hot spot" within the site boundary. During the operation of groundwater pump-and-treat from 1982 to 1988, a total of 93 285 pounds of VOCs were removed from groundwater.

SVE system was operated between 1987 and 1990 to treat vadose zone soil contamination. Vapor treatment system consists of vacuum pumps (4500 cfm at 0.67 atm), a dehumidifier, and five GAC adsorption units to capture vapor (two 3000-lb GAC beds operating in parallel followed by a second set of GAC beds and then a final 3000-lb GAC bed). SVE operated five days per week. At any particular time, the system was operated with a maximum of 25 of the 39 extraction wells. It was estimated that SVE removed a total of 16 000 lb solvents from soil during the 16 months of operation. Extraction rate increased to a peak, and then decreased. The concentrations decreased for TCA, DCE, but increased in some soil borings due to the variations of contaminant concentrations in the soils (common to real-world problem). The maximum rate of removal was achieved at 130 lb/day at the early stage of operation (~2 months) and then the rate decreased (<10 lb/day after 8 months; <4 lb/day after 16 months).

In total, 146 000 pounds of VOCs were removed from the site through soil excavation, P&T, and SVE. No active remediation has been performed at the site since 1998. The site used monitored natural attenuation as a polishing step. The capital cost for SVE was $2.1 million US dollars, excluding slurry wall and P&T cost. O&M costs totaled $1.8 million dollars for 16 months of operation. This corresponds to $240 per lb of contaminant removed and $93 per cubic yard of soil treated.

Bibliography

Bear, J. (1979). *Hydraulics of Groundwater*. New York, NY: McGraw-Hill.

Beckett, G.D. and Huntley, D. (1994). Characterization of flow parameters controlling soil vapor extraction. *Groundwater* 32 (2): 239–247.

Bedient, P.B., Rifai, H.S., and Newell, C.J. (1999). *Ground Water Contamination: Transport and Remediation*, 2e. Upper Saddle River, NJ: Prentice Hall PTR.

Bird, R., Stewart, W., and Lightfoot, E. (2002). *Transport Phenomena*. New York, NY: John Wiley.

Boudouch, O., Ying, Y., and Benadda, B. (2009). The influence of surface covers on the performance of a soil vapor extraction system. *Clean (Weinh)* 37 (8): 621–628.

Clement, T.P., Kim, Y.-C., Gautam, T.R., and Lee, K.-K. (2004). Experimental and numerical investigation of DNAPL dissolution processes in a laboratory aquifer model. *Ground Water Monit. Remidiat.* 24 (4): 88–96.

Cole, G.M. (1994). *Assessment and Remediation of Petroleum Contaminated Sites*. Boca Raton, FL: CRC Press.

DePaoli, D.W. (1996). Design equations for soil aeration via bioventing. *Sep. Tech.* 6: 165–174.

Dickson, J.R. and Stenson, R. (2011). Insufficient source area remediation results in the rebound of TCE breakdown products in groundwater. *Remediation* Winter: 87–103.

DiGiulio, D. (1992). Evaluation of soil venting application. *J. Haz. Materials* 32 (2–3): 279–291.

Dixon, K.L. and Nichols, R.L. (2006). Soil vapor extraction system design: a case study comparing vacuum and pore-gas velocity cutoff criteria. *Remediation* Winter: 55–67.

Essaid, H.I., Bekins, B.A., Herkelrath, W.N., and Delin, G.N. (2011). Crude oil at the Bemidji site: 25 years of monitoring, modeling, and understanding. *Ground Water* 49 (5): 706–726.

Garges, J.A. and Baehr, A.L. (1998). Type curves to determine the relative importance of advection and dispersion for solute and vapor transport. *Ground Water* 36 (6): 959–965.

Grasso, D. (1993). *Hazardous Waste Site Remediation – Source Control*. Lewis Publishers.

Johnson, P.C., Kemblowski, M.W., and Colthart, J.D. (1990a). Quantitative analysis of the cleanup of hydrocarbon-contaminated soils by *in-situ* soil venting. *Ground Water* 28: 413–429.

Johnson, P.C., Stanley, C.C., Kemblowski, M.W. et al. (1990b). A practical approach to the design, operation and monitoring of *in situ* soil-venting systems. *Ground Water Monitoring Rev.* 10 (2): 159–178.

Kuo, J. (1999). *Practical Design Calculations for Groundwater and Soil Remediation*. Boca Raton, FL: Lewis Publishers.

Lundegard, P.D. and LaBrecque, D.J. (1998). Geophysical and hydrologic monitoring of air sparging flow behavior: comparison of two extreme sites. *Remediation* Summer: 59–71.

McAlary, T.A. and Barker, J.F. (1987). Volatilization losses of organics during ground water sampling from low permeability materials. *GWMR* 7 (4): 63–68.

Mobile, M.A., Widdowson, M.A., and Gallaher, D.L. (2012). Multicomponent NAPL source dissolution: evaluation of mass-transfer coefficient. *Environ. Sci. Technol.* 46 (18): 10047–10054.

Nyer, E.K. (1993). *Practical Techniques for Groundwater and Soil Remediation*. Boca Raton, FL: Lewis Publishers, CRC Press, Inc.

Nyer, E.K., Palmer, P.L., Carman, E.P. et al. (2001). *In Situ Treatment Technology*, 2e. Boca Raton, FL: Lewis Publishers.

Pichtel, J. (2007). *Fundamentals of Site Remediation*. Government Institute.

Thomson, N. and Flynn, D.J. (2000). Soil vacuum extraction of perchloroethylene from the Borden aquifer. *Ground Water* 38 (5): 673–688.

USEPA (1991a). *Soil Vapor Extraction Technology: Reference Handbook*, EPA/540/2-91/003. Cincinnati, OH: Office of Research and Development.

USEPA (1991b). *Guide for Treatability Studies Under CERCLA: Soil Vapor Extraction*, EPA/540/2-91/019A. Washington, DC: Office of Emergency and Remedial Response.

USEPA (1993). *Decision-Support Software for Soil Vapor Extraction Technology Application: HyperVentilate*, EPA/600/R-93/028. Cincinnati, OH: Office of Research and Development.

USEPA (1995). *Review of Mathematical Modeling for Evaluating Soil Vapor Extraction Systems*, EPA/540-R-95/513. USEPA.

USEPA (2010). *Green Remediation Best Management Practices: Soil Vapor Extraction and Air Sparging*, EPA 542-F-10-007. USEPA.

USEPA (2012). *A citizen's guide to soil vapor extraction and air sparging*, EPA 542-F-12-018. USEPA.

USEPA (2017). *How to Evaluate Alternative Cleanup Technologies for Underground Storage Tank Sites: A Guide for Corrective Action Plan Reviewers*, EPA 510-B-17-003. USEPA.

Weiner, E.R. (2000). *Application of Environmental Chemistry: A Practical Guide for Environmental Processional*. Washington, D.C: Lewis Publishers.

Wisconsin Department of Natural Resources (DNR) (2014). *Guidance for Design, Installation and Operation of Soil Venting Systems*, PRB-BR-185. DNR.

Wong, J.H.C., Lim, C.H., and Nolen, G.L. (1997). *Design of Remediation System*. Lewis Publishers.

Questions and Problems

1. What compounds are not amenable to IAS/SVE: BTEX, dioxins, heating oils, lubricating oils, PCBs?

2. Describe how the following factors affect SVE in the vadose zone: humic contents (f_{oc}), soil moisture, fine-grained sands, clays, and coarse-grained sands.

3. Determine which of the following is unfavorable for SVE: (a) groundwater table at 4 ft, (b) groundwater table at depth of 200 ft, (c) groundwater table fluctuates considerably.

4 Determine which of the following is unfavorable for IAS: (a) a free LNAPL phase is present on the groundwater table, (b) contamination underneath a building, (c) presence of sewage lines, (d) contamination at 5 ft BGS, (e) an unconfined aquifer.

5 Explain how an upwelling affects SVE performance and how to alleviate this problem.

6 What are generally the minimal Henry's constant (atm), vapor pressure (mmHg), maximum boiling point (°C), and minimal intrinsic permeability (cm^2) for air sparging to be an appropriate technology?

7 What are generally the minimal Henry's constant (atm), vapor pressure (mmHg), maximum boiling point (°C), and minimal intrinsic permeability (cm^2) for soil vapor extraction to be an appropriate technology?

8 Describe how the air permeability is affected by heterogeneity, porosity, and moisture.

9 Hydraulic conductivity and intrinsic permeability to water are two terms interchangeably used in the literature, but they are different. List several major differences between the two. What are the typical range of hydraulic conductivity to water flow and intrinsic permeability to airflow in a sandy soil?

10 During a site characterization stage, a contaminated soil was determined to have a hydraulic conductivity of 1.8×10^{-8} m/s and air permeability of 5×10^{-12} cm^2. Determine the intrinsic permeability to water in the unit of cm^2 and darcy. What is the air permeability in the unit of darcy? Given: μ (water) = 0.01 g/cm-s, ρ (water) = 1 g/cm^3, and $g = 981$ cm/s^2.

11 If the intrinsic permeability of water is 1.5×10^{-8} cm^2, what would be its saturated hydraulic conductivity in cm/s? If this soil is completely depleted of water (dry), then what would be its permeability to air?

12 Describe the difference between soil–air permeability and soil–water permeability.

13 How can intrinsic permeability to airflow be determined in lab and *in situ*? Provide key equations when needed.

14 Describe the typical system components for (a) SVE and (b) IAS.

15 Describe the functions of wells used in SVE and IAS: vacuum wells, injection wells, passive wells, and monitoring wells.

16 Which is the predominant contaminant removal pathway in SVE: advection, desorption, volatilization, or diffusion?

17 Dry soils may strongly adsorb VOCs, whereas wet soils facilitate desorption. How could this phenomenon limit the effectiveness of SVE in southwestern states in the United States?

18 Use Raoult's law to calculate the vapor pressure of an aqueous solution containing 40 g/L of glucose (MW = 180). The vapor pressure of pure water at 25 °C is 23.8 mmHg.

19 If we mix 15 g of cyclohexane ($p^0 = 81$ torr, MW = 84) with 20 g of ethanol ($p^0 = 52$ torr, MW = 92), what are the partial pressures of cyclohexane and ethanol in the saturated vapor?

20 An aqueous solution has saturated benzene of 1780 mg/L at 25 °C. Use Raoult's law to calculate how the dissolved benzene affects the vapor pressure. The vapor pressures of pure water and benzene at 25 °C are 23.8 and 95 mmHg, respectively.

21 A site is contaminated with 50% benzene and 50% xylene by weight. Calculate the saturated vapor concentrations of extracted benzene and xylene at 25 °C. Benzene has a molecular weight of 78 and vapor pressure of 95 mmHg, and xylene has a molecular weight of 106 and vapor pressure of 10 mmHg.

22 Pump-and-Treat (P&T) versus Soil Vapor Extraction (SVE)

- **a.** Briefly summarize (tabulate) the difference between P&T and SVE with regard to the media, chemicals treated, contaminant form, physicochemical principle, cleanup time, cost, and other aspects that might be important.
- **b.** P&T is very effective as a containment strategy for plume control, but not cost-effective in restoring contaminated aquifer. Explain why and why not.
- **c.** SVE is the most commonly used "innovative" remediation technique with success in many contaminated sites in the United States. Explain why.

23 Use the SVE nomograph to predict the likelihood of success of a soil vapor extraction under the following sets of conditions: (a) benzene released in a coarse sandy soil two to three weeks ago, (b) benzene released in a fine sandy soil three months ago, and (c) benzene released in a clay soil two to three weeks ago.

24 For the above question, what if the contaminant is: (a) toluene, (b) a phenolic compound, (c) a low-molecular-weight PAH such as naphthalene?

25 The normal atmospheric partial pressure of oxygen at sea level is about 0.2 atm. The partial pressure of oxygen in vadose zone is lower and assumed to be 0.063 atm due to the oxygen consumption by soil bacteria. Henry's law constant (H) of oxygen is 26 (unitless) or 635 L-atm/mol at 20 °C.

- **a.** Calculate the oxygen concentration in mg/L in soil pore water.
- **b.** Is this oxygen concentration sufficiently high to support the growth of aerobic bacteria if a minimum requirement for dissolved oxygen is 4.0 mg/L?

26 An underground storage tank containing toluene ruptured and released pure toluene into a vadose zone sandy soil with a very low moisture content. What would be its concentration in saturated soil vapor? Toluene has a molecular weight of 92.14, vapor pressure of 0.037 atm, and dimensionless Henry's law constant of 0.28 at 20 °C.

27 Both SVE and IAS have a minimal vapor pressure of 0.5 mmHg for the compound to be effectively removed. (a) What is the equivalent equilibrium concentration reported in mol/L? (b) If methane (CH_4) is reported, then what is the concentration in mg/L?

28 Atrazine is an herbicide commonly used in the United States. The Henry's law constant of atrazine is 3×10^{-6} atm-L/mol or 1×10^{-7} (dimensionless) at 20 °C. The vapor pressure is 4×10^{-10} atm.

a. Calculate the solubility of atrazine in soil pore water in mg/L_{water} (*Hint*: use molecular weight of 216 g/mol to convert mol/L into mg/L)

b. Calculate the concentration of atrazine in soil air in mg/L_{air}

c. What is the % of total atrazine mass in the soil gas? Assume the aquifer contaminated with atrazine has a residual gas saturation (i.e. the % of soil pores occupied by gas) of 15%.

d. Can SVE be used to remediate atrazine-contaminated vadose zone? Justify your answer as to why or why not.

29 Evaluate how an exceedingly high vacuum affects: (a) advection vs. diffusion based contaminant removal, (b) cost-effectiveness of SVE.

30 Estimate the mass of VOC removed by soil vapor extraction in kg/day if the vapor contains 1 mg/L (C) of contaminant at a constant vapor flow of 25 scfm (Q).

31 Perform the same calculation as the above question, but assume $C = 5$ mg/L and $Q = 50$ scfm.

32 Given the following conditions: air permeability (k) of 10^{-7} cm^2, radius of extraction well (R_w) = 5.1 cm, and radius of vacuum influence (R_I) = 12 m, vacuum pressure (P_w) = 0.90 atm (i.e. 9.09×10^5 g/cm-s^2), and air viscosity of 1.8×10^{-4} g/cm-s. (a) Find the airflow rate per unit thickness of well screen Q/H by using Figure 8.6 and applying Eq. (8.19) to verify, (b) If the well is screened to the entire contaminated zone of 5 m, what is the airflow rate in cfm? Does this flow fall in the ideal range of 100–1000 cfm?

33 Use Figure 8.6 to find the Q/H for a medium sand with air permeability of 10 darcy, $P_w = 0.8$ atm, $R_I = 12$ m, and $R_w = 5.1$ cm. What will be the Q if the screen section of the extraction well is 10 m?

34 Propose some best management practices (BMPs) for sustainable remediation using SVE and IAS during design, construction, and operation and monitoring phases.

35 At the most, how many possible phases of NAPLs can be present in (a) vadose zone (b) saturated zone?

Chapter 9

Bioremediation and Environmental Biotechnology

LEARNING OBJECTIVES

1. Understand the basic microbiology and the roles that various microorganisms play in bioremediation and why bacteria are particularly important in bioremediation
2. Comprehend the electron tower theory and the interactions between bacteria and contaminant substrates regarding electrons, carbon, and energy requirements
3. Perform stoichiometric calculations for oxygen and nutrient demands by bacteria
4. Perform calculation of the cleanup time using the first-order biodegradation kinetics
5. Learn biodegradation potentials for various contaminant groups and bacteria's strategies (metabolic pathways) under aerobic and anaerobic conditions
6. Understand the optimal conditions for microbe-based bioremediation, including hydrological parameters and soil/groundwater properties such as pH, temperature, DO, moisture, and toxicity
7. Delineate applications, pros and cons of various *in situ* and *ex situ* bioremediation technologies, their enhancement such as biostimulation and bioaugmentation, and the use of oxygen-releasing compounds
8. Delineate how and what types of plants can be employed in phytoremediation of selected environmental contaminants as a part of the green remediation technology
9. Describe general design factors for some common bioremediation techniques including monitored natural attenuation, bioventing, biosparging, composting, and landfill
10. Know the cost-effectiveness of various bioremediation techniques relative to other physical–chemical methods through the case study examples

Bioremediation has become the third most common innovative remediation technology for the cleanup of contaminated soil and groundwater, since its first commercial application in 1972 by Sun Tech in Amber, Pennsylvania (Norris and Figgins 2005). At present, bioremediation is one of the major environmental applications of biotechnology, which uses living organisms to improve, modify, or produce products or processes for environmental purposes. This chapter introduces the basic principles, applications, designs, and case studies of bioremediation. Basic principles include microbial growth, stoichiometry, reaction kinetics, and biochemical pathways of detoxification and biodegradation. The comprehension of these principles will be critical to determine the feasibility and design any bioremediation system, such as the estimation of nutrient requirement, rate of oxygen delivery, cost-effective monitoring, and cleanup time. Bioremediation applications include those conventional biotechnologies (biofilm, activated sludge, lagoons, nutrient removal, and anaerobic processes) developed traditionally for wastewater treatment, but our focus will be exclusively on soil and groundwater bioremediation techniques, such as *in situ* and *ex situ* bioremediation, monitored natural attenuation, bioventing, composting, landfarming, slurry-phase bioreactors, and phytoremediation.

Soil and Groundwater Remediation: Fundamentals, Practices, and Sustainability, First Edition. Chunlong Zhang.

Companion website: www.wiley.com/go/Zhang/Remediation_1e

9.1 Principles of Bioremediation and Biotechnology

This section is intended for readers to review some basic microbiology essential to the understanding of bioremediation and environmental biotechnology (Box 9.1). Since bacteria are the major players in bioremediation and environmental biotechnology, our emphasis will be the bacterial growth, bacteria-mediated reaction kinetics, and pathways relevant to bioremediation. Fungi- and plant-based remediation will be only briefly described for special applications.

Box 9.1 Environmental biotechnology

Biotechnology has various applications in fields of agriculture, medicine, food, drug, energy, industrial production, and environment protection. **Environmental biotechnologies** are the applications of primary microorganisms and sometimes plants and their products in the prevention and control of environmental pollution through biotreatment of solid, liquid, and gaseous wastes, and biomonitoring of environment and treatment processes.

In a broad sense, environmental biotechnologies include the well-established field of **biological treatment** of wastewater such as activated sludge system that was invented and first employed in England in 1914, and the microbe-based landfill system such as the Fresno Municipal Sanitary Landfill, opened in Fresno, California in 1937, which is considered to be the first modern sanitary landfill in the United States. Environmental biotechnologies also include centuries-old composting technology that was used to dispose of solid wastes and later evolved to treat hazardous wastes.

Although the subject of environmental biotechnology is loosely defined in the literature, it is in general consensus that **bioremediation** falls in one of the major applications of environmental biotechnologies. The term bioremediation is commonly reserved for the treatment of soil and groundwater. While bioremediation can be referred to microbe-based oil spills in surface water remedy, it is rarely used to refer to conventional technologies for the biological treatment of domestic and industrial wastewater. An example of these bioremediation techniques include, but are certainly not limited to, *in situ* and *ex situ* soil and groundwater treatment methods listed in Table 6.4.

Environmental biotechnologies rely on indigenous or genetically modified microorganisms of various types to degrade and detoxify contaminants, such as the use of anaerobic, anoxic, and aerobic microorganisms. The use of native and genetically modified higher plants capable of metabolizing contaminants also belongs to the realm of environmental biotechnologies, such as phytoremediation and wetland-based treatment systems. In many cases, environmental biotechnologies rely on the stimulation or enhancement of soil and groundwater conditions because of the abundant nature of indigenous bacteria or bacterial consortia in the environment. These optimal conditions include oxygen, electron acceptors, nutrients, soil moisture, pH, etc.

Another area of environmental biotechnological application is the biologically based monitoring. Biosensors are essential tools in biomonitoring of environment and treatment processes. Combinations of biosensors in array can be used to measure concentration or toxicity of various chemical substances. Microarrays for simultaneous qualitative or quantitative detection of different microorganisms or specific genes in environmental samples are also useful in the monitoring of microorganisms in bioremediation systems.

9.1.1 Microorganisms and Microbial Growth

9.1.1.1 Types of Microorganisms

Herein, we first introduce the classifications (taxonomy) of life and explain why bacteria are particularly important in bioremediation systems. As illustrated in Figure 9.1, bacteria, archaea, and eucarya comprise three major domains of all organisms.

Bacteria are microscopic scale unicellular prokaryotic organisms characterized by the lack of a membrane-bound nucleus and membrane-bound organelles. **Archaea** were not recognized as distinctly different from bacteria until molecular methods became available. Archaea differ biochemically from bacteria due to the arrangement of bases in their ribosomal RNA, and in the composition of the plasma membranes and cell walls. Although they are similar to bacteria in size, biologists have determined that chemically, they have more in common with plants and animals than they do with bacteria. Archaea can usually survive in the extreme environment, such as high temperature, oxygen-depleted, and high-salinity conditions. An example of Archaea important to environmental remediation is **methanogens**, which produce methane as a metabolic by-product under anoxic conditions. Eucaryota (eukaryota) is an older name for the domain **Eucarya**, which contains all the eukaryotic kingdoms, including plants, animals, protists, and fungi.

Microorganisms are a diverse group of organisms, including bacteria, fungi, archaea, and protists; microscopic plants (green algae); and animals such as plankton and the planarian (nonparasitic, freshwater worm). Some microbiologists also include viruses, but others

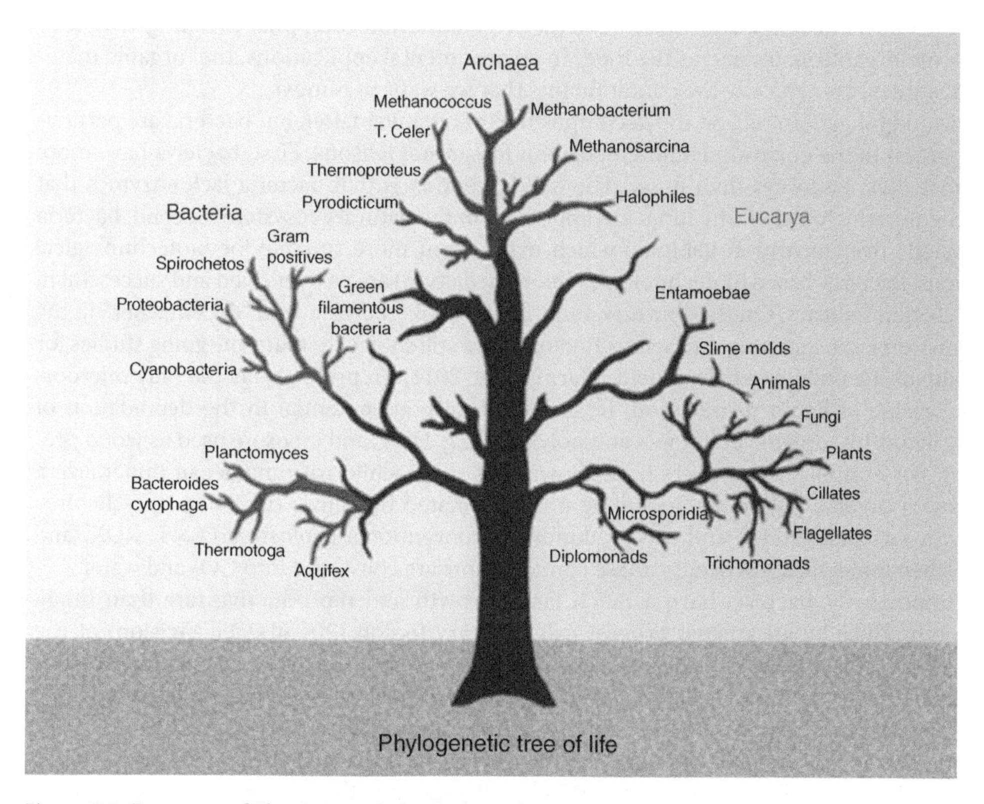

Figure 9.1 Taxonomy of life: phylogenic (evolutionary) tree.

consider viruses as nonliving. Most microorganisms are unicellular (single-celled); however, this is not universal since some multicellular organisms are microscopic. Some unicellular protists and bacteria, such as *Thiomargarita namibiensis* (0.75 mm in diameter), are macroscopic and even visible to the naked eye.

Table 9.1 Types of microorganisms based on carbon and energy sources.

	Energy source	
Carbon sources	**Chemical**	**Photochemical (light)**
Organic matter	*Chemoheterotrophs:* All fungi and protozoans, and most bacteria. Use organic sources for both energy and carbon.	*Photoheterotrophs:* A few specialized bacteria that use photoenergy, but are dependent on organic matter for a carbon source.
Inorganic carbon CO_2, HCO_3^-	*Chemoautotrophs:* Use CO_2 for biomass and oxidize substances such as H_2, NH_4^+, and S for energy.	*Photoautotrophs:* Algae, cyanobacteria, and photosynthetic bacteria. Use light energy to convert CO_2, HCO_3^- to biomass by photosynthesis.

There are four different types of microorganisms based on their ability to use carbon and energy sources (Table 9.1): chemoheterotrophs, photoheterotrophs, chemoautotrophs, and photoautotrophs. All fungi and most bacteria are **chemoheterotrophs**, meaning that they must rely on "organic materials" as the food. In environmental applications, the "organic materials" (substrates) are typically the contaminants that we want to remove.

Bacteria and fungi both rely on the decomposition of organic matter, but bacteria are particularly important in the environmental remediation for several reasons. First, bacteria have more diverse metabolic pathways than fungi. The only exception is that bacteria lack enzymes that degrade lignin, which is also why fungi are important in the natural ecosystem. Second, bacteria prefer aquatic over terrestrial habitats, which make them more suitable for biotechnological applications. Bacteria-based biotechnologies (bioremediation) are well received and successful in the remediation industry. On the contrary, fungi are a major "recycler" and "decomposer" in the natural environment, and fungal-based technologies are still subject to many on-going studies for the cleanup of the polluted environment (Harms et al. 2011). In principle, as the only microorganisms capable of lignin degradation, fungi have enormous potential in the degradation of lignins or lignin-like organic pollutants and polymers (e.g. DDT, and creosote used as wood preservative). For example, researchers have shown that some white rot fungus can mineralize a wide array of organic pollutants, including polychlorinated biphenyls (PCBs), polycyclic aromatic hydrocarbons (PAHs), and the predominant conventional explosives (TNT, RDX, and HMX). When mineralization occurs, these contaminants are converted into CO_2 and water.

More importantly, bacteria have a much faster growth and reproductive rate than fungi. Bacteria reproduce by an asexual process called binary fission (20–30 s/cell division) at the number of 1, 2, 4, 8, 16, 32, ... if they are unrestricted in growth. Bacterial population doubles approximately every 45 minutes. It is so fast that a single bacterium (10^{-13} g) can reproduce 10^{16} bacteria in a single day and would have a dry weight of about 18 kg!

The reason why bacteria can consume various chemicals is because they have various types of metabolic enzymes. **Enzymes** are biological "catalysts," but they differ from conventional chemical catalysts in that biochemical oxidation occurs at low temperature and pressure in the presence of enzymes. Enzymes are proteins made of amino acids, and are typically named after what they catalyze and the addition of "-ase." For example, hydrolytic enzymes and respiratory enzymes in bacterial cells entail oxido-reductases, transferases, lysases, isomerases, and ligases. Hydrolases are generally located outside cells (extracellular enzymes), whereas respiratory enzymes (desmolases) are intracellular enzymes (inside the cell). The activities of enzymes depend on temperature, pH, oxidation–reduction potential (Eh), and substrate concentration.

Box 9.2 Superbugs and genetically modified bacteria

The name of superbugs is commonly perceived by the public as the antibiotic resistant bacterial strains. These are the harmful microorganisms that will threaten the public health and safety. In the environmental arena, superbugs have been sought for many years in a hope that the bacterial strains are desired to detoxify and degrade multiple contaminants including those recalcitrant ones such as PCBs, DDTs, and dioxins. From extensive research efforts, there are many successful isolations of bacterial strains that are capable of degrading one contaminant or a group of structurally similar contaminants. These isolates are either from the enriched culture in the laboratory or from polluted environment where certain bacteria strains have been adapted to and even metabolize the contaminants.

An alternative way is to develop **genetically engineered microbes (GEMs)** or the so-called superbugs which combine the desired genetic characteristics from several bacteria. Biotechnologists were able to transfer specific plasmids into a single bacterium, making it a superbug bacterium. These **plasmids** are extra chromosomal genetic elements of bacteria containing certain genetic traits that can be transferred to other bacteria and gain the capability of reproducing independently into the new ones and share the traits with them.

In the 1970s, Chakrabarty and coworkers made the first claim for their development of a superbug capable of degrading a number of oil-spill related contaminants, including octane, hexane, xylene, toluene, and naphthalene. Consequently in 1980, a US patent was granted for their invention of this genetically engineered microbe (Chakrabarty 1981). As the first living organism to be patented, this superbug, *Pseudomonas putida*, can degrade about three-fourths of the oil pollution. Its plasmids contain the genes responsible for oil degradation from four different species of bacteria. This superbug was then used in the cleanup of oil spills in Texas in 1990.

A true superbug bacterium able to degrade all persistent contaminants is probably nonexistent. However, genetic engineers are actively constructing strains of microorganism with a broad spectrum of catabolic potential ideal for bioremediation applications in both aquatic and terrestrial environments.

Bioremediation via bacteria is by far the most important application in environmental biotechnology. It is the use of bacteria (either native or genetically engineered, see Box 9.2) to biodegrade contaminants. The success of bioremediation technology depends on our state-of-the-art understanding of the interactions between bacteria, contaminants, and the environmental conditions in soil and groundwater. The contents present below are devoted to the growth and energetics of bacteria when contaminants are present as the substrate. Energetics quantifies how much energy bacteria can gain from the degradation of organic contaminants from the thermodynamic standpoint. Bacterial degradation kinetics (the rate of degradation related to the cleanup time) and biochemical pathways (bacteria's strategies in degrading and detoxifying contaminants) are elaborated in the next section.

9.1.1.2 Cell Growth on Contaminant

Contaminants (substrates) often serve as electron donors when they are degraded by bacteria. An **electron donor** is a chemical that donates electrons to another compound. Since it donates electrons, it itself is oxidized in the process. By virtue of its electron donation, this chemical is a reducing agent. In the biodegradation process, a portion of the electrons (f_e) is transferred to the electron acceptor to provide energy (**catabolism**), and another portion of electrons (f_s) is used for cell synthesis (**anabolism**) (Figure 9.2). For example, values of f_e = 0.703 and f_s = 0.297, as shown in Figure 9.2, indicate that 70.3% of electrons from the degradation of 2,4-DNT is used for energy, whereas 29.7% of the electrons is used for cell synthesis (Zhang and Hughes 2004).

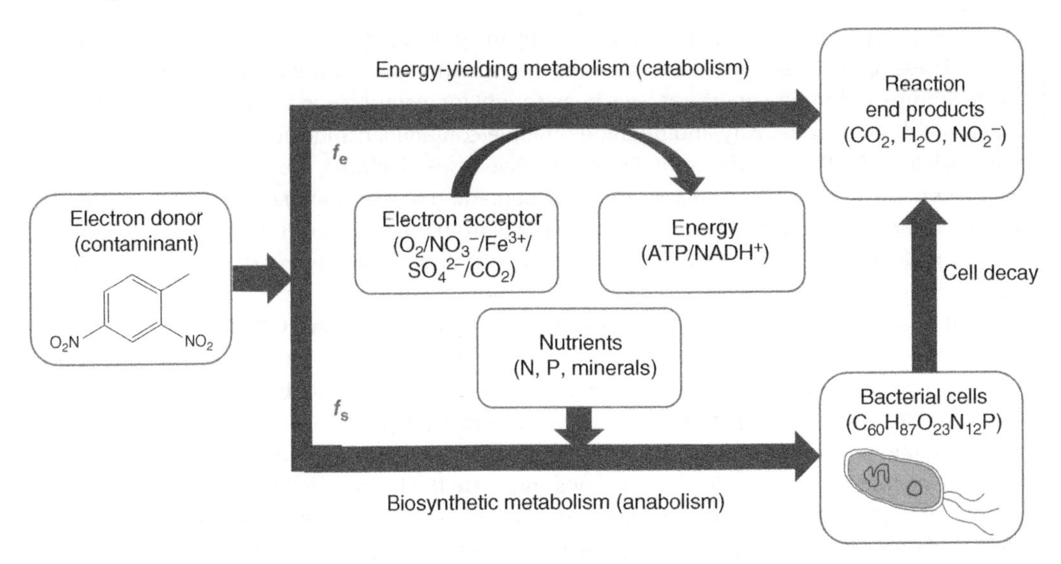

Figure 9.2 Bacterial degradation and cell growth on contaminant as the electron donor. For 2,4-dinitrotoluene, the portion of electrons transferred to an electron acceptor is f_e = 0.703 (70.3%), whereas the portion of electrons used for bacterial cell growth and reproduction is f_s = 0.297 (29.7%).

Cell synthesis refers to the making of bacterial cells, which have an empirical chemical formula of $C_5H_7O_2N$ or $C_{60}H_{87}O_{23}N_{12}P$. Depending on the availability of the electron acceptor, biodegradation occurs by various bacterial pathways (shown as the **electron tower theory** from Eqs. (9.1) to (9.5)). An **electron acceptor** is a chemical that accepts electrons from another compound, such as O_2 (oxygen), NO_3^- (nitrate), Fe^{3+} (ferric iron), SO_4^{2-} (sulfate), CO_2 (carbon dioxide). Bacteria would obtain as much energy as possible from the oxidation of a given "food" (substrate). Hence, the reaction with the highest free energy production (i.e. the most negative ΔG) will be the most thermodynamically favorable reaction.

- Aerobic oxidation (aerobic bacteria):

$$\frac{1}{4}O_2 + H^+ + e^- \rightarrow \frac{1}{2}H_2O \qquad \Delta G^0 = -220.9\frac{kJ}{mol} \qquad \text{Eq. (9.1)}$$

- Denitrification (anaerobic bacteria):

$$\frac{1}{5}NO_3^- + \frac{6}{5}H^+ + e^- \rightarrow \frac{1}{10}N_2 + \frac{3}{5}H_2O \qquad \Delta G^0 = -210\frac{kJ}{mol} \qquad \text{Eq. (9.2)}$$

- Iron reduction (iron-reducing bacteria):

$$Fe^{3+} + e^- \rightarrow Fe^{2+} \qquad \Delta G^0 = -100\frac{kJ}{mol} \qquad \text{Eq. (9.3)}$$

- Sulfate reduction (sulfate-reducing bacteria):

$$\frac{1}{8}SO_4^{2-}+\frac{19}{16}H^+ + e^- \rightarrow \frac{1}{16}H_2S+\frac{1}{16}HS^- +\frac{1}{2}H_2O \qquad \Delta G^0 = -20\frac{kJ}{mol} \qquad \text{Eq. (9.4)}$$

- Methanogenesis (methanogenic bacteria):

$$\frac{1}{8}CO_2 + H^+ + e^- \rightarrow \frac{1}{8}CH_4 + \frac{1}{4}H_2O \qquad \Delta G^0 = -15\frac{kJ}{mol} \qquad \text{Eq. (9.5)}$$

where ΔG^0 is the standard Gibbs free energy change that accompanies the formation of 1 mol of this chemical from its constituent elements at standard state (1 bar at 25 °C). From the ΔG^0 values of the above reactions (Rittmannand McCarty 2001), it is clear that bacteria would benefit the most when using oxygen as an electron acceptor (Eq. 9.1). Accordingly, the order of preference for the electron acceptor is as follows: $O_2 > NO_3^- > Fe^{3+} > SO_4^{2-} > CO_2$. This also means that the aerobic bacteria (which use oxygen) are the most energy-efficient. The energy available and the rate of process are both in the decreasing order of: aerobic, denitrifying, iron reducing, sulfate reducing, and methanogenic bacteria.

Typical electron acceptors utilized by microorganisms in soil and groundwater are O_2, NO_3^-, Fe^{3+}, SO_4^{2-}, and CO_2. When oxygen is utilized as the electron acceptor, microbial respiration is termed aerobic. When other electron acceptors are utilized, it is termed anaerobic. Depending on the mode of respiration, microbes can be classified into three categories: (a) aerobes, (b) anaerobes, and (c) facultative bacteria. **Aerobes** thrive only in oxygenated environments using dissolved oxygen as an electron acceptor. Strict **anaerobes** grow only under highly reduced conditions, where oxygen is effectively absent. Strict anaerobes use electron acceptors such as sulfate or carbon dioxide. Many microorganisms are able to adapt to both aerobic and anaerobic conditions, but are typically more active in the presence of oxygen. These organisms are termed **facultative**, and most microbes utilizing nitrate as an electron acceptor tend to be facultative.

Figure 9.3 schematically shows the order of electron acceptor utilization in a contaminated plume. As depicted, the respiratory conditions of the plume vary from highly reactive aerobic conditions at the outmost of the contaminated plume, through anoxic nitrate and iron reduction, to highly reduced sulfate and methanogenic conditions at the contaminant source. The order of electron acceptor agrees well with the utilization order predicted by the electron tower theory.

The development of redox zones shown in Figure 9.3 has been recognized in many contaminated sites (Cozzarelli et al. 2011). The field evidence also suggests that *in situ* bioremediation may employ different electron acceptors at various locations throughout a given site. Figure 9.4 is an example of the field data on the distribution of electron acceptors and reduced products in a kerosene-contaminated site near Berlin, Germany. Groundwater flow is mainly east–west by a perched aquifer in the unsaturated zone through a silt unit, whereas a small area in the center has a flow direction of south–north. As shown in Figure 9.4a, anoxic conditions with dissolved oxygen (DO) concentration below 1 mg/L exist beneath the source zone; and DO levels rise to 3–4 mg/L at 200–300 m downstream from the source zone. DO is also found in the center of the site where water interchange between the perched aquifer and the main aquifer. Fe(II) is found at high concentrations (>30 mg/L) beneath the entire source zone, and its concentration pattern corresponds to the contaminant concentrations. The distribution of Fe(II) clearly indicates a significant biodegradation process in the source zone and contaminated plume. Elevated Fe(II) concentrations were associated with low sulfate concentrations and high methane concentrations, which indicate the co-occurrence of sulfate reduction and methanogenesis. Detailed analysis of electron acceptor in this site can be found in Miles et al. (2008).

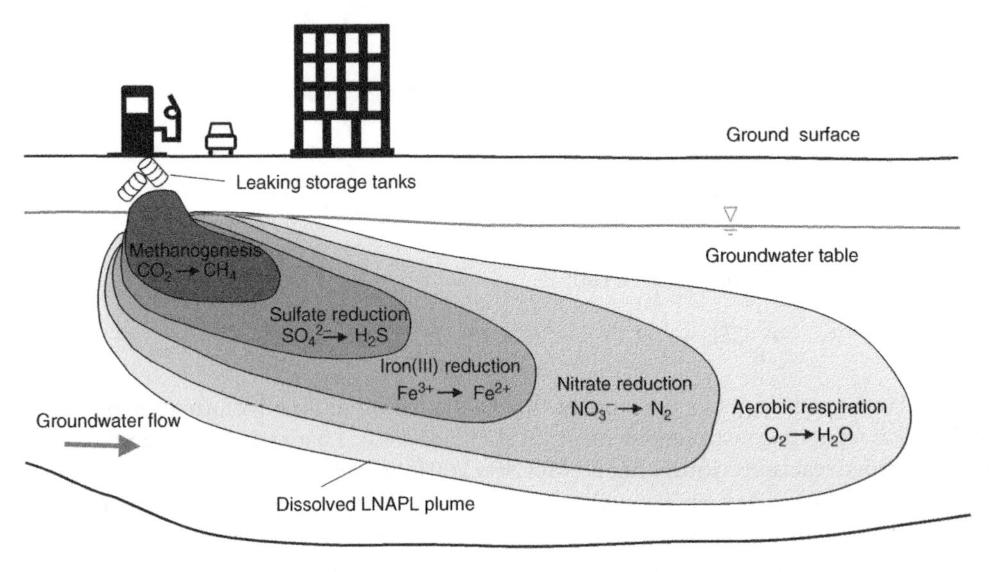

Figure 9.3 Schematics of electron accepting conditions in a dissolved contaminant plume.

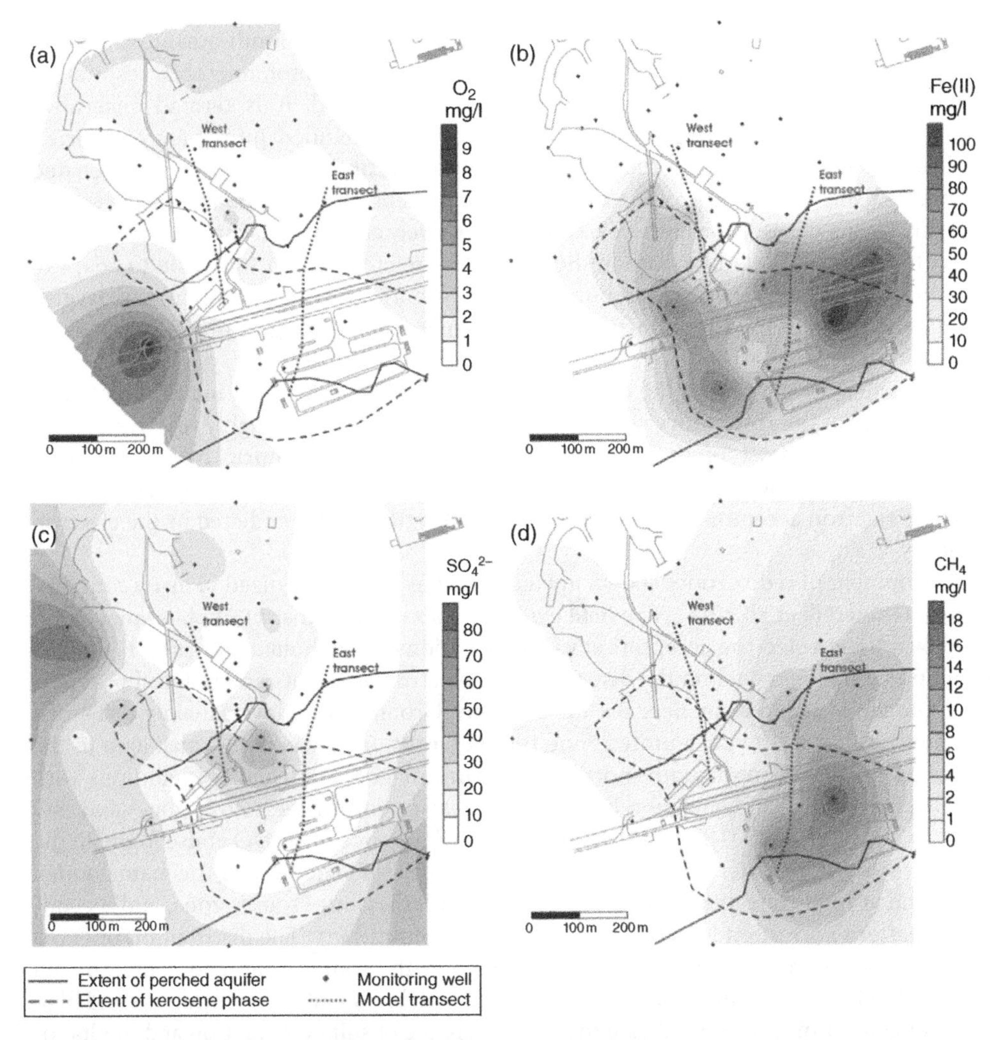

Figure 9.4 Field evidence of the distribution of electron acceptors and reduced products in a kerosene-contaminated site near Berlin, Germany. (a) dissolved oxygen, (b) Fe(II), (c) sulfate, and (d) methane. Nitrate (data not shown) is not present at significant concentrations at the site. The source of kerosene spill was from the tank farm in the Southeast area, and the extent of the kerosene spill is also shown. *Source:* Miles et al. (2008); Reprinted with permission from John Wiley & Sons.

9.1.2 Reaction Stoichiometry and Kinetics

Stoichiometry is the molar relationship between reactants and products in a chemical or microbe-mediated reaction. It is the basis for the mass balance calculation which is essential for the engineering design of processes based on a chemical or biological reaction. With the molar relationship, for example, one can estimate the oxygen demand or nutrient requirement (as the reactants), the biomass production, or methane production during an anaerobic reaction. For example, using Gibbs free energy data, Zhang et al. (2001) derived the following microbial reaction for the degradation of 2,4-dinitrotoluene (2,4-DNT, molecular formula: $C_7H_6N_2O_4$) in a slurry-phase bioreactor:

$$C_7H_6N_2O_4 + 5.62O_2 \rightarrow 5.17CO_2 + 1.63NO_2^- + 1.63H^+ + 0.90H_2O + 0.37C_5H_7O_2N \quad \text{Eq. (9.6)}$$

The stoichiometry of the above reaction indicates that for every mole of 2,4-DNT (MW = 182), the aerobic bacteria need 5.62 mol of oxygen (MW = 32). In the meantime, the bacteria produce 5.17 mol of CO_2, 1.63 mol of NO_2^-, 1.63 mol of H^+, and 0.90 mol of H_2O. In addition, 0.37 mol of bacterial cells (empirical formula: $C_5H_7O_2N$, MW = 113) will be synthesized for every mole of degraded 2,4-DNT. The mass balance of the above reaction for the chemicals of the same order is as follows:

$$182\,g\,C_7H_6N_2O_4 + 180\,g\,O_2 = 227\,g\,CO_2 + 75\,g\,NO_2^- + 1.63\,g\,H^+ + 16\,g\,H_2O + 42\,g\,C_5H_7O_2N \quad \text{Eq. (9.7)}$$

The bacteria employed in this study (Zhang et al. 2001) were capable of using 2,4-DNT as the sole carbon and nitrogen source without additional supply of nitrogen. It should be noted that another nutrient element of phosphorous is always required. Since a simplified empirical formula of bacterial cells ($C_5H_7O_2N$) is used, phosphorous is not included in the above reaction. Another commonly used empirical formula for bacterial cell is $C_{60}H_{87}O_{32}N_{12}P$ if phosphorous needs to be included (Tchobanoglous et al. 2014). The exact constituents of bacteria involved in a specific chemical reaction can be experimentally determined by elemental analysis of the biomass samples.

The use of stoichiometry to estimate the required amount of oxygen, nitrogen, and phosphorous is illustrated in Examples 9.1 through 9.3. For practical purposes, or when the stoichiometric reaction is not available, the C:N:P mass ratios of 100:10:1 to 100:1:0.5 can be used to estimate the nutrient (N and P) requirements. Moreover, an average N/P mass ratio of 6:1 can be used to estimate the required P if the amount of required N is already known.

Example 9.1 Calculate stoichiometric dissolved oxygen (DO) requirement

The groundwater is contaminated due to a recent accidental spill of benzene from an underground storage tank. Assume free-phase benzene on the groundwater table has been removed, and the dissolved phase concentration of benzene is 50% saturated in the aqueous phase. The solubility of benzene = 1780 mg/L. Dissolved oxygen (DO) at a typical groundwater temperature of 13 °C is 10 mg/L. Is DO concentration sufficient for aerobic biodegradation of benzene at 50% of its saturated concentration?

Solution

We simplify the calculation without incorporating cell mass during the biodegradation of benzene using the stoichiometric reaction: $C_6H_6 + 7.5O_2 \rightleftharpoons 6CO_2 + 3H_2O$.

The molecular weight of C_6H_6 and O_2 are 78 and 32, respectively. The mass ratio of benzene to $O_2 = 1 \times 78 : 7.5 \times 32 = 78 : 240 = 1 : 3.08$. The dissolved phase benzene concentration at 50% saturation = 1780 mg/L × 50% = 890 mg/L, which requires 890 × 3.08 = 2740 mg/L dissolved oxygen – far exceeding the saturated DO in groundwater. This example indicates that groundwater is typically oxygen limited for biodegradation. Fortunately, O_2 will be replenished by oxygenated groundwater flowing through the contaminant plume. However, this could always be problematic because of the slow-flowing groundwater in a less permeable aquifer. For an aquifer low in hydraulic conductivity and a saturated DO concentration of 10 mg/L in groundwater, only 10/3.08 = 3.2 mg/L of benzene can be degraded without the replenishment of DO.

Example 9.2 Calculate nutrients (N, P) requirement given stoichiometry

Estimate the electron acceptor and nutrient (N, P) needs for bioremediation of 3000 yd^3 soil contaminated with 7000 mg/kg of total hydrocarbon. Assume: (i) The total hydrocarbon containing BTEX, *n*-pentane, 2,2-dimethylheptane, benzene 1,2-diol, and butyric acid has the average chemical structural formula of $C_6H_{10}O$; (ii) NH_3 is used as the nitrogen source according to: $C_6H_{10}O + 4O_2 + 0.8\,NH_3 \rightleftharpoons 2CO_2 + 0.8C_5H_7O_2N + 3.4H_2O$; (iii) Soil density = 115 lb/ft^3.

Solution

Molecular weight: $C_6H_{10}O$ = 98; O_2 = 32; NH_3 as N = 14; CO_2 = 44. Since phosphorous (P) is not included in the stoichiometric reaction, we assume that P is 1/6 of the required nitrogen by mass. The mass ratios of $C_6H_{10}O:O_2:N:P:CO_2 = 98:4\times32:0.8\times14:1/6\times0.8\times14:2\times44$. Divided by 98, we get the ratio = 1 : 1.31 : 0.114 : 0.019 : 0.90.

Mass of total hydrocarbons = volume of soil × soil density × concentrtaion

$$\text{Mass of total hydrocarbons} = 3\,000\,\text{yd}^3 \times \frac{115\,\text{lb}}{\text{ft}^3} \times \frac{7\,000\,\text{lb}}{10^6\,\text{lb}} \times \frac{27\,\text{ft}^3}{\text{yd}^3} = 65206\,\text{lb}$$

Note that 7000 mg/kg is the same as 7000 parts per million (ppm), which is the same as the mass ratio in any unit, such as 7000 lb/10^6 lb or 7000 kg/10^6 kg. The requirements for O, N, and P are as follows:

Oxygen requirement = 65 205 × 1.31 = 85 420 lb
Nitrogen requirement = 65 205 × 0.114 = 7 430 lb as N
Phosphorus requirement = 65 205 × 0.019 = 1 240 lb as P

Note in the above calculation that the atomic weight of N (14) instead of the molecular weight of NH_3 (MW = 17) is used. This is because both N and P compositions in fertilizers commonly used are reported as per N or P basis, which can be easily converted into the mass of the parent compounds such as ammonium sulfate ($[NH_4]_2SO_4$) or tripolyphosphate ($K_3P_3O_{10}$).

Example 9.3 Calculate nutrients (N, P) requirements without stoichiometry

A leaking underground storage tank (LUST) has an estimated contaminated soil volume of 90 000 ft^3. The average TPH concentration in the contaminated soil is 1000 mg/kg, and the soil bulk density is 50 kg/ft^3 (1.75 g/cm^3). Estimate the nutrient requirement for the bioremediation of this contaminated site.

Solution

The mass of contaminated soil is equal to the product of volume and bulk density:

$$\text{Soil mass} = 90000\,\text{ft}^3 \times 50\frac{\text{kg}}{\text{ft}^3} = 4.5\times10^6\,\text{kg}$$

The mass of contaminants (TPH) is the product of the mass of contaminated soil and the average TPH concentration in the contaminated soil (note 1 kg = 2.2 lb), hence:

$$\text{Contaminant}(\text{TPH})\,\text{mass} = 4.5\times10^6\,\text{kg}\times1000\frac{\text{mg}}{\text{kg}} = 4.5\times10^9\,\text{mg} = 4500\,\text{kg} = 10000\,\text{lb}$$

Unlike Example 9.2, we are not provided the stoichiometric reaction, so an empirical C:N:P ratio of 100:10:1 is used to estimate N and P requirement. Herein, we assume that the total carbon mass is the same as the total mass of TPH. This is rather a conservative approximation by assuming that the total mass of hydrocarbon in the soil represents the mass of carbon available for biodegradation. This simplifying assumption could be valid because the carbon content of the petroleum hydrocarbons commonly encountered at UST sites is approximately 90% carbon by weight (USEPA 1994).

Using the C:N:P mass ratio of 100:10:1, the required mass of nitrogen would be 1000 lb (450 kg), and the required mass of phosphorous would be 100 lb (45 kg). After converting these masses into concentration units (i.e. 450 kg per 4.5×10^6 kg soil or 100 mg/kg for nitrogen; 45 kg per 4.5×10^6 kg soil or 10 mg/kg for phosphorous), they can be compared with the results of the soil analyses to determine if nutrient addition is necessary. Since 100 mg N/kg is markedly higher than the background concentration of nitrogen in most soils, nitrogen addition is necessary and slow release sources should be preferred. An important consideration for slow release is the lowering of soil pH after the nitrogen is added to the soil.

Reaction stoichiometry entails the *amount* (mass) of reactants and products involved in a reaction at thermodynamic equilibrium. However, the *kinetic rate* of reaction must also be revealed if the rates of reactant utilization (e.g. contaminant degradation, oxygen utilization) or the rates of product formation must be known. For practical purposes, the rate of biodegradation can be approximated by the **first-order reaction**, i.e. the reaction rate (dC/dt) is directly proportional to the concentration (C) of one reactant:

$$\frac{dC}{dt} = -kCt \qquad \text{Eq. (9.8)}$$

where C is the concentration at time t, and k is the first-order rate constant. Integrating Eq. (9.8) will result in

$$C_t = C_o e^{-kt} \qquad \text{Eq. (9.9)}$$

where C_0 = original concentration of organic compounds, C_t = concentration after degradation at time t (days), k = first-order rate constant (day^{-1}). The rate constant (k) can be estimated if a half-life ($t_{1/2}$) is known:

$$t_{1/2} = \frac{0.693}{k} \qquad \text{Eq. (9.10)}$$

In some cases, the percent biodegradation (or in general, percent removal = $1 - C_t/C_0$) is given, and the cleanup time (t) can be estimated:

$$t = \frac{\ln\left(1 - \%\,\text{degradation}\right)}{-k} \qquad \text{Eq. (9.11)}$$

Equations (9.9) through (9.11) are quite useful to estimate the cleanup time in many remediation systems in which first-order reaction can oftentimes be approximated, regardless of whether the reaction is physical (thermal and radioactivity decay), chemical, or biological. We will revisit the use of these equations in future chapters, but Example 9.4 is employed here for the bioremediation system.

Example 9.4 Calculate oxygen delivery rate based on biodegradation kinetics

The half-life for the biodegradation of naphthalene is estimated to be 35 days in a lab study. For a contaminated site, a cleanup goal of less than 100 mg/kg must be achieved, and this remediation corresponds to a >99% removal of the naphthalene. (a) Estimate the cleanup time in days. (b) What is the percentage of naphthalene remaining after the first two weeks of remediation? (c) If the total oxygen requirement is calculated to be 100 000 lb, what is the average oxygen delivery rate during the first two weeks? and the second two weeks? (adapted from Cookson 1995)

Solution

a) We first use the half-life ($t_{1/2}$ = 35 days) to calculate the first-order rate constant according to Eq. (9.10):

$$k = \frac{0.693}{t_{1/2}} = 0.0198\,\text{day}^{-1}$$

The cleanup time for a 99% removal can be calculated from Eq. (9.11):

$$t = \frac{\ln\,(1 - 0.99)}{-0.0198} = 233\text{ days} \quad (8\,\text{months})$$

Note that the cleanup time is independent of the actual concentrations, but depends on the percent removal, i.e. the *ratio* of contaminant concentration (C_t/C_0) at a given time (t) to the initial concentration at $t = 0$.

b) By rearranging Eq. (9.9), we can calculate percent remaining after the first two weeks at $t = 14$ days:

$$\%\text{remaining} = \frac{C_t}{C_0} = e^{-kt} = e^{-0.0198 \times 14} = 0.758\,(75.8\%)$$

c) At the end of first two weeks, the percent degradation = 1 − 0.758 = 0.242 (24.2%). This is equivalent to a total of 24.2% × 100 000 lb = 24 210 lb of oxygen consumed during the first two weeks. The average rate during the first two weeks = 24 210 lb/14 days = 1700 lb/day.

Now we do the same calculation at the end of second two weeks (Day 15 to Day 28):

$$\% \text{ Remaining at 28 days} = \frac{C_t}{C_0} = e^{-kt} = e^{-0.0198 \times 28} = 0.574\,(57.4\%)$$

The percent degradation at the end of second two weeks = 1 − 0.574 = 0.426 (42.6%). This is equivalent to 42.6% × 100 000 lb = 42 558 lb of O_2 required at the end of second two weeks.

The oxygen needed during the second two weeks of biodegradation (Day 15 to Day 28) = 425 58 − 24 210 = 18 348 lb.

Again if we take the average during the second two weeks, then the rate of oxygen delivery = 18 348/14 = 1 300 lb/day. This is smaller than the average O_2 delivery during the first two weeks (17 000 lb/day). The oxygen utilization rate is decreased over time, implying that adjustment should be made for a cost-effective implementation of the bioremediation.

9.1.3 Biodegradation Potentials and Pathways

The technical feasibility of bioremediation depends on the rate and extent of contaminant biodegradation, or in the general term, the **biodegradability** or the biodegradation potential. Another needed piece of information to assess the technical feasibility of bioremediation and monitor bioremediation progress is the metabolic (biochemical) pathways that bacteria employed to consume corresponding compounds. Ideally, contaminants are mineralized into CO_2, H_2O, and other innocuous end products. However, this is not always the case. Sometimes daughter compounds can be more toxic than the parent compounds (see an example in Box 2.2).

A great deal of research work has been done on the biodegradation potential and pathways of a variety of contaminants. In general, studies show that some environmental contaminants can be biodegraded only by aerobic bacteria, whereas other chemicals can be biodegraded only by anaerobic bacteria. Some chemicals can be biodegraded by either type of bacteria, although the strategies will be different. Moreover, some chemicals need the participation of a consortium of both aerobic and anaerobic bacteria. Bacteria capable of degrading common contaminants are usually abundant in the natural soil and shallow groundwater. The diversity of the bacteria in the natural environment is essential, because no single bacterial species (so-called superbug) is capable of degrading all synthetic (human-made) chemicals. Table 9.2 is a preview of biodegradability of several major contaminant classes. These are further elaborated below according to the classes of contaminants commonly detected in soil and groundwater.

Table 9.2 Chemical classes and their susceptibility to bioremediation.

Chemical class	Examples[a]	Biodegradability
Aliphatic hydrocarbons	Pentane, hexane	Aerobic
Aromatic hydrocarbons	Benzene, toluene	Aerobic and anaerobic
Ketones and esters	Acetone, MEK	Aerobic and anaerobic
Petroleum hydrocarbons	Fuel oils	Aerobic
Chlorinated solvents	TCE, PCE	Anaerobic (reductive dechlorination)
Polyaromatic hydrocarbons	Anthracene, B(a)P, creosote	Aerobic
Polychlorinated biphenyls (PCBs)	Arochlors	Some evidence; not readily biodegradable
Metals	Cadmium	Not degradable; biosorption

Source: Adapted from Baker and Herson (1994).
[a] MEK = Methyl ethyl ketone; TCE = trichloroethene; PCE = tetrachloroethylene; B(a)P = Benzopyrene.

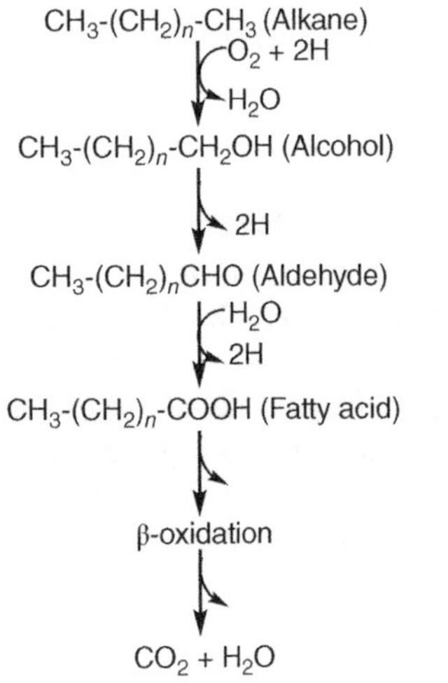

CH_3-$(CH_2)_n$-CH_3 (Alkane)
O_2 + 2H
H_2O
CH_3-$(CH_2)_n$-CH_2OH (Alcohol)
2H
CH_3-$(CH_2)_nCHO$ (Aldehyde)
H_2O
2H
CH_3-$(CH_2)_n$-COOH (Fatty acid)
β-oxidation
CO_2 + H_2O

Figure 9.5 Metabolism of an alkane by aerobic microorganisms.

9.1.3.1 Biodegradation of Petroleum Aliphatic Hydrocarbons

Under the normal standard temperature and pressure conditions (NSTP, 1 atm, 20 °C), alkanes in the range of CH_4 (methane) to C_4H_{10} (butane) are gaseous; from C_5H_{12} (pentane) to $C_{17}H_{36}$ are liquids; and after $C_{18}H_{38}$ are solids. Alkanes in gasoline that contain 4–10 carbons per molecule of saturated aliphatic hydrocarbons are seldom found in most contaminated sites because of their high volatility. Longer carbon chain alkanes from gasoline spills can be biodegraded by aerobic bacteria into CO_2 and water (i.e. mineralization), as shown in Figure 9.5. Biodegradation potential is a function of the carbon chain length, degree of saturation, and the extent of branch of the alkanes. Longer-chain alkanes are degraded more readily than shorter-chain alkanes. This appears counterintuitive that shorter-chain alkanes are less readily biodegradable than the longer-chain alkanes. It is because alkanes with less than nine carbons in length tend to be toxic as a result of their higher solubility. Saturated alkanes are more susceptible to aerobic bacterial attack than unsaturated alkanes (i.e. alkenes, alkynes). Alkanes with branched chains are degraded less readily than straight-chain alkanes.

As shown in Figure 9.5, the most common aerobic pathway for alkane degradation is the oxidation of the terminal methyl group into carboxyl acid through an alcohol intermediate, and eventually the complete mineralization through **β-oxidation** (Leahy and Colwell 1990; Cookson 1995). The β-oxidation is so named because the oxidation occurs at the carbon of the β-position to the carboxyl carbon by the removal of two carbons at a time as acetyl coenzyme A (CoA). These reactions occur in the mitochondria. It is unknown whether physiologically and phylogenetically distinct anaerobes possess the similar biological mechanisms. However, recent data with *n*-hexane utilizing denitrifying isolate pointed to a different pathway involving an initial enzymatic attack by fumarate ($^-$OOC—H=CHCOO$^-$) addition in a manner similar to that for toluene as discussed later. It is apparent that bioremediation of aliphatic hydrocarbons should be performed as an aerobic rather than anaerobic process. While natural soils normally contain adequate alkane-degrading microorganisms, additional oxygen must be supplied.

9.1.3.2 Biodegradation of Single-Ring Petroleum Aromatic Hydrocarbon (BTEX)

While saturated alkanes in oil spills are less of a health concern, petroleum aromatic hydrocarbons with one benzene ring (i.e. BTEX) are more important because of the potential health effects. Petroleum aromatic compounds can be degraded via the initial microbial attack leading to the formation of dihydrodiol, which is typical for single-, double-, and triple-ring aromatic hydrocarbons. The next step in the aerobic biodegradation is an oxidative attack on the aromatic nucleus leading to the formation of catechol followed by the cleavage of the aromatic carbon ring, as shown in Figure 9.6. This will ultimately lead to the mineralization of benzene into CO_2 and H_2O.

Under anaerobic conditions, certain bacteria are capable of mineralizing single-ring aromatic hydrocarbons (BTEX) into CO_2 and H_2O. The biochemical pathways, however, are different from that of aerobic conditions. Figure 9.7 shows the general strategies of anaerobic bacteria in degrading four major gasoline components of environmental concern, i.e. benzene, toluene, ethylbenzene, and xylene (BTEX). Variations of pathways, however, exist since various bacteria depend on different electron acceptors corresponding to differing redox conditions (Chakraborty and Coates 2004). Although a complete mineralization has been reported for all BTEX compounds except *p*-xylene, research thus far can only elucidate the initial enzymatic reactions, as shown in Figure 9.7.

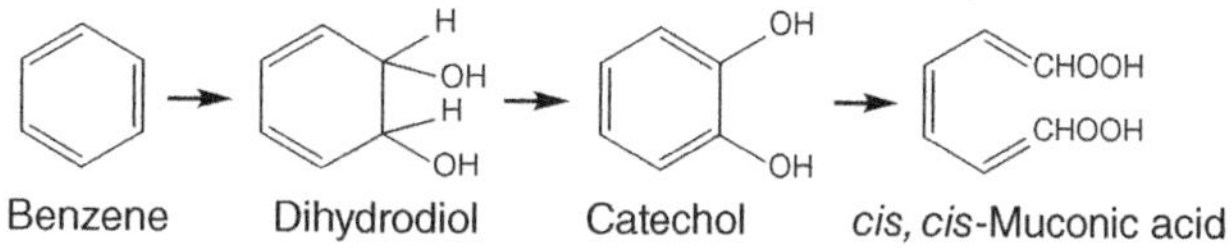

Figure 9.6 Initial steps for the aerobic biodegradation of benzene.

A striking difference from aerobic mechanisms is the introduction of O_2 through H_2O to form less stable oxygenated monoaromatic compounds which are susceptible to further ring cleavage. Also shown in Figure 9.7 is benzoyl coenzyme A (benzoyl-CoA), the common intermediate for all BTEX compounds. Benzoyl-CoA is formed through the addition of fumarate to BTEX compounds through the enzymatic action of benzylsuccinate synthase (BSS) (for toluene) or methylbenzylsuccinate synthase (for *o*- and *m*-xylene). Benzoyl-CoA is transformed to 1,5-diene-1-carboxyl-CoA through a key enzyme benzoyl-CoA reductase. After a series of hydration and dehydrogenation steps, 3 mol of acetyl-CoA and 1 mol of CO_2 is formed from each mole of BTEX compound (Mogensen et al. 2003).

Successful degradation of BTEX has been demonstrated under all anaerobic conditions: nitrate respiration, sulfate reduction, and methanogenic fermentation. Anaerobic processes are often less favorable thermodynamically than aerobic processes because of the longer degradation time. The rate of denitrifying reactions appears to be 50% lower compared with oxygen as the electron acceptor. However, the cost of using nitrate as an electron acceptor was reported to be 50% less per unit soil volume as compared with oxygen supplied with H_2O_2 as the electron acceptor (Cookson 1995).

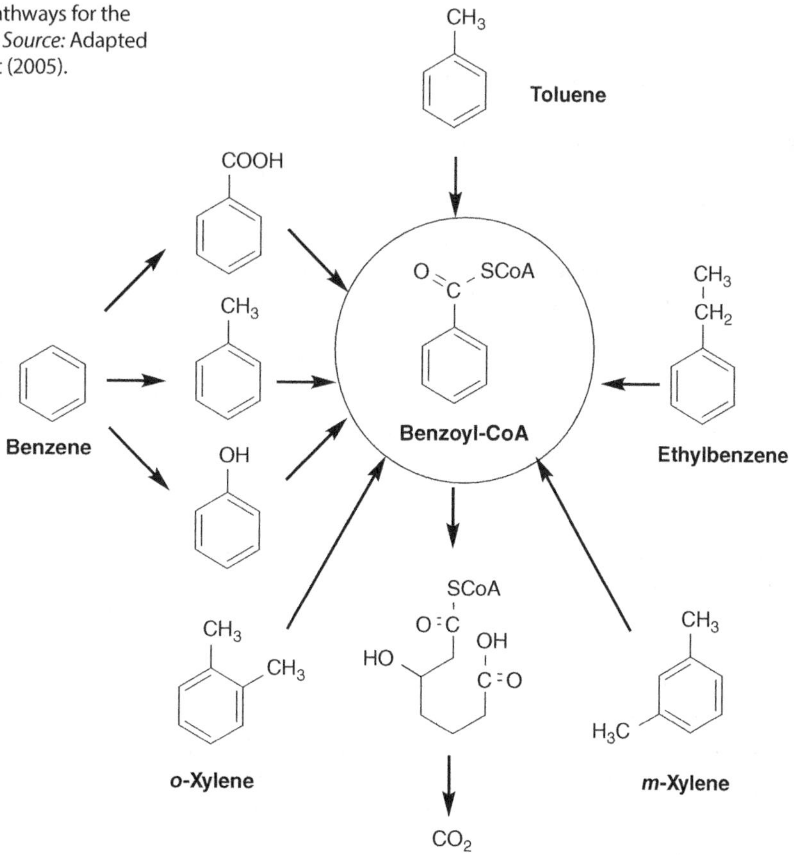

Figure 9.7 Anaerobic pathways for the biodegradation of BTEX. *Source:* Adapted from Zhang and Bennett (2005).

9.1.3.3 Biodegradation of Fuel Additives (MTBE)

Methyl *tert*-butyl ether (MTBE) has been used to replace tetraacetyl lead to increase octane rating of gasoline and comply with the Clean Air Act mandates. It was approved by the USEPA for use in 1979 and was added to gasoline during the 1980s at approximately 2–5% by volume as an octane booster. In 1992, it was blended at 10–15% by volume for use in some areas in the wintertime oxygenated fuel program. In addition, several other ethers have been used as oxygenates in gasoline such as tertiary amyl methyl ether (TAME) and ethyl tertiary butyl ether (ETBE). Because their use has not been as widespread, the threat is unlikely to be as great as MTBE at most sites.

MTBE is fairly volatile (vapor pressure 245 mmHg, and dimensionless Henry's law constant 0.02 at 25 °C); however, it persists in groundwater due to its high water solubility (51 g/L at 25 °C) and resistance to microbial attack. Since MTBE can be quickly spread in groundwater, the cleanup strategy for MTBE-laden petroleum spills will be different from the non-MTBE-laden petroleum release. The ether bond is structurally hindered, thereby resisting to abiotic or biotic hydrolysis. The metabolic pathways of MTBE under both aerobic and anaerobic conditions are not yet well understood. Available studies have shown that aerobes attack MTBE by a monooxygenase leading to the cleavage of the recalcitrant ether bond. With anaerobic bacteria, the cleavage involves methyl transferases and tetrahydrofolate for the degradation of lignin (a naturally occurring ether compound) and hydroxyl group addition during the fermentation of polyethylene glycols ($-O-CH_2-CH_2OH$).

There have been reported cases of MTBE bioremediation. Several pilot tests and limited field applications show that MTBE biodegradation is often significant at sites where monitored natural attenuation is the remediation process (Davis and Erickson 2004). In the past decade, the number of species that degrade MTBE and the reported rates of biodegradation have increased significantly. Until further work is done it will not be possible to determine the extent to which bioaugmentation may be beneficial at field sites. Both aerobic and anaerobic strategies may be applicable; however, aerobic biodegradation rates are higher similar to other contaminants.

9.1.3.4 Biodegradation of Polycyclic Aromatic Hydrocarbons (PAHs)

Bacteria-mediated biotransformation and mineralization have been demonstrated under both aerobic and anaerobic conditions for PAHs and substituted PAHs. Under aerobic conditions, two- or three-ring PAHs are readily biodegradable. PAHs with four or more aromatic rings are significantly more difficult and some may be recalcitrant. The biodegradability of PAHs depends on the solubility of PAHs (which in turn is dependent on the number of aromatic rings), the alkyl substitution (number, type, and position), and the soil-to-water partitioning coefficient (K_d). Alkyl substitutions reduce the biodegradation of PAHs. A less hydrophobic PAH will tend to be more bioavailable as a result of the faster desorption from soil particles. For PAHs with four or more rings, cometabolism coupled with analog substrate enrichment may be the efficient way to achieve cleanup of high-molecular-weight PAHs. **Cometabolism** is the simultaneous degradation of two compounds, in which the degradation of the second compound (the secondary substrate) depends on the presence of the first compound (the primary substrate). An example of **analog substrate enrichment** is the use of lower-molecular-weight PAHs (e.g. naphthalene) to induce enzyme production that will facilitate the degradation of higher-molecular-weight PAHs.

Although the detailed pathways and enzymes involved are still the active research, it is well known that the rate-limiting step for aerobic PAH degradation is the initial ring oxidation by attaching the functionality of OH and the epoxide (a cyclic ether with a three-atom ring) to one of the carbons in the aromatic ring. The initial attack is also rate-limiting for PAH-degrading anaerobes. Figure 9.8 shows anaerobic bacteria's strategies in degrading naphthalene and a substituted PAH of a similar structure. Similar to aerobic processes, the benzene ring must be opened up first such that bacteria can transform it into CO_2 and water. Two- and

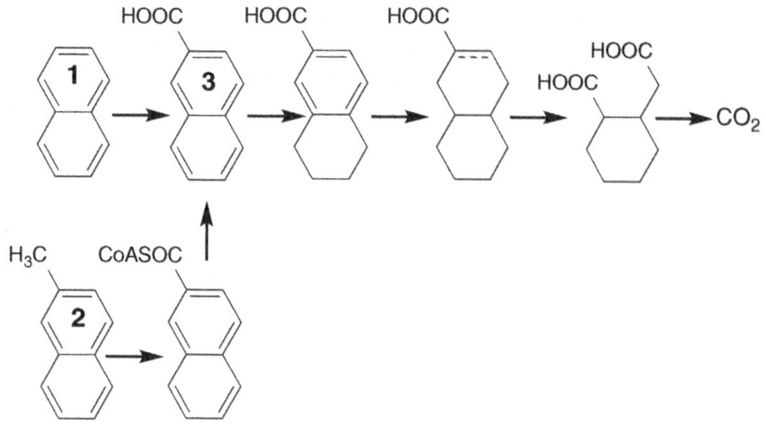

Figure 9.8 Anaerobic pathways for the biodegradation of PAHs (1 = naphthalene; 2 = 2-methylnapthalene, 3 = 2-naphthoic acid). *Source:* Adapted from Zhang and Bennett (2005).

three-ring aromatic compounds have been transformed under denitrifying, sulfate-reducing, fermentation, and methanogenic conditions.

9.1.3.5 Biodegradation of Chlorinated Aliphatic Hydrocarbons (CAHs)

Three major biodegradation pathways exist for chlorinated (or halogenated in general) aliphatic hydrocarbons containing one to two carbons, such as the frequent groundwater contaminants PCE and TCE. These pathways include the aerobic oxidation (chlorinated alkanes used as a sole energy source), cometabolism under aerobic conditions, and **reductive dechlorination** (or **dehalogenation** in general) under anaerobic conditions. In bioremediation applications, reductive dechlorination is the most important.

Aerobic transformation uses oxygen as the electron acceptor for one- to three-halogen atom–substituted aliphatic compounds (e.g. vinyl chloride, DCE, and TCE). Enzymes involved for such aerobic transformation are oxygenases, dehalogenases, or hydrolytic dehalogenases. The oxygenase enzyme transforms CAHs into alcohols, aldehydes, or epoxides. Dehalogenase transforms CAHs into aldehyde and glutathione, the latter being a cofactor for the nucleophilic substitution by this enzyme. Hydrolytic dehalogenase hydrolyzes CAHs into alcohols.

Figure 9.9 shows anaerobic bacteria's strategies in degrading PCE and TCE into the unharmful product ethene. However, the process should be well controlled so that the accumulation of toxic intermediate vinyl chloride (VC) can be avoided. The reductive transformation is carried out by the electrons released from the oxidation of the primary substrate such as hydrogen gas (H_2) as the electron donors (H_2 is oxidized to H^+, hence it is an electron donor). Since CAHs serve as the electron acceptors during the reductive dehalogenation, the presence of other competing electron acceptors such as nitrate, sulfate, and CO_2 will adversely affect the reductive dehalogenation. The reductive process is usually through cometabolism, for example, in the presence of methanotrophic microorganisms.

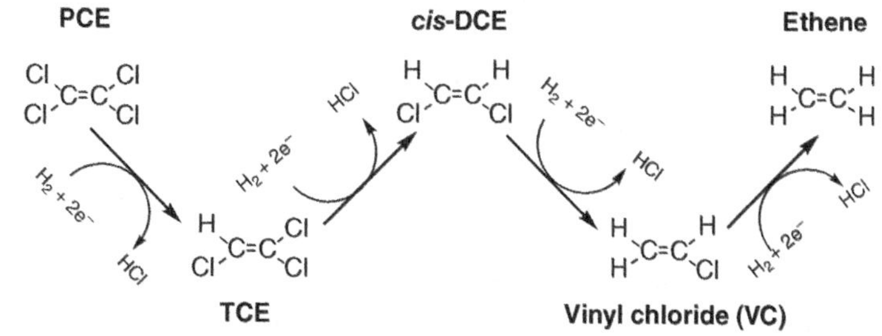

Figure 9.9 Anaerobic pathways for the biodegradation of chlorinated aliphatic tetrachloroethylene (PCE). *Source:* Adapted from Rittmann and McCarty (2001).

9.1.3.6 Biodegradation of Chlorinated Aromatic Compounds

Compounds of this category include a diverse group of environmentally significant contaminants such as chlorobenzenes, chlorophenols, chloronitrophenols, some pesticides (e.g. DDT), PCBs, and dioxins. Hence, a detailed discussion regarding biodegradation potential and pathways is not appropriate here, and only some general remarks are made below.

For some single-ring chlorinated aromatic compounds, such as 3-chlorobenzoate and halocatchols (e.g. 3-methyl-5-chlorocatechol), mineralization can be readily achieved by aerobes after the benzene ring is opened up. The replacement of the halide with the hydroxyl group is essential for the initial enzymatic attack of a typical aerobic biotransformation process. The rate of elimination of halogens under aerobic condition decreases as halogen substitution increases. Interestingly, the opposite is true for anaerobic transformation, that is, reductive dechlorination increases as the number of chlorine substitution increases. Another important aspect for chlorinated aromatics is the need of consortium bacteria in order to degrade some highly chlorinated aromatic compounds such as PCBs. Highly Cl-substituted PCBs are not susceptible to aerobic bacteria. However, once chlorines have been removed, aerobic transformation is likely. A sequential anaerobic–aerobic process, therefore, should be employed for such complex halogenated aromatics. The anaerobic stage promotes reductive dechlorination, whereas a subsequent aerobic stage further oxidizes less chlorinated metabolic intermediates into CO_2, Cl^-, and H_2O (Figure 9.10). These two complementary processes (anaerobic and aerobic) for the biodegradation of PCBs are known to occur in the Hudson River sediment and many other contaminated sites (Abramowicz 1995).

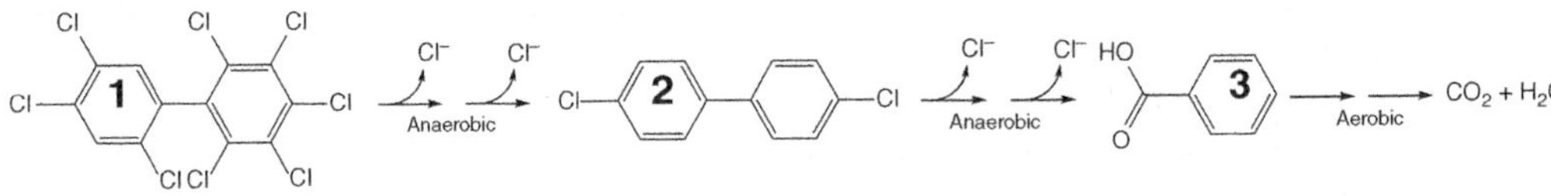

Figure 9.10 Proposed sequential anaerobic–aerobic processes for the biodegradation of PCBs(1 = 2,2′,3,4,4′,5,5′,6-octachlorobiphenyl, 2 = 4,4′-dichlorobiphenyl, 3 = benzoic acid).

An important consideration should be given to the effect of electron donors and acceptors on the reductive dechlorination of aromatic hydrocarbons. Since halogens are being reduced by gaining electrons from electron donors, the addition of electron donor such as butyrate, propionate, ethanol, or acetone will likely increase the rate and extent of biodegradation. On the other hand, electron acceptors such as sulfate (from sulfate-reducing bacteria) will adversely impact dehalogenation, because dehalogenating bacteria and sulfate-reducing bacteria will compete for suitable electron donors.

9.1.3.7 Biodegradation of Explosive Compounds

In Chapter 6 (Section 6.1.3), several common explosives contaminants were briefly introduced regarding their biodegradation potential for a technological screening purpose. These compounds include TNT, 2,4-DNT, 2,6-DNT, RDX, and HMX (see Figure 2.4 for the structure). TNT and DNTs are nitroaromatic compounds (NACs), and both RDX and HMX are nonaromatic heterocyclic compounds that contain C and N in their ring structure.

TNT has three oxidized NO_2 groups at the 2,4,6-positions. Because of the highly oxidized state, aerobic oxidation is unlikely to destabilize this molecule. Conversely, because of the electrophilic nature, these external NO_2 groups are amenable to enzymatic reduction. In the meantime, since π-electrons in the benzene ring are shielded by four functional groups (three NO_2 and one CH_3) due to steric hindrance, the aromatic structure is very stable, preventing ring cleavage from any enzymatic attack. This unique chemical structure explains, to a large extent, why **biotransformation** of TNT occurs rapidly but appreciable **mineralization** into CO_2 and H_2O has not been achieved in either aerobic or anaerobic systems. Not only can TNT not be biodegraded all the way into CO_2 and water, but also some of the nitro-reduction intermediate products are even more toxic than TNT, such as the hydroxylamino (NHOH) derivatives of TNT.

With one nitro-group less than TNT, the biodegradation potentials of 2,4-DNT and 2,6-DNT are considerably improved. Under optimal aerobic conditions, a complete and rapid mineralization was demonstrated in a pilot-scale bioslurry reactor system (Zhang et al. 2000). The major metabolic pathways were elucidated (Figure 9.11). Under optimal pH, nutrients, and mixing conditions, both DNTs are completely mineralized into nitrite (NO_3^-), CO_2, and H_2O. The overall stoichiometric reaction has been given previously in Eq. (9.1).

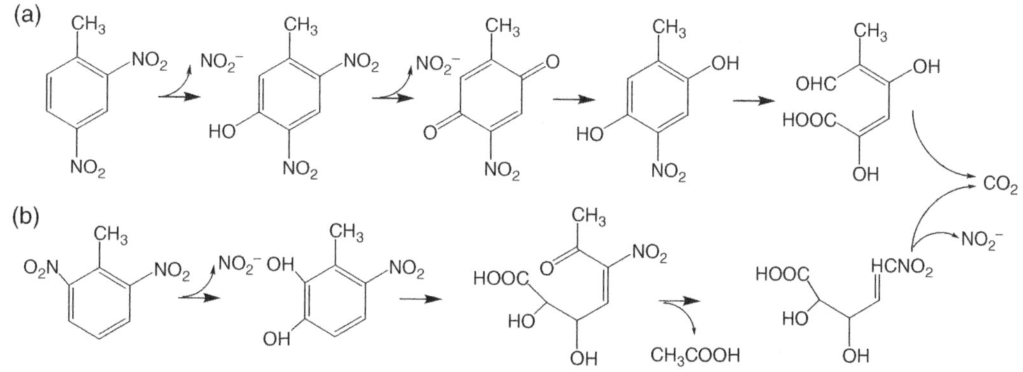

Figure 9.11 The aerobic pathway leading to the oxidative mineralization of (1) 2,4-DNT by *Burkholderiacepacia* JS872, and (2) 2,6-DNT by *Burkholderiacepacia* JS850 and *Hydrogenophagapalleronii* JS863. *Source:* Adapted from Nishino et al. (2000).

Under anaerobic condition, however, both DNTs can only undergo transformation rather than mineralization (see Box 2.2). As shown in Figure 9.12, the nitro group (NO_2) is reduced to hydroxylamino group (NHOH) which is more reactive and toxic than its parent compounds. The NHOH intermediates can further form diaminotoluene as the end products. The aforementioned aerobic versus anaerobic pathways clearly indicate that bioremediation of DNTs can only be achieved satisfactorily under the aerobic conditions.

9.1.4 Optimal Conditions for Bioremediation

A compound's biodegradability is a prerequisite for the success of any bioremediation technology. However, merely having a desired biodegradability from literature or laboratory test data does not warrant its success in the field where bioremediation is employed. Some fundamental questions should be addressed during the technology screening process to evaluate the applicability of bioremediation. In a broad sense, one should ensure (a) abundant and right bacteria in the contaminated zone of the subsurface (right bacteria); (b) site conditions are optimal for bacterial growth and biodegradation (right environmental conditions); and (c) intimate contacts are maintained through engineering or hydrogeologic approaches (e.g. mixing, advective flow) between bacteria, contaminants, nutrients, and electron donors/acceptors (right contact).

In the discussion that follows, we group these parameters into four categories: (a) hydrogeologic parameters, (b) soil/groundwater physicochemical parameters, (c) microbial presence, and (d) contaminant characteristics. The general trends for various bacteria-based bioremediation

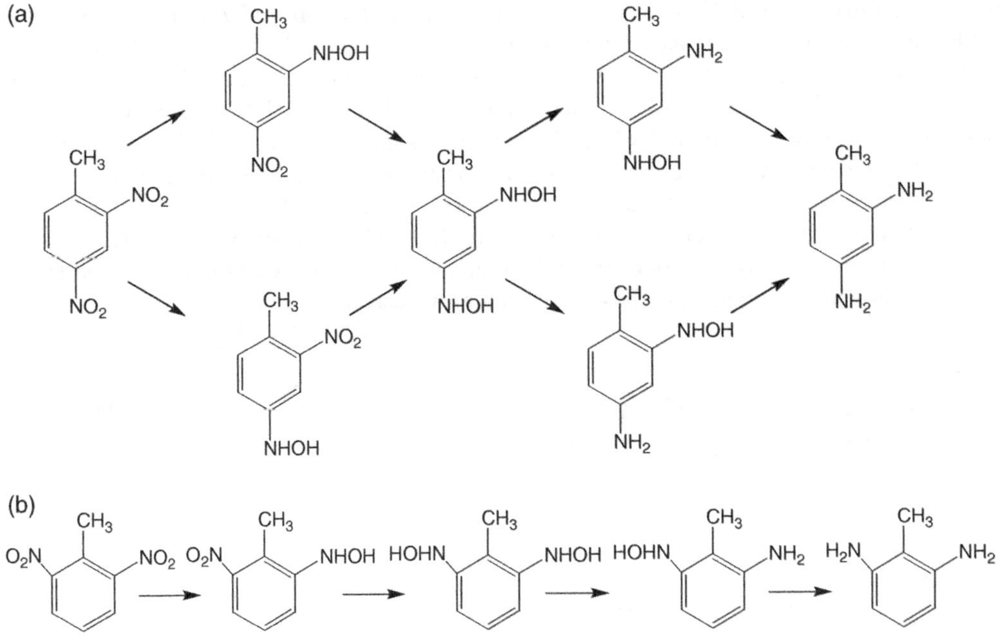

Figure 9.12 The anaerobic pathway leading to the reductive transformation of (a) 2,4-DNT and (b) 2,6-DNT in cell culture and cell extract of *Clostridium acetobutylicum*. *Source:* Adapted with permission from Hughes et al. (1999). Copyright (1999) American Chemical Society.

processes will be introduced in Section 9.2. Readers should be cautious that exceptions exist. For example, the environmental factors (e.g. temperature, soil moisture) that affect biopiles (composting) will be very different from *in situ* groundwater bioremediation. Likewise, parameters affecting vadose zone bioremediation will differ from those related to saturated zone bioremediation. Furthermore, plant-based bioremediation (i.e. phytoremediation) will have very different process parameters from bacteria-based parameters. We will discuss these process-specific parameters in Section 9.2.

9.1.4.1 Hydrogeologic Parameters

Hydraulic conductivity of the contaminated aquifer controls the transport and distribution of contaminants, electron donors/acceptors, water, and nutrients in the subsurface – all are critical for the bioremediation. Bioremediation is generally effective in permeable (e.g. sandy, gravel) aquifer media. In less permeable silty or clay media, *in situ* bioremediation is not likely to be very effective and it will require longer time to clean up than a more permeable medium. It is generally considered that a hydraulic conductivity of greater than 10^{-4} cm/s is needed for bioremediation to be effective. For sites with lower hydraulic conductivities (e.g. 10^{-4}–10^{-6} cm/s), the technology can also be effective, but it must be carefully evaluated, designed, and controlled.

Another important consideration for the effective *in situ* bioremediation is soil structure and stratification. Structural characteristics such as microfracturing can result in a higher permeability than expected for certain soils (e.g. clays). Stratified soils with different permeabilities can considerably increase the lateral flow of groundwater in the more permeable strata. As a result, a **preferential flow** reduces the flow in the less permeable portion of the strata, thereby increasing the time for site cleanup.

9.1.4.2 Soil/Groundwater Physicochemical Parameters

Soil moisture and organic contents, as well as groundwater dissolved oxygen (DO), pH, temperature, osmotic pressure, salinity, nutrients, dissolved iron, and electron donors/acceptors, are the factors affecting bioremediation. Serving as the terminal electron acceptors (TEA), dissolved oxygen is commonly the limiting factor for the aerobic degradation of petroleum hydrocarbons and other readily biodegradable oxygen-consuming organics. However, the presence of dissolved oxygen will become detrimental to anaerobic bacteria. Extreme pH values (i.e. lower than 5 or higher than 10) are generally unfavorable for microbial activity. Typically, optimal microbial activity occurs under neutral pH conditions (i.e. in the range of 6–8). Only aggressive microbes can survive at lower pH conditions outside this range (e.g. 4.5–5) in natural systems. If contaminant degradation will alter the pH, a close monitoring of pH change should be conducted. Rapid changes of more than one or two units can inhibit microbial activity and may require an extended acclimation period before the microbes resume their activity. When the pH of the groundwater is too low, lime or sodium hydroxide can be added to increase the pH. When the pH is too high (too alkaline), a suitable acid (e.g. hydrochloric, muriatic acid) can be added to reduce the pH. An example effect of pH and phosphorous on the biodegradation of 2,4-DNT (as measured by CO_2 production) is shown in Figure 9.13.

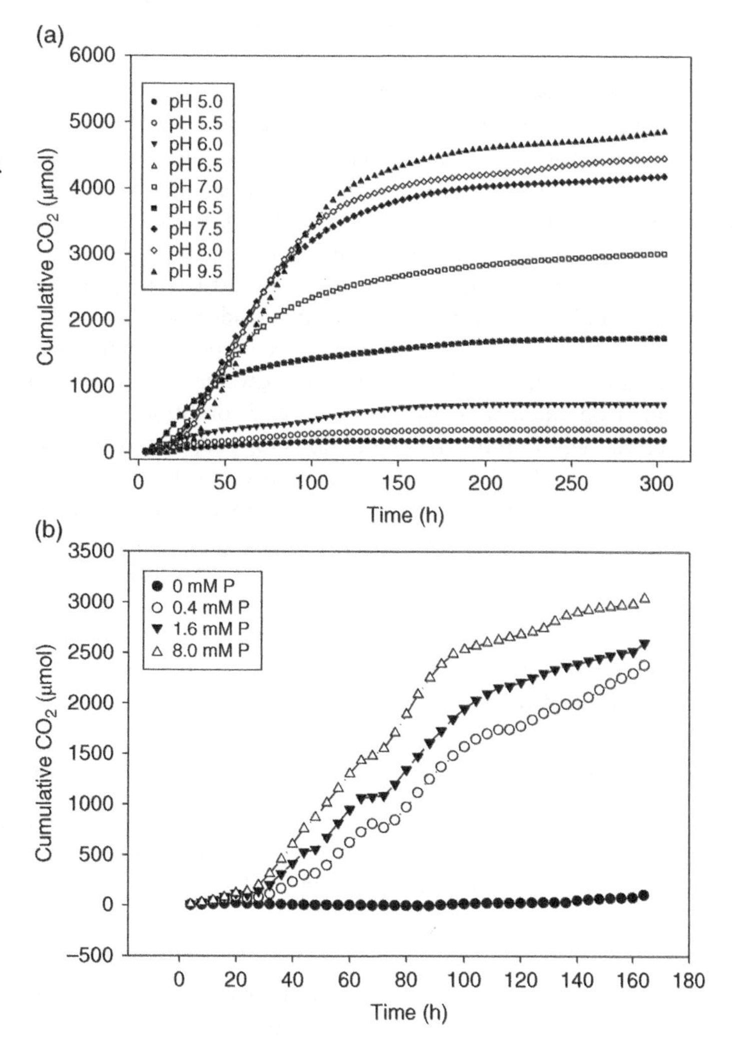

Figure 9.13 Effects of (a) pH and (b) phosphorous on the aerobic degradation of 2,4-DNT as measured by cumulative CO_2 production in a well-controlled batch study using a respirometer. *Source:* Adapted with permission from Zhang and Hughes (2004). Reproduced with permission of John Wiley & Sons.

Bacterial growth rates and microbial activities increase with the increase in temperature. Within the range of 10–45 °C, the rate of microbial activity typically doubles for every 10 °C rise in temperature. In most cases of *in situ* groundwater bioremediation, the bacteria living in the subsurface are likely to experience relatively stable temperatures with only slight seasonal variations. In most areas of the United States, the average groundwater temperature is about 13 °C, but groundwater temperatures may be somewhat lower or higher in the extreme northern and southern states. Subsurface microbial activity has been shown to decrease significantly at temperatures below 10 °C and essentially to cease below 5 °C. Microbial activity of most bacterial species also diminishes at temperatures greater than 45 °C.

Excessively high mineral contents (Ca^{2+}, Mg^{2+}, Fe^{2+}) will react with dissolved carbon dioxide (partially produced from microorganism as an end product), causing scaling and therefore the difficulty in operation of bioremediation systems. Calcium and magnesium (commonly measured as the hardness) will also react with phosphate, which is typically supplied as a nutrient in the form of tripolyphosphate, to form phosphate precipitate. This makes phosphate unavailable to microbial use as a nutrient. This effect can be minimized by using tripolyphosphate in a mole ratio of greater than 1:1 tripolyphosphate to total minerals (Ca^{2+}, Mg^{2+}). At this ratio, the tripolyphosphate acts as a sequestering agent to keep Ca^{2+} and Mg^{2+} in solution.

In reduced groundwater environment, iron exists in its soluble divalent form as ferrous iron (Fe^{2+}). Oxidation of Fe^{2+} occurs when it is exposed to oxygen, particularly when oxygen is introduced as the electron acceptor at the injection wells. The trivalent oxidation product of ferric iron (Fe^{3+}) exists as $Fe(OH)_3$, which then forms an insoluble ferric oxide (Fe_2O_3) precipitate. This precipitate can be deposited in aquifer flow channels, reducing permeability. Since iron is very abundant in soil, the effects of iron precipitation tend to be most noticeable around injection wells. It was suggested that injection wells require periodic testing and may need periodic cleaning or replacement if Fe^{2+} is in the range of 10–20 mg/L. *In situ* bioremediation is not recommended if Fe^{2+} is greater than 20 mg/L.

Nitrogen is usually the primary limiting nutrient in groundwater bioremediation. However, excessive addition of nitrogen should be avoided because of the potential in lowering groundwater pH. A slow-releasing nitrogen source is preferred because of the possibility of exceeding the nitrate standard of 40 mg/L in groundwater. In fact, *in situ* groundwater bioremediation should be operated at near nutrient-limited condition.

Bacteria require moist soil conditions for proper growth. Excessive soil moisture, however, reduces the availability of oxygen by restricting the flow of air through soil pores. The ideal range for soil moisture is between 40 and 85% of the **water-holding capacity** of the soil (see Example 9.5). Generally, water-saturated soils prohibit airflow and oxygen delivery to bacteria, while dry soils lack the moisture necessary for bacterial growth. In bioventing process, for example, the moisture content of soils within the capillary fringe may be too high. If so, depression of the groundwater table by pumping may be necessary to biovent soils. Bioventing promotes dehydration of moist soils through an increased air in soil. On the other hand, excessive dehydration hinders bioventing performance and extends cleanup time.

Example 9.5 Estimate soil moisture requirement for bioremediation

Surrounding a recently ruptured underground gasoline tank, there was an estimated total of 500 yd^3 of soil need to be biologically treated. The soil has a porosity of 40%, and soil moisture at 15% of its water-holding capacity. How much water needs to be spread?

Solution

The total volume of soil pores = 500 $yd^3 \times 0.40 = 200\ yd^3$. Since the ideal range of soil moisture is 40–85% of the water-holing capacity, additional water is needed for the optimal condition of bioremediation. If we select the near middle value of 60%, the additional amount of water = $200\,yd^3 \times (0.60 - 0.15) = 90\,yd^3 = 18\,180\,gal\ (1\,yd^3 = 202\,gal)$.

Note that the soil water saturation needs to be frequently monitored, and the amount of make-up water depends on the local climate at the contaminated site.

9.1.4.3 Microbial Presence

In regard to the ubiquitous, diverse, and large amount of microorganism in soils, additional bacteria inocula oftentimes are not necessary for *in situ* bioremediation. **Bioaugmentation**, a process of introducing a group of natural or foreign microbial strains to treat contaminated soil or water, is only provided on as needed basis, such as in the case of some special xenobiotic contaminants. Except in coarse-grained, highly permeable aquifers, microbes tend not to move very far past the point of injection, therefore, the effectiveness of bioaugmentation could be limited in extent. Bioaugmentation is frequently used in bioreactors and in *ex situ* systems.

To determine the presence and population density of naturally occurring bacteria capable of degrading target contaminants, laboratory analysis of soil samples from the site should be conducted. Plate counts for **total heterotrophic bacteria** (i.e. bacteria that use organic compounds as an energy source) and if possible, contaminant-degrading bacteria can be used as indicators. Although heterotrophic bacteria are normally present in all soil environments, plate counts of greater than 1000 colony-forming units (CFU)/g of soil could indicate the good microbial presence for *in situ* bioremediation. A CFU/g soil of less than 1000 indicates potential depletion of oxygen or other essential nutrients or the presence of toxic constituents. However, concentrations as low as 100 CFU/g of soil can be stimulated to an acceptable level, assuming toxic conditions (e.g. exceptionally high concentrations of heavy metals) are not present.

9.1.4.4 Contaminant Characteristics

The contaminant characteristics important to biodegradation include, but are not limited to, contaminant's chemical structure, concentrations, and solubility. In the preceding section, the reliance of biodegradability and pathways on the inherent chemical classes and structures are delineated. Hence, only the effects of contaminant concentrations and solubilities are discussed here.

High concentrations of organic and inorganic contaminants in soils can be toxic and inhibit the growth and reproduction of bacteria responsible for biodegradation. For example, concentrations of petroleum hydrocarbons (measured as total petroleum hydrocarbons, or TPH) in excess of 50 000 mg/kg, organic solvent concentrations in excess of 7 000 mg/kg, or heavy metals in excess of 2 500 mg/kg in the aquifer medium are generally considered inhibitory and/or toxic to aerobic bacteria. The concentration effects should be part of the treatability study. Figure 9.14 shows how 2,4-DNT concentrations affect the biodegradation of two isomers of dinitrotoluenes (2,4-DNT and 2,6-DNT) in a treatability study. Such information is important because high concentrations are found in explosives-contaminated soils and two isomers coexisted in these soils.

In addition to maximum concentrations, very low concentrations of organic material will result in diminished levels of bacterial activity. One should also consider the cleanup concentrations proposed for the treated soils. This is because bacteria cannot obtain sufficient carbon from degradation of the constituents to maintain adequate biological activity if the concentrations are below a certain "threshold" level. The threshold level determined from laboratory treatability studies may be much lower than what is achievable in the field under less than optimal conditions. Although the threshold limit varies greatly depending on bacteria-specific and constituent-specific features, constituent concentrations below 0.1 mg/L in the total aquifer matrix may be difficult to achieve. However, concentrations in the groundwater for these

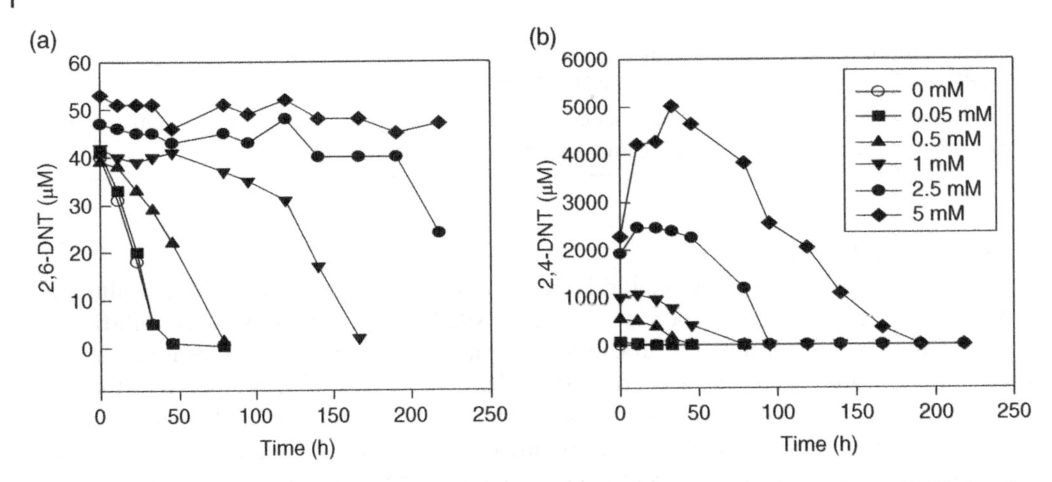

Figure 9.14 Effects of 2,4-DNT concentrations on the biodegradation of (a) 2,6-DNT and (b) 2,4-DNT in batch shake flask studies. 2,4-DNT and 2,6-DNT are often coexisted at explosives-contaminated sites. *Source:* Adapted with permission from Zhang et al. (2000). Copyright (2000) American Chemical Society.

specific constituents may be below the detection limits. Experience has also shown that reductions in contaminants such as petroleum hydrocarbon concentrations greater than 95% can be very difficult to achieve because of the presence of "resistant" or nondegradable petroleum constituents.

Contaminants that are highly soluble have a tendency to dissolve into the groundwater and are more bioavailable to bacteria for biodegradation. Conversely, chemicals that have low water solubilities tend to remain in the adsorbed phase and will be biodegraded more slowly. In general, lower molecular weight constituents tend to be more soluble and biodegrade more readily than do higher-molecular-weight or heavier constituents. In the field, aqueous contaminant concentrations rarely approach its solubility because dissolved phase concentrations tend to be reduced through competitive dissolution of other constituents and degradation processes such as biodegradation, dilution, and adsorption.

9.2 Process Description of Bioremediation and Biotechnologies

There are basically two biodegradation-based technologies (i.e. biotechnologies) from its development perspectives. The **conventional biological processes** including biofilm, activated sludge, lagoons, denitrification, phosphorous removal, and anaerobic processes. These are the biological applications traditionally developed for the treatment of domestic and industrial wastewater. These conventional biological processes are the realms of traditional environmental engineering textbooks. An excellent monograph covering these topics is by Rittmann and McCarty (2001). In addition to the conventional biological processes, the innovative biotechniques have been developed for the treatment of hazardous wastes, and the cleanup of contaminated soil and groundwater, such as *in situ* and *ex situ* bioremediation (including bioventing, biosparging), monitored natural attenuation, composting, landfarming, landfill, *ex situ* bioslurry reactors, and phytoremediation. Refer to Table 6.1 in Chapter 6 for an overview of these bioremediation technologies. Several important bioremediation and biotechnology processes are further elaborated in the following section.

9.2.1 *In Situ* Bioremediation

In situ bioremediation is by far the most widely used innovative remediation technology, next to the application of soil vapor extraction in contaminated sites. It has many variations, including bioventing, biosparging, and enhanced bioremediation for *in situ* biological treatment of soil, as well as enhanced bioremediation and monitored natural attenuation for the *in situ* treatment of contaminated groundwater. Bioremediation can also be enhanced in a variety of ways, for example, bioaugmentation and biostimulation. As we mentioned previously, bioaugmentation is the addition of bacterial cultures to a contaminated medium; frequently used in bioreactors and *ex situ* systems. Lippincott et al. (2015) also succeeded bioaugmentation in a deep aquifer of 82–92 ft in depth using enriched bacterial strain *Rhodococusruber* for the degradation of 1,4-dioxane. **Biostimulation** is the stimulation of indigenous microbial populations in soils and/or groundwater by the addition of various forms of rate-limiting nutrients and electron acceptors, such as nitrogen, phosphorous, and oxygen. Biostimulation may be done *in situ* or *ex situ*.

Bioventing is the method of treating contaminated soils by drawing air through the soil to stimulate microbial growth and activity in the *unsaturated* zone. Bioventing is preferred for low-vapor-pressure heavy hydrocarbons, while hydrocarbons with higher vapor pressure tend to lose into air through volatilization. Low airflow rates are usually used for bioventing to minimize volatilization. When volatilization is high, bioventing is frequently applied in cooperation with, and usually following soil vapor extraction (SVE) for vadose zone remediation. Bioventing is relatively noninvasive, and it is especially valuable for treating contaminated soils at military bases, industrial complexes, and gas stations, where structures and utilities cannot be disturbed (Figure 9.15).

Unlike bioventing in the vadose zone, **biosparging** is an *in situ* remediation technology that uses indigenous microorganisms to biodegrade organic constituents in the *saturated* zone. Although contaminants adsorbed to soils in the unsaturated zone can also be treated by biosparging, bioventing is typically more effective for this situation. The biosparging process is similar to air sparging described in Chapter 8. However, while air sparging removes contaminants primarily through volatilization, biosparging promotes biodegradation of constituents rather than volatilization (generally by using lower flow rates than are used in air sparging).

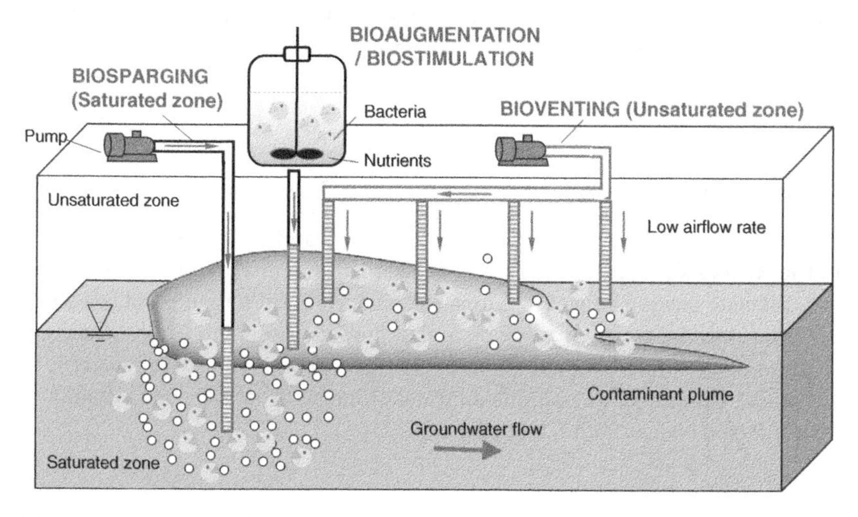

Figure 9.15 Schematic diagram of various bioremediation processes: a biosparging system in saturated zone, a bioventing process in unsaturated zone, and a bioaugmentation/biostimulation process.

In practice, some degree of volatilization and biodegradation occurs when either air sparging or biosparging is used.

Aerobic bioremediation can be enhanced through the injection of air, pure oxygen, or the use of oxygen-releasing compounds – all termed as the **enhanced aerobic bioremediation**. With **pure oxygen injection**, dissolved oxygen concentrations can reach up to 40–50 mg/L, in contrast to 8–10 mg/L when the saturated zone is aerated using atmospheric air containing approximately 20.95% oxygen (by volume).

Commonly used **oxygen-releasing compounds** include calcium and magnesium peroxides that are introduced to the saturated zone in solid or slurry phases. These peroxides release oxygen to the aquifer when hydrated by groundwater as the peroxides are ultimately converted to their respective hydroxides. **Magnesium peroxide** has been more commonly applied in field applications than calcium peroxide because magnesium peroxide has a lower solubility and consequently a prolonged release of oxygen (Eqs. 9.12 and 9.13).

$$2MgO_2 + 4H_2O \rightarrow 2Mg(OH)_2 + 2H_2O_2 \quad \text{Eq. (9.12)}$$

$$2H_2O_2 \rightarrow 2H_2O + O_2 \quad \text{Eq. (9.13)}$$

Magnesium peroxide formulations placed in the saturated zone during a short-term injection event can release oxygen to groundwater over a four- to eight-month period. Significant quantities of magnesium peroxide are required based on stoichiometry and the fact that 72% of the weight of the compound is not oxygen. Oxygen amounting to approximately 28% of the weight of magnesium peroxide is released to the aquifer over the active period.

Introducing **hydrogen peroxide** (a chemical oxidant) to the saturated zone can also significantly augment existing oxygen levels as demonstrated in Example 9.6. H_2O_2 naturally decomposes rapidly into oxygen. For every 2 mol of hydrogen peroxide introduced to the groundwater, 1 mol of oxygen can be produced (Eq. 9.13). Hydrogen peroxide has the potential of providing much higher levels of available oxygen to contaminated groundwater relative to other enhanced aerobic bioremediation technologies because it is infinitely soluble in water. In theory, 10% hydrogen peroxide could provide 50 000 mg/L of available oxygen.

Lastly, **ozone injection** is both a chemical oxidation technology and an enhanced aerobic bioremediation technology. This is because ozone is a strong oxidizing agent as well as an oxygen-releasing compound. Ozone is over 10 times more soluble in water than is pure oxygen (109 vs. 8.0 mg/L at 25 °C). Because of its instability, however, ozone is usually generated on site, and air containing up to 5% ozone is injected into wells.

Example 9.6 Delivery of dissolved oxygen to groundwater using O_2-releasing H_2O_2

(a) Estimate the dissolved oxygen (DO) concentration (mg/L) in groundwater at equilibrium with air in soil voids. The Henry's law constant of oxygen at 25 °C is 756.7 atm/(mol/L). For a comparison purpose, also calculate oxygen concentration in soil air and report it in mg/L of soil air. Assume that air in soil voids has the same composition as that in the atmosphere (i.e. 20.95% O_2 and 79.05% N_2). (b) As an oxygen-releasing compound, hydrogen peroxide in groundwater is often kept below 1000 mg/L because of its potential biocidal property. What DO concentration can H_2O_2 provide if the injected concentration is 1000 mg/L H_2O_2?

Solution

a) Dissolved oxygen in groundwater can be calculated using Henry's law $H = p/C$, where p = partial pressure of oxygen (atm) and C = mole concentration (mol/L) of oxygen dissolved in groundwater.

A 20.95% oxygen is the same as 1 atm × 0.2095 = 0.2095 atm for the partial pressure of oxygen in the soil air, hence DO in groundwater is as follows:

$$C = \frac{p}{H} = \frac{0.2095\,\text{atm}}{\left[\dfrac{756.7\,\text{atm}}{\left(\dfrac{\text{mol}}{\text{L}}\right)}\right]} = 0.000277\frac{\text{mol}}{\text{L}}$$

Convert mol/L into mg/L by multiplying the molecular weight of oxygen of 32 g/mol or 32 000 mg/mol, we have:

$$C = 0.000277\frac{\text{mol}}{\text{L}} \times \frac{32000\,\text{mg}}{1\,\text{mol}} = 8.86\frac{\text{mg}}{\text{L(groundwater)}}$$

This is the maximum concentration of molecular oxygen that can be dissolved in water at 25 °C. This low DO concentration confirms that O_2 is only sparingly soluble in water. Now, let us convert 0.2095 atm into mg/L using the ideal gas law:

$$C = \frac{n}{V} = \frac{p}{RT} = \frac{0.2095\,\text{atm}}{0.0821\dfrac{\text{atm} \times \text{L}}{\text{mol} \times \text{K}} \times (273 + 25)\text{K}} = 0.008563\frac{\text{mol}}{\text{L}}$$

Again, using the conversion factor of 32 000 mg/mol, we can calculate the oxygen concentration in soil air:

$$C = 0.008563\frac{\text{mol}}{\text{L}} \times \frac{32000\,\text{mg}}{1\,\text{mol}} = 274\frac{\text{mg}}{\text{L(soil air)}}$$

b) For stoichiometric calculations, H_2O_2 in the unit of mg/L must be converted into its molar concentration. The molecular weight of H_2O_2 is 34 g/mol or 34 000 mg/mol, so the molar concentration of 1 000 mg/L H_2O_2 is: 1 000 mg/L/(1 mol/34 000 mg) = 0.0294 mol/L.

From Eq. (9.13), we know that every 2 mol of H_2O_2 will produce 1 mol of dissolved oxygen if we assume a complete dissociation of H_2O_2. Hence, 0.0294 mol/L will produce 0.0147 mol/L O_2 or 0.0147 × 32 000 = 471 mg/L O_2 in the groundwater. The DO concentration following the introduction of H_2O_2 is over 50 times higher than the saturated DO concentration in groundwater as previously calculated (8.86 mg/L), demonstrating the effectiveness of H_2O_2 as an oxygen-releasing compound.

Natural attenuation, also referred to as intrinsic bioremediation, is a remediation approach that relies on naturally occurring processes such as dispersion, sorption, evaporation, and biodegradation to control the migration of contaminants dissolved in groundwater. However, natural attenuation, by its definition, is not solely a biological process, but also a physico-chemical approach. Natural attention cannot be misinterpreted as a "doing nothing approach." Instead, active monitoring is required, hence it is commonly referred to as the **monitored natural attenuation** (MNA). Some advantages of MNA include the following: (a) Reduced wastes generation and exposure as with other *in situ* technologies, (b) Less intrusive, (c) Lower overall remediation cost. The disadvantages of MNA include the following: (a) Longer time-frames for cleanup, (b) Need for long-term monitoring, (c) Continued contaminant migration, (d) Complex and costly site characterization.

Overall, the *in situ* bioremediation techniques have the following advantages: (a) Less disturbance than other physicochemical methods, (b) No soils need to be excavated and no groundwater need to be extracted, (c) No waste to be stockpiled or disposed of, (d) Smaller footprint for the aboveground component of the treatment system, (e) Safe and less exposure risk associated with hazardous waste transports and handling, (f) Complete restoration to basic constituents rather than simply transferring waste to another medium (such as the atmosphere), (g) Appealing to the general public because of the idea of utilizing naturally occurring microorganisms to degrade toxics into harmless products, (h) Often less expensive than conventional technologies in the capital cost. The long-term costs and low maintenance dominate over the relatively high initial costs.

General disadvantages of *in situ* bioremediation are several folds: (a) High up-front costs to conduct site characterization and determine the feasibility and parameters for system design, (b) It is difficult to convince the responsible party that up-front sampling with detailed analysis is necessary and that the associated costs are justified, (c) Require high technical skills and high level of training for remediation personnel to perform the sampling prior to installation and after the system becomes operational, (d) It is generally difficult to prove that bioremediation is the mechanism responsible for the declined concentrations of contaminants in the subsurface.

9.2.2 *Ex Situ* Biological Treatment

The *ex situ* biological techniques include biopiles, composting, landfarming, and slurry-phase bioreactors for the treatment of contaminated soil, as well as *ex situ* bioslurry reactors and constructed wetlands for the treatment of contaminated groundwater. The constructed wetland systems will be briefly described in the section of phytoremediation (Section 9.2.4).

9.2.2.1 Biopiles and Composting

Biopiles, also known as biocells, bioheaps, biomounds, and compost piles, are used to reduce contaminants in excavated soils through the use of biodegradation. This technology involves heaping contaminated soils into piles (or "cells") and stimulating aerobic microbial activity within the soils through the aeration and/or addition of minerals, nutrients, and moisture. The enhanced microbial activity results in degradation of adsorbed contaminants through microbial respiration. A typical biopile cell is shown in Figure 9.16.

More commonly known as composting, the biopile is actually a century-old technology for the disposal of yard waste in India and China. Composting is an established technology for the management of municipal solid wastes, and is an innovative technology for the treatment of hazardous wastes. It is the aerobic thermophilic (>45 °C) decaying of organic matter by bacteria, fungi, and lower form of animals. The product of composting, known as "compost," is partially decomposed, humus-rich material with the appearance of dark, easily crumbled, and an earthy aroma smell. Compost from municipal wastes can be utilized as a soil fertilizer, and it also reduces the volume of wastes sent to landfill.

A healthy composting system involves a diversity of organisms. The primary players are microorganisms including fungi, bacteria, nematodes, and actinomycetes (fungi-like bacteria that are filamentous). Actinomycetes are the primary decomposers of recalcitrant plant materials such as bark, newspaper, and woody stems. They are especially effective at attacking raw plant tissues (cellulose, chitin, and lignin), softening them up for other organisms. Nematodes are tiny (400 μm–5 mm long), cylindrical, often transparent microscopic worms that are the most abundant of the physical decomposers.

There are some types of invertebrates such as mites, millipedes, snails, slugs, earthworms, sowbugs, and white worms. These invertebrates shred the plant materials, creating more sur-

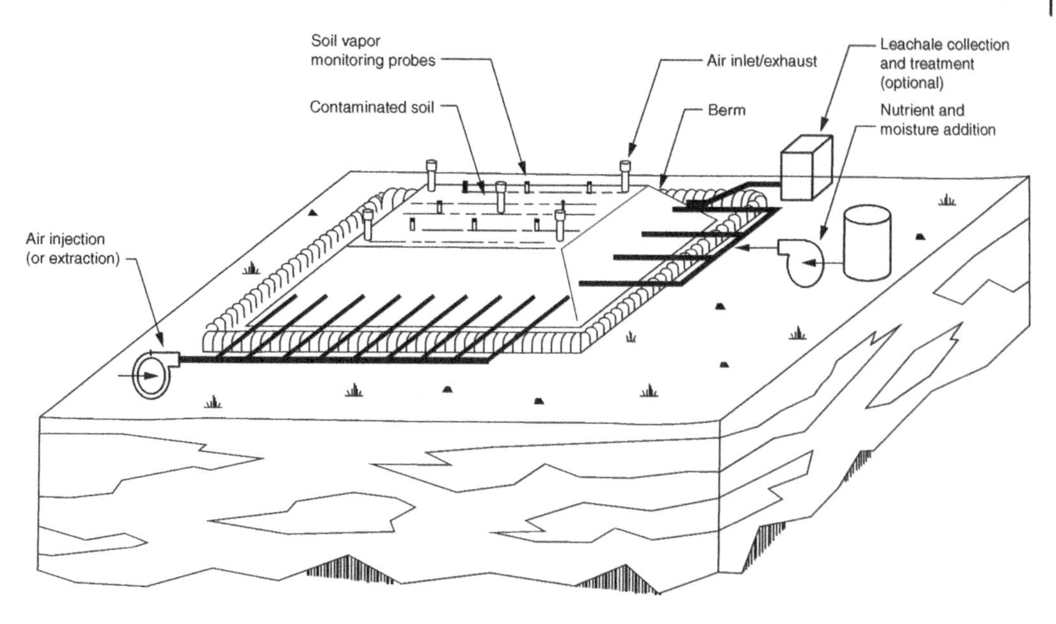

Figure 9.16 A typical biopile system (USEPA 1994).

face area for action by fungi, bacteria, and actinomycetes. The invertebrates can be consumed by higher forms of organisms such as mites and springtails. Sowbugs are relatives of crabs and lobsters. Their powerful mouth-parts are used to fragment plant residue and leaf litter. Millipedes possess two pairs of legs on each body segment. They are the shredders, which chew up dead plant matter as they eat bacteria and fungi on the surface of the plant matter. Earthworms are critical during the decomposition process. They tunnel and feed on dead plants and decaying insects. Tunneling aerates the compost and enables water, nutrients, and oxygen to filter down. As soils and organic matters pass through earthworm's body, materials are finely ground and then digested.

Composting for hazardous wastes has been applied to wastes containing TNT, petroleum sludge, and certain pesticides. Typically, hazardous wastes are mixed with bulking agents such as sawdust, wood chip, and baled straw. Composting provides an inexpensive and technologically straightforward solution for managing hazardous industrial waste streams and for remediating soil contaminated with toxic organic compounds. Costs are usually significantly lower than conventional technologies: \$20–\$40/ton (windrow) and \$60–\$120/ton (static pile). Generally, more than 50% saving can be achieved using composting over conventional technologies such as incineration.

There are three types of composting systems from the design standpoint: (a) Windrow composting: The material to be composted is piled on an impervious platform such as concrete or asphalt. The pile is aerated by turning the compost mixture, either manually or mechanically. (b) Static pile process: Forced aeration is utilized in static piles, and no turning is required. A system of perforated pipes is laid out at the base of pile and aeration can be achieved in a positive mode by pushing air through the pile or in a negative mode by applying vacuum on the pile. (c) In-vessel composting: Unlike windrow composting or static pile, the compost mixture is placed in an enclosed reactor (drum, solo, concrete-lined trench, or similar equipment) and mixing and aeration are achieved by turning and forced aeration.

9.2.2.2 Landfarming

Landfarming is a solid-phase treatment system for contaminated soils, which may be done *in situ* when only the upper 1 ft of soil is contaminated, or in a constructed soil treatment cell (prepared bed) for soil depth greater than 1 ft following excavation (Cookson 1995). Landfarming is similar to biopiles in that they are both aboveground, engineered systems that use oxygen, generally from air, to stimulate the growth and reproduction of aerobic bacteria involved in degradation of organic contaminants adsorbed to soil. While biopiles are aerated most often by forcing air through slotted or perforated piping placed throughout the pile, landfarming systems are aerated by typical agricultural operations such as tilling or plowing.

A critical component of constructed soil treatment cell is the protective layer called "liner," which is made of impermeable materials for the protection of groundwater from leaching contaminants (Figure 9.17). The bed consisting of a base of sand or soil is used to protect the clay or liner from equipment.

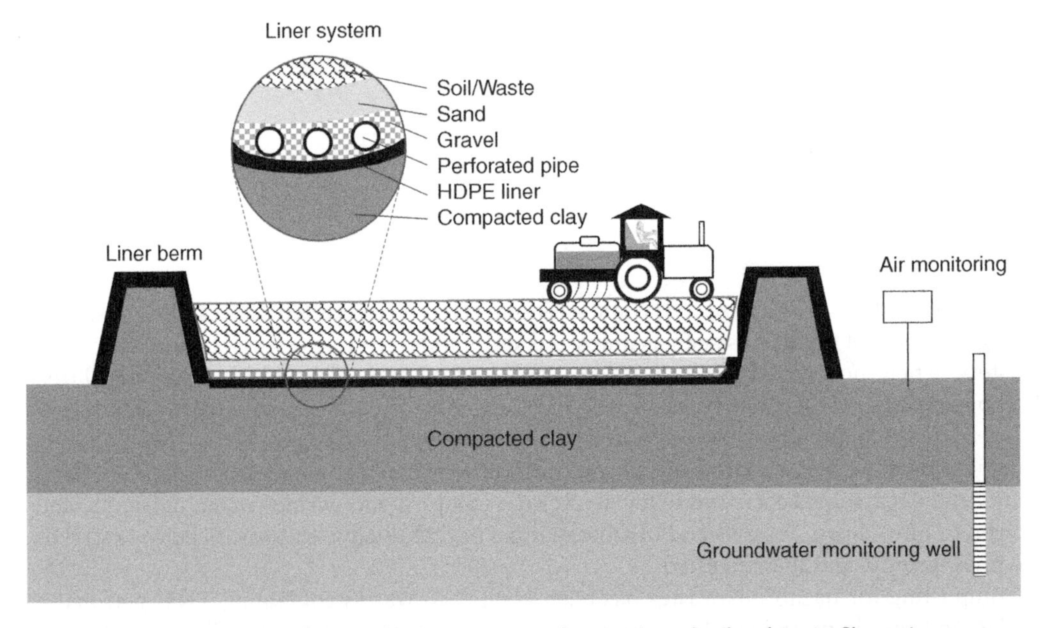

Figure 9.17 Schematic of landfarming for the treatment of contaminated soil and waste. Shown is a cross-section of a land treatment unit, and a close-up of a land treatment unit liner and leachate collection system.

The rate of contaminant degradation in landfarming is enhanced by fertilizing, irrigating, and tilling soil to increase the availability of nutrients, moisture, and oxygen to the soil microorganisms. Usually with indigenous bacteria, landfarming is effective for upper soil layers (zone of incorporation, 18–24 in thick). Landfarming is a well-established technique especially at petroleum refinery sites, and creosote-contaminated sludge and soils. Its approximate cost is \$50–\$80/yd^3 in general; \$30–\$50/yd^3 for petroleum-contaminated sites containing greater than 1000 yd^3 of soils.

Landfarming is low in cost and simple to implement. It is particularly applicable to slowly biodegradable contaminants which could render bioreactor approaches infeasible due to large bioreactor sizes and costs. Landfarming is usually employed for moderate to low contaminant concentration. There might be problems related to toxic and dust generation. It requires bottom lining if leaching is a concern, and it requires pretreatment if toxic metals are present. Landfarming can only be used where environmental temperatures are favorable during most of the year. Since it is aerobic, it requires a large area of land. It usually is less rapid compared to other bioremediation techniques.

9.2.2.3 Bioslurry Reactors

Bioslurry reactors can be soil slurry reactors operated in the batch draw-and-fill mode (Figure 9.18), or in a slurry lagoon system operated in a continuous mode. The enclosed bioslurry reactor systems offer a high degree of control through optimized bacterial degradation, pH, temperature, nutrient, O_2, mixing, and contact. As such, bioslurry reactors can be highly effective and fast. Although the cost may be higher than landfarming and composting, it is lower than conventional remediation such as incineration and solidification/stabilization. One key operational parameter is the maximal percent of solid concentration that can be loaded in the reactor system to maintain the slurry state for mechanical operation (e.g. pumping). Solid concentrations may range 5–50% depending on physical properties of soils. Highly viscous contaminants such as tars and certain oils may not be appropriate for bioslurry reactors.

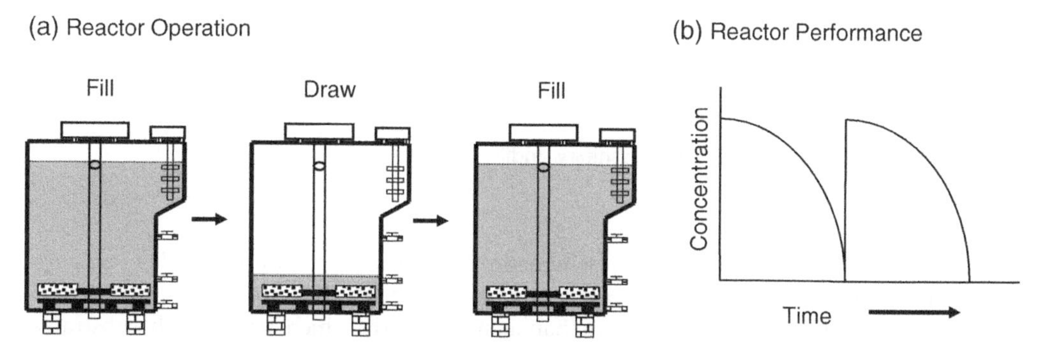

Figure 9.18 Schematic diagram of a pilot-scale bioslurry reactor. (a) Draw-and-fill operation mode: In drawing operation, a certain percentage of the total soil slurry volume is retained as the bacterial seeds. (b) Reactor performance as indicated by the contaminant concentrations as a function of time.

Bioslurry reactors have been used in RCRA-listed hazardous wastes such as wood treating wastes, dissolved air flotation wastes, oil emulsion solids, and oil separator sludge.

9.2.3 Sanitary Landfills

Landfill is listed as a containment method by the EPA screening matrix (Table 6.1, Chapter 6). However, it is a biologically based process that utilizes initially the aerobic process and subsequently the anaerobic process for the degradation of hazardous wastes and hazardous materials. Landfills are a system that predominantly uses anaerobic bacteria to degrade organic contaminants in the solid wastes. Anaerobes convert contaminants into methane gas, which can be used as an energy source. A modern **sanitary landfill** is not just a simple open dump, because the microbiology and engineering control of a sanitary landfill is rather complicated.

Modern day landfills are subject to stringent regulatory control. The requirements posted by these regulations typically entail, for example, landfill siting above the groundwater table; **liner** and clay on the bottom and sides; **leachate** collection system; monitoring of groundwater; daily cover and final cover with soil; and venting and collection of methane. All landfills should be carefully situated where clay deposits and other land features act as natural buffers between the landfills and the surrounding environment. They all sit above the groundwater table (5 ft minimum between bottom liner to seasonal high groundwater; 10 ft minimum between bottom liner and bedrock). The bottom and sides of modern landfills are lined with layers of clay or plastic to keep the leachate from entering into the soil and groundwater. Modern landfills have leachate collection and gas recovery systems to further prevent the pollution to groundwater

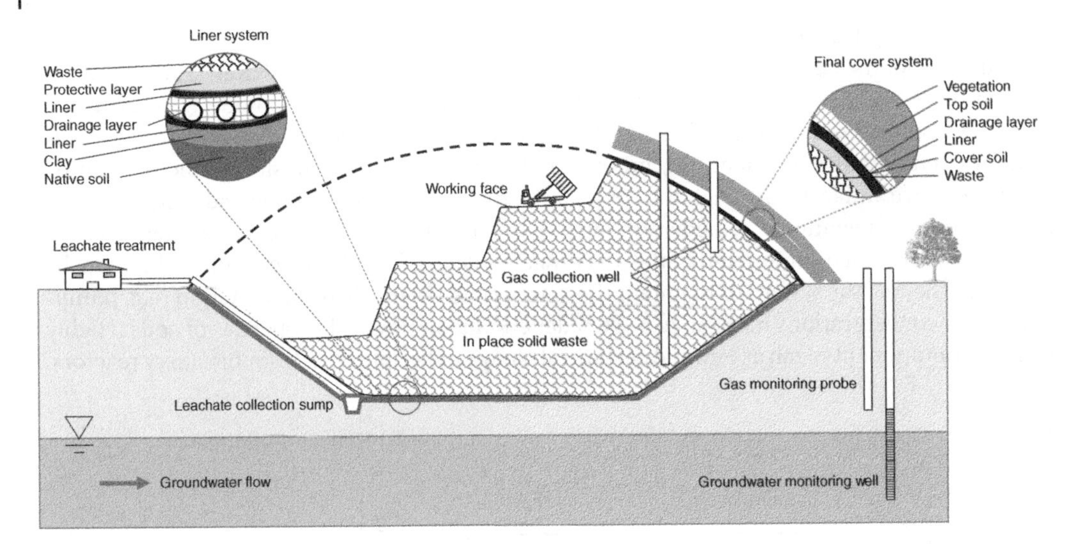

Figure 9.19 Schematic diagram of a modern sanitary landfill.

and atmosphere (see Figure 9.19 for a schematic diagram of a landfill). The laws require groundwater be monitored at least 20 years or longer after the landfill closure.

The main anaerobes of landfills are methanogenic bacteria which use growth substrates directly from solid wastes, such as sugar, proteins, lipids, and various contaminants of concerns. Before they can function, other bacteria must first convert these organic chemicals into acetate, H_2, CO_2, formate, methanol, methylated amines, and dimethyl sulfide that methanogenic bacteria can use. This conversion is a series of decomposition processes, such as aerobic degradation, sulfate reduction, and acetogenesis (the formation of acetate through the reduction of CO_2 and electron sources). Methanogenic bacteria can only grow under anoxic (oxygen-free) or anaerobic conditions. Oxygen must be depleted in an initial aerobic stage in the landfill before methanogenic bacteria can take over. Methanogens are usually coccoid (spherical) or bacilli (rod shaped). There are over 50 described species of methanogens, which do not form a monophyletic group, although all methanogens belong to the domain of Archaea (Figure 9.1).

9.2.4 Phytoremediation and Constructed Wetland

Phytoremediation uses plants (hence **rhizobium bacteria** associated with plants) to clean up contaminated soil, groundwater, wastewater, or sediment. Plants have been historically used for wind and soil erosion control, but the use of plants to clean up the contaminated environment is a relatively new biotechnological application. It is still a subject of numerous on-going studies and field tests in both the scientific community and the remediation industry. Constructed wetland system is related to phytoremediation. **Constructed wetland** is an artificial land system with water-saturated soil/sand/rock and the prevalence of vegetations adapted for growth in the water-saturated soils. Familiar examples of natural wetlands are swamps, marshes, and bogs. When wastewater flows through constructed wetland systems, contaminants are removed via several mechanisms, including phytoremediation as well as other mechanisms such as physical filtration, biofilms, etc.

Depending on the types of contaminants, the mechanisms involved in phytoremediation may include phytoextraction, phytovolatilization, and phytodegradation. **Phytoextraction** is the uptake and storage of pollutants in the plant stems or leaves. Some plants, called **hyperaccumulators**, draw a significant amount of pollutants through the roots. After the accumulation of pollutants in stems and leaves, the plants are harvested, and then either burned or disposed of. Even if the plants cannot be used, incineration and disposal of the plants is still less expensive than traditional remediation methods. As a comparison, it is estimated that a site containing 5000 tons of contaminated soil will produce only 20–30 tons of ash (Black 1995). This method is particularly useful when remediating metals or radionuclides (see Box 9.3). **Phytovolatization** is the uptake and vaporization of pollutants by a plant. This mechanism takes a solid or liquid contaminant and transforms it to an airborne vapor. Either the vapor can be a pure pollutant or the pollutant can be metabolized by the plant before it is vaporized, as in the case of mercury, lead, and selenium.

Box 9.3 Phytoremediation in Chernobyl nuclear plant, Soviet Union

The Chernobyl nuclear accident occurred from a ruptured nuclear reactor vessel in Ukraine in 1986. An unprecedented amount of radioactive materials was released from this catastrophe, resulting in an estimated 50 000 excess cancer cases and 25 000 excess cancer deaths. Radioactivity was about 14 exabecquerel (1EBq = 10^{18} Bq), where 1 Bq is defined as the activity of radioactive material in which one nucleus decays per second. The order of magnitude of this release can be better perceived by comparing its equivalency with the global inventory of carbon-14 at an estimated 8.5×10^{18} Bq (8.5 EBq). Radioactive substances from this accidental release include iodine (I), cesium (Ce), strontium (Sr), and plutonium (Pu) radioisotopes. The majority of radionuclides had short half-lives and have decayed (8 days for I-131), so the radioactive chemicals with longer half-life, Cs-137 (39 years) and Sr-90 (29 years), become the most significant at the present time.

Following the release, contamination in the air reduced significantly in the first few years, and most radioactive materials eventually deposited into soil or sediment bed in waters. Therefore, the Chernobyl accident mostly affected extensive agricultural systems with soils containing a high organic content (peat soil) and not plowed pastures.

The removal and disposal at the Chernobyl site is not feasible, because of the vast impacted area and the radioactive hazard not permitted in landfills. Hydraulic methods such as pump-and-treat is also not appropriate because radionuclides are fixed in soil/sediment matrix. The basic remediation strategy would be to isolate, contain, and stabilize radioactive materials in the impacted environment such that the exposure near the inhabited land is kept at minimum.

Bioremediation is a sustainable, renewable, and clean technology. Two technologies, i.e. phytoremediation and mycoremediation, have been tested to remediate this area. For example, cesium and strontium are easily absorbed by plants because of the analogues between the radionuclides and plant essential minerals (Sr vs. Ca; Cs vs. K). The Chernobyl Sunflower project was created in 1994 by a Phytotech (a remediation company from New Jersey, USA) to use sunflower as a radionuclide collector.

The test project was located 1 km from the Chernobyl reactor on the area of a 75 m^2 pond. Results showed that almost 95% of radionuclides were removed in 10 days (mature plants) with the majority of Cs-137 located in roots and the majority of Sr-90 in shoots. Sunflower can be planted in floating Styrofoam rafts to uptake radioactive substance from water. To adapt to a toxic environment, plants have evolved sophisticated sequestration and metabolic mechanisms. The cost was approximately \$0.50–\$1.6 per thousand liters treated. The plants did not metabolize the radionuclides, but after the incineration of sunflowers, the amount of radionuclide waste can be deposited safely.

Mycoremediation using fungi is another bioremediation technology that has been proposed and used in Chernobyl area. Fungi are typically more tolerant than bacteria to a very high level of toxicity and especially useful for radionuclide remediation. The black mold was found covering the walls in Chernobyl closed reactor. It was discovered that melanin contained in some types of fungi played a similar role as chlorophyll in plants, and fungi could transform gamma rays into chemical energy. Hyperaccumulating mycorrhizal mushrooms were reported to absorb and concentrate Cs-137 approximately 10 000 times more than ambient background level. Following harvest, mushrooms can be incinerated and the radioactive ash can later be safely stored.

The Chernobyl area still remains heavily contaminated and affects millions of people within Europe. Bioremediation using plant and fungi for contaminated water and soils appear to be a more viable and sustainable alternative to conventional remediation options.

Phytodegradation is the metabolism of pollutants by plants. After the contaminant has been uptaken into the plant, it is assimilated into plant tissue, where the plant then degrades the pollutant. This metabolism by plant-derived enzymes, such as nitroreductase, lactase, dehalogenase, and nitrilase, has yet to be fully documented, but has been demonstrated in field studies (Boyajian and Carreira 1997). The daughter compounds can be either volatized or stored in the plant. If the daughter compounds are relatively benign, the plants can still be used in traditional applications. If the daughter compounds are less harmful than the parent compound, but not benign, then the plant biomass can be burned or used in alternate applications.

Phytoremediation has a good public perception as a green technology. It has gained its popularity among the general public because it is low in cost and esthetically pleasant. For example, phytoremediation was tested in pilot scale to remediate radionuclides at the Chernobyl site (Box 9.3). However, it is also subject to public scrutiny and controversy regarding the recent development of genetically modified plants for faster and more efficient remediation. Advantages of phytoremediation are several folds. It is solar driven and low in maintenance, and it works for metals and slightly hydrophobic compounds, including many organics. This *in situ* method can stimulate bioremediation in soil closely associated with plant roots. Plants can stimulate microorganisms through the release of nutrients and the transport of oxygen to their roots. Phytoremediation is relatively inexpensive – it can cost as little as \$10–\$100/yd^3, whereas metal washing can cost \$30–\$300/yd^3 (Watanabe 1997). Having ground cover on property reduces exposure risk to the community (i.e. lead). In addition, planting vegetation on a site also reduces erosion by wind and water, which can leave usable topsoil intact.

However, like any bacteria-based bioremediation, phytoremediation can take many growing seasons to clean up a site. Phytoremediation works for contaminants at low to medium concentrations in shallow contaminated aquifers. Plants have short roots, so phytoremediation is appropriate to clean up near surface soil or groundwater *in situ*, typically 3–6 ft. Further design work is needed if plants are used to remediate deep aquifers. Trees have longer roots and can clean up slightly deeper contamination than plants, typically 10–15 ft. Tree roots grow in the capillary fringe, but do not extend deep into the aquifer. This makes remediating DNAPLs *in situ* with plants and trees not recommended. Also, plants that absorb toxic materials may contaminate the food chain. Volatilization of compounds can transform a groundwater pollution problem to an air pollution problem. Phytoremediation is less efficient for hydrophobic contaminants which bind tightly to soil.

In practice, the selection of candidate plants for phytoremediation is critical (Table 9.3). The plants should tolerate the contaminant, and absorb a lot of target contaminant (the so-called hyperaccumulator). Ideally, the plants are native (not invasive species), and they

Table 9.3 Plants used to remove selected contaminants in phytoremediation.

Contaminants removed	Plants used
Heavy metals (Cu, Ni, Hg, Pb, etc.)	Indian mustard, hybrid poplar, cottonwood
Chlorinated solvents (TCE, etc.)	Hybrid poplar, eastern cottonwood
Petroleum hydrocarbons	Alfalfa, poplar juniper, fescue
Polycyclic aromatic hydrocarbons (PAHs)	Ryegrasses, mulberry
Explosives (TNT, etc.)	Duckweed, parrot feather, poplar tree
Radionuclides (cesium, uranium, etc.)	Sunflowers, Indian mustard, cabbage

grow fast with a large amount of biomass (roots, leaves, stems) productions. Plants having deep-roots and fibrous root tissues are preferred because of a great surface area. The selected plant species should uptake a lot of water (for example, a poplar tree can transpire 50–300 gal of water per day), and cannot be eaten by insects or animals which are parts of the food chains. Thus the use of crop plants and pasture grasses may pose a potential risk to animals and humans.

The potential use of transgenic plants in phytoremediation has been studied. Transgenic plants are genetically modified plants following the transfer of non-native genes (transgenes) with certain beneficial or desirable traits (e.g. expression of enzyme to breakdown contaminants). Because they are genetically modified, they more likely have a poor public image.

The success of phytoremediation depends on a number of other factors related to contaminants, soil, and climate conditions. For example, the contaminant needs to be bioavailable for plant to uptake. If the compound is too hydrophobic, the compound will remain in the lipid bilayer of the cell membrane. If the compound is too hydrophilic, the compound will not be able to cross the cell wall. Soil properties affect contaminant bioavailability, hence soil conditions could become critical for the success of phytoremediation. Soils rich in organics tend to bind hydrophobic compounds, such that the compound cannot be taken up by the plant. Climate also plays an important role since warmer climate gives longer growing season and the longer biological processes. Phytodegradation needs optimal temperature for plant enzymes to remain active.

Example 9.7 calculates the required cleanup time for phytoremediation through harvesting plant biomass accumulating metals from soil. The reliance on the choice of available plant hyperaccumulator is clearly demonstrated.

Example 9.7 Estimate the required cleanup time using phytoremediation

Cadmium (Cd) ranks first among all the metals in the agricultural soils in China. It was reported that 7% of soil (equivalent to approximately 9 million ha of agricultural soils) in China exceeded the Cd standard of the Ministry of Environmental Protection (Zhao et al. 2015). Cd has a Class II Chinese Soil Environmental Quality Standard of 0.3 mg/kg for pH $<$ 6.5 and 0.6 mg/kg for pH $>$ 7.5. To reduce Cd from a hypothetical average Cd concentration of 0.8 mg/kg to a cleanup goal of 0.3 mg/kg, (a) estimate the required cleanup time in years if rice is used to cleanup soil, assuming biomass of grain and straw = 12 000 kg/ha-yr, and rice can accumulate Cd as high as 0.2 mg/kg (equal to China's food standard for Cd); and (b) estimate the required cleanup time in years if a hyperaccumulative plant such as Indian mustard is planted, fertilized, and harvested for phytoextraction, assuming its concentration is 0.01% of its dried biomass and the biomass productivity is

2500 kg/ha-yr. Also assume that contaminated surface soil has a depth of 0.2 m and a soil bulk density of 2000 kg/m^3. There are no additional sources of Cd from atmospheric deposition and fertilizer use during the period of phytoremediation. Note that 1 ha = 10 000 m^2.

Solution

a) We first calculate contaminated soil mass based on unit soil area of 1 ha and depth of 0.2 m.

$$\text{Soil}(\text{kg}) = \text{area} \times \text{depth} \times \text{density} = \frac{10000\,\text{m}^2}{1\,\text{ha}} \times 0.2\,\text{m} \times 2000\frac{\text{kg}}{\text{m}^3} = 4 \times 10^6\frac{\text{kg}}{\text{ha}}$$

The mass of Cd removed from 0.8 to 0.3 mg/kg per ha of surface soil (0.2 m) is as follows:

$$\text{Cd}\left(\frac{\text{g}}{\text{ha}}\right) = \text{Cd concentration in soil}s \times \text{soil mass} = (0.8 - 0.3)\frac{\text{mg}}{\text{kg}} \times 4 \times 10^6\frac{\text{kg}}{\text{ha}}$$
$$= 2 \times 10^6\frac{\text{mg}}{\text{ha}} = 2000\frac{\text{g}}{\text{ha}}$$

The uptake rate of Cd by rice plant is as follows:

$$\text{Cd uptake by rice} = \text{Cd concentration in rice} \times \text{biomass procution rate}$$
$$= 0.2\frac{\text{mg}}{\text{kg}} \times 12000\frac{\text{kg}}{\text{ha-yr}} = 2400\frac{\text{mg Cd}}{\text{ha-yr}} = 2.4\frac{\text{g Cd}}{\text{ha-yr}}$$

The cleanup time is estimated using the mass of Cd removal and the Cd uptake rate:

$$\text{Clean-up time} = \frac{2000\dfrac{\text{g Cd}}{\text{ha}}}{2.4\dfrac{\text{g Cd}}{\text{ha-yr}}} = 833\,\text{years}$$

b) For Indian mustard, Cd concentration of 0.01% is equivalent to 100 mg/kg of Cd. The uptake rate of Cd by Indian mustard is as follows:

$$\text{Cd uptake by mustard} = \text{Cd concentration in mustard} \times \text{biomass procution rate}$$
$$= 100\frac{\text{mg}}{\text{kg}} \times 2500\frac{\text{kg}}{\text{ha-yr}} = 2.5 \times 10^5\frac{\text{mg Cd}}{\text{ha-yr}} = 250\frac{\text{g Cd}}{\text{ha-yr}}$$

The cleanup time is estimated using the same mass of Cd removal as calculated above and the Cd uptake rate:

$$\text{Clean-up time} = \frac{2000\dfrac{\text{g Cd}}{\text{ha}}}{250\dfrac{\text{g Cd}}{\text{ha-yr}}} = 8\,\text{years}$$

This simplified example illustrates the importance of properly selecting a hyperaccumulative plant and the required high biomass production of the plant native to the contaminated soil. Note that the required cleanup time was calculated based on the contaminant concentration in the plant biomass. This approach is valid if the plant biomass is harvested as a mechanism of contaminant removal. However, for trees that will stay on the contaminated sites and remove organic contaminants primarily through evapotranspiration process, a slightly different approach should be used. This second approach is illustrated in Example 9.8 according to Schnoor (1997). First, we assume that the contaminant concentration untaken by plant is related to the concentration (*C*) in the groundwater by a **transpiration stream concentration factor** (TSCF, dimensionless). If the rate of evapotranspiration (ET in L/day) is known, then the removal rate of contaminant (m in mg/day) is

$$m = \mathrm{ET} \times (C \times \mathrm{TSCF}) \qquad \text{Eq. (9.14)}$$

where TSCF correlates with the $\log K_{ow}$ of the contaminant, and the typical values are provided accordingly (e.g. TCE 0.74, atrazine: 0.74, and RDX: 0.25). Second, we assume first-order kinet-

Example 9.8 Phytoremediation of RDX using a hybrid poplar tree through transpiration of contaminated groundwater

RDX has contaminated the surface soil (2 m) nearby a former manufacturing plan. From the lysimeter samples, the RDX in unsaturated soil water samples was determined to be 2 mg/L. In a proposed phytoremediation project, 1000 poplar trees per acre will be planted, and the plants are expected to transpire an average 1 in of soil water per acre per year. Estimate the cleanup time if the goal is to set up a 90% removal RDX from 2 to 0.02 mg/L or from 0.5 to 0.05 kg/acre within the surface soil.

Solution

With TSCF = 0.25, C = 10 mg/L, and ET = 1 in-acre/yr (=103 m^3/yr = 1.03×10^5 L/yr), we can estimate the RDX removal per year using Eq. (9.14):

$$m = 1.03 \times 10^5 \frac{\mathrm{L}}{\mathrm{yr}} \times 10 \frac{\mathrm{mg}}{\mathrm{L}} \times 0.25 = 51500 \frac{\mathrm{mg}}{\mathrm{yr}} = 0.0515 \frac{\mathrm{kg}}{\mathrm{yr}}$$

Applying Eq. (9.15), we have:

$$k = \frac{0.0515 \frac{\mathrm{kg}}{\mathrm{yr}}}{0.5\,\mathrm{kg}} = 0.2575\,\mathrm{yr}^{-1}$$

The cleanup time for the 90% removal can be estimated using Eq. (9.11):

$$\text{Clean-up time} = \frac{\ln(1 - \%\text{degradation})}{-k} = \frac{\ln(1 - 0.9)}{\frac{-0.2575}{\mathrm{yr}}} = 8.9\,\text{years}$$

ics for the transpiration; consequently the required rate constant (k in day^{-1}) can be calculated if the initial mass of contaminant in soil (M in mg) is known according to:

$$k = \frac{m}{M} \qquad \text{Eq. (9.15)}$$

Note that m and M have the units of mg/day and mg, respectively. With the calculated k value from Eq. (9.15), we then can use Eq. (9.11) to estimate the required cleanup time as long as we have a defined cleanup goal, such as the percent removal or the initial and the final concentrations of the contaminant (C_t/C_0).

9.3 Design Considerations and Cost-Effectiveness

In Section 9.1.2, we have illustrated some calculations that one may encounter during the design of bioremediation, including the oxygen and nutrient demands and cleanup time. There are many other aspects of design-related considerations. However, with many *in situ* and *ex situ* bioremediation technologies, it is prudent not to include any specific design detail in this section. Instead, only design elements and rationales are present here for several selected bioremediation technologies.

9.3.1 General Design Rationales

Readers should perceive this section as a basic design consideration, rather than guides for conceptual or detailed engineering design. Detailed manuals and handbooks should be consulted when specific engineering designs are engaged.

9.3.1.1 Design for *in Situ* Groundwater Bioremediation

The following elements are presented in the order in which design information might typically be collected. The design principles of some of these components are the same as those for the extraction/injection well systems discussed in Chapter 7. Other design components, such as oxygen/nutrient delivery and cleanup time, have been introduced previously in this chapter.

- Volume and area of aquifer to be treated
- Initial concentrations of constituents of concern
- Required final constituent concentrations
- Estimates of electron acceptor and nutrient requirements
- Layout of injection and extraction wells
- Design area of influence
- Groundwater extraction and injection flow rates
- Site construction limitations
- Electron acceptor system
- Nutrient formulation and delivery system
- Bioaugmentation
- Extracted groundwater treatment and disposition
- Remedial cleanup time
- Ratio of injection/infiltration to extraction
- Free-phase NAPL recovery system

9.3.1.2 Design for Bioventing

For bioventing systems, design components entail well placements, orientation, and construction for extraction/injection/monitoring wells, piping, vapor treatment, and blower specification. The design should include the following information:

- Design radius of influence (ROI)
- Wellhead pressure
- Induced vapor flow rate
- Initial constituent vapor concentrations
- Required final constituent concentrations
- Required remedial cleanup time
- Soil volume to be treated
- Pore volume calculations
- Discharge limitation and monitoring requirements
- Site construction limitation
- Nutrient formulation and delivery rate

9.3.1.3 Design for Biosparging

Critical components for biosparging will be the sparging well orientation, placement, construction, manifold piping, compressed air equipment, and monitoring and control equipment. Since biosparging may be combined with nutrient delivery and SVE, additional design considerations shall be given as well. Factors that should be considered for biosparging systems are the following:

- Radius of influence for sparging wells
- Sparging airflow rate
- Sparging air pressure
- Nutrient formulation and delivery rate (if needed)
- Initial contaminant concentrations
- Initial concentrations of oxygen and CO_2 in the saturated zone
- Required final dissolved constituent concentrations in the saturated zone
- Required remedial cleanup time
- Saturated zone volume to be treated
- Discharge limitations and monitoring requirements
- Site construction limitations (e.g. building locations, utilities, buried objects, residences)

9.3.1.4 Design for Biopiles and Composting

Important technical parameters will be the aeration, moisture contents, and nutrient requirements. The optimal soil moisture for biopiles is between 40 and 85%, and the C : N ratio is 20 : 1 to 40 : 1.

- Land requirements
- Biopile layout
- Biopile construction
- Aeration equipment
- Water management systems for control of runon and runoff
- Soil erosion control from wind or water
- pH adjustment, moisture addition, and nutrient supply methods
- Site security
- Air emission controls (e.g. covers or structural enclosures)

9.3.1.5 Design of Landfill

Some important design considerations for a sanitary landfill are as follows:

- Landfill capacity and area
- Landfill siting relative to airport, floodplains, and nearby properties
- Landfill construction (bottom) – double liner system
- Landfill construction (top) – daily soil cover, final cover, and slope
- Landfill cell and lift requirement
- Leachate collection and recovery system
- Landfill gas composition and the amount of methane estimation

9.3.2 Cost Effectiveness Case Studies

Overall, bioremediation is the lowest in total cost as compared to physicochemical and thermal technologies (Figure 9.20). In this figure, bioremediation refers to a wide array of *in situ*, on-site, and off-site techniques, such as landfilling, biopilling, phytoremediation, bioslurry reactor, bioslurping, landfarming, composting, and soil venting. Likewise, cost data for physicochemical remediation include *in-situ*, on-site, and off-site techniques such as soil washing, immobilization, air sparging, chemical oxidation, and reactive barriers. Thermal technologies refer mainly to incineration, vetrification, and several thermally enhanced techniques. The unit cost will be site specific and vary considerably depending on the hydrogeologic settings, amendments, types, and amount of contaminants. The unit cost will also be increased with the decreasing volumes of soil and groundwater to be treated. Regardless, Figure 9.20 indicates that the cost factor is very favorable for the use of these various bioremediation technologies. The case studies below regard the cost-effectiveness of several specific bioremediation approaches.

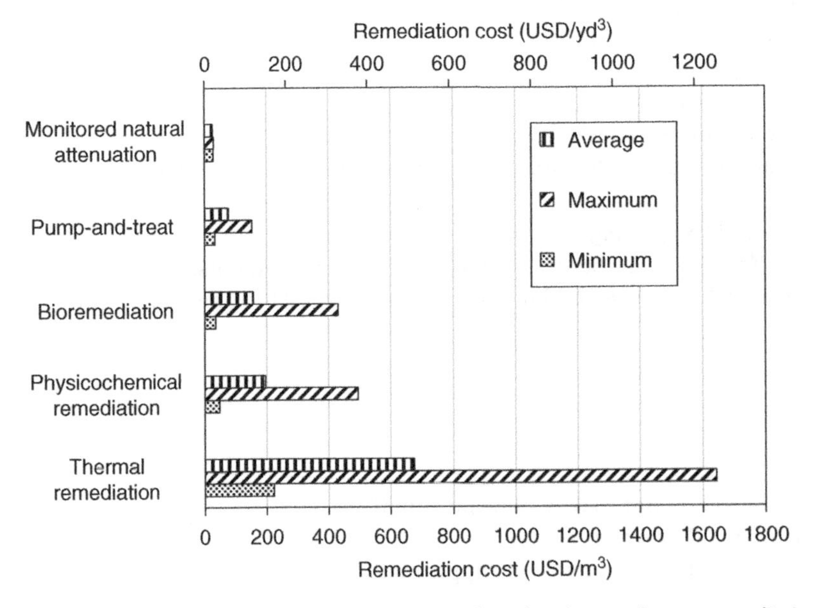

Figure 9.20 Relative cost comparison of available soil and groundwater remediation technologies. Costs reflect remediation technology in the United Kingdom and Europe from 2001 to 2005. Cost data in Euro from Summersgill (2006) were converted to US dollars (USD) in 2018 assuming 5% interest rate.

Case Study 1 Bioventing at the Hill Air Force Base, Salt Lake City, Utah, USA

The Hill Air Force Base in Salt Lake City, Utah had contaminated soils due to spills of 27 000 gal of JP-4 jet fuel oils. The fuel contamination migrated to a depth of approximately 65 ft in the vadose zone consisting primarily of sands and gravels.

The site was initially constructed as a SVE at high airflow rates. After several months, lower airflow rates were used to favor biodegradation over volatilization during the period of 1991 to 1995. Approximately 1 500 lb of hydrocarbon fuel was removed through volatilization, whereas 93 000 lb of fuel was removed through biodegradation. Most of the biodegradation was attributed to the bioventing phase of the cleanup process. Soil sampling conducted in December 1994 revealed that TPH and BTEX levels had declined at all (from 400 to <5 mg/kg) but one soil depth within a 25-ft radius of the injection well. Only at a depth of 90–100 ft (at the capillary fringe) did TPH and BTEX levels fail to decline, suggesting that the capillary fringe was not adequately aerated. It was also reported that the addition of moisture resulted in a significant increase in contaminant biodegradation, whereas addition of nutrients (N and P) did not enhance the removal.

Case Study 2 Pilot-scale bioslurry reactors for explosives remediation

This *ex situ* pilot-scale slurry reactor system used contaminated soils from two heavily contaminated soils from former army ammunition plants in Tennessee and Wisconsin. Soils contained up to 8 940–10 890 mg/kg of 2,4-DNT and 480–870 mg/kg of 2,6-DNT. The 80-L slurry bioreactors were fed with soil slurries after soils were prewashed to remove sands and then inoculated with DNT-degrading bacteria. The reactor systems were operated at well-controlled pH and nutrient condition at the draw-and-fill batch mode.

Results showed that a single reactor system removed 2,4-DNT efficiently but only partially for 2,6-DNT. The dual reactors operated in a sequential mode removed both 2,4-DNT and 2,6-DNT to acceptable levels (Table 9.4). Due to the toxicity at high concentrations and longer period for the lack of substrate, reinoculation and effluent dilution were needed from time to time.

Table 9.4 Reactor performance and mass balance (Zhang et al. 2001).

	VAAP soil at 20% loading[a]		**BAAP soil at 40% loading**[b]	
	2,4-DNT	**2,6-DNT**	**2,4-DNT**	**2,6-DNT**
Contaminated soil (mg/kg)	10 890	870	8 940	480
Washed soil (mg/kg)[c]	97	19	28	0.4
First reactor:				
Initial concentration (mg/L)	1 434	189	2 011	71
Effluent slurry (mg/L)	39	179	2	56
Effluent solids (mg/kg)	224	84	92	21
Second reactor:				
Initial concentration (mg/L)[d]	21	57	6	58
Effluent slurry (mg/L)	3.5	4.9	ND[e]	2.7
Effluent solids (mg/kg)	NA[f]	NA	71	4

Source: Reproduced with permission of Elsevier.

[a, b] Average of four feeding cycles at a 20% solids loading, and two feeding cycles at a 40% solids loading, respectively.

[c] Soil washing was performed at water/soil ratio of 17 L/kg.

[d] The second reactor was fed a mixture of effluent from the first reactor. The effluent was diluted prior to use.

[e] Not detectable.

[f] Data not available.

Case Study 3 Landfarming of kerosene spilled soil, New Jersey, USA

Dibble and Bartha (1979) reported landfarming treatment of 1.5 ha of an agricultural land due to 1.9 million liters of kerosene spill from a pipeline leak in New Jersey. After the removal of 200 m^3 of heavily contaminated soil, the land was amended to a depth of 117 cm with 6350 kg lime, 200 kg N, 20 kg P, and 17 kg K per hectare (twice). The soil underwent periodic tilling and mixing to promote aerobic biodegradation. After 24 months, contaminants in the top 30-cm soil were reduced from 8700 mg/kg to a very low level, and contaminants in soil at 30–45 cm depth were decreased to 3000 mg/kg.

Case Study 4 Phytoremediation at the Naval Air Station, Fort Worth, Texas, USA

Phytoremediation was used to clean up low-level trichloroethylene (TCE) from a shallow oxic groundwater (<3.7 m deep) in Fort Worth, Texas. The site was planted with eastern cottonwoods (*Populusdeltoides*) above shallow groundwater. The microbial populations under the newly planted cottonwood trees had not yet matured to an anaerobic community that could dechlorinatetrichloroethene (TCE); however, the microbial population under a mature (22-year-old) cottonwood tree had an established anaerobic population capable of dechlorinating TCE to DCE, and DCE to vinyl chloride (VC). Compounds consistent with the degradation of root exudates and complex aromatic compounds, such as phenol, benzoic acid, acetic acid, and a cyclic hydrocarbon, were identified in sediment samples under the mature cottonwood trees. Elsewhere at the site, transpiration and degradation by the cottonwood trees appear to be responsible for the loss of chlorinated ethenes.

Like what we have discussed in pump-and-treat and vapor extraction (Chapters 7 and 8), many sustainability practices can be incorporated into various types of bioremediation projects. Box 9.4 elaborates some of the sustainability practices in bioremediation.

Box 9.4 Green remediation best management practices for bioremediation

In principle, BMPs discussed for pump-and-treat (Box 7.3) and soil vapor extraction (Box 8.3) can also be adopted in bioremediation systems, partially because some bioremediation techniques employ the same infrastructure such as wells for pumping, injection, and monitoring. Herein, we only provide a sampling of success measures for green bioremediation that are more unique to bioremediation.

During the design stage, site characterization data and modeling efforts can play a crucial role in reducing environmental footprints. For example, a good evaluation of the radius of influence will ensure injected substrates reach the zone of bioremediation. The BIOPLUME model (Borden and Bedient 1986), one of the most widely used bioremediation models, can simulate microbial growth and decay, as well as the transport of microorganisms, oxygen, and hydrocarbons.

Prior to the implementation of any bioremediation system for an optimized full-scale operation, a bench-scale treatability test using site soil, groundwater, and microbes should be conducted to determine the site-specific contaminant removal efficiency, the degradation mechanisms and products, timeframe, required substrates, and amendments.

Construction-related BMPs for bioremediation involve the installation and testing of wells used to deliver the reagents such as oxygen, nutrients, and non-native bacteria. Direct push technologies and horizontal wells are the examples of BMPs. For land-based bioremediation systems, examples of BMPs include the construction of retention ponds within a bermed treatment area and the reclamation of clean or treated water from other site activities.

BMPs during the operation and monitoring of bioremediation systems include the use of gravity feed in existing wells, pulsed rather than continuous injections when delivering air, and use of solar energy. Another example is the use of local agricultural and industrial by-products to provide bacteria and enzymes, such as manure, woodchips, and biodigestor wastes for composting. Environmentally preferable purchasing should also be considered, such as the use of reagents in a high quality or concentrated form, and the use of biodegradable and recyclable reagents.

Bibliography

Abramowicz, D.A. (1995). Aerobic and anaerobic PCB biodegradation in the environment. *Environ. Health Perspect.* 103: 97–99.

Adler, T. (1996). Botanical clean up crews. *Sci. News* 150 (3): 42–45.

Baker, K.H. and Herson, D.S. (1994). *Bioremediation.* McGraw Hill.

Black, H. (1995). Absorbing possibilities: phytoremediation. *Environ.Health Perspect.* 103 (12): 1106–1108.

Borden, R.C. and Bedient, P.B. (1986). Transport of dissolved hydrocarbons influenced by oxygen-limited biodegradation: 1. Theoretical development. *Water Resources Res.* 22 (13): 1973–1982.

Boyajian, G. and Carreira, L.H. (1997). Phytoremediation: a clean transition from laboratory to marketplace? *Nat. Biotechnol.* 15: 127–128.

Burken, J.G. and Schnoor, J.L. (1998). Predictive relationships for uptake of organic contaminants by hybrid poplar trees. *Environ. Sci. Technol.* 32 (21): 3379–3385.

Chakrabarty, A.M. (1981). Microorganisms having multiple compatible degradative energy-generating plasmids and preparation thereof. US Patent US4259444A.

Chakraborty, R. and Coates, J.D. (2004). Anaerobic degradation of monoaromatic hydrocarbons. *Appl. Microbiol. Biotechnol* 64 (4): 437–446.

Cookson, J.T. Jr. (1995). *Bioremediation Engineering Design and Application.* McGraw Hill.

Cooney, C.M. (1996). Sunflowers remove radionuclides from water in ongoing phytoremediation field tests. *Environ. Sci. Technol.* 30 (5): 194.

Cozzarelli, I.M., Böhlke, J.K., Masoner, J. et al. (2011). Biogeochemical evolution of a landfill leachate plume, Norman, Oklahoma. *Ground Water* 49 (5): 663–687.

Davis, L.C. and Erickson, L.E. (2004). A review of bioremediation and natural attenuation of MTBE. *Environ. Progress* 23 (3): 243–252.

Dibble, J.T. and Bartha, R. (1979). Rehabilitation of oil-inundated agricultural land: a case history. *Soil Sci.* 128 (1): 52–60.

Dushenkov, S., Mikheev, A., Prokhnevsky, A. et al. (1999). Phytoremediation of Radiocesium-contaminated soil in the vicinity of Chernobyl, Ukraine. *Environ. Sci Technol.* 33 (3): 469–475.

Evans, G.M. and Furlong, J.C. (2011). *Environmental Biotechnology: Theory and Application*, 2e. Wiley-Blackwell.

Eweis, J.B., Ergas, S.J., Chang, D.P.Y., and Schroeder, E.D. (1998). *Bioremediation Principles.* McGraw Hill.

Fesenko, S.V., Alexakhin, R.M., Balonov, M.I. et al. (2007). An extended critical review of twenty years of countermeasures used in agriculture after the Chernobyl accident. *Sci. Total Environ.* 383: 1–24.

Fesenko, S., Jacob, P., Ulanovsky, A. et al. (2010). Justification of remediation strategies in the long term after the Chernobyl accident. *J. Environ. Radioact.* 119: 39–47.

Fortner, J.D., Zhang, C., Spain, J.C., and Hughes, J.B. (2003). Soil column evaluation of factors controlling biodegradation of DNT in the vadose zone. *Environ. Sci. Technol.* 37 (15): 3382–3391.
Harms, H., Schlossor, D., and Wick, L.Y. (2011). Untapped potential: exploiting fungi in bioremediation of hazardous chemicals. *Appl. Industrial Microbiol.* 9: 177–192.
Hughes, J.B., Wang, C.Y., and Zhang, C. (1999). Anaerobic biotransformation of 2,4-dinitrotoluene and 2,6-dinitrotoluene by *Clostridium acetobutylicum*: a pathway through dihydroxylamino intermediates. *Environ. Sci. Technol.* 33 (7): 1065–1070.
Hughes, J.B., Neale, C.N., and Ward, C.H. (2000). *Bioremediation, Encyclopedia of Microbiology*, 7e, vol. 1, 587–610. Boston: Academic Press, Inc.
Leahy, J.G. and Colwell, R.R. (1990). Microbial degradation of hydrocarbons in the environment. *Microbiol. Rev.* 54 (3): 305–315.
Lippincott, D., Streger, S.H., Schaefer, C.E. et al. (2015). Bioaugmentation and propane biosparging for in situ biodegradation of 1,4-dioxane. *Groundwater Monitoring Remediation* 35 (2): 81–97.
Miles, B., Peter, A., and Teutsch, G. (2008). Multicomponent simulation of contrasting redox environments at an LNAPL site. *Ground Water* 46 (5): 727–742.
Mogensen, A.S., Dolfing, J., Haagensen, F., and Ahring, B.K. (2003). Potential for anaerobic conversion of xenobiotics. *Adv. Biochem. Eng. Biotechnol.* 82: 69–134.
Nishino, S.F., Paoli, G.C., and Spain, J.C. (2000). Aerobic degradation of dinitrotoluenes and pathway for bacterial degradation of 2,6-dinitrotoluene. *Appl. Environ. Microbiol.* 66 (5): 2139–2147.
Norris, R.D. and Figgins, S. (2005). Editor's perspective – guest column: apersonal view of Richard L. Raymond and his influence on bioremediation. *Remediation* Autumn: 1–4.
Rittmann, B.E. and McCarty, P.L. (2001). *Environmental Biotechnology: Principles and Applications*. McGraw Hill.
Schnoor, J.L. (1997). Phytoremediation, Technology Evaluation Report, Ground-Water Remediation Technologies Analysis Center, TE-98-01.
Summersgill, M. (2006). Remediation Technology Costs in the UK & Europe; Drivers and Changes from 2001 to 2005, 5th ICEG Environmental Geotechnics: Opportunities, Challenges and Responsibilities for Environmental Geotechnics, Cardiff, UK, 26–30 June 2006.
Tchobanoglous, G., Stensel, H.D., Tsuchihashi, R., and Burton, F. (2014). *Wastewater Engineering: Treatment and Resource Recovery*, 5e. McGraw-Hill.
USEPA (1994). Chapter V: landfarming. In: *How to Evaluate Alternative Cleanup Technologiesfor Underground Storage Tank Sites: AGuidefor Corrective Action Plan Reviewers*, EPA 510-B-17-003. USEPA.
USEPA (2017a). Chapter III: bioventing. In: *How to Evaluate Alternative Cleanup Technologies for Underground Storage Tank Sites: A Guide for Corrective Action Plan Reviewers*, EPA 510-B-17-003. USEPA.
USEPA (2017b). Chapter IV: biopiles. In: *How to Evaluate Alternative Cleanup Technologies for Underground Storage Tank Sites: A Guide for Corrective Action Plan Reviewers*, EPA 510-B-17-003. USEPA.
USEPA (2017c). Chapter X: *in-situ* groundwater bioremediation. In: *How to Evaluate Alternative Cleanup Technologiesfor Underground Storage Tank Sites: A Guide for Corrective Action Plan Reviewers*, EPA 510-B-17-003. USEPA.
USEPA (2017d). Chapter XII: enhanced aerobic bioremediation. In: *How to Evaluate Alternative Cleanup Technologiesfor Underground Storage Tank Sites: A Guide for Corrective Action Plan Reviewers*, EPA 510-B-17-003. USEPA.

Watanabe, M.E. (1997). Phytoremediation on the brink of commercialization. *Environ. Sci. Technol.* 31 (4): 182–186A.
Zhang, C. and Bennett, G.N. (2005). Biodegradation of xenobiotics by anaerobic bacteria. *Appl. Microbiol. Biotechnol.* 67: 600–618.
Zhang, C. and Hughes, J.B. (2003). Biodegradation pathways of hexahydro-1,3,5-trinitro-1,3,5-triazine (RDX) by *Clostridium acetobutylicum*cell-free extract. *Chemosphere* 50 (5): 665–671.
Zhang, C. and Hughes, J.B. (2004). Bacterial energetics, stoichiometry and kinetic modeling of 2,4-dinitrotoluene biodegradation in a batch respirometer. *Environ. Toxicol. Chem.* 23 (12): 2799–2806.
Zhang, C., Hughes, J.B., Nishino, S.F., and Spain, J. (2000). Slurry-phase biological treatment of 2,4-dinitrotoluene and 2,6-dinitrotoluene: role of bioaugmentation and effects of high dinitrotoluene concentrations. *Environ. Sci. Technol.* 34 (13): 2810–2816.
Zhang, C., Hughes, J.B., Daprato, R.C. et al. (2001). Remediation of dinitrotoluene contaminated soils from former ammunition plants: soil washing efficiency and effective process monitoring in bioslurry reactors. *J. Haz. Materials* B87: 139–154.
Zhao, F.-J., Ma, Y., Zhu, Y.-G. et al. (2015). Soil contamination in China: current status and mitigation strategies. *Environ. Sci. Technol.* 49 (2): 750–759.

Questions and Problems

1 Which domain (bacteria, archaea, eucarya) dothe following organisms belong to: (a) protists, (b) mold, (c) methanogen, (d) algae, (e) cyanobacteria, (f) alfalfa, and (g) denitrifying bacteria?

2 What are the important types of microorganisms important in landfills?

3 What are the important types of microorganisms important in composting?

4 What do the important roles bacteria or fungi play in (a) the natural environment and (b) the polluted environment?

5 What makes bacteria especially more important than fungi in bioremediation?

6 Do fungi belong to (a) chemoheterotroph, (b) photoheterotroph, (c) chemoautotroph, or (d) chemoautotroph?

7 What are the electron acceptors and electron donors for the degradation of organic compounds by (a) aerobes, (b) nitrifying bacteria, (c) iron reducing bacteria, and (d) sulfate-reducing bacteria?

8 Describe the electron tower theory. Why the metabolism of aerobes is more favorable thermodynamically than anaerobes?

9 Describe the following terms: aerobic, anaerobic, facultative, and anoxic. Give examples of aerobes, anaerobes, and facultative bacteria.

10 The aerobic degradation of 2,4-DNT is as follows (Eq. 9.6):

$$C_7H_6N_2O_4 + 5.62O_2 \rightarrow 5.17\ CO_2 + 1.63NO_2^- + 1.63H^+ + 0.90H_2O + 0.37C_5H_7O_2N$$

If 500 lb of soil is contaminated with 800 mg/kg 2,4-DNT (soil density = 125 lb/ft^3), and a batch-scale slurry reactor is used to degrade 2,4-DNT, how much of the stoichiometric nitrogen, oxygen, and phosphorous are needed? Assume that oxygen demand comes only from 2,4-DNT, and no significant other organic compounds are present in this soil.

11 When incorporating phosphorous in the reaction, the stoichiometric reaction describing the biodegradation of 2,4-DNT ($C_7H_6N_2O_4$) is experimentally obtained as follows:

$$C_7H_6N_2O_4 + 5.621O_2 + 0.0296PO_4^{3-} \rightarrow 5.227CO_2 + 1.645NO_2^- + 1.557H^+ + 0.936H_2O + 0.0296C_{60}H_{87}O_{23}N_{12}P$$

a. Check the mass balance and report error for each element, if any.
b. Check how well the charge is balanced in the above equation.
c. Calculate the stoichiometric requirements of oxygen, nitrogen, and phosphorous, and compare the results calculated using Eq. (9.6) that does not incorporate phosphorous in the equation. Report the requirement as per gram of 2,4-DNT.
d. Calculate the biomass production per gram of 2,4-DNT.

12 Equation (9.6) reveals that pH will be lowered as a result of 2,4-DNT degradation. If no pH buffer is present, how much pH drop will be observed for every 500 mg/L of 2,4-DNT being degraded? Note that pH = – log [H$^+$].

13 Many two- and three-ring PAHs such as the ones given in the following table can be biodegraded under aerobic conditions.

Chemicals	Naphthalene	Phenanthrene
Structure		
Chemical formula	$C_{10}H_8$	$C_{14}H_{10}$
Molecular weight	128.16	178.23
Solubility in water (mg/L)	31.7	1.29

a. For each of the above chemical that is saturated in water at 20 °C, estimate the oxygen requirement (mg/L) in order to achieve a complete biodegradation. Neglect the bacterial growth.

$$C_{10}H_8 + 12O_2 \rightarrow 10CO_2 + 4H_2O$$

$$C_{14}H_{10} + 33/2O_2 \rightarrow 14CO_2 + 5H_2O$$

b. Water (20 °C) in equilibrium with atmosphere contains 8 mg/L dissolved oxygen. Comment the results calculated in (a).
c. Estimate the total theoretical oxygen demand (kg) to degrade a tank of liquid containing 500 m^3 of water contaminated with 5 mg/L of naphthalene.

14 The half-life for the complete dechlorination of TCE to ethene was found to be 14 days from a lab study. (a) Estimate the first-order rate constant. (b) If the remediation goal is to achieve 95% removal of TCE in a contaminated site, what is the cleanup time in days? (c) What is the % TCE remaining after the first, second, third, and fourth week of remediation?

15 The half-life for the aerobic biodegradation of kerosene is reported to be 28 days in a lab study. For a contaminated site, a cleanup goal of 50 ppm must be achieved, and this remediation corresponds to a 98% removal of the naphthalene. (a) Estimate the cleanup time in days.(b) What is the percentage of kerosene remaining after the first two weeks of remediation? (c) If the total oxygen requirement is calculated to be 50 000 lb, what is the average oxygen delivery rate during the first two weeks? and the second two weeks? Use $C_{12}H_{26}$ as the molecular formula to represent kerosene.

16 Kerosene contains a mixture of hydrocarbon liquids ranging from $C_{12}H_{26}$ to $C_{15}H_{32}$. (a) Use molecular formula of $C_{15}H_{32}$ (for an approximate but more conservative estimate) to develop a balanced equation for the aerobic oxidation of kerosene using NH_3, i.e. $C_{15}H_{32}O + ?O_2 + ?\ NH_3 \rightleftharpoons ?\ CO_2 + ?\ C_5H_7O_2N + 3.4H_2O$. (b) Estimate the electron acceptor and nutrient needs for bioremediation of 5000 yd^3 soil contaminated with 2000 mg/kg of kerosene. The contaminated soil has a density of 100 lb/ft^3.

17 A leaking underground storage tank has an estimated contaminated soil volume of 100 000 ft^3. The average TPH concentration in the contaminated soil is 5000 mg/kg, and the soil bulk density is 50 kg/ft^3 (1.75 g/cm^3). Estimate the nutrient requirement for the bioremediation of this site.

18 Three compounds of interest were present in a Superfund site in Louisiana. Their concentrations in the extracted wastewater stream along with their properties are listed in the table below:

Chemicals	Concentration (mg/L)	Solubility (mg/L)	Aerobic half-life	Rate constant
Hexachlorobenzene	0.005	0.005	?	0.165 year^{-1}
Naphthalene	0.32	31.7	20 days	?
Tetrachloroethylene	6.93	150	?	0.077 month^{-1}

a. Calculate the half-life in years for hexachlorobenzene.
b. Calculate the rate constant for naphthalene, and estimate the time needed for a 99% removal.
c. The EPA's maximum contaminant level of tetrachloroethylene in water is 0.5 μg/L. If this is also the cleanup goal, how long will be the cleanup time in years?
d. Evaluate the applicability of bioremediation to each of these compounds based on the above calculation.

19 Estimate the electron acceptor, nitrogen, and phosphorous nutrient needs (report nutrients in lb as N or P) for bioremediation of 5000 yd^3 soil contaminated with 4000 mg/kg of total hydrocarbons. The soil has a density of 125 lb/ft^3. Assume that the total hydrocarbon containing BTEX, *n*-pentane, 2,2-dimethylheptane, benzene 1,2-diol, and butyric acid has the average chemical structural formula of $C_6H_{10}O$. Also assume that NO_3^- was selected as the nutrient for nitrogen source:

$$C_6H_{10}O + 4O_2 + 0.57\ NO_3^- \rightleftharpoons 3.1CO_2 + 0.57C_5H_7O_2N + 3.3H_2O$$

20 It was reported that a 10% (w/v) H_2O_2 addition in an enhanced aerobic bioremediation could provide approximately 50 000 mg/L of available dissolved oxygen. Using Eq. (9.13), i.e. $2H_2O_2 \rightarrow 2H_2O + O_2$, to double check whether this statement is true or not. H_2O_2 has a MW of 34 g/mol.

21 Estimate the amount of water needed to be supplemented for 750 yd^3 of contaminated soil that is considered for *in situ* bioremediation. The soil has a porosity of 30% and current soil moisture is 20% of its water-holding capacity. Assume that 60% of water-holding capacity is the desired level of soil moisture content.

22 Composting for yard wastes and municipal solid wastes can be accomplished, for example, using chicken manure to supply N source and straw as the bulking materials. The design of composting technology is the control of C:N ratio = 20:1 to 40:1. The following data are available:

Material	Moisture	N (% dry mass)	C:N ratio (dry mass basis)
Chicken manure	70%	6.3	C:N = 15
Oat straw	20%	1.1	C:N = 48

How many kg of oat straw should be mixed with 1 kg chicken manure to give a final C:N of 30:1?

23 Is bioaugmentation always needed in bioremediation? Explain why or why not.

24 Which of the following is feasible using aerobic bioremediation process based on our current understanding of their biodegradation pathways: (a) BTEX, (b) Arochlors, (c) TCE, (d) lower-molecular-weight PAHs (two to three rings), and (e) dinitrotoluenes?

25 Which of the following is feasible using anaerobic bioremediation process based on our current understanding of their biodegradation pathways: (a) PCBs, (b) TCE, (c) TNT, and (d) alkanes?

26 Which of the following is best with sequential anaerobic–aerobic or aerobic–anaerobic process strategy: (a) PAHs, (b) PCBs, (c) TCE, and (d) RDX?

27 Three separate industries in Houston area are potentially responsible for contamination of a municipal well supply with chlorinated aliphatic hydrocarbons (CAHs). Industry records indicate that Industry A was a large user of toluene, Industry B was a large user of tetrachloroethene (PCE), and Industry C was a large user of PCBs. As far as can be determined, each industry used only one chlorinated solvent. The only CAHs found present in the municipal well supply are vinyl chloride, 1,1-dichloroethene, and 1,1-dichloroethane. Since none of the industries used these compounds, they all claim they are innocent.

You are called in by a state environmental agency as an expert witness in this case to help shed some light on the possible source or sources of these contaminants.

- Does the evidence suggest the identified industries are innocent as claimed and that some unidentified industry (*D*) is responsible, or does it support the potential that one or all of the identified industries may be responsible?
- Who is the most likely culprit? Give technically sound reasons for your judgments.

28 For the following contamination cases, briefly and clearly state whether or not the given bioremediation technologies will be a viable option:
  a. Natural attenuation: Complains have been filed by the local residents regarding PCBs in a well serving as drinking water purpose.
  b. Composting: A hot spot located in a former ammunition plant where high concentration of 2,4-dinitrotoluene was found.
  c. Phytoremediation: Metals in soils from the discharge of mining wastes.

29 Based on the current understanding, which of the following contaminant (or contaminant group) has been reported for its complete mineralization: (a) alkanes, (b) BTEX, (c) TNT, (d) PCBs, and (e) dioxins?

30 Why is the groundwater remediation of MTBE from spilled gasoline particularly troublesome? Why must a fast remediation response be in place immediately after a spill with MTBE?

31 What chemical structural feature (bonding and functional group) or properties (solubility, sorption) causes the recalcitrance (resistance to biodegradation) of the following compounds: (a) TNT, (b) MTBE, (c) PCBs, (d) PAHs, and (e) metals?

32 Studies have revealed the different strategies that bacteria use to degrade BTEX under aerobic and anaerobic conditions. What is the main difference for bacteria to incorporate oxygen into the aromatic ring prior to the ring cleavage?

33 How does K_d value affect the bioremediation of PAHs in soil?

34 Use example(s) to explain: (a) cometabolism, (b) analogy substrate enrichment, and (c) dehalogenation.

35 How can the number of chlorine atoms affect differently on the aerobic and anaerobic degradation of chlorinated aromatic hydrocarbons?

36 When contaminant concentration is either too high or too low, why could it be unfavorable to microbial degradation?

37 List the potential toxic daughter compounds (metabolic intermediates or end-products) that their toxicities are more potent than their corresponding parent compounds from the degradation of (a) TNT, (b) PCE, and (c) PAH.

38 How is 2,4-dinitrotoluene biodegraded differently under aerobic and anaerobic condition? What is its implication to bioremediation?

39 Explain in general why it is important for *in situ* bioremediation to succeed with (a) right bacteria, (b) right condition, and (c) right place.

40 Explain how hydrogeologic conditions and soil stratification affect *in situ* bioremediation.

41 Explain the types of effects and optimal range for bioremediation to occur for the following parameters in groundwater: (a) DO, (b) seasonal variations in temperature, and (c) pH.

42 How does groundwater hardness affect the bioavailability of nutrient? Give a specific example.

43 How can soil moisture contents in a vadose zone affect bioremediation?

44 Why can ferrous iron in groundwater be detrimental to the implementation of bioremediation?

45 What is the major difference between bioaugmentation and biostimulation?

46 What are the common approaches to increase dissolved oxygen concentration for *in situ* bioremediation?

47 How can total heterotrophic bacteria be used as an indicator for microbial presence in soil and groundwater remediation?

48 What are the differences between bioventing and biosparging?

49 List some environmentally important contaminants that have been shown successful using composting as the remediation approach.

50 In landfarming and landfills, what is the liner used for?

51 Describe some unique applications of bioslurry reactors in soil remediation.

52 What are some important regulatory requirements for the design of a sanitary landfill?

53 What are the pros and cons of phytoremediation?

54 What are the major remediation mechanisms for (a) phytoremediation and (b) constructed wetlands for contaminant removal?

55 What is a hyperaccumulator? List some typical plants used in phytoremediation with their corresponding contaminants.

56 List some key design parameters for landfills.

57 What are some key operational parameters for a healthy compositing system?

58 Given the following scenarios, what bioremediation technique would you choose: (a) metal-contaminated surface soil of less than 3 ft deep in a remote rural area; (b) a light fraction petroleum spill that has been infiltrated to a saturated zone, cleanup time is not essential; (c) 1000 yd^3 of surface soil soaked with nonvolatile compounds? Concentrations are relatively high, but still in the aerobic biodegradable range; and (d) an already excavated pile of soil contaminated with mostly three-ring aromatic PAHs in a rural area where animal manure and straw are available from a nearby farm. Land area is not a constraint.

59 Phytoremediation was selected to remediate lead (Pb) in surface soil at a historical battery-recycling center. The current lead concentration from a composite soil sample of this site was 1250 mg/kg, and the cleanup goal is 300 mg/kg. Estimate the required cleanup time in years if a hyperaccumulative plant in the local area is planted, fertilized, and harvested for phytoextraction, assuming that its Pb concentration is 0.03% of its dried biomass and the biomass productivity is 3000 kg/ha-yr. Also assume that the contaminated surface soil has a depth of 0.5 m, and a soil bulk density of 2000 kg/m^3. There are no additional sources of Pb from atmospheric deposition and fertilizer use during the period of phytoremediation. Note that 1 ha = 10 000 m^2.

60 In a former army ammunition plant, surface soil (1.5 m) within the property boundary was found to contain elevated TNT and the saturated soil water in this surface soil contain 5 mg/L of TNT from a lysimeter measurement. In a proposed phytoremediation project, 1500 poplar trees will be planted per acre, and the plants are expected to transpire an average of 1.5 in of soil water per acre per year. Estimate the cleanup time if the goal is to set up a 95% TNT removal from 2 to 0.10 mg/L in pore water or 5 to 0.25 kg/acre within the surface soil. TSCF for TNT in poplar tree was assumed to be 0.46 (Burken and Schnoor 1998). Note that 1 in-acre/yr = 103 m^3/yr.

Chapter 10

Thermal Remediation Technologies

LEARNING OBJECTIVES

1. Become acquainted with several high- and low-temperature thermal remediation technologies including incineration, vitrification, and various thermal enhancements
2. Calculate combustion efficiency and thermal destruction efficiency in relation to the four nines and six nines rules by the USEPA
3. Estimate the heating value using Dulong's equation and calculate stoichiometric and excess oxygen (air) requirement in combustion reactions
4. Delineate the importance of the three T's (time, temperature, and turbulence) in the design of combustion/incineration processes
5. Delineate the key components of an incineration system and pros and cons of four types of combustion chambers
6. Delineate key design considerations for an incinerator (dimensions and flow rate) as well as regulatory and siting considerations
7. Comprehend how an increased temperature affects physicochemical and biological properties (e.g. viscosity, solubility, vapor pressure, Henry's constant, sorption, diffusion, and biodegradation rate) and how these can be applied in a thermally enhanced remediation system
8. Delineate the applicability of hot air, steam, and hot water in remediation
9. Understand how electrical resistance heating, thermal induction heating, and radio frequency heating can be used in soil remediation

Thermal remediation described in this chapter refers to incineration and vitrification (thermal destruction) as well as thermal enhancement. **Incineration** is the conversion of organic contaminants into CO_2, H_2O, and other combustion products at the high temperature. **Vitrification** is a thermal treatment process that melts soil or sludge into a glass-like material. **Thermally enhanced remediation** is the enhanced volatilization due to an increased vapor pressure and volatilization rate of VOC and SVOCs, and enhanced mobility due to the reduced viscosity and residual saturation of VOCs and SVOCs. Thermal remediation technologies, mainly the on-site and off-site incineration, and various thermally enhanced processes, are listed as the third most used remediation technology in the United States for VOCs and SVOCs removal. A properly designed incineration system is capable of the highest overall degree of destruction and control for the broadest range of hazardous wastes among all of the "terminal" treatment technologies. However, incineration is also a highly controversial technology. This chapter focuses on the incineration process. Since some of the thermally enhanced remediation technologies belong to the processes we have described in previous chapters such as soil vapor extraction, air sparging, and bioremediation, our focus will be on the effects of heat

Soil and Groundwater Remediation: Fundamentals, Practices, and Sustainability, First Edition. Chunlong Zhang.

Companion website: www.wiley.com/go/Zhang/Remediation_1e

transfer and temperature on physicochemical and biological properties that are the foundation of these processes using hot air, steam, hot water, and electro-heating techniques.

10.1 Thermal Destruction by Incineration

The underlying principles, key system components, design calculations, and regulatory and siting considerations of incineration are presented in this section.

10.1.1 Principles of Combustion and Incineration

Modern understanding of the incineration of solid and hazardous waste destruction started in the mid-1970s, when incineration was found to be effective for the disposal and treatment of wastes. For an effective design and operation, an understanding of the basic principles of combustion is essential. Generally, three types of reaction mechanisms are involved in the combustion/incineration process: oxidation, pyrolysis, and free radical reactions. We'll introduce the unique principles of thermal processes, including thermal oxidation, heating values, three-T's (time, temperature, and turbulence) of the incinerator systems.

10.1.1.1 Combustion Chemistry and Combustion Efficiency

Combustion or incineration of hazardous wastes is a thermal oxidation process. Its basic principle is similar to that of conventional combustion of fossil fuels. For example, when a hydrocarbon fuel with a general composition of C_aH_b is completely oxidized, the reaction is as follows:

$$C_aH_b + \left(a + \frac{b}{4}\right)O_2 \rightarrow aCO_2 + \frac{b}{2}H_2O \qquad \text{Eq. (10.1)}$$

Pure oxygen (O_2) is introduced through air, which is a mixture of 20.95% O_2 and 79.05% N_2 (a mole ratio of 1–3.77). Although N_2 is generally inert, it is always useful to carry this inert gas into the combustion calculation. Eq. (10.1) now becomes:

$$C_aH_b + \left(a + \frac{b}{4}\right)\left(O_2 + 3.77N_2\right) \rightarrow aCO_2 + \frac{b}{2}H_2O + 3.77\left(a + \frac{b}{4}\right)N_2 \qquad \text{Eq. (10.2)}$$

Equation (10.2) states that, for every mole of fuel C_aH_b combusted, $a + b/4$ mol of O_2 and 3.77 $(a + b/4)$ mol of N_2, or 4.77 $(a + b/4)$ mol of air are required. Thus the **fuel/air ratio** for stoichiometric combustion of hydrocarbon C_aH_b is $1/[4.77\ (a + b/4)]$.

Now we consider a waste containing C, H, O, Cl, S, and N with a general formula of $C_aH_bO_cCl_dS_eN_f$. The stoichiometric reaction can be written as (Santoleri et al. 2000)

$$\begin{aligned} &C_aH_bO_cCl_dS_eN_f + \left(a + g + e - \frac{1}{2}c\right)O_2 + \frac{79}{21}\left(a + g + e - \frac{1}{2}c\right)N_2 \\ &\rightarrow aCO_2 + 2gH_2O + dHCl + eSO_2 + \left[\frac{1}{2}f + \frac{79}{21}\left(a + g + e - \frac{1}{2}c\right)\right]N_2 \end{aligned} \qquad \text{Eq. (10.3)}$$

where a, b, c, d, e, and f = number of moles of C, H, O, Cl, S, and N, respectively, present in the waste-fuel mixture; $g = ¼(b - d)$ when $b > d$; $g = 0$ when $b \leq d$. For example, the complete combustion of octane (C_8H_{18}), where a = 8, b = 18, and $c = d = e = f = 0$, can be written as follows:

$$C_8H_{18} + 12.5O_2 + 12.5 \times 3.77N_2 \rightarrow 8CO_2 + 9H_2O + 47.125N_2 \qquad \text{Eq. (10.4)}$$

The combustion reactions described in Eqs. (10.1) through (10.4) are the examples of oxidation. The second process **pyrolysis** is the thermal degradation of organic matter in the absence of oxygen at a high temperature. Although incineration requires about 50–100% excess air, oxygen in certain area of the combustion chamber may be limited due to inadequate mixing. For example, when cellulose ($C_6H_{10}O_5$) and PCBs ($C_{12}H_7Cl_3$) undergo pyrolysis, the incomplete combustion products include carbon monoxide (CO) and element carbon (C). The carbon (C) or char produced will stay in the solid phase.

$$C_6H_{10}O_5 \rightarrow 3C + 2CO + CH_4 + 3H_2O$$

$$C_{12}H_7Cl_3 \rightarrow 12C + 3HCl + 2H_2$$

The third reaction process in thermal combustion and incineration is the **radical attack**. A radical is an atom or group of atoms with one or more unpaired electrons. For example, the most common free radical OH• (hydroxyl radical) is composed of an oxygen atom and a hydrogen atom with a total of seven valence electrons. This radical needs one more electron to saturate the valence shell, and therefore is unstable and very reactive because of this unpaired electron. Common radicals in incineration around 1000 °C include: atomic oxygen (O), atomic hydrogen (H), atomic chlorine (Cl), hydroxyl radical (OH•), methyl radical (CH_3•), and chloroxy radical (ClO•). The decomposition of chemicals in the waste is facilitated by the radical attack, which usually induces a chain of reactions.

Ideally, a complete combustion of a waste containing C, H, N, Cl, S, and P should be converted into CO_2, H_2O, N_2, HCl, SO_2, and P_2O_5, respectively. Under incomplete combustion conditions, unwanted by-products are produced and emitted from the combustion chamber. These by-products, termed the **products of incomplete combustion** (PICs), includecarbon monoxide (CO), SO_2 (1–5% SO_3), NO_x (NO + NO_2; ~95% NO), Cl_2 (less favored than HCl), polychlorinated dioxin/furan, and metals. Note that metals cannot be destroyed during incineration. Alkali metals such as K and Na will become the hydroxides or carbonates. Non-alkaline metals will become the metal oxides, such as CuO and Fe_2O_3. After combustion, most metals become residue in bottom ash, some volatile metals will vaporize and diffuse into the exhaust gas stream. Dioxins and furans refer to families of 75 related chemical compounds known as polychlorinated dibenzo-*p*-dioxin (PCDDs) and 135 related chemical compounds known as polychlorinated dibenzofurans (PCDFs), respectively. The presence of toxic PCDD/F in the municipal and hazardous incinerators has been the major concern of this supposedly well-established technology. Monitoring data have shown PCDD concentrations of 1–10 700, 117–450, and ND – 16 ng/Nm^3 in municipal, medical, and hazardous wastes, respectively, whereas the concentrations of PCDF in the respective wastes are 2–37 500, 52–30 300, and ND – 56 ng/Nm^3, where ND = not detected and ng/Nm^3 stands for nanogram per unit normal cubic meter at the standard temperature and pressure (25 °C and 1 atm) (Steverson 1991).

The efficiency of thermal destruction is termed **destruction and removal efficiency** (DRE), which is calculated as:

$$DRE = \frac{m_i - m_o}{m_i} \times 100\% = \left(1 - \frac{m_o}{m_i}\right) \times 100\% \qquad \text{Eq. (10.5)}$$

where m can be reported as the mass or concentration, and the subscripts "i" and "o" denote the mass or concentration in the feed and the flue gas emitted directly to the air, respectively. Unlike the low-temperature thermal enhanced technologies, which usually have an average removal efficiency of approximately 65%, the DRE in a well-designed and operated incinerator is very high. The USEPA's rules on the treatment efficiency for incineration are as follows: (a) 99.99% (**the four nines rule**) for Principal Organic Hazardous Constituents (POHCs) including halogenated, nonhalogenated, aliphatic, aromatic, and polynuclear organic compounds; and (b) 99.9999% (**the six nines rule**) for PCBs and dioxins (Example 10.1).

Example 10.1 Destruction efficiency of a hazardous incinerator

If the mass of PCBs in the emitted air of a hazardous incinerator must not exceed 0.001 g/h, what must be the maximum rate of PCBs in the feed to the incinerator as per the USEPA's requirement?

Solution

For PCBs, the "six nines" rule applies.

$$\mathrm{DRE} = 1 - \frac{m_o}{m_i}$$

$$0.999\ 999 = 1 - \frac{0.001\frac{\mathrm{g}}{\mathrm{h}}}{m_i}$$

$$m_i = \frac{0.001\frac{\mathrm{g}}{\mathrm{h}}}{1 - 0.999\ 999} = 1000\frac{\mathrm{g}}{\mathrm{h}} = 1\frac{\mathrm{kg}}{\mathrm{h}}$$

The DRE should not be confused with a related operational parameter describing the efficiency of combustion. Since carbon monoxide is the product of incomplete combustion, it is used as an indicator for the **combustion efficiency** (CE). That is, the more CO_2 (or the less CO), the higher the combustion efficiency. Theoretical combustion efficiency can be calculated from a given stoichiometric reaction. The actual combustion efficiency can be calculated based on the ratio of carbon dioxide to total carbon species measured in the flue gas effluent samples:

$$\mathrm{CE} = \frac{\mathrm{CO_2}}{\mathrm{CO_2} + \mathrm{CO}} \times 100\% \qquad \text{Eq. (10.6)}$$

where the combustion efficiency can be calculated either by volume or by mass (see Example 10.2 for the calculation of volume-based and mass-based CE).

Example 10.2 Combustion efficiency in an incinerator

Estimate the combustion efficiency by volume and by mass under the following two conditions: (a) The incomplete combustion of propane (C_3H_8) in an oxygen-deficit condition, given: $C_3H_8 + 4O_2 \rightarrow 2CO + CO_2 + 4H_2O$. (b) The flue gas in an incinerator stack contains 15% CO_2 and 10 ppm CO.

Solution

a) For every mole of C_3H_8 combusted, the molar ratio of CO and CO_2 is 2 : 1. Since the volume ratio is equal to the molar ratio, the volume-based combustion efficiency is as follows:

$$\mathrm{CE}\left(\text{by volume}\right) = \frac{\text{moles of CO}_2}{\text{moles of CO}_2 + \text{moles of CO}} = \frac{1}{1+2} = 0.333(33.3\%)$$

The mass-based combustion efficiency can be calculated in a similar way based on the 2:1 ratio of CO (MW = 28) and CO_2 (MW = 44)

$$\text{CE(by mass)} = \frac{\text{mass of } CO_2}{\text{mass of } CO_2 + \text{mass of CO}} = \frac{44}{44 + 2 \times 28} = 0.44(44\%)$$

b) Both % and ppm are measured on the volume basis (v/v), so ppm can be converted to %. A concentration of 10 ppm is equal to $10/10^6$ (v/v) or 0.001/100, i.e. 0.001%.

$$\text{CE(by volume)} = \frac{15}{15 + 0.001} = 0.999\ 933(99.9933\%)$$

Again, since the volume ratio is equal to the molar ratio, the mass-based ratio is as follows:

$$\text{CE(by mass)} = \frac{15 \times 44}{15 \times 44 + 0.001 \times 28} = 0.999958(99.9958\%)$$

The above-calculated combustion efficiency in (a) is too low for propane, but it is sufficient in (b) for POHCs in the incinerator, as stated in the four nines rule.

10.1.1.2 Heating Values of Fuels/Wastes

The **heating value** (HV) of a waste is a measure of the energy released when the waste is burned and is commonly expressed in the unit of British Thermal Units per pound (1 Btu/lb = 2.326 J/g). The analysis of heating value is important to determine whether a waste is suitable for incineration and how much of the auxiliary fuel should be provided to sustain combustion in the incinerator. The heating value is equal to the enthalpy of combustion but of the opposite sign. Thus, an exothermic reaction with a negative ΔH should be reported as a positive HV.

$$\Delta H^0 = \left(\sum_i H^0\right)_{\text{products}} - \left(\sum_j H^0\right)_{\text{reactants}} \qquad \text{Eq. (10.7)}$$

where H is the enthalpy of either product i or reactant j in the combustion reaction. The heating value depends on the phase of the water, i.e. condensed water versus water vapor present in the combustion products. A **higher heating value** (HHV) is the heating value when water is in its condensed form, whereas a **lower heating value** (LHV) is the heating value when water is in its vapor form. The difference between the HHV and the LHV is the latent heat of vaporization of water, which is 10 520 kcal/kg-mol at 20 °C.

The approximate heating value of a chemical can be estimated using **Dulong's equation** if we know the composition (mass percentage) of the waste (fuel).

$$\frac{\text{Btu}}{\text{lb}} = 14540C + 62000\left(\text{H} - \frac{\text{O}}{8}\right) + 4100\text{S} + 1000\text{N} \qquad \text{Eq. (10.8)}$$

where C, H, O, S, and N represent the weight percentage of the respective element (i.e. 0.25 for 25%). The term H − O/8 represents the hydrogen (H) combined with oxygen as moisture as well as H and O that are combined in other forms (see Example 10.3).

Typical heating values for some common materials are given in Table 10.1. Note that a minimum heating value of approximately 5 000 Btu/lb (11 630 J/g) is required to sustain combustion, and a heating value of greater than 2 500 Btu/lb is required for a waste to be treated using a thermal incinerator.

Example 10.3 Heating values of fuels/wastes using Dulong's equation

Polyvinyl chloride (PVC) is a plastic that has the chemical formula of $(CH_2 = CHCl)_n$. Estimate the heating value using Dulong's formula.

Solution

Since the weight percentage of each element in PVC is independent of the value of *n*, we use the formula of its single unit of polymer CH_2=CHCl or C_2H_3Cl to calculate the weight percentages. The formula weight of C_2H_3Cl is $12\times2+1\times3+35.5\times1 = 62.5$.

The weight percentage of C = 12×2/62.5 = 0.384 (38.4%).
The weight percentage of H = 1×3/62.5 = 0.048 (4.8%).
The weight percentages of O, S, and N are all zero.

By substituting the numbers into the Dulong's equation, we have:

$$\text{Btu/lb} = 14\,540\times0.384+62\,000\times(0.048-0)+4\,100\times0+1\,000\times0 = 8\,559\,\text{Btu/lb}$$

It is clear that plastic has a heating value high enough to sustain its own combustion in an incinerator without auxiliary fuels.

Table 10.1 Typical heating values for some common materials.

Materials	Heating value (Btu/lb)	Materials	Heating value (Btu/lb)
Wood	8300	Scrape tires	12000–16000
Coal (high sulfur)	10500	Gasoline	20200
Coal (low sulfur, Eastern)	13400	RFD from MSW	3000–6000
Coal (low sulfur, Western)	8000		

RFD = Refuse derived fuel; MSW = Municipal solid waste.

10.1.1.3 Oxygen (Air) Requirement

Ambient air is often used to sustain combustion reactions that require oxygen, which contains approximately 79.05% N_2 and 20.95% O_2 by volume (Table 10.2). The amount of oxygen required under ideal or "perfect" conditions to burn all of the organic materials into CO_2 and H_2O with no oxygen left is called the **stoichiometric (or theoretical) oxygen requirement** and the process is termed **complete combustion**. Typically, an incinerator is operated at 1.5–2 times the stoichiometric oxygen demand (i.e. 150–200% of the stoichiometric oxygen, or 50–100% excess oxygen or air). **Excess air** (EA) is the air supplied in excess of what is necessary to burn a compound completely and it appears in the products of combustion (flue gas). The amount of EA in a typical range of 1.5–2.0 is normally expressed as a percentage of the theoretical air required for a complete combustion. Note that in the reaction, O_2 is the chemical involved, not "air." However, since air has a constant percentof oxygen and nitrogen, the volume of air can be calculated if we know the volume of oxygen.

From Table 10.2, one can multiply the volume of O_2 by 3.77 to get the volume of N_2, or multiply by 4.77 to get the volume of air. In practice, a volume ratio of 79/21 = 3.76 is used in place of 3.77 for N_2/O_2 or 4.77 for air/O_2 ratio. To be consistent, we will use 3.77 for N_2/O_2 or 4.77 for air/O_2 ratio in this text. The calculations of theoretical air, excess air, and actual air requirements are illustrated using methane CH_4 (Example 10.4) and propane C_3H_8 (Example 10.5) as simplified examples.

Table 10.2 Ratio of N_2 and air to O_2 in air.

	By weight	By volume
O_2 in air	23.15%	20.95%
N_2 in air	76.85%	79.05%
N_2/O_2 ratio in air	3.32 (=76.85/23.15)	3.77 (=79.05/20.95)
Air/O_2 ratio in air	4.32 (= 1 + 3.32)	4.77 (=1 + 3.77)

Example 10.4 Theoretical, excess, and actual air requirements during combustion

The combustion of methane can be written as: $CH_4 + 2O_2 \rightarrow CO_2 + 2H_2O$. Determine (a) the theoretical combustion air, (b) the excess combustion air at 50% excess rate, and (c) the actual combustion air. Report the amount per lb-mol of CH_4 and per lb of CH_4.

Solution

Prior to solving this problem, one should be familiar with the lb_m or "lb-mol" unit in the US unit system in performing stoichiometric calculation of such combustion reactions. In this text, we use the mass notation lb_m instead of lb-mole so it will not be misread as the product of lb and mole. The unit of lb_m is analogous to the regular gram-mole. One gram-mole (i.e. the regular 1 mol) of CH_4 is 16 g, whereas 1 lb-mol (lb_m) of CH_4 is 16 lb.

a) We rewrite the balanced equation by adding N_2 into the reaction using the N_2/O_2 volume ratio of 3.77 in air.

$$CH_4 + 2O_2 + 2(3.77)N_2 \rightarrow CO_2 + 2H_2O + 2(3.77)N_2$$

Since 1 lb_m of CH_4 (or 16 lb CH_4) is involved in the above reaction, the air requirement can be expressed as per mole or per 16 lb. CH_4. Thus:
Using N_2/O_2 ratio in air (Table 10.2), the theoretical combustion air (mol/mol CH_4) = 2 + 2(3.77) mol/mol CH_4 = 9.54 mol/mol CH_4.
The theoretical combustion air (lb/lb CH_4) = 2(32) + 2(3.77)(28) = 275.12 lb/16 lb CH_4 = 17.2 lb/lb CH_4

b) The excess combustion air (mol/mol CH_4) = 0.5(9.54) = 4.77 mol/mol CH_4.
The excess combustion air (lb/lb CH_4) = 0.5 × 17.2 lb/lb CH_4 = 8.6 lb/lb CH_4.

c) The actual combustion air (mol/mol CH_4) = 9.54 + 4.77 mol/mol CH_4 = 14.3 mol/mol CH_4.
The actual combustion air (lb/lb CH_4) = 17.2 + 8.6 lb/lb CH_4 = 25.8 lb/lb CH_4.

Example 10.5 Oxygen requirements and flue gas composition

Calculate (a) the oxygen requirement (O_2 lb/lb fuel) and (b) the volumetric ratio of CO_2 in the flue gas of propane (C_3H_8) if an excess air of 20% is used for the combustion of propane: $C_3H_8 + 5O_2 \rightarrow 3CO_2 + 4H_2O$. The molecular weight of C_3H_8 = 44.

Solution

a) At 120% of the stoichiometric amount (i.e. 20% excess air):

Moles of O_2 = 5 × 120% = 6

Moles of $N_2 = 6 \times 3.77$ (the volume ratio of N_2/O_2) = 22.6

Re-write the equation by including N_2 in the air:

$$C_3H_8 + 6O_2 + 22.6N_2 \rightarrow 3CO_2 + 4H_2O + O_2 + 22.6N_2$$

So the O_2 requirement in terms of mass of O_2 per unit mass of propane = 6×32/44 = 4.36 lb O_2/lb propane. We then use the mass ratio of air/O_2 in the air (Table 10.2) of 4.32 to convert from O_2 into air, the air requirement = 4.36×4.32 = 18.8 lb air/lb propane.

b) According to the molar ratio in the above equation, we can calculate the volumetric ratio of CO_2 in the flue gas = mole of CO_2/(mole of CO_2 + mole of all gases) = 3/(3+4+1+22.6) = 0.098 (9.8%).

An alternative way of calculating stoichiometric oxygen, nitrogen, and air is to directly use the following formula:

$$O_2 = 2.67C + 8\left(\frac{H - Cl}{35.6}\right) + S - B_o \qquad \text{Eq. (10.9)}$$

$$N_2 = 3.32O_2 \qquad \text{Eq. (10.10)}$$

$$Air = O_2 + N_2 \qquad \text{Eq. (10.11)}$$

where C, H, Cl, and S are the mass of C, H, Cl, and S per unit mass (not percent) of the waste (fuel), respectively, and B_o is the mass of **bound oxygen** per unit mass of waste (fuel) that is associated with waste or fuel. It reacts with reactants and is a negative value because it contributes to the O_2 from the ambient air. For example, B_o for a waste containing $C_{59}H_{93}O_{37}N$ is equal to $37 \times 16/(59 \times 12 + 93 \times 1 + 37 \times 16 + 14 \times 1) = 0.421$. The coefficient 2.67 is the mass ratio of O_2/C (32/12) for the combustion of C ($C + O_2 \rightarrow CO_2$). For sulfur, this coefficient is equal to 1, because of the ratio of O_2/S (32/32) for the combustion of S ($S + O_2 \rightarrow SO_2$). H – Cl/35.5 is equal to the H left after Cl reacts with H, and the coefficient 8 is equal to the ratio of $0.5O_2/H_2$ ($0.5 \times 32/2 = 8$) for the combustion of each H atom ($H_2 + 0.5O_2 \rightarrow H_2O$). Example 10.6 illustrates the use of Eqs. (10.9)–(10.11) as an alternative method to solving Example 10.4.

Example 10.6 Stoichiometric oxygen requirement during a complete combustion

Use Eqs. (10.9)–(10.11) to confirm the calculation made in Example 10.4 for the stoichiometric oxygen needed during the complete combustion of methane: $CH_4 + 2O_2 \rightarrow CO_2 + 2H_2O$. Report the results on lb O/lb CH_4 basis.

Solution

For every lb of CH_4, there are 12/16 lb C and 4/16 lb H. Applying Eq. (10.9), we have:

$$O_2 = 2.67C + 8\left(\frac{H - Cl}{35.6}\right) + S - B_o$$

$$= 2.67 \times \frac{12}{16} + 8 \times \left(\frac{4}{16} - \frac{0}{35.5}\right) + 0 - 0 = 4.00 \frac{lbO}{lbCH_4}$$

Applying Eq. (10.10), we can calculate the required N_2:

$$N_2 = 3.32\ O_2 = 3.32 \times 4.00 = 13.28 \frac{\text{lb N}}{\text{lb CH}_4}$$

Using Eq. (10.11), the amount of air needed = $O_2 + N_2 = 4.00 + 13.28 = 17.28$ lb air/lb CH_4. This stoichiometric amount of air is the same as that calculated in Example 10.4a.

10.1.1.4 Three T's of the Combustion/Incineration

Besides the essential heating values of wastes and sufficient oxygen, an efficient combustion/incineration should meet the requirement of the **three T's**: time, temperature, and turbulence (mixing). Here, time refers to the **residence time** (also called **retention time, dwell time**) of waste staying in the incinerator. A longer residence time in combustion chamber generally improves destruction efficiency. However, a longer residence time will require a large volume of the incinerator. Temperature is a function of the heating values of the waste and air flow rate. A higher temperature results in a higher removal efficiency. However, too high temperature should be avoided for the safety of the refractory material (e.g. ceramics) in the incinerator. **Turbulence** promotes mixing between the waste compounds, oxygen, and auxiliary fuel if provided. A greater turbulence will increase combustion efficiency. Turbulence can be measured by the Reynolds number (see Eq. 3.13, Chapter 3).

If we know the kinetic rate of the combustion, the residence time can be calculated. The residence time (t) relates the size of an incinerator to the flow rate by

$$t = \frac{V}{Q} \qquad \text{Eq. (10.12)}$$

where V = incinerator (combustion chamber) volume, Q = volumetric flow rate (m^3/s). Since the volume of gas depends on temperature, the volumetric flow rate Q at one temperature can be converted to the value at another temperature using Charles' law (Box 10.1, Example 10.7).

Box 10.1 Charles' law – temperature–volume relationship at constant mass

Charles' law describes how gases tend to expand when heated. Charles' law states that the volume of a given mass of an ideal gas is directly proportional to its temperature on the absolute temperature scale (Kelvin or Rankine) if the pressure and the amount of gas remain constant (Eq. 10.13). Charles' law is particularly useful when we need to convert the gas flow rate from one temperature to another temperature:

$$\frac{Q_2(\text{acfm})}{Q_1(\text{scfm})} = \frac{T_2}{T_1} \qquad \text{Eq. (10.13)}$$

where Q_1 = the standard flow rate (e.g. scfm) at standard temperature T_1, Q_2 = the actual flow rate (e.g. acfm) at T_2. For most environmental engineering applications, standard conditions are 60 °F (15 °C) or 32 °F (0 °C) and 1 atm. In designing the equipment, one should always use the actual conditions rather than the standard temperature and pressure. For example, the ratio of acfm (2000 °F) to scfm (60 °F) in Example 10.7b is 4.73. The reader is also cautioned on the use of absolute temperature and pressure when using the Charles' law. For the unit of T in Eq. (10.13), either Kelvin (°K) or Rankie (°R) should be used. The same answer will result because T (°K)/T (°R) = 5/9.

Example 10.7 Use of Charles' law to calibrate gas flow rate at different temperatures

(a) A hazardous waste incinerator operates at 2200 °F with a combustion flow rate of 45 000 **actual cubic foot per minute (acfm)**. If the incinerator is 10 ft wide, 15 ft deep, and 20 ft high, what is the maximum residence time in the incinerator? (b) Another incinerator has a gas flow rate of 4500 **standard cubic foot per minute (scfm)** at 15 °C (60 °F or 288 °K). It operates at 2000 °F (1366°K). If the minimum residence time of 1.5 seconds is required, calculate the required volume of this incinerator (adapted from Reynolds et al. 2007)

Solution

a) The volume of the incinerator = width × length × depth = 10 × 15 × 20 = 3000 ft^3
The maximum residence time = V (volume)/Q (flow rate) = 3 000 ft^3/(45 000 ft^3/min) = 0.0667 minutes = 4 seconds.

b) The flow rate at the standard temperature condition (60 °F) should be converted into the flow rate at the actual condition at its operating temperature of 2000 °F (1366°K). According to the ideal gas law, volume is proportional to the temperature for a given mass (mole) of the chemical. Using temperature conversion: T (°R) = T (°F) + 460, where T (°F) is the actual temperature in °F, and the addition of 460 is to convert the temperature from °F to the thermodynamic (absolute) temperature degree Rankine T (°R), we have:

$$Q_2 = Q_1 \times \frac{T_2}{T_1} = 4500 \times \frac{2000 + 460}{60 + 460} = 4500 \times 4.73 = 21285\,\text{acfm}$$

The required volume (V) of this incinerator = $Q \times t$ = 21 285 ft^3/min × (1.5 s × 1 min/60 s) = 532 ft^3. In the above calculation, the absolute temperature in the unit of °R is used. If absolute temperature in the unit of °K is used, the ratio of T_2/T_1 is the same (i.e. 1366/288 = 4.73).

10.1.2 Components of Hazardous Waste Incinerator Systems

In this section, the general applications of hazardous waste incinerators and their pros and cons are briefly described. This is followed by the description of incineration system components, with special detail of four commonly used combustion chambers.

10.1.2.1 General Applications, Pros and Cons

Incineration has been employed for the disposal of industrial chemical wastes (hazardous wastes) since the 1960s. The early incineration units, based on technologies for municipal wastes using grate type units, had poor performance and adaptability. This led to the subsequent development of many rotary kiln facilities in West Germany. One of the first kiln units in the United States was at the Dow Chemical Company facility in Midland, Michigan. Incineration has now become the well-established technology for the disposal and treatment of both municipal and hazardous wastes. It is applicable for numerous waste streams including hazardous wastes, municipal wastes, toxic wastes, laboratory wastes, infectious (medical) wastes, spent pesticides, wastewater sludge, contaminated soils, and liquids/wastewater. In the United States, 14% of the municipal solid wastes (MSWs) are treated by incineration, compared to 73% by landfill and 13% by recycling. Incineration was reported to be employed at 150 Superfund sites in the United States.

Some general advantages of thermal destruction (incineration) are as follows:

- Hazardous organic compounds are destroyed
- Both volume and weight are significantly reduced
- Air emissions can be well controlled

- Heat can be recovered
- Possibility of on-site operation
- Applicable to numerous waste types
- Generally accepted by regulations
- Fast due to short residence times (can treat several hundred tons of wastes per day)

Incineration is an environmentally and technologically superior method, but it can be highly controversial and expensive. Some general disadvantages of thermal destruction (incineration) are as follows:

- Chemicals such as metals are not incinerable
- Some organics generate toxic Products of Incomplete Combustion (PICs), such as dioxins which were not present in the feed
- High capital and operating cost (~$500/ton)
- Skilled operators are required
- Supplemental fuel is often required
- Public opposition, the so-called NIMBY (not in my back yard) attitude

10.1.2.2 Incinerator System Components

Despite the many variations of the incineration facilities employed in hazardous waste disposal and treatment, the major subsystems of an incineration system typically include waste preparation and feeding, combustion chamber(s), air pollution control, and residue/ash handling components (Figure 10.1). The most common air pollution control system involves combustion gas quenching followed by a venturi scrubber (for particulate removal), a packed tower absorber (for acid gas removal), and a mist eliminator. Residue and ash from incineration facilities are typically used for construction materials or disposed of in landfills. Figure 10.1 illustrates the normal orientation of these subsystems along with typical process component options.

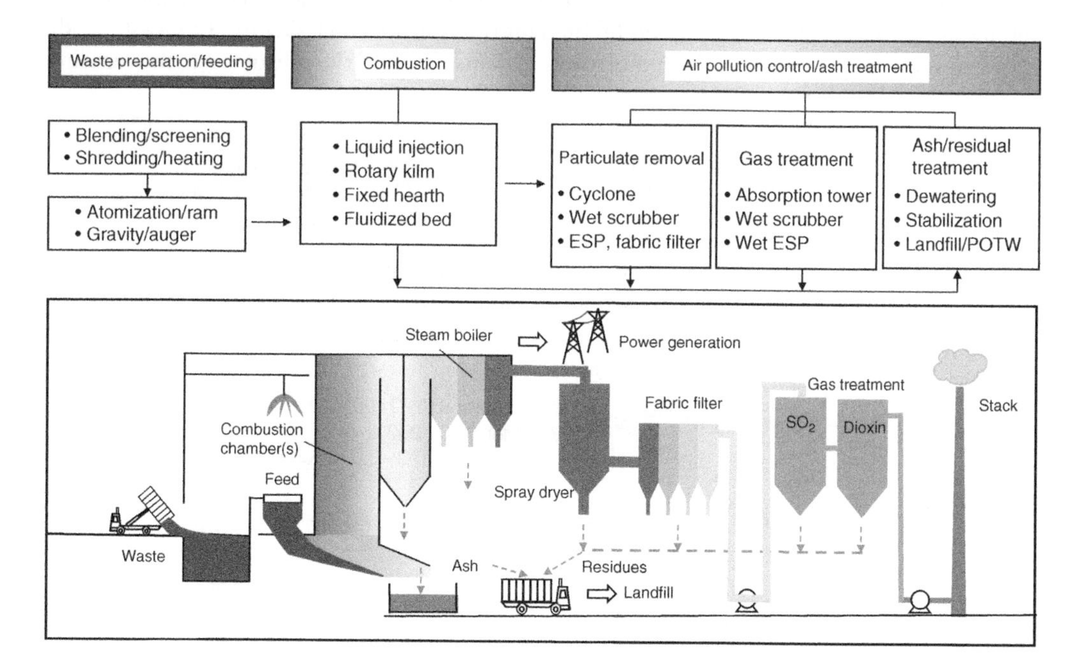

Figure 10.1 General orientation of incineration subsystems and typical process component options (ESP = Electrostatic Precipitator; POTW = Publicly Owned Treatment Works).

10.1.2.3 Four Types of Combustion Chambers

The key component of an incineration system is the combustion chamber. In the order of their usages, the four most common configurations of combustion chambers are: liquid injection incinerators (sometimes combined with fume incineration), rotary kilns, fixed hearths, and fluidized bed incinerators. These are depicted below with some brief description of operational principles.

Liquid injection incinerators are applicable almost exclusively for pumpable liquid waste with viscosity <10 000 SUS (typical of the viscosity of honey; Saybolt universal seconds, defined as the time needed to let 60 cm^3 of oil to flow through a calibrated orifice of a Saybolt universal viscometer at a controlled temperature). These incinerators are usually simple, refractory-lined cylinders (either horizontally or vertically aligned) equipped with one or more waste burners (Figure 10.2 for a horizontally aligned system). Liquid wastes are injected through the burner(s), atomized to fine droplets through a nozzle, and burned in suspension. Burners as well as separate waste injection nozzles may be oriented for axial, radial, or tangential firing. An improved utilization of combustion space and higher heat release rates, however, can be achieved with the utilization of swirl or vortex burners or designs involving tangential entry. Vertically aligned liquid injection incinerators are preferred when wastes are high in inorganic salts and fusible ash content, while horizontal units may be used with low ash content. The typical capacity of liquid injection incinerators is approximately 2.8×10^7 Btu/h heat release.

Rotary kiln incinerators (Figure 10.3) are more versatile incinerators because they are applicable to the destruction of solid wastes, slurries, and containerized waste as well as liquids. These units are most frequently used in commercial off-site incineration facilities. Wastes are loaded into a rotary kiln followed by an afterburner. The **rotary kiln** is an inclined cylindrical shell lined with refractory materials. The primary function of the kiln is to convert solid wastes to gases, which occurs through a series of volatilization, pyrolysis, and partial combustion reactions. Rotation of the shell provides for transportation of waste through the kiln as well as for enhancement of waste mixing. The residence time of waste solids in the kiln is generally 1–1.5 hours. This is controlled by the kiln rotation speed (1–5 rpm), the waste feed rate and, in some instances, the inclusion of internal dams to retard the rate of waste movement

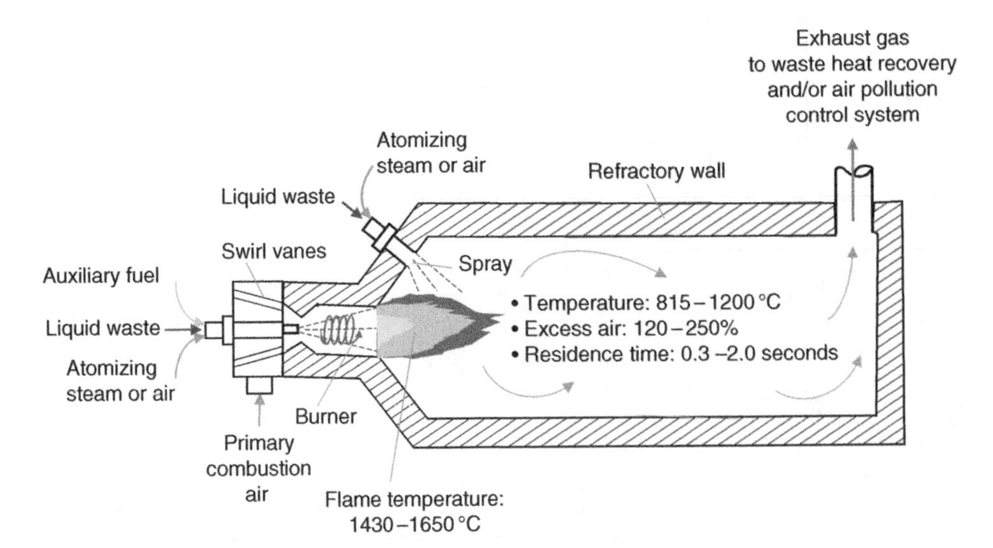

Figure 10.2 A typical liquid injection incinerator with a horizontally aligned liquid injection. *Source:* Adapted from Dempsey and Oppelt (1993).

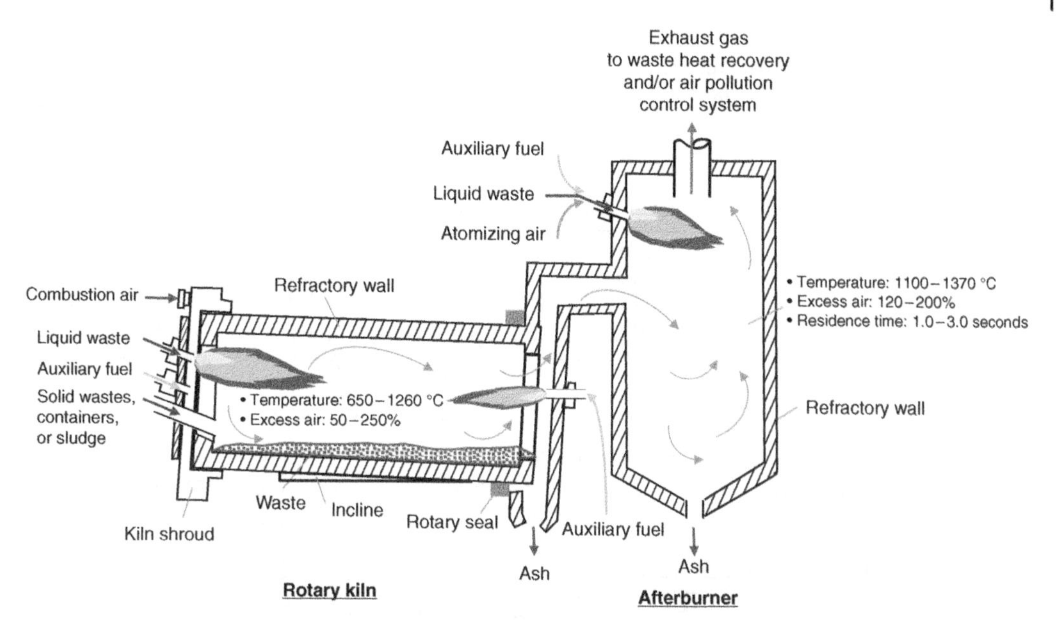

Figure 10.3 A typical rotary kiln incinerator. *Source:* Adapted from Dempsey and Oppelt (1993).

through the kiln. An internal dam in a rotary kiln works much like a dam in a flowing stream, so that material builds up behind the dam, forcing retention time to increase. The feed rate is generally adjusted to limit the amount of waste being processed in the kiln to at most 20% of the kiln volume.

An **afterburner** is necessary to complete the gas phase combustion reactions. It is connected directly to the discharge end of the kiln, whereby the gases exiting the kiln turn from a horizontal flow path upward to the afterburner chamber. The afterburner itself may be horizontally or vertically aligned, and essentially functions much on the same principles as a liquid injection incinerator. In fact, many facilities also fire liquid hazardous waste through separate waste burners in the afterburner. Both the afterburner and kiln are usually equipped with an auxiliary fuel firing system to bring the units up to and maintain the desired operating temperatures. Rotary kilns have been designed with a heat release capacity as high as 9×10^7 Btu/h in the United States. On average, however, units are typically 6×10^7 Btu/h.

Fixed hearth incinerators (also called controlled air incinerators, starved air incinerators, or pyrolytic incinerators) are the third major incinerator in current use for hazardous wastes (Figure 10.4). This type of incinerator is a "fixed" hearth and air is controlled in two stages – a primary combustion in an oxygen-deficient ignition chamber, followed by an oxygen-rich gaseous phase combustion. Waste is ram fed into the first stage, or the primary chamber, and burned at roughly 50–80% of stoichiometric air requirements. This starved air condition destroys most of the volatile fraction pyrolytically, with the required endothermic heat provided by the oxidation of the fixed carbon fraction. Fixed carbon is the part of total carbon remained in the solid residue (other than ash) after pyrolysis of the waste (fuel). The resultant smoke and pyrolytic products, consisting primarily of volatile hydrocarbons and carbon monoxide, along with products of combustion, pass to the second stage, or secondary chamber. Here, additional air is injected to complete the combustion either spontaneously or through the addition of supplementary fuels. The primary chamber combustion reactions and turbulent velocities are maintained at low levels by the starved-air conditions to minimize particulate carryover. With the addition of secondary air, total excess air for fixed hearth incinerators is in the 100–200% range.

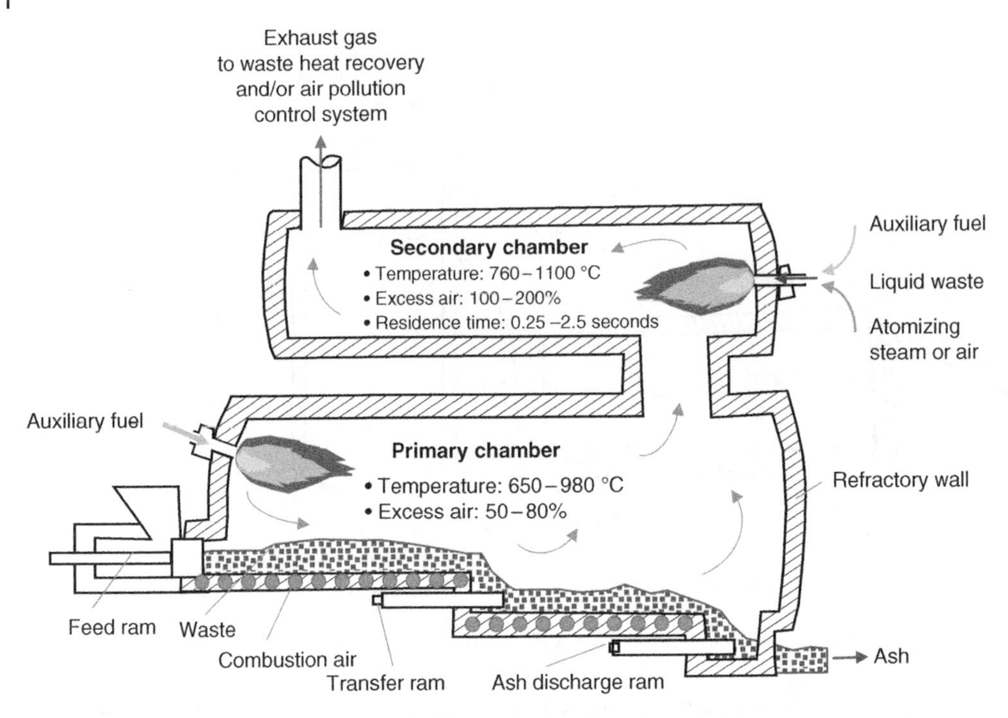

Figure 10.4 A typical fixed hearth incinerator. *Source:* Adapted from Dempsey and Oppelt (1993).

Fixed hearth units tend to be of smaller capacity than liquid injection or rotary kiln incinerators because of physical limitations in ram-feeding and transporting large amounts of waste material through the combustion chamber. These lower relative capital costs and potentially reduced particulate control requirements make them more attractive than rotary kilns for smaller on-site installations.

Fluidized bed incinerators may be either bubbling or circulating bed designs. Both types consist of a single refractory-lined combustion vessel partially filled with particles of sand, alumina, sodium carbonate, or other materials. Combustion air is supplied through a distributor plate at the base of the combustor (Figure 10.5) at a rate sufficient to fluidize (bubbling bed) or circulate the bed material (circulating bed). In the circulating bed design, air velocities are higher and the solids are blown overhead, separated in a cyclone, and returned to the combustion chamber. Operating conditions are normally maintained in the range of 760–1100 °C and excess air requirements from 100 to 150%.

Fluidized bed incinerators are primarily used for sludges or shredded solid materials. To allow for good distribution of waste within the bed and removal of solid residues from the bed, all solids generally require prescreening or crushing to a size less than 2 inch in diameter. Fluidized bed incinerators offer several advantages, such as high gas to solids ratios, high heat transfer efficiencies, high turbulence in both gas and solid phases, uniform temperatures throughout the bed, and the potential for *in situ* acid gas neutralization by lime or carbonate addition. However, fluidized beds also have the potential for solids agglomeration in the bed if salts are present in waste feeds and may have a low residence time for fine particulates.

10.1.3 Design Considerations for Incineration

Some important design considerations and calculations for incineration are presented here. More detailed calculation examples can be found in references (e.g. Lee and Lin 2007).

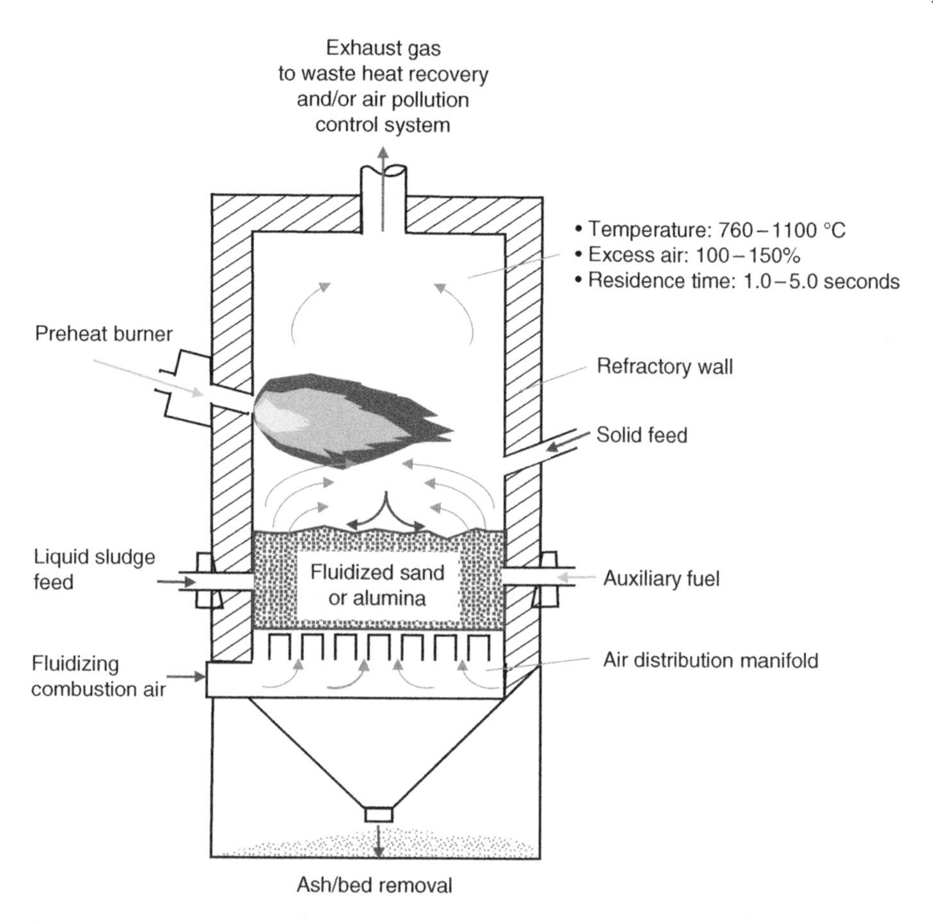

Figure 10.5 A typical fluidized-bed combustion chamber. *Source:* Adapted from Dempsey and Oppelt (1993).

10.1.3.1 Incinerator Size and Dimensions

For vapor incinerators (thermal oxidizer, afterburner), the residence time (t) can be estimated from its assumed first-order rate constant (k) as follows:

$$t = \frac{1}{k}\ln\frac{C_0}{C} \qquad \text{Eq. (10.14)}$$

where C_0 is hazardous chemical concentration in the feed, and C is the exit concentration of the afterburner. For hazardous waste, oftentimes the minimal residence time (t) is specified in the regulations (typically between 0.2 and 2 seconds). If we can determine the total volumetric flow rate (Q) of the flue gas, we can also estimate the incinerator volume according to:

$$V = t \times Q \qquad \text{Eq. (10.15)}$$

For example, if we know the stoichiometric reaction for waste containing C_aH_b (refer to Eq. (10.2), Section 10.1.1):

$$C_aH_b + \left(a + \frac{b}{4}\right)\left(O_2 + 3.77N_2\right) \rightarrow aCO_2 + \frac{b}{2}H_2O + 3.77\left(a + \frac{b}{4}\right)N_2 \qquad \text{Eq. (10.2)}$$

If the flow rate (mol/h) of the contaminant (C_aH_b) is given, then the total number of moles (n) of the flue gas can be calculated using the above stoichiometric relationship:

$$n = n_{CO_2} + n_{H_2O} + n_{N_2} + n_{O_2}(\text{excess}) \qquad \text{Eq. (10.16)}$$

where a 50% excess air is assumed in the above equation. The volumetric flow rate (Q in l/h) can be converted from the flow rate of the contaminants in mol/h (n) using the ideal gas law:

$$Q = \frac{nRT}{P} \qquad \text{Eq. (10.17)}$$

where P = pressure in atm, T = temperature in °K, n = mol/h, R = 0.082 atm-l/mol × °K. The length (L) of the vapor incinerator can be estimated from the gas flow velocity (v) and the residence time (t) as follows:

$$L = v \times t \qquad \text{Eq. (10.18)}$$

The diameter of the vapor incinerator is as follows (Cooper and Alley 2011):

$$d = \sqrt{\frac{4Q}{\pi v}} \qquad \text{Eq. (10.19)}$$

Example 10.8 illustrates the use of some of these above equations in designing an incinerator.

Example 10.8 Design of a thermal oxidizer (afterburner)

An afterburner receives 800 mol/h of propane (C_5H_{12}). The afterburner is operated at 50% excess of air and assumed isothermal condition at 1450 °F (1061 K) and 1 atm at the flue gas exit. The desired residence time is 1.2 seconds and the desired linear velocity is 8.2 ft/s (2.5 m/s). Determine the length and diameter of the afterburner. Assume a length-to-diameter (L/d) ratio of 3.5.

Solution

Write a balanced combustion reaction by referencing Eq. (10.2), where a = 5 and b = 12 for C_5H_{12}. Hence, the balanced reaction is as follows:

$$C_5H_{12} + 8O_2 + 30.2N_2 \rightarrow 5CO_2 + 6H_2O + 30.2N_2$$

At 150% of stoichiometric amount (i.e. 50% excess air), O_2 = 8 × 150% = 12 and N_2 = 12 × 3.77 = 45.12. Thus, the balanced reaction at 50% excess air becomes:

$$C_5H_{12} + 12O_2 + 45.2N_2 \rightarrow 5CO_2 + 6H_2O + 45.2N_2 + 4O_2$$

Using Eq. (10.16), we obtain the total moles of the flue gas from 800 mol/h of C_5H_{12}:

$$n = n_{CO_2} + n_{H_2O} + n_{N_2} + n_{O_2}(\text{excess}) = 800 \times (5 + 6 + 45.2 + 4) = 48\,160\frac{\text{mol}}{\text{h}}$$

Converting 48 160 mol/h into volumetric flow rate in m³/s using Eq. (10.17), we have:

$$Q = \frac{nRT}{P} = \frac{48\,160\frac{\text{mol}}{\text{h}} \times 0.082\frac{\text{atm-L}}{\text{mol-°K}} \times 1\,061\text{°K}}{1\,\text{atm}} = 4.19 \times 10^6\frac{\text{L}}{\text{h}} = 4.19 \times 10^3\frac{\text{m}^3}{\text{h}} = 1.16\frac{\text{m}^3}{\text{s}}$$

The diameter of the afterburner can be calculated from Eq. (10.19):

$$d = \sqrt{\frac{4Q}{\pi v}} = \sqrt{\frac{4 \times 1.16 \frac{m^3}{s}}{\pi \times 2.5 \frac{m}{s}}} = 0.77\,m$$

The length of the reaction chamber is as follows (Eq. 10.18):

$$L = v \times t = 2.5 \frac{m}{s} \times 1.2 s = 3.0\,m$$

If we use the L/d ratio of 3.5, then $L = 3.5 \times 0.77\,m = 2.7\,m$.

10.1.3.2 Factors Affecting Incinerator Performance

These important factors are in three groups: (a) combustion air, (b) combustion temperature, and (c) waste characteristics such as heating values, chlorine content, ultimate composition, and moisture content. In the preceding section, we have discussed about combustion air and heating values. Below are some additional considerations that are important to the design.

a) *Combustion air:* At the theoretical (stoichiometric) air condition, complete combustion is attained. In reality due to a nonideal turbulent mixing condition in the combustion chamber, an excess air at 150–200% of stoichiometric ratio should be provided. The maximum temperature is achieved at the stoichiometric air condition. At the level below the stoichiometric point, some of the organic compounds are not reacted, and pollutants are emitted as a result of incomplete combustion. On the other hand, when air is overly in excess, the temperature in the incinerator drops because energy is used to heat the extra combustion air.
b) *Combustion temperature:* Temperature should be maintained at the level specified by incinerator manufacturers. The minimum temperature (residence time) was set to be >1800 °F (t > 2 seconds) for hazardous waste incinerator, and > 1600 °F (t > 1 second) for hospital (biomedical) waste by the USEPA. High temperature is essential for complete combustion, but excessively high temperature will waste supplemental fuel as well as cause the damage of incinerator. The supplemental fuel required to compensate for the heat loss and to maintain the needed temperature can be determined by the heat balance:

$$\text{Heat Loss} = \text{Heat In}\left(C_aH_b + \text{Air}\right) - \text{Heat Out}\left(\text{Air} + CO_2 + H_2O\right) \qquad \text{Eq. (10.20)}$$

c) *Waste characteristics:* These factors are the heating values, chlorine contents, ultimate composition, and moisture. If the heating value is below the minimally required value of 5000 Btu/lb, auxiliary fuel should be provided to sustain the combustion. A heat content of greater than 2500 Btu/lb is needed for incineration to be applicable for hazardous wastes.

The heating value of the waste is therefore needed to calculate the total heat input to the incinerator according to the following equation:

$$\text{Heat input}\left(\frac{\text{Btu}}{\text{h}}\right) = \text{Feed rate}\left(\frac{\text{lb}}{\text{h}}\right) \times \text{Heating value}\left(\frac{\text{Btu}}{\text{lb}}\right) \qquad \text{Eq. (10.21)}$$

Ultimate composition (C, H, O, Cl, S, etc.) of wastes should be characterized to estimate the stoichiometric air requirement (Examples 10.4–10.6). This information is also important to determine the ash content that need to be disposed of. Special precaution should be exercised for the chlorine content in the waste. Chlorine content in the wastes should be below 1000 ppm halogens since the production of HCl is corrosive to the downstream components of the incinerator. A scrubbing system should be installed to remove the corrosive HCl. Finally, wastes high in moisture are detrimental because it uses energy, thereby reducing the temperature of incinerator. The water vapor will pass through the incinerator system, increasing the gas flow rate which in turn reduces combustion gas residence time.

10.1.4 Regulatory and Siting Considerations

Incineration has received its regulatory acceptance with its technologically sound performance on the disposal of solid and hazardous wastes. In the United States, several laws enforced its use, such as RCRA (subtitle C for hazardous waste, subtitle D for municipal waste, subtitle J for medical waste), TSCA, CWA (for sludge), CERCLA (for Superfund wastes), SARA (for extremely hazardous substances), and FIFRA (for pesticides). In particular, the 1984 amendments and reauthorization of RCRA, termed the Hazardous and Solid Waste Act of 1984 (HSWA), established a strict timeline for the land disposal of untreated hazardous wastes. Regulatory standards for incinerators have been specified regarding the performance (e.g. requirements for acceptable levels of combustion efficiency, destruction efficiency, halogen removal efficiency, and an emission limit for particulate matter), as well as operation (e.g. requirements of semicontinuous monitoring of process variables such as CO, specific minimum temperature, and combustion gas residence time).

The RCRA performance standards require that in order for an incinerator facility to receive a RCRA permit, it must attain the designated performance levels. As stated in Section 10.1.1., RCRA requires 99.99% destruction and removal efficiency (DRE) for each principal organic hazardous constituent (POHC) in the waste feed. The so-called "dioxin rule" further requires all "certified" incinerators achieve 99.9999% DRE for chlorinated dioxins or similar compounds (certain chlorinated dibenzo-*p*-dioxins, chlorinated dibenzofurans, and chlorinated phenols).

The TSCA also requires a DRE of 99.9999% for PCBs and PCDD/F. For liquid wastes, TSCA requires 1200 ± 100 °C, 2 seconds dewell time, 3% excess oxygen in stackgas; or 1600 ± 100 °C, 1.5 seconds dewell time, 2% excess oxygen in stackgas. For nonliquid wastes, TSCA requires mass emission to be less than 0.001 g POHC emitted/kg POHC introduced.

One of the major barriers to increased incineration capacity is public opposition to the permitting and siting of new facilities, especially for the off-site commercial incinerator necessary to handle the increasing amount of the solid and hazardous wastes. Public opposition to the permitting of new thermal destruction operations has been strong for many years. The normal time required for permitting new incineration facilities is three years (Dempsey and Oppelt 1993). On-site facilities that directly serve a single waste generator have greater public acceptance than off-site commercial incinerators that serve multiple generators in a large market area. Off-site facilities cannot often be justified by the sufficient economic benefits to offset the risks associated with the introduction of wastes from other areas to the local community. On-site facilities are more clearly perceived as being linked to business opportunities that are important to the local economy, and are generally not perceived as being importers of hazardous wastes.

10.2 Thermally Enhanced Technologies

In principle, thermally enhanced remediation is totally different from thermal destruction as described above. Under much lower-temperature conditions in the thermal enhancement, chemical contaminants are not structurally modified. Instead, their physicochemical and/or biological properties are modified to favor their removal by a subsequent technology, such as P&T, SVE, air sparing, or bioremediation. Strictly speaking, these thermally enhanced technologies belong to each of these remediation technologies as we described in the previous chapters. In this section, we first discuss the principles of thermally enhanced contaminated removal by examining how temperature affects physicochemical and biological parameters of contaminants. Heat transfer mechanisms relevant in soil and the engineering approaches of applying heat in subsurface will then be introduced.

10.2.1 Temperature Effects on Physicochemical and Biological Properties

The temperature-dependent physicochemical and biological properties include, but are not limited to, density, viscosity, solubility, vapor pressure, Henry's law constant, sorption coefficient, diffusion coefficient, and biodegradation rate.

Generally, when a compound is heated, its density is reduced, its vapor pressure is increased, its adsorption onto soil organic matter is decreased, and its molecular diffusion in the aqueous and gaseous phase is increased. The viscosity of a *liquid* will decrease as the temperature is increased, but the viscosity of *gases* will increase with temperature. The biodegradation rates should be increased as long as the temperature remains in the range that the growth and reproduction of bacteria are not adversely affected. These are the general rules, but exceptions sometimes exist (a notable exception is water, which has the maximum density at 4 °C). The net effects of heat on the enhanced recovery of a particular contaminant depend mostly on the properties of the contaminant and the dominant mechanism that limits the removal rate of the contaminant in the particular circumstance.

In theory, the temperature dependence can be derived from Clausius–Clapeyron equation that relates vapor pressures (P) to temperatures (see Eq. 8.2, Chapter 8), or van't Hoff equation that relates equilibrium constant (K) to temperature (Mackay et al. 2006).

$$\ln P = C + \frac{\Delta H}{R}\frac{1}{T} \qquad \text{Eq. (10.22)}$$

where P can be any physicochemical or biological property as described above, R = universal gas constant, T = temperature in Kelvin, and C = constant. ΔH is the enthalpy of phase change, such as the heat of evaporation from pure state for vapor pressure, heat of dissolution from pure state into water for aqueous solubility, or the enthalpy change from air–water transition in the case of Henry's law constant. As described below, the use of Eq. (10.22) will lead to a linear correlation between ln P and $1/T$ (or more commonly $1000/T$).

Temperature Effects on Viscosity: Liquid chemicals generally expand approximately 0.1% per degree Celsius. Thus, increasing the temperature by 100 °C will increase the liquid volume by approximately 10%. The expansion of liquids with increasing temperature causes a reduction in the interaction between molecules, and thus a reduction in its viscosity. Common liquid organic chemicals can generally be assumed to have about 1% reduction in viscosity per degree Celsius temperature increase. Thus, the higher the viscosity of the liquid at ambient temperatures, the greater the reduction in viscosity as the temperature is increased. The reduced viscosity in response to increasing temperatures is provided in Figure 10.6 for several petroleum products in reference to water. As can also be seen, the experimental data on the temperature-dependent viscosity of liquids fit well to the linear relationship between logarithm of viscosity and $1/T$ (or $1000/T$ shown in Eq. 10.22). From the contaminant remediation perspective, a reduced viscosity will increase the mobility and potential to be pumped out from the subsurface.

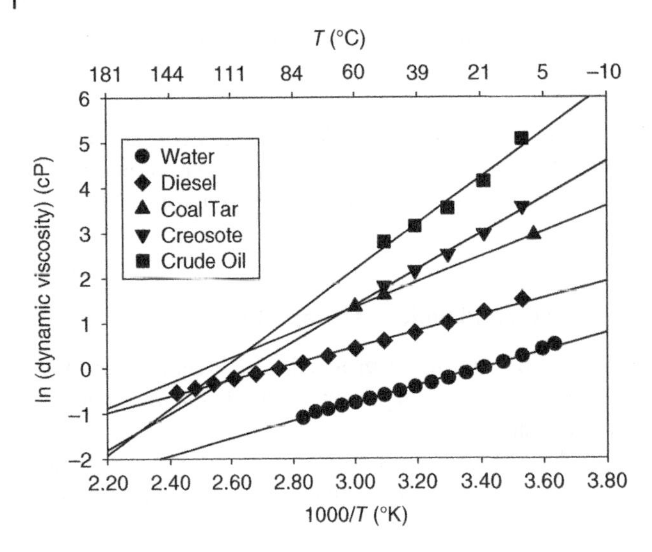

Figure 10.6 Temperature effects on the dynamic viscosity of water and several petroleum products (diesel, coal tar, creosote, and crude oil). Data sources: Esteban et al. (2012); USEPA 1997, EPA/540/S-97/502.

For gaseous compounds, since the volume of a gas is directly proportional to the temperature given in Kelvin, a 100 °C increase in temperature, e.g., a temperature change from 25 to 125 °C which is equivalent to temperature increase in Kelvin = (398 − 298)/298 ≈ 30%, will cause approximately a 30% increase in the gas volume. The viscosity of gases at ambient temperatures is approximately one to two orders of magnitude lower than the viscosity of liquids. However, the increase in the velocity of gas molecules with temperature is such that it causes greater interaction between molecules as the temperature increases, causing an increase in viscosity with temperature. This increase is proportional to the temperature in degrees Kelvin, so that a 100 °C increase in temperature will increase the viscosity of a gas by approximately 30%.

Temperature Effects on Solubility: The effect of temperature on solubility is dependent on the chemical. Increasing temperature will reduce the water–water, water–solute, and solute–solute interactions, so the net effect of temperature on solubility will vary depending on which interactions are affected to the greatest extent (Yalkowsky and Banerjee 1992). Most liquid chemicals such as TNT, RDX, and chlorobenzenes show increasing solubility with temperature (Figure10.7a), whereas others (e.g. MTBE) show decreasing solubility with temperature (Figure 10.7b). Maximum or minimum solubilities with temperature have also been observed for some chemicals. For example, toluene and ethylbenzene exhibit minima in solubility at about room temperature (Figure 10.7c). Regardless, the change in solubility of common organic chemicals in the ambient temperature range is generally less than a factor of 2.

Note also that the liner relationships shown in Figure 10.7 reflect the van't Hoff plots. Specifically, Eq. (10.22) can be rewritten as follows:

$$\ln S = C + \frac{\Delta H}{R}\frac{1}{T} \qquad \text{Eq. (10.23)}$$

where S is the aqueous solubility (e.g. mol/l) and ΔH is the enthalpy of solution (kJ/mol). This linear relationship is analogous to the temperature dependence of viscosity.

Temperature Effects on Vapor Pressure: Vapor pressures always increase with temperature. For the organics having a boiling point of less than 100 °C, the vapor pressure increases by a factor of 5–7 as the temperature increases from 10 to 50 °C. For compounds having a boiling point greater than 100 °C, their vapor pressure generally increases by a factor of 40–50 by raising the temperature from 10 to 100 °C. Limited data on the desorption of organics from soils

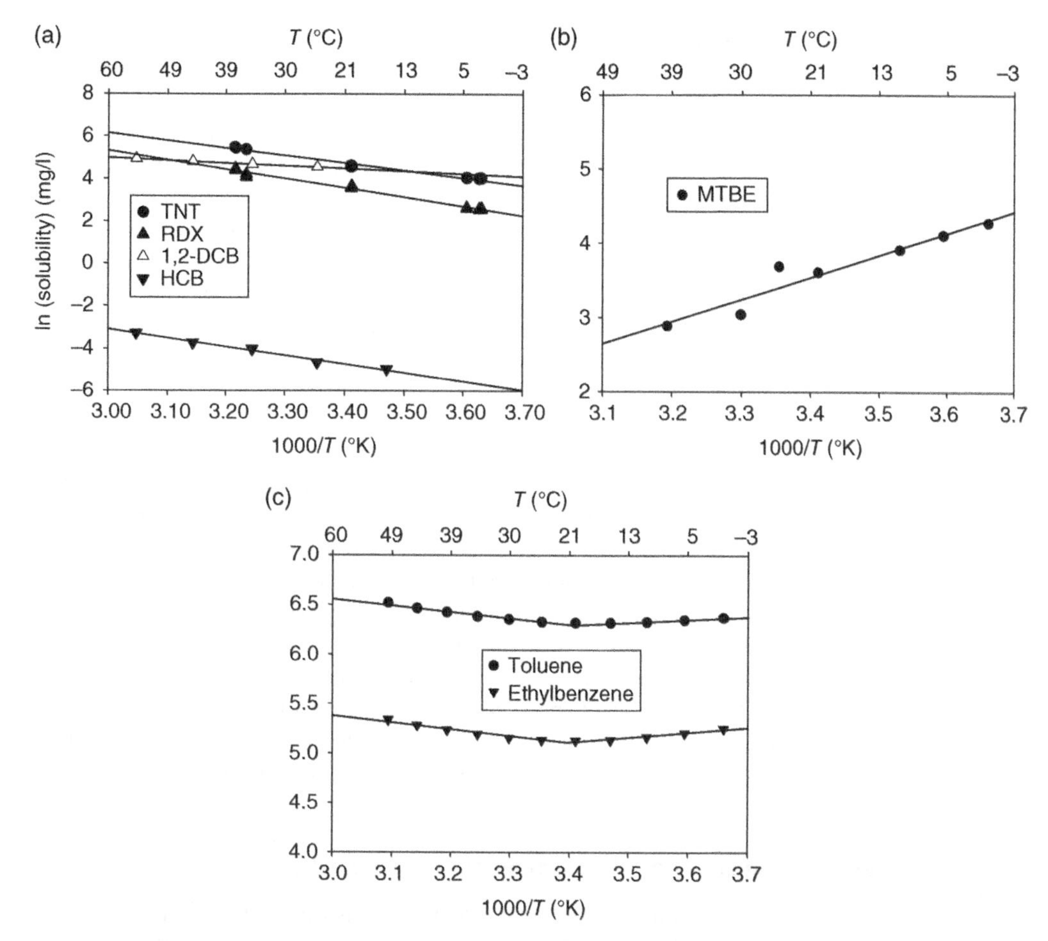

Figure 10.7 Temperature effects on solubility: (a) TNT, RDX, 1, 2-dichlorobenzene (1,2-DCB), hexachlorobenzene (HCB) (b) MTBE (c) Toluene and ethylbenzene. Data sources: Lynch et al. (2001), Horst et al. (2018), Fischer et al. (2004), Oleszek-Kudlak et al. (2004), and Sawamura et al. (2001).

show that the exponential increase in the vapor pressure with temperature also holds when the organic chemical is presentin soils.

As we have discussed in Chapter 8 on soil vapor extraction, the quantitative relationship describing the dependence of vapor pressure on temperature is the Clausius–Clapeyron equation (see Eq. 8.2, Chapter 8). Because vapor pressure is the determining factor for the success of soil vapor extraction, temperature control through hot water, air, and steam is an effective means to improve the performance of SVE for the removal of some semivolatile compounds with low volatility.

Temperature Effects on Henry's Constant: Recall from Eq. (2.3) in Chapter 2 that Henry's constant (H) is the ratio of vapor pressure to solubility. The combination of only small changes in solubility with temperature but large increases in vapor pressure results in an increased Henry's constant as a function of temperature. The temperature dependency of Henry's law constant can also be described by the van't Hoff equation (Eq. 10.22). However, very limited measurements of Henry's constants for chemicals of environmental concern as a function of temperature have been made, and most of these measurements are over a limited temperature range. Heron et al. (1998a) calculated and measured H values for TCE as a function of temperature and found an order of magnitude increase when the temperature was raised from 10 to 95 °C. This very dramatic temperature effect on the Henry's law, hence the large effects on volatility, is expected during thermally enhanced remediation. For the more soluble compounds such as dichloromethane or 2-butanone, and the water-miscible compounds such as acetone and methanol, H values may not be influenced significantly by a temperature increase.

Temperature Effects on Adsorption/Desorption: Adsorption is generally exothermic (heat releasing), thus adsorption will decrease as the temperature increases (Le Châtelier's principle, see Box 10.2). There are exceptions of some systems in which adsorption may increase with temperature increase over a narrow temperature range. The magnitude of the effect of temperature depends on the particular chemical, the soil, and the water content, as these factors will determine the mechanism causing the adsorption. Based on heats of sorption, adsorption from the aqueous phase onto soils can be expected to decrease by a factor of approximately 2.2 as the temperature is increased from 20 to 90 °C. Adsorption from the vapor phase onto dry soils generally has a larger heat of sorption, which leads to an even greater influence of temperature on the adsorption process. For example, TCE had approximately an order of magnitude decreased adsorption onto dry soil as the temperature was increased from 20 to 90 °C (Heron et al. 1998b).

Temperature Effects on Diffusion: Experimental data have shown that the diffusion coefficient (refer to Fick's second law, Eq. 8.8, Chapter 8) in liquids is proportional to temperature in degrees Kelvin. Increasing the temperature from 10 to 100 °C will increase the diffusion of a solute in the aqueous phase by approximately 30%. The diffusion coefficient for gases is also dependent on temperature. Observation of the theoretical equation for diffusivity in the gas phase developed for mixtures of nonpolar gases or of a polar with a nonpolar gas shows that the diffusivity varies almost as $T^{3/2}$ (Treybal 1980). Increasing the temperature from 10 to 100 °C

Box 10.2 Le Châtelier's principle: How does temperature affect equilibrium

In a general sense, Le Châtelier's principle states that "when a stress is applied to a system at equilibrium, the equilibrium shifts to relieve the stress." In a chemical system at equilibrium, if it experiences a change in concentration, temperature, or partial pressure, then the equilibrium shifts to counteract the imposed change until a new equilibrium is established.

Le Chatelier's principle can be used to explain many of the observed effects of temperature on chemical equilibrium such as adsorption, desorption, volatilization, and diffusion. For example, increasing temperature generally reduces chemical adsorption on to soils, whereas the opposite is true for the temperature effect on desorption.

Take desorption as an example. Increasing the temperature will increase chemical desorption thereby favoring contaminant removal through thermal enhancement. This is because desorption is generally endothermic, meaning that heat must be added into the system to desorb the chemical from the soil. When the temperature is increased, then heat is added. According to Le Chatelier's principle, the equilibrium must shift from the sorbed state to desorbed state to consume (counteract) some of the added heat. This will increase the amount of chemical in the desorbed state and will increase the desorption of this chemical.

More important is the thermal desorption process from the remediation standpoint as compared to the adsorption process. This is particularly true for hydrophobic organic contaminants such as PAHs and PCBs, because desorption can always become the rate-limiting step in soil remediation. A large fraction of PCBs remains adsorbed to the soil at ambient temperatures, but significantly higher temperatures (300–400°C) may be required for desorption to occur (Uzgiris et al. 1995).

will increase diffusion in the air phase by approximately 50%, while increasing the temperature from 10 to 300 °C will increase diffusion by approximately 200%.

Temperature Effects on Biodegradation Rate: It is generally assumed that biochemical reaction velocity including biodegradation rate constant (k_b) increases two to three times for every 10 °C rise in the normal temperature range. **Arrhenius equation,** $k = C\exp(-E/RT)$, describes how temperature affects the rate constants for both chemical and biological reactions. The logarithmic form of the Arrhenius equation results in the following equation analogous to Eqs. (10.22) and (10.23) for temperature-dependent biodegradation rate (Thompson et al. 2011):

$$\ln k_b = C + \frac{E}{R}\frac{1}{T} \qquad \text{Eq. (10.24)}$$

where k_b is the biodegradation rate constant at temperature T, E is the average activation energy (kJ/mol) in the temperature range, C is a constant, and R is the ideal gas law constant. An example plot of the experimental data on the temperature effects on biodegradation rates of PCE and *tert*-butyl alcohol (TBA) is shown in Figure 10.8. These results show that bioremediation in low-temperature groundwater (~5 °C) in the winter may remain inactive in shallow aquifer systems (Greenwood et al. 2007), unless the predominant microbial consortia are shifted from mesophilic to psychrophilic with an optimum operating temperature ranges of 20–40 and 0–20 °C, respectively.

As a summary of the above analysis regarding temperature effects, it becomes clear that essentially all of these changes with temperature can aid in the recovery of contaminants from the subsurface. The thermal expansion of a liquid with its accompanying decrease in viscosity will allow the heated liquid to flow more readily. For gases, the expansion with temperature will be largely offset by the increase in viscosity. However, since the viscosity of gases is approximately two orders of magnitude lower than the viscosity of liquids, conversion of a liquid to a gas will greatly increase its mobility. The expansion itself will facilitate the movement of the fluids out of the pore space, with the greatest effects coming from the vaporization of a liquid to a gas. The increased diffusion of contaminants as the temperature increases in both the aqueous and the gaseous phases will help to move contaminants from areas of low permeability to areas of high permeability and speed their recovery.

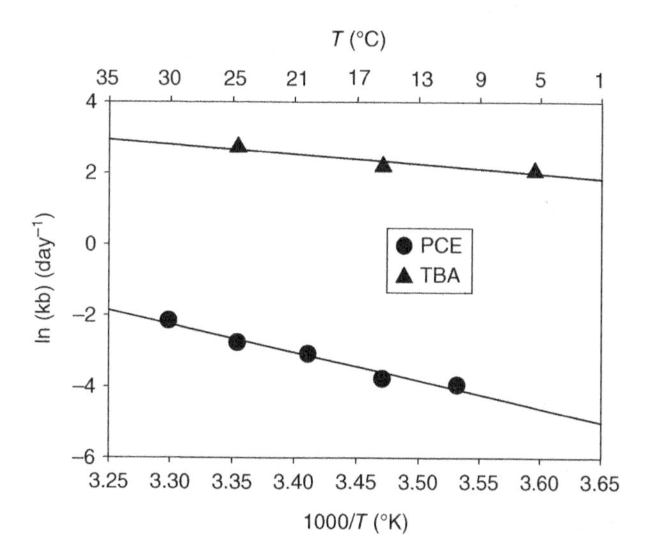

Figure 10.8 Temperature dependence of biodegradation rates of tetrachloroethylene (PCE) and *tert*-butyl alcohol (TBA). Data sources: Yagi et al. (1992) and Greenwood et al. (2007).

10.2.2 Heat Transfer Mechanisms in Soil and Groundwater

Increased subsurface temperature can be accomplished through one of the four technologies, hot air injection, steam injection, electro-heating, and radio frequency heating. Their applications are as follows: (i) Hot air injection for the removal of volatiles (gasoline, solvents, and jet fuel); (ii) Steam injection for the removal of semivolatiles and nonvolatiles (diesel fuel, heavy oils); (iii) Electro-heating for the removal of semivolatiles and nonvolatiles (diesel fuel, heavy oils) in low-permeability clays and silts; (iv) Radio frequency (RF) heating for the removal of semivolatiles and nonvolatiles (diesel fuel, heavy oils) at low temperatures (<100 °C) and removal of nonvolatiles (certain PAHs, PCBs, and pesticides) at high temperatures (100–400 °C).

Heat is transferred through one or more of the following three mechanisms: (i) convection, (ii) conduction, and (iii) radiation. In soil and groundwater, heat transfer by **radiation** should not be relevant because radiation is the heat transfer through space by electromagnetic waves without the requirement for an intervening media. Additional explanations on heat transfer by convection and conduction in soil are further described as follows.

Heat Transfer by Convection: **Convection** is the transfer of heat by the mixing motion of one fluid with another or between a surface and an adjacent fluid. There are two modes of convective heat transfer. In a **forced convection**, thermal energy is transferred from a hot fluid moving through the soil matrix under the influence of a blower or a compressor. The heat is transferred from the fluid by contact to the surrounding soil grains, pore water, and soil vapor. In a **natural convection**, the result of motion is caused by variation in density caused by differential temperature under the influence of gravity. The convective *heat transfer* is analogous to *mass transfer* which is dependent on the concentration gradient in place of temperature gradient. Mathematically, the convection can be described as follows:

$$\frac{dq}{dA} = -h_c\left(T_2 - T_1\right) \qquad \text{Eq. (10.25)}$$

where dq/dA = heat flux per unit area (W/m^2), A = area perpendicular to flow direction, h_c = heat transfer coefficient (W/m^2-K), T_1 = ambient temperature, and T_2 = temperature at the interface of soil surface and liquid media. Equation (10.25) indicates that the rate of convective heat transfer depends on the temperature difference as well as the heat conducting medium (i.e. soil and groundwater) through which heat flows.

Heat Transfer by Conduction: **Conduction** is the transfer of heat by means of a temperature gradient without the displacement of adjacent matter. Conductive heating is analogous to diffusion and can be described by an equation similar to Fick's second law for the mass transfer (Eq. 8.8, Chapter 8). Heat is transferred from the high-temperature region to the low-temperature region of the soil. The mathematical equation describing the conductive heat transfer is **Fourier's law:**

$$\frac{dq}{dA} = -K\frac{dT}{dL} \qquad \text{Eq. (10.26)}$$

where K = thermal conductivity (W/m-K), T = temperature, and L = distance. The above equation is also analogous to the Darcy's law (Eq. 2.17, Chapter 2) for the flow of fluid under a hydraulic gradient.

10.2.3 Required Heat-Up Time and Radius of Influence

Heat capacity is the required change in heat content of a unit volume of soil per unit change in temperature, and has a unit of cal/(cm^3 × °C). As soil is a mixture of various minerals, organics, moisture, and gaseous compounds, the heat capacity of soils (C_p) is a function of the heat capacity of all of these components in the soil system. Soil heat capacity (C_p) can be described by

$$C_p = f_1C_1 + f_2C_2 + f_3C_3 + f_4C_4 \quad \text{Eq. (10.27)}$$

where f = volumetric fraction of each soil componentand the subscripts 1, 2, 3, and 4 denote soil mineral, soil organic, soil water, and soil gas, respectively. The heat capacity due to soil gas f_4C_4 can be neglected.

The time needed to heat up soil to a target temperature through steam or hot water can be derived from the heat balance between the heat adsorbed by soil (Q) and the heat released from steam (Q'):

$$Q = C_p \times (V \times \rho) \times \Delta T \quad \text{Eq. (10.28)}$$

$$Q' = \Delta H_v \times (1 - f) \times m \times t \quad \text{Eq. (10.29)}$$

where C_p = heat capacity of soil (Btu/lb-F), $V \times \rho$ = volume of heated soil × soil density, ΔT = the difference between the target temperature and the ambient temperature (°F), ΔH_v = heat of vaporization of water (970 Btu/lb), f = heat loss fraction, m = mass of steam injection rate (lb/h), and t = heat-up time (hours) using steam. By equating Eqs. (10.28) and (10.29) (i.e. $Q = Q'$), we can solve for the required heat-up time:

$$t = \frac{C_p \times (V \times \rho) \times \Delta T}{\Delta H_v \times (1 - f) \times m} \quad \text{Eq. (10.30)}$$

If we substitute $\pi r^2 h$ for soil volume (V), then the radius of steam influence (r) can be estimated from the equation below when steam is injected through an injection well:

$$r = \sqrt{\frac{\Delta H_v \times (1 - f) \times m \times t}{\pi h C_p \times \rho \times \Delta T}} \quad \text{Eq. (10.31)}$$

where h is the depth of heated soil zone, and all other parameters are the same as described previously.

10.2.4 Use of Hot Air, Steam, Hot Water, and Electro-Heating

Generally, three groups of methods can be used to inject or apply heat to the subsurface to enhance remediation: (a) injection of hot gases such as steam or air, (b) injection of hot water, and (c) electrical or electromagnetic energy heating. All of these methods were first developed by the petroleum industry for enhanced oil recovery, and have more recently been adapted to soil and aquifer remediation applications with various degrees of success, particularly steam injection, electrical resistance heating, and thermal conductive heating (also termed *in situ* thermal desorption).

10.2.4.1 Hot Air, Steam, Hot Water, and Electro-Heating

Hot air at temperature as high as 425 °C (800 °F) can be obtained from a heat exchanger in thermal oxidizer exhaust stack or a wellhead heater. When hot air is injected into vadose zone, soil temperature can be raised to 65–80 °C (150–180 °F). Enhanced remediation is achieved by the increased vapor pressure, decreased viscosity, or decreased soil interfacial tension. When hot air is combined with soil vapor extraction (see Figure 8.1), it enhances the efficiency of vapor removal. Hot air can be supplied at a very high temperature, but its field use is limited because of the very low heat capacity of air (~1 kJ/kg °C). For situations that require dried soil, hot air injection may be complementary to recover contaminants.

The main mechanism for enhanced recovery using **hot water** is generally through a reduced viscosity and interfacial tension. Improved permeability and reduced residual saturation may also aid in the recovery of nonvolatile contaminants. Because hot water does not vaporize contaminants, only liquid recovery systems are needed without the use of a vapor recovery system (CDM Smith 2017).

Steam-enhanced extraction (SEE) uses steam (temperature at 100 °C or 212 °F) from an on-site generator or a portable trailer. It is injected into the vadose and saturated zones to heat the soils up to 93–104 °C (200–220 °F) by forced convection (Figure 10.9). Steam enhances contaminant recovery by increasing SVOCs' vapor pressure and decreasing viscosity, interfacial tension, and residual saturation. Unlike hot air, steam has a heat capacity of approximately four times that of air (~4 kJ/kg °C), and the heat of evaporation of more than 2000 kJ/kg. Steam can thus be used more effectively to heat soils and aquifers. The drawback of steam is the residual water from condensed steam, which leaves water-soluble contaminants at high concentrations in this residual water. This issue can be resolved by co-injecting air with steam; for example, DNAPL vapor from steam injection could remain suspended in the air, preventing DNAPL condensation and downward migration (Kaslusky and Udell 2005).

Hot air, steam, and hot water differ in the applicable contaminant phases, their volatilities, etc. As summarized in Table 10.3, hot air injection has been used to recover contaminants only in the *vapor phase*, whereas steam injection will displace mobile contaminants in front of the steam as well as vaporize volatile residual contaminants, and therefore can recover volatile contaminants in *both the liquid phase and the vapor phase*. **Hot water** injection generally recovers contaminants only in the *liquid phase*. In terms of volatilities, hot air is applicable to VOCs and SVOCs that are water soluble, whereas steam injection is applicable to VOCs and SVOCs that are immiscible with water. In comparison, hot water injection is applicable for contaminants that have low volatility and very low solubility in water.

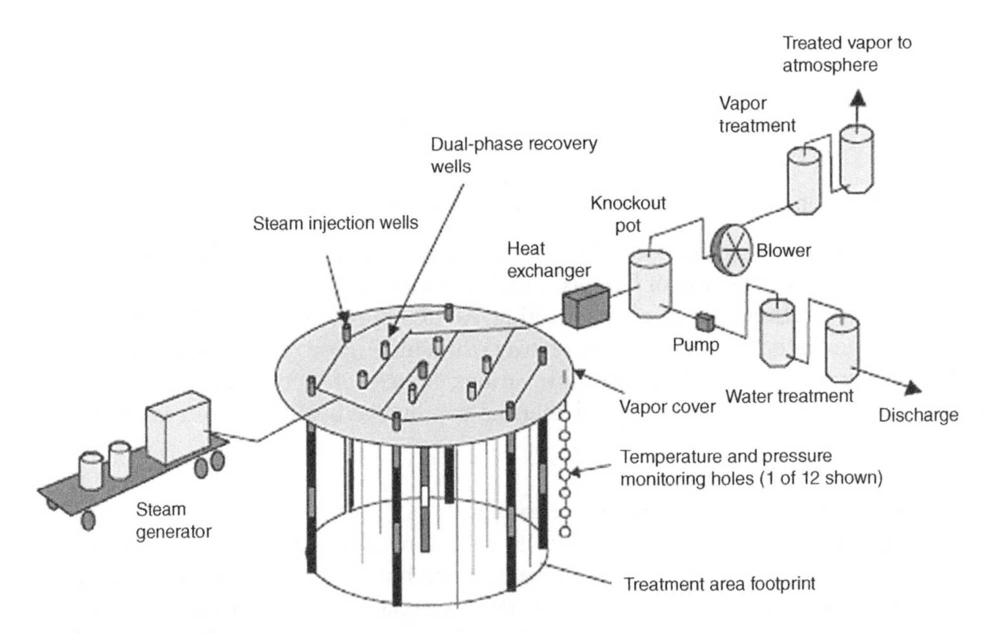

Figure 10.9 Schematics of steam-enhanced extraction. *Source:* Environmental Security Technology Certification Program (2009).

Table 10.3 Applicability of hot air, steam, and hot water in remediation.

	Hot air	Steam	Hot water
Contaminant phase:			
Vapor	A	A	N
Liquid	N	A	A
Contaminant type:			
VOCs	A[a]	A[b]	N
SVOCs	A[a]	A[b]	N
NVOCs	N	N	A[c]

A = Applicable; N = Not applicable
[a] Water-soluble VOCs and SVOCs.
[b] Water-immiscible VOCs and SVOCs.
[c] NAPLs having low volatility and very low solubility in water.

Hot water injection is most likely effective only when the *nonaqueous phase* is present in quantities greater than the residual saturation, as the main recovery mechanism is the physical displacement of the nonaqueous phase. Hot water injection may be most effective for light contaminants (e.g., LNAPLs) that are floating on top of the water table, as the lower-density hot water has a tendency to rise if injected below the water table. For contaminants that are more dense than water at ambient temperatures but less dense than water at the displacement temperature, heating of the subsurface by hot water injection may help to float these oils, which may aid in their recovery. Steam injection has a definite advantage over hot water injection when the contaminants have a low boiling point and are present as an immiscible phase, and thus can be steam distilled at the temperatures achieved by steam injection.

Electrical resistance heating (ERH) uses an electrical current to heat less permeable soils such as clays and fine-grained sediments. Electrodes are placed typically 3–7 m (14–24 ft) apart directly into the less permeable soil matrix and activated so that electrical current passes through the soil, creating a resistance which then heats the soil (saturated or unsaturated zone). These electrodes are typically constructed of steel pipe or copper plates, and installed in a similar manner as the installation of monitoring wells. When the temperature reaches the boiling point of water, steam is generated which strips contaminants from the soils and enables them to be extracted from the subsurface by SVE. The heat also causes fracture, which make the soil more permeable for an enhanced removal of contaminants through SVE.

The laboratory results of two-dimensional temperature and contaminant (TCE) profiles in a homogeneous silty soil tank are shown in Figure 10.10. Since ERH does not rely on advective flow of groundwater, it works well in heterogeneous and low-permeability subsurface with silts and clays (Beyke and Fleming 2005). ERH has proven success for the cleanup of highly impacted sites, including those with DNAPL and LNAPL sources, heterogeneous lithologies, and sites with low-permeability silt and clays (Kingston et al. 2012). ERH technology can be operated under buildings with limited interference from the business operations or the public (McGee 2003).

Thermal conduction heating (TCH) resembles ERH in that both involve simultaneous application of electrically generated heat and vacuum to extract vapor from subsurface soils, followed by the aboveground vapor treatment. In TCH, a heating element is situated inside a non-perforated pipe running down the length of the casing of heater/vacuum wells. Heat is dissipated through an array of vertical heater/vacuum wells and the vapor is captured through these well systems (Figure 10.11). The dissipation of heat to the contaminated region is largely through thermal conduction, i.e. the transfer of heat due to temperature gradients between the

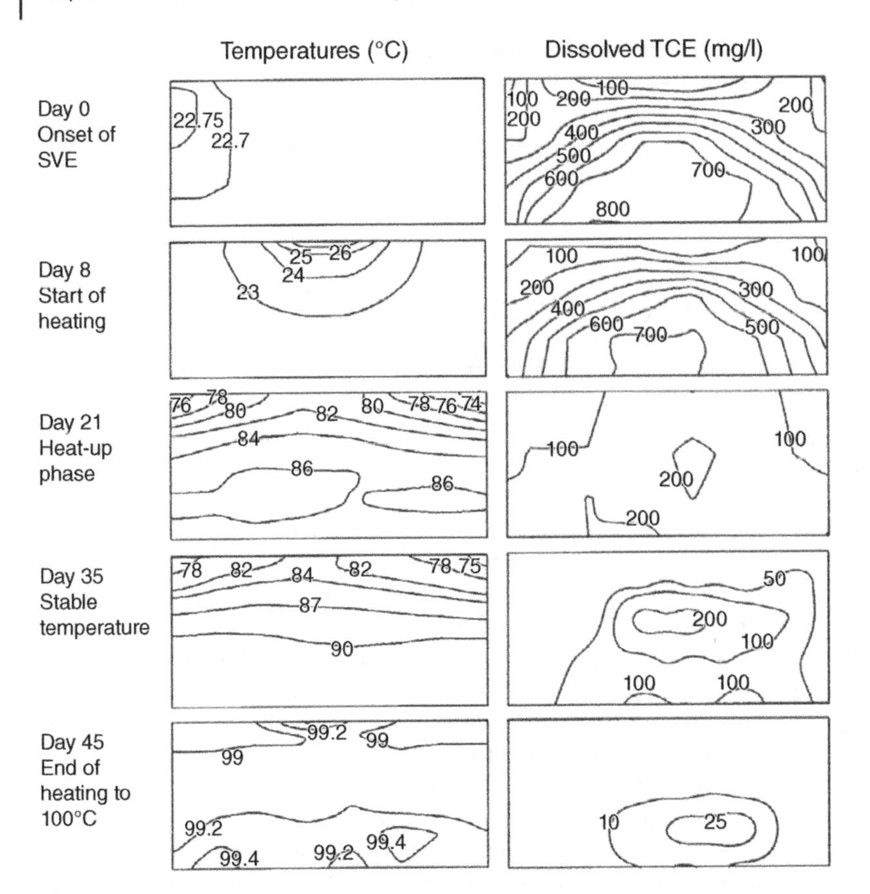

Figure 10.10 Distribution of temperature and dissolved trichloroethylene (TCE) in a two-dimensional electrical heating – soil vapor extraction reactor at selected times. The box was 120 cm wide, 60 cm high, and 12 cm deep. *Source:* Reprinted with permission from Heron et al. (1998b). Copyright (1998) American Chemical Society.

heater wells and the surrounding aquifer. Convective heat transfer may also occur during the early stage of steam formation from pore water. Because TCH can achieve elevated soil temperatures (in excess of 500 °C) compared to the maximum of 100 °C (water boiling point) in the ERH, a significant portion (up to 99%) of organic contaminants either oxidize (if sufficient air is present) or pyrolize once high soil temperatures are achieved in TCH (USEPA 2004).

Due to the different operating temperatures, ERH is suitable for VOC removal and TCH is suitable for SVOC removal. Because of this complementary nature in contaminant removal and the similarities in system components, technology has become available to provide both ERH and TCH in a single borehole for better cost-effective remediation. The major drawback is that both ERH and TCH will not function well when groundwater flow is high (>1 ft/day). High groundwater flow may result in a significant heat loss (USEPA 2014; CDM Smith 2017).

Radio frequency heating (RFH) is an *in situ* process that uses electromagnetic energy to heat soil to over 300 °C. RHF enhances soil vapor extraction (SVE) through a decreased viscosity, increased volatility, and improved permeability by drying. RFH technique works the same way as a microwave oven except at a different frequency (wavelength). RFH affords its ability to penetrate deeply even dense and large-size materials.

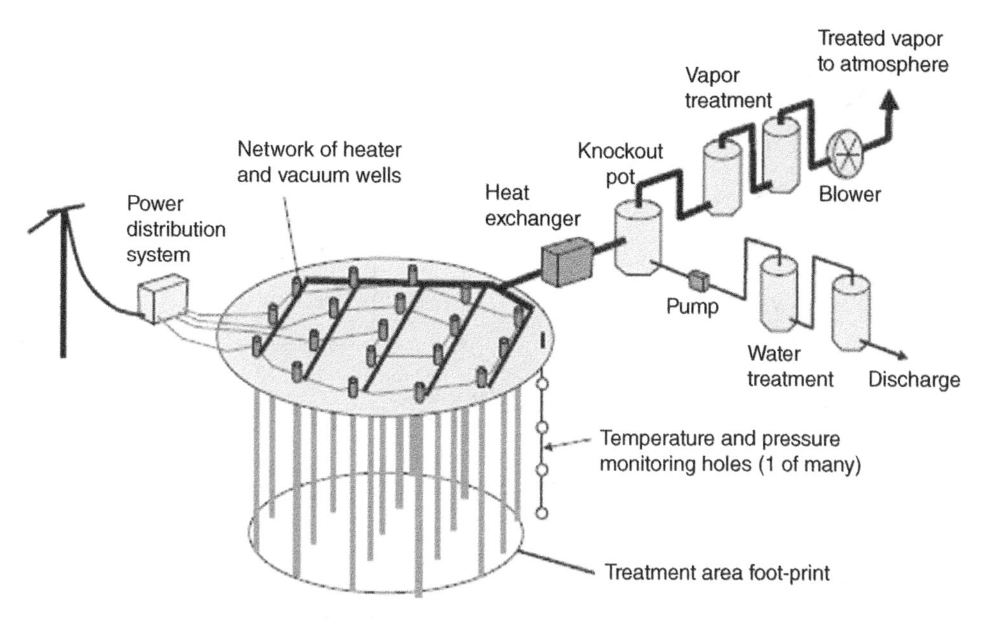

Figure 10.11 Typical thermal conduction heating configuration. *Source:* Environmental Security Technology Certification Program (2009) and USEPA (2014).

When **electromagnetic heating** is used through a downhole antennas, the water in the pore spaces of the soil absorbs essentially all the applied energy, so the evaporation of water limits the transport of energy in the soil and, therefore, limits the heating process. Thus, for the **low-frequency methods**, the boiling point of water (100 °C) is the highest temperature that can be achieved. For semivolatile organic contaminants, the vapor pressure at 100 °C may not be adequate to effectively recover the contaminants. Compared to the low-frequency methods (60 Hz), the **high-frequency methods** use RF energy (3 kHz–300 GHz, analogous to a household microwave with an average 2.45 GHz) from a radio transmitter trailer. The electromagnetic energy is used to heat the vadose zone and top of the saturated zone. The RF energy can be absorbed by the soil itself, and thus there is no fluid injected into the soil. RFH can provide more uniform heating independently from soil condition (Bientinesi et al. 2015). At present, commercial success in the field scale RFH is still limited.

10.2.4.2 Flow Chart to Select Thermal Processes

Each of the above-mentioned thermally enhanced methods is generally applicable only to certain types of contaminated sites. The choice of technique must be based on the characteristics of both the subsurface (permeability, heterogeneity) and the contaminants to be recovered (volatility, adsorption, and solubility). Steam, hot air injection, and electrical heating techniques are generally applicable for VOCs and SVOCs, whereas hot water injection is generally applicable for nonvolatile oils such as automatic transmission fluids, coal tar, creosote, and crude oil. Electrical heating may be favored in low-permeable media and when there is a significant heterogeneity. For highly solublecontaminants, drying the soil may be necessary and, thus, hot air or RF heating may be more applicable. Because desorption can be a slow process, higher temperatures and/or longer remediationtimes may be necessary when adsorption is significant. Figure 10.12 can be used as a quick guide to determine which of the thermally enhanced techniques would likely be applicable for a given situation; in some cases, more than one technique may be applicable.

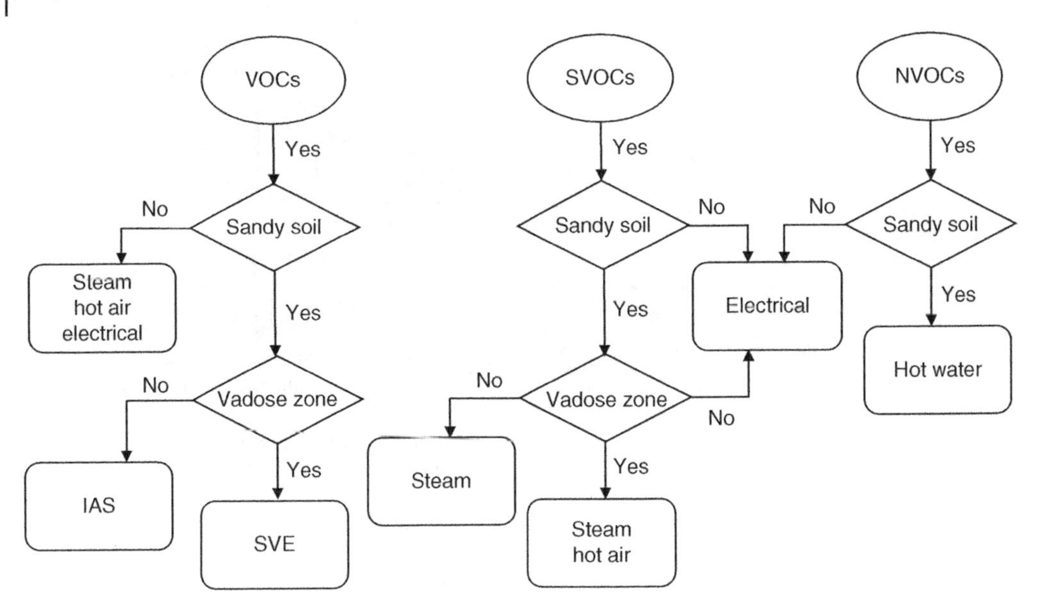

Figure 10.12 Flow chart to indicate which of the thermally enhanced techniques may be applicable for a particular site. *Source:* Adapted from USEPA (1997).

10.3 Vitrification

Vitrification uses electric power to create the heat needed to melt soil in the temperature range of 1590–2010 °C (2900–3650 °F). Soils can be vitrified either *in situ* or *ex situ*. Four rods of graphite electrodes are drilled in the polluted area. Under high voltage (4000 V to start up), an electric current is passed between the electrodes, melting the soil between them (Figure 10.13). Melting starts near the ground surface and moves down. As the soil melts, the electrodes sink further into the ground causing deeper soil to melt. When the power is turned off, the melted soil cools and vitrifies, which means it turns into a solid block of glass-like material and the electrodes become part of the block. This material comprises fused inorganic oxides, notably silica oxides that immobilize nonvolatile metals within the glass. When vitrified, the volume of soil will shrink and the vitrified soil remains on site. As such, clean soil will be needed to fill the sunken area.

Vitrification permanently traps hazardous wastes in a solid block of a glass-like material, which is left in place. This prevents rainfall, groundwater flow, and wind from transporting the chemicals off-site. In the vitrification process, the heat used to melt the soil can also destroy some of the organic contaminants and cause others to evaporate. Organic compounds are pyrolyzed (in the absence of oxygen) or combusted (in the presence of oxygen) during the vitrification. The evaporated chemicals (including small amount of volatile metals such as Pb, Cd, and Zn) rise through the melted soil to the ground surface, and are subsequently collected by a hood that covers the heated area. These chemicals are sent to an off-gas treatment system where they are cleaned up.

The time it takes for *in situ* vitrification to clean up a site depends on several factors, including size and depth of the polluted area, types and amounts of chemicals present, and how wet the soil is (wet soil must be dried, which takes more time). In general, *in situ* vitrification offers faster cleanup times than most methods. The cleanup can take from weeks to months, rather than years. Vitrification has been successfully used at sites where a small amount of inorganic

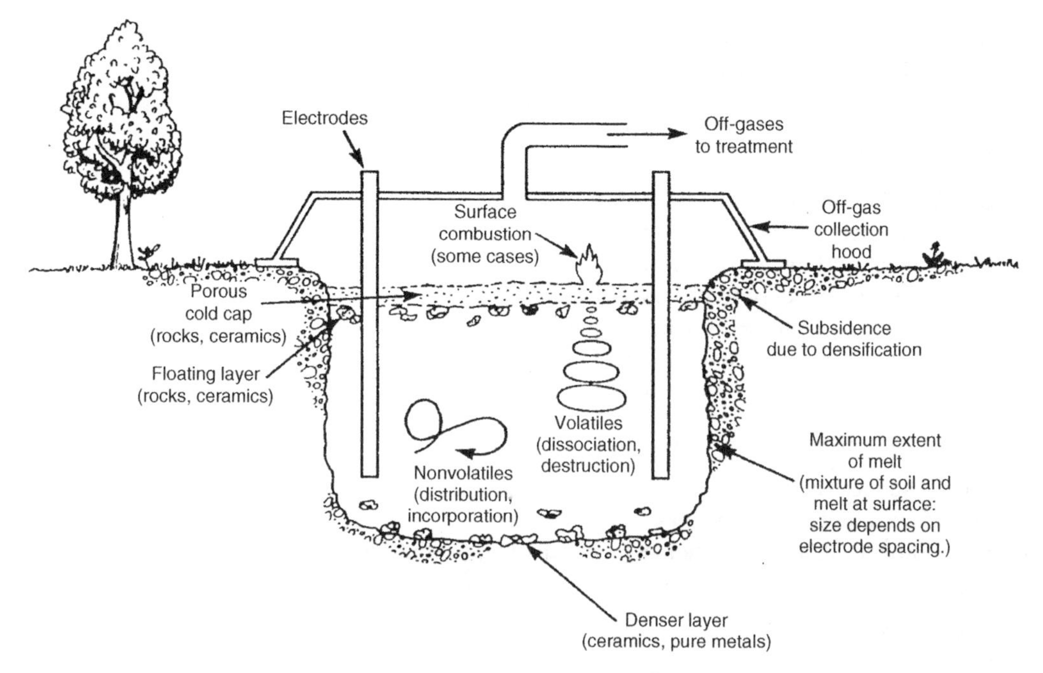

Figure 10.13 Schematic of an *in situ* vitrification process (USEPA 1992).

wastes is of particular concern, such as radioactive wastes, metal sludge, asbestos-containing wastes, or soil and ash contaminated with metals. Vitrification can clean up several types of chemicals and soils. By cleaning up soil in place, it avoids the expense of digging up soil or trucking it to a landfill for disposal. When used properly, vitrification can be quite safe if volatile chemicals released during vitrification process are controlled.

Vitrification solidifies inorganic metals and nuclides. Like incineration, vitrification will also destruct organic contaminants at its extreme high temperature. Both processes require a high input of energy from either an electrical power or a fuel source. On the other hand, with much lower temperature, *in situ* thermal enhancement relies on existing remediation infrastructure such as vapor extraction, flushing, or bioremediation, and the heat source can be acquired through the use of waste heat such as hot air, steam, and water if available on site. Various green remediation practices are possible for many *in situ* thermal enhancement technologies (Box 10.3).

Box 10.3 Green remediation for *in situ* thermal enhancement technologies

In situ thermal systems have been increasingly used to remediate contaminated sites, including Superfund sites, RCRA facilities, brownfield sites, or military installations needing accelerated cleanup for the time frame of months instead of years. Since all *in situ* thermal systems are energy intensive, the maximized use of best management practices is essential.

Opportunities to reduce the environmental footprint of *in situ* thermal applications correlate to the common cleanup phases of design, construction, operation and maintenance, and monitoring (USEPA 2014). The U.S. EPA provides a checklist for such BMPs.

Design Stage:

- Establish a conceptual site model
- Maximize use of high-resolution imagery techniques

- Consider a phased heating approach
- Integrate sources of renewable energy
- Establish a baseline on resource consumption and waste generation

Construction Stage:

- Consider colocating wells with heating equipment
- Choose materials with recycled contents
- Employ direct-push technology wherever feasible
- Screen drill cuttings for potential on-site reuse
- Integrate techniques to lower or buffer noise
- Reclaim treated or clean pumped water for on-site use or return to the aquifer
- Employ cleaner fuels, clean emission technologies, and fuel conservation techniques

Operation and Maintenance Stage:

- Maintain surface seals
- Modify flow rates to meet changing site conditions
- Continuously evaluate the potential for downsizing or shutting down equipment as cleanup progresses

Monitoring Stage:

- Maximize automated and remote monitoring capabilities
- Use field test kits whenever feasible
- Include data collection from areas immediately beyond the target area

In the following cases studies, only thermal enhancement technologies are given, because incineration is a well-established remediation technology. For more case studies, readers can refer to reports by the USEPA (1995a–c).

Case Study 1 *In situ* steam enhancement in Huntington Beach, CA, USA

An *in situ* thermal enhanced recovery process was tested at full-scale at the Rainbow Disposal site in Huntington Beach, CA. A total of 100 000 gal of diesel fuel was leaked from a piping rupture, impacting soil and groundwater to approximately 40 ft below ground. Thirty-five injection wells and 38 extraction wells were arranged in an overlapping pattern of alternating two types of wells over a treatment area of 2.3 acres. The spacing between wells was about 45 ft for wells of opposite type and 60 ft for wells of the same type. The liquid contaminants were gravimetrically removed from the water by an oil/water separator, and the vapor was treated in a thermal oxidizer unit.

Steam injection was conducted over a two-year period between August 1991 and August 1993. Monitoring data revealed that 700 gal of diesel was recovered, and 15 400 gal of diesel was treated by thermal oxidizer unit, accounting for 12–24% of the initial amount of spilled diesel. Steam injection removed a significant amount of spilled diesel, but it did not meet the site cleanup objective of 1000 mg/kg TPH. *In situ* biodegradation by aerobic bacteria was used to reach cleanup goals after termination of the steam injection process. Site closure was issued by the California Regional Water Quality Control Board in 1994.

Case Study 2 Radio frequency heating with SVE, Volk Air National Guard Base, WI, USA

A radio frequency heating was tested for an enhanced soil vapor extraction (SVE) at the Volk Air National Guard Base, Wisconsin. The field test took place at a former fire training pit contaminated with JP-4 jet fuel spilled during routine fire training exercises. The soil was 98% silica sand. Nineteen cubic yards of sandy soil were heated to an average depth of 7 ft. The area of the heated zone was 72 square feet. The soil was heated to a temperature range of 150–160 °C in eight days with a 40 kW RF power source. The treatment temperature was maintained for four days. Vapors rising from the treatment zone were captured and channeled to a vapor treatment system consisting of an air-cooled heat exchanger (for condensation of steam and contaminant vapors) followed by a separator (to remove the condensate from the vapor stream) and carbon adsorbers. At least 99% of the volatile hydrocarbons and 94–99% of the semivolatile hydrocarbons were removed from the site. Hexadecane, with a normal boiling point of 289 °C, was used as a target compound to represent the semivolatile aliphatic compounds. On the average, 83% removal of hexadecane was achieved.

Bibliography

Beyke, G. and Fleming, D. (2005). *In situ* thermal remediation of DNAPL and LNAPL using electrical resistance heating. *Remediation* Summer: 5–22.

Bientinesi, M., Scali, C., and Petarca, L. (2015). Radio frequency heating for oil recovery and soil remediation. *IFAC (International Federation of Automatic Control)-Papers Online* 48-8: 11198–11203.

Smith, C.D.M. (2017). *Draft White Paper on Thermal Remediation Technologies for Treatment of Chlorinated Solvents: Santa Susana Field Laboratory*. California: Simi Valley.

Cooper, C. and Alley, F.C. (2011). *Air Pollution Control: A Design Approach*, 4e. Prospect Heights, IL: Waveland Press, Inc.

Davis, E. (1998). Steam injection for soil and aquifer remediation. *Ground Water Issue* 16, EPA/540/S-97/505.

Dempsey, C.R. and Oppelt, E.T. (1993). Incineration of hazardous waste: a critical review update. *Air and Waste* 43 (1): 25–73.

Environmental Security Technology Certification Program (ESTCP) (2009). State-of-the-practice overview: critical evaluation of state-of-the-art *in situ* thermal treatment technologies for DNAPL source zone treatment. *ESTCP Project* ER-0314.

Esteban, B., Riba, J.-R., Baquero, G. et al. (2012). Temperature dependence of density and viscosity of vegetable oils. *Biomass Bioenergy* 42: 164–171.

Federal Remediation Technologies Roundtable (FRTR)(n.d.), Remediation technologies screening matrix and reference guide, version 4.0, 4.10. Thermal Treatment, http://www.frtr.gov/matrix2/section4/4-9.html. (accessed December 2018).

Fischer, A., Muller, M., and Klasmeier, J. (2004). Determination of Henry's law constant for methyl *tert*-butyl ether (MTBE) at groundwater temperatures. *Chemosphere* 54: 689–694.

Grasso, D. (1993). *Hazardous Waste Site Remediation – Source Control*. Lewis Publishers.

Greenwood, M.H., Sims, R.C., McLean, J.E., and Doucette, W.J. (2007). Temperature effect on tert-butyl alcohol (TBA) biodegradation kinetics in hyporheic zone soils. *BioMedical Eng. OnLine* 6: 34–42.

Heron, G., Christensen, T.H., and Enfield, C.G. (1998a). Henry's law constant for trichloroethylene between 10 and 95 °C. *Environ. Sci. Technol.* 32 (10): 1433–1437.

Heron, G., van Zutphen, M., Christensen, T.H., Enfield, C. G. (1998b), Soil heating for enhanced solvents: a laboratory study on resistive heating and vapor extraction in a silty, low-permeable soil contaminated with trichloroethylene, *Environ. Sci. Technol.*32 (10): 1474–1481.

Horst, J., Flanders, C., Klemmer, M. et al. (2018). Low-temperature thermal remediation: gaining traction as a green remedial alternative. *Groundwater Monitoring & Remediation* 38 (3): 18–27.

Kaslusky, S.F. and Udell, K.S. (2005). Co-injection of air and steam for the prevention of the downward migration of DNAPLs during steam enhanced extraction: an experimental evaluation of optimum injection ratio predictions. *J. Contaminant Hydrol.* 77: 325–347.

Kingston, J.L., Dahlen, P.R., and Johnson, P.C. (2012). Assessment of groundwater quality improvements and mass discharge reductions at five in situ electrical resistance heating remediation sites. *Ground Water Monitoring Remediation* 32 (3): 41–51.

Lee, C.C. and Lin, S.D. (2007). *Handbook of Environmental Engineering Calculations,* 2e. McGraw Hill.

Lynch, J.C., Myers, K.F., Brannon, J.M., and Delfino, J.J. (2001). Effects of pH and temperature on the aqueous solubility dissolution rate of 2,4,6-trinitrotoluene (TNT), hexahydro-1,3,5-trinitro-1,3,5-triazine (RDX), and octahydro-1,3,5,7-tetranitro-1,3,5,7-tetrazocine (HMX). *J. Chem. Eng. Data* 46: 1549–1555.

Mackay, D., Shiu, W.Y., Ma, K.-C., and Lee, S.C. (2006). Handbook of physical-chemical properties and environmental fate for organic chemicals. In: *Introduction and Hydrocarbons,* 2e, vol. 1. Taylor & Francis Group.

McGee, B.C.W. (2003). Electro-thermal dynamic stripping process for in situ remediation under an occupied apartment building. *Remediation* Summer: 67–79.

Oleszek-Kudlak, S., Shibata, E., and Nakamura, T. (2004). The effects of temperature and inorganic salts on the aqueous solubility of selected chlorobenzenes. *J. Chem. Eng.* 49: 570–575.

Reynolds, J.P., Jeris, J., and Theodore, L. (2007). *Handbook of Chemical and Environmental Engineering Calculation.* John Wiley & Sons.

Santoleri, J.J., Reynolds, J., and Theodore, L. (2000). *Introduction to Hazardous Waste Incineration,* 2e. NY: Wiley-Interscience.

Sawamura, S., Nagaoka, K., and Machikawa, T. (2001). Effects of pressure and temperature on the solubility of alkylbenzenes in properties of hydrophobic hydration. *J. Phys. Chem., B* 105 (12): 2429–2436.

Sellers, K. (1998). *Fundamentals of Hazardous Waste Site Remediation.* Lewis Publishers.

Steverson, E.M. (1991). Provoking a Firestorm: Waste Incineration. *Environ. Sci. Technol.* 25 (11): 1808–1814.

Theodore, L. and Reynolds, J. (1987). *Introduction to Hazardous Waste Incineration.* New York, NY: Wiley-Interscience.

Thompson, K., Zhang, J., and Zhang, C. (2011). Use of fugacity model to analyze temperature-dependent removal of micro-contaminants in sewage treatment plants. *Chemosphere* 84 (8): 1066–1071.

Treybal, R.E. (1980). *Mass Transfer Operations,* 3e. McGraw-Hill.

US Army Corps of Engineers (2014). Design: *In Situ* Thermal Remediation Engineer Manual, EM-200-1-21, 2014.

USEPA (1992). *Handbook: Vitrification Technologies for Treatment of Hazardous and Radioactive Waste,* EPA/625/R-92/002. USEPA.

USEPA (1995a). *Site Technology Capsule: In Situ Steam Enhanced Recovery Process,* EPA/540/R-94/510a. USEPA.

USEPA (1995b). *IITRI Radio Frequency Heating Technology: Innovative Technology Evaluation Report,* EPA/540/R-94/527. USEPA.

USEPA (1995c). *In Situ Remediation Technology Status Report: Thermal Enhancement,* EPA/542/K-94/009. Washington, DC: EPA, OSWER.

USEPA (1997). *Issue Paper: How Heat Can Enhance In Situ Soil and Aquifer Remediation*, EPA 540-S-97-502, Washington, DC: Office of Research and Development.

USEPA (1999). Cost and Performance Report: Six-Phase Heating (SPH) at a Former Manufacturing Facility, Skokie, Illinois.

USEPA (2004). *In Situ Thermal Treatment of Chlorinated Solvents: Fundamentals and Field Applications*, EPA 542-R-04-010. USEPA.

USEPA (2012). *Green Remediation Best Management Practices: Implementing In Situ Thermal Technologies*, EPA 542-F-12-029. USEPA.

USEPA (2014). *In Situ* Thermal Treatment Technologies: Lessons Learned, 18 pp.

Uzgiris, E.E., Edelstein, W.A., Philipp, H.R., and Iben, I.E.T. (1995). Complex thermal desorption of PCBs from soil. *Chemosphere* 30 (2): 377–387.

Yagi, O., Uchiyama, H., and Iwasaki, K. (1992). Biodegradation rate of chloroethylene in soil environment. *Wat. Sci. Technol.* 25: 419–424.

Yalkowsky, S.H. and Banerjee, S. (1992). *Aqueous Solubility: Methods of Estimation of Organic Compounds*. NY: Marcel Dekker Inc.

Questions and Problems

1 Which thermal process does not involve pyrolysis: (a) incineration, (b) thermal enhancement, (c) vitrification?

2 Develop an oxidation reaction for C_3H_5OCl using the general reaction in Eq. (10.3).

3 Describe the typical PICs and POHCs from the incineration of hazardous wastes.

4 List the factors that affect the efficiency of combustion. How is combustion efficiency mearured?

5 Why the heating value of waste/fuel is important for incineration? How heating values can be used to design the combustion/incineration?

6 If the mass flow rate of a principal organic hazardous constituent (POHC) in a rotary kiln incinerator contains 200 kg/h, what is the maximum outlet mass flow rate from the incinerator allowed by the RCRA? What if this chemical is a PCB with a feed flow rate of 2 kg/h?

7 The flue gas in a hazardous waste incinerator contains 13% CO_2 and 25 ppm CO. What is the combustion efficiency?

8 An incinerator operates at 1950 °F, its gas flow rate is 5300 standard cubic foot per minute (scfm) at 60 °F. If the minimum residence time of two seconds is required, calculate the required volume of this incinerator.

9 Use Charles' law to convert the flow rate of benzene from 4500 scfm at 6 °F to acfm at 1100 °F at a constant pressure of 1 atm.

10 Use Le Chatelier's principle to explain how the temperature affects volatilization.

11 Why is an excess air (1.5–2 times of the stoichiometric oxygen) needed in an incinerator? Why overly excessive air is not essential, and why oxygen deficiency should be avoided?

12 Calculate the mass in lb for 5 lb-mol of the following compounds: (a) benzene, (b) octane, (c) ethanol, and (d) butyl acetate.

13 If a refuse-derived fuel made of paper, plastic, and textile requires 1000 L oxygen per kg of waste, what is the needed amount of air in L/kg waste?

14 In the equation used to estimate stoichiometric oxygen (Eq. 10.9), the coefficients of 2.67, 8, 1, and −1 are used to convert C, H, S, and bound oxygen into oxygen demand. Derive these coefficients.

15 In the equation used to estimate stoichiometric nitrogen (Eq. 10.10), the coefficient 3.32 is used to convert from oxygen into nitrogen. Why?

16 The incomplete combustion of gasoline component octane has the following equation: $C_8H_{18} + 12O_2 \rightarrow 9H_2O + 7CO_2 + CO$. Calculate the combustion efficiency (CE) (a) by volume and (b) by mass.

17 The combustion of propane can be written as: $C_3H_8 + 5O_2 \rightleftharpoons 3CO_2 + 4H_2O$. Determine (a) the theoretical combustion air, (b) the excess combustion air at 100% excess rate, and (c) the actual combustion air. Report the amount per lb-mole of C_3H_8 and per lb of C_3H_8.

18 Determine the heating value of gasoline in Btu/lb. Gasoline is a complex mixture of many components, use octane's formula (C_8H_{18}) for the approximate calculation.

19 Calculate the stoichiometric amount of oxygen, nitrogen, and air required for the complete combustion of cellulose ($C_6H_{10}O_5$) at a feed rate of 500 lb/h. (*Hint*: Use Eqs. (10.9)–(10.11)).

20 For the above questions, calculate the actual amount of oxygen, nitrogen, and air if an excess air is at 150% of the stoichiometric amount (i.e. 50% excess).

21 Given the balanced equation for the complete combustion of cellulose: $C_6H_{10}O_5 + 6O_2 \rightleftharpoons 6CO_2 + 5H_2O$, develop a balanced equation by adding N_2 into the reaction under the condition of (a) 20% excess air and (b) 100% excess air.

22 The flow rate of air at the standard temperature condition (60 °F) in an incinerator is 4000 scfm (standard cubic foot per minute). What is the flow rate when it is converted into the flow rate expressed as acfm (actual cubic foot per minute) at the actual condition at its operating temperature of 1800 °F?

23 Explain why (a) vapor pressure, (b) Henry's law constant, (c) diffusion coefficient, and (d) biodegradation rate increase with the increase of temperature.

24 Explain why an increase in temperature does not always increase the following parameters: (a) aqueous solubility, (b) viscosity, and (c) sorption.

25 Which heat transfer process is analogous to the mass transfer due to a concentration gradient? Which heat transfer process is analogous to the Darcy's law due to a hydraulic gradient?

26 Use Eqs. (10.30) and (10.31) to explain the factors that affect the soil heat-up time using steam.

27 Describe the pros and cons of the incinerators. What is NIMBY?

28 What are the specifications (limits) in RCRA regarding the hazardous waste incinerator: (a) minimum heating value, (b) maximum halogen content, and (c) DRE for POHCs?

29 List the four most common types of the incinerator systems, and describe the typical subsystems of an incinerator system.

30 In a fixed hearth incinerator, what is the benefit to have a primary combustion chamber operated under the starved oxygen condition?

31 Which type of the incinerator is the most versatile in disposal of various forms of hazardous wastes (solid, liquid, and gas)?

32 Which type of the incinerator can be commonly used in a small-scale for on-site installations for the treatment of hazardous wastes?

33 For which of the incinerator, the waste feed materials must be prescreened and crushed into smaller sizes for incineration?

34 Provide the typical temperature and residence time for fluidized bed incinerators. What are the required minimum temperature and residence time for hazardous waste incinerator (HWI) and hospital (biomedical) waste incinerator by the U.S. EPA?

35 Describe the limitations of thermal enhancement using hot air and steam.

36 Describe in general the applicable range of each thermally enhanced remediation technologies in respect to contaminant phase (liquid, vapor), volatility (VOCs, SVOVs, NVOCs), and type of soil matrix (sandy and clayed).

37 What are the main differences between low-frequency thermal methods and high-frequency thermal methods?

38 Which of the thermal enhancement technique (e.g. hot air, steam, hot water, electro-heating) is applicable given: (a) VOC in a clay soil, (b) SVOVs in a sandy soil under unsaturated condition, (c) NVOCs in a sandy soil, and (d) NVOCs in a clay soil?

39 Under what circumstances, vitrification can be used as a remediation approach for contaminated soils.

Chapter 11

Soil Washing and Flushing

LEARNING OBJECTIVES

1. Describe the similarities and differences between soil washing and in situ soil flushing
2. State how soil texture affects the applications of soil washing, and how the volume reduction can be achieved in soil washing
3. Know why various chemical additives are used in soil washing and soil flushing for the removal of metals and organic contaminants, including acids, chelates, surfactants, and cosolvents
4. Delineate the differences between solubilization and mobilization, and some essential terms associated with these mechanisms such as apparent solubility, interfacial tension, and capillary number
5. Describe how cosolvents differ from surfactants for the recovery of NAPLs
6. Understand the pros and cons of soil washing and in situ flushing, and site-related or chemical-related factors that affect the successful applications of in situ flushing
7. Be able to calculate the volume reduction and percent contaminant present in coarse-grained sands and fine-grained fractions after soil washing
8. Understand the selection of various surfactants and the problems associated with the recycle and disposal of surfactant-laden washing/flushing liquid streams
9. Be aware of some recent research and development in surfactant-enhanced aquifer remediation of NAPLs, including, but not limited to, foam and the combined use of surfactants with cosolvents/oxidants

This chapter addresses soil washing and soil flushing technologies that are innovative in nature. Being innovative technologies, they are not yet conventional or commonly used, but have been sufficiently developed with full-scale applications. Soil washing is an *ex situ* technology, whereas soil flushing is *in situ*. The key to these two technologies is the use of plain water or other chemical additives (e.g. acids, chelating compounds, surfactants, and cosolvents) to increase contaminant solubility and mobility for the removal of contaminants. Since contaminants are separated but not destroyed, soil washing and flushing technologies are usually used in conjunction with other treatment technologies. For example, *in situ* soil flushing using surfactants is virtually the enhanced pump-and-treat for the remediation of contaminant source zone and NAPLs (Chapter 7). This chapter will focus on the scientific principles of contaminant solubilization and mobilization, general process description, applicability, and limitations. Significant research and development work has been done on soil washing and *in situ* soil flushing since the late 1980s. As such, some recent progress of these technologies will also be surveyed.

Soil and Groundwater Remediation: Fundamentals, Practices, and Sustainability, First Edition. Chunlong Zhang.

Companion website: www.wiley.com/go/Zhang/Remediation_1e

11.1 Basic Principles of Soil Washing and Flushing

The basic principles of two fundamentally similar processes, soil washing and soil flushing, are briefly described in this section. This is followed by the discussion on how surfactants and cosolvents can be used to enhance the cleanup of contaminants from subsurface soils through two primary mechanisms, i.e. solubilization and mobilization.

11.1.1 Overview of Soil Washing and Flushing

In soil washing, contaminated soils must be excavated first from the site (that is why soil washing is typically an *ex situ* technology). The excavated soil undergoes **soil washing,** which is a mechanical process of using liquids (usually water, sometimes combined with chemical additives) to scrub soils. This "scrubbing" removes hazardous contaminants from soil and concentrates them into a smaller volume attached to fine-grained particles (silt and clay), called **fines**. In the subsequent separation step, fine-grained particles are separated from coarse-grained particles (sand and gravel) through various mechanical equipment such as trammel and hydrocyclones (Figure 11.1).

The principle of soil washing is based on the assumption that contaminants are typically associated with fine particles because of the larger surface area and sorption coefficients (K_d, see Eq. 2.9 in Chapter 2) for silt and clay particles relative to sands and gravels. Consequently, a significant volume reduction of contaminated soil is achievable if the content of silt and clay in the treated soil is minor, whereas sand and gravel are the predominant soil components. It is, therefore, critical to have a prior knowledge of soil particle distribution for the soil being treated. Soil particles are divided into gravel (>2 mm), coarse sand (0.2–2 mm), fine sand

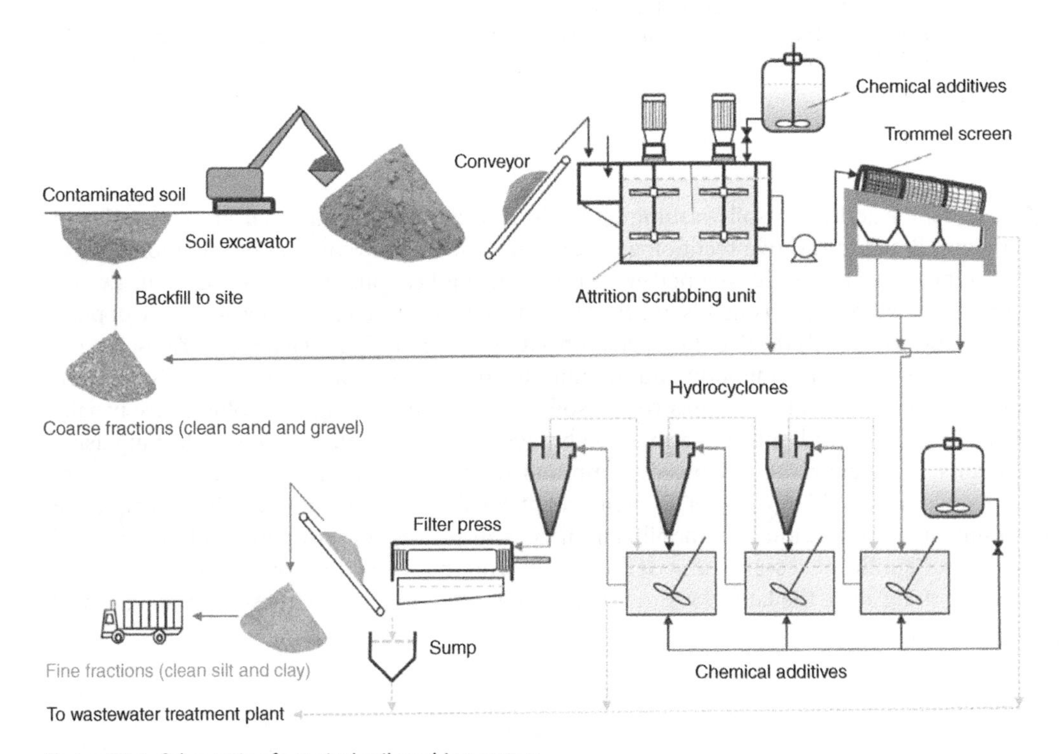

Figure 11.1 Schematic of a typical soil washing system.

(0.02–0.2 mm), silt (0.002–0.02 mm), and clay (<0.002 mm). The proportion of sand, silt and clay sized particles that make up the mineral fraction of the soil is termed **soil texture**. The commonly used 12 textural classifications are described in the textural triangle diagram (see Figure 3.2 in Chapter 3).

Soil washing is a relatively low-cost alternative for separating wastes, as well as minimizing the volume of soil required for subsequent treatment, such as incineration or bioremediation. Soil washing units can be brought to the site as a mobile system, so soil washing is a transportable technology.

In situ **soil flushing** involves flooding contaminated soils with a solution that moves the contaminants to an area where they are discharged or treated. The process generally begins with the drilling of injection and extraction wells into the ground on the contamination site. The soil flushing equipment pumps the flushing solution into the injection wells. In addition to wells, surface spraying and infiltration are also possible. The solution then passes through the soil, carries contaminants as it moves toward the extraction wells (or trench). The extraction wells collect the flushing solution containing the contaminants. The number, location, and depth of the injection and extraction wells should be decided based on several geological factors and engineering considerations, as we have discussed in pump-and-treat (Section 7.2, Chapter 7). After the solution–contaminant mixture is pumped out of the ground through the extraction wells, the mixture is then treated by a wastewater treatment system to remove the contaminants and reclaim the flushing chemicals whenever possible (Figure 11.2). Therefore, other equipment typical for wastewater treatment must be transported to or built on the site.

Soil flushing differs from soil washing in that soil flushing treats soil in place using an injection/recirculation process, and soil flushing requires the drilling of injection and extraction wells on site. As such, soil flushing can be combined/incorporated into pump-and-treat systems.

Both soil washing and soil flushing processes are effective for soils with low silt or clay contents. Soil washing and soil flushing processes also share some common features of using solutions containing chemical additives. The mechanisms for contaminant removals include one or more of the following processes, depending on the contaminants and additives employed: solubilization, mobilization, emulsion formation, and chemical reactions. Soil washing or flushing accelerates one or more geochemical reactions that alter contaminant concentrations in soil and groundwater, such as adsorption/desorption, acid/base, oxidation/

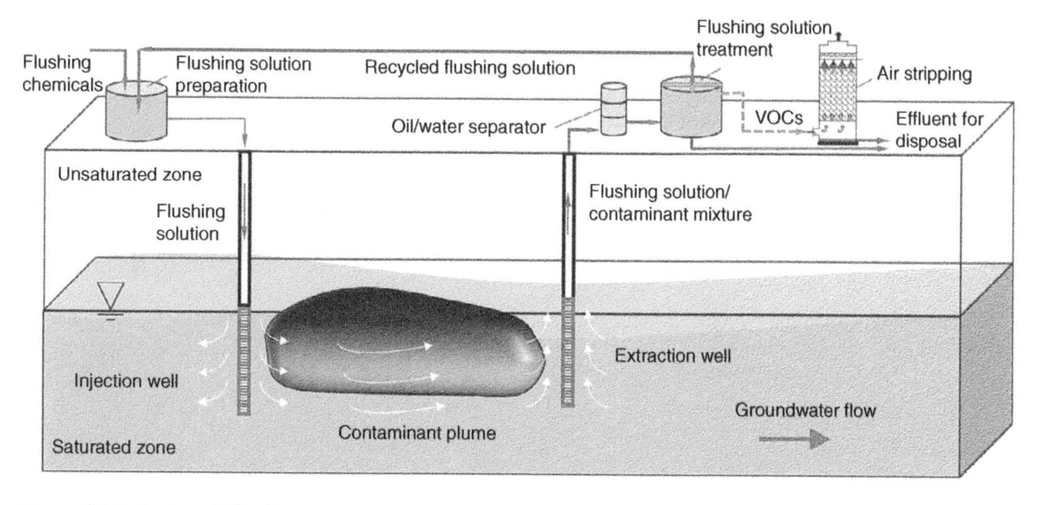

Figure 11.2 *In situ* soil flushing process.

reduction, dissolution/precipitation, ion paring or complexation, and biodegradation. Accelerated transport processes might include mechanisms such as advection, dispersion, molecular diffusion, depletion via volatilization, or solubilization.

The washing/flushing solution is typically one of two types of fluids: (i) water only, or (ii) water plus additives such as surfactants, cosolvents, acids, bases, oxidants, and chelants. Acids and bases are employed to enhance the removal of metals and phenols, respectively. **Chelants** (or chelators, chelating agents) are usually organic compounds that have high affinity to bind metal ions through coordinate bonds. For the enhanced removal of hydrophobic organic compounds, surfactants and cosolvents are the most commonly used based on the principles of enhanced solubilization and mobilization, which will be the focus in the subsequent discussions.

11.1.2 Surfactant-Enhanced Contaminant Solubilization

Surfactants (surface active agents) are chemicals with both hydrophilic (water-soluble) and lipophilic (oil-soluble) structures that can concentrate on the interface and reduce surface tension of the interface thereby increasing its spreading and wetting properties. Surfactants are classified according to the nature of the hydrophilic portion of the molecule (Table 11.1). The head group may carry a negative charge (anionic), a positive charge (cationic), both negative and positive charges (zwitterionic), or no charge (nonionic). Differences in the chemistry of surfactants due to the nature of the hydrophobic tails (degree of branching, carbon number, and aromaticity) are usually less pronounced than those due to the hydrophilic head group.

Anionic surfactants account for 60% of all the surfactants produced in the United States, mainly sulfonates such as linear alkylbenzene sulfonates (LAS), alkyl or alcohol ether sulfate,

Table 11.1 Examples of four surfactant types.

Ionic type	Surfactant example	Molecular structure
Anionic	Sodium dodecylsulfate	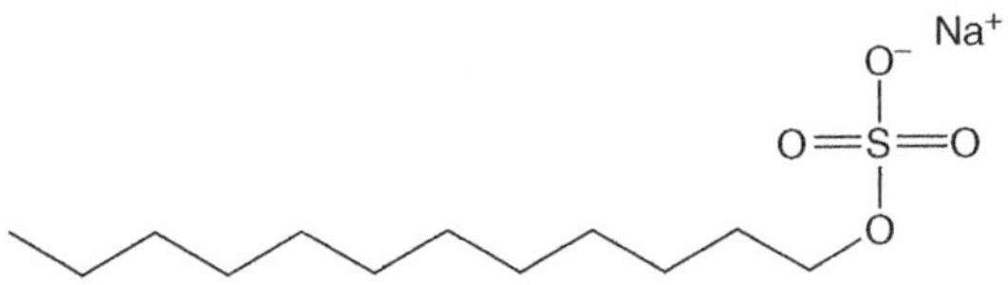 Na^+ O^- O=S=O O
Cationic	Benzyltrimethyl-ammonium bromide	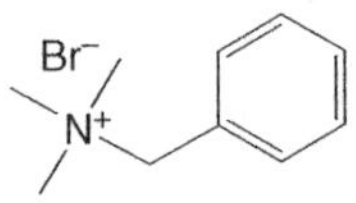 Br^- N^+
Nonionic	Triton X-100	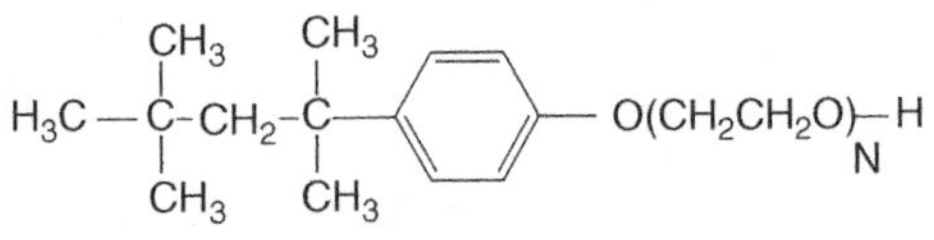 CH_3 CH_3 $H_3C-C-CH_2-C-$ $O(CH_2CH_2O)_N-H$ CH_3 CH_3
Zwitterionic	*n*-Hexadecyl phosphocholine	 N^+ O $^-O-P-O$ O

and carboxylates (soap). **Nonionic surfactants** account for 30% of all surfactants in the United States, mainly alcohol and alkylphenol ethoxylates. The **cationic surfactants** are primarily the alkyl amines and quaternary ammonium compounds (e.g. $R\text{-}N[CH_3]_3Br^-$). The **zwitterionic (amphoteric)** surfactants account for less than 1% of all the surfactants in the United States. For remediation applications, anionic and nonionic surfactants are the most commonly used.

Surfactants have been used in the petroleum industry for many years to enhance the oil recovery. Injection of surfactants in the oil field reduces the interfacial tension thus improving the mobility of oil. The use of surfactants in soil and groundwater remediation is called **surfactant-enhanced aquifer remediation** (SEAR). As we have discussed in Chapter 7, traditional pump-and-treat has a limited success because adsorbed phase and residual saturation are not amenable to pumping. Surfactants enhance pump-and-treat by two factors (mechanisms): (i) increasing solubilities of contaminants in groundwater, a process called **solubilization**; (ii) reducing interfacial tension between a liquid contaminant (e.g. NAPL) and water, a process termed **mobilization**. It is important to note that it is the interfacial tension that is responsible for the trapping of NAPLs (residual saturation) in the porous media. We will describe solubilization in this section, and introduce mobilization in the subsequent section.

A unique feature that distinguishes surfactants from cosolvents is the formation of micelles. **Micelles** are the aggregated structure of surfactant molecules in which the hydrophobic tails of the surfactant molecules point toward the interior of the micelles and the hydrophilic heads of the surfactant molecules orient toward the bulk aqueous solution. In a dilute surfactant solution, surfactant molecules exist as numerous single units called **surfactant monomer**. These surfactant monomers start forming micelles when monomer concentration increases up to the **critical micelle concentration** (CMC). At concentrations at or above the CMC, the number of monomers remains constant and the excess surfactant molecules aggregate to form micelles (Figure 11.3). The concentration required to form micelles (CMC) of typical aqueous-based surfactants is in the range of mg/L to g/L. The hydrophobic nature inside the micelles makes it suitable for NAPL to reside, as shown in Figure 11.3. The mechanism attributed to the removal of NAPL through micelles is defined as **micellar solubilization**, or solubilization in short.

At above the CMC, a contaminant's solubility (**apparent solubility**) is a linear function of the surfactant concentrations according to

$$S = S_0 + K_m (C_S - CMC) \tag{Eq. (11.1)}$$

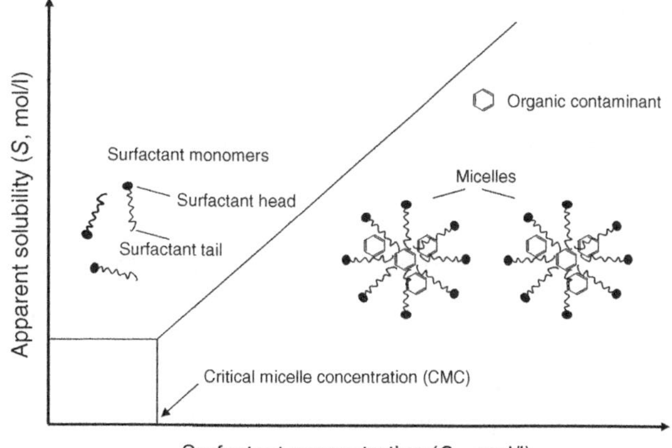

Figure 11.3 Formation of micelles from surfactant monomers above the critical micelle concentration (CMC), and micellar solubilization showing the liner relationship between apparent solubility of an organic hydrophobic contaminant and surfactant concentrations.

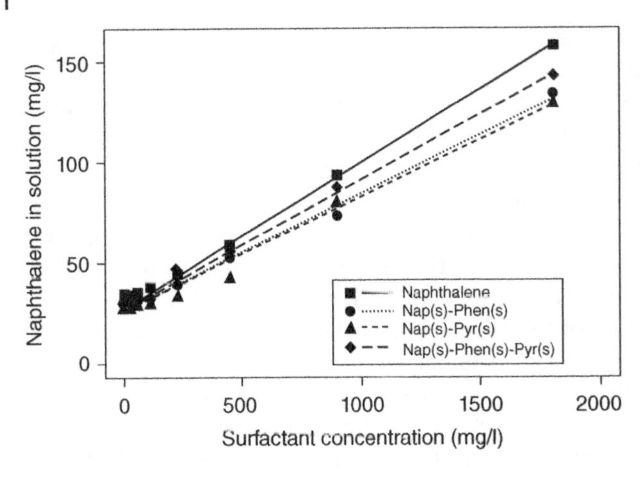

Figure 11.4 Enhanced solubility of naphthalene by increasing concentrations of nonionic surfactant Triton X-100 (octylphenol polyoxyethylene). Data also show how the solubilities of naphthalene are affected by other PAHs in the solution. Nap(s) = naphthalene; Phen(s) = phenanthrene; and Pyr(s) = pyrene. Source: Reprinted with permission from Guha et al. (1998). Copyright (1998) American Chemical Society.

where S = the apparent solubility (mol/L) of contaminant in a surfactant solution, S_0 = the aqueous solubility (mol/L) of dissolved contaminant in the absence of a surfactant, K_m = solubilizing capacity defined as the number of moles of solubilized organic compounds per mole of micellized surfactant (mol/mol), C_s = surfactant concentration (mol/L), and CMC = critical micellar concentration (mol/L). Solubilizing capacity (K_m), commonly termed the micelle/water partition coefficient, is a particular type of equilibrium partition coefficient in describing how a hydrophobic contaminant is distributed between micellar phase and bulk aqueous phase.

In the case of multiple components such as the case of NAPLs, the micelle/water partition coefficient of an individual NAPL contaminant may be reduced, unaffected, or increased. Generally speaking, solubilization favors more hydrophobic contaminants that could displace less hydrophobic contaminants. Thus, the solubility enhancement of naphthalene is slightly reduced in the presence of more hydrophobic phenanthrene and/or pyrene (Figure 11.4), and the solubility of phenanthrene is greatly enhanced in both binary and ternary mixtures with two other PAHs (data not shown, see Guha et al. 1998).

Surfactants are also characterized by their **hydrophile–lipophile balance** (HLB), which is an indication of the relative strength of the hydrophilic and hydrophobic portions of the surfactant molecules. The HLB scale ranges from 0 to 20. Surfactants with HLB values in the range of 3.5–6.0 are more suitable for use in W/O emulsions (water drops in oil), and surfactants with HLB values in the range of 8–18 are most commonly used in O/W emulsions (oil drops in water). Hydrophilic groups in surfactant molecules can be $-SO_4^-Na^+$, $-COO-K^+$, $-COOH$, $-OH$, whereas hydrophobic groups can be $-CH-$, $-CH_2-$, CH_3-, or $=CH-$. A high HLB value indicates a higher water solubility and less affinity for the NAPL (Sabatini et al. 1995). When choosing an optimal surfactant to remove a given contaminant composition, an important factor to consider is that the HLB of the surfactant should be as close as possible to that of the contaminant (Rosen and Kunjappu 2012).

11.1.3 Surfactant-Enhanced Contaminant Mobilization

In the presence of surfactant, mobilization occurs mainly by reducing the interfacial tension (IFT) between NAPL and surfactant. For example, the water-immiscible NAPL in the form of residual saturation (see Chapter 7) is trapped in soil pores due to **capillary forces**. These capillary forces are proportional to the interfacial tension at the NAPL–water interface (West and Harwell 1992). When a surfactant is used for an enhanced aquifer remediation, surfactant molecules accumulate at the NAPL–water interface, and IFT is reduced between the two phases because of the amphiphilic nature of the surfactant. If the buoyancy forces (related to the density difference) and viscous forces overcome capillary forces, then the DNAPL migrates in the direction of the net force and the phenomenon is termed **mobilization**.

Interfacial tension can be thought of as a measure of the increased free energy required to enlarge an interfacial area. This energy is called **surface tension** if the interface is between a gas and a liquid or **interfacial tension** if it is between two liquids (solvent and water). The reason for an increased free energy as a result of increased interfacial area can be easily explained by a bottle of water containing oil drops subject to shaking. First, all oil drops tend to stay in a spherical shape in water or any solid surface, because the spherical shape keeps its minimum surface area. This minimal surface area corresponds to the minimal energy that is thermodynamically stable. If we then let the solution stand still for a while, we observe many small oil drops start sticking together and become fewer and larger ones. For the same reason, this process is spontaneous because larger oil drops have smaller surface area and lower energy.

By analogy, we now can perceive a similar interfacial phenomenon will occur for NAPLs in the groundwater. For a NAPL phase to be dispersed in groundwater, it will take some free energy to maintain the interfacial area. In other words, free energy is required to enlarge the interfacial area. This required energy is the interfacial tension between NAPLs and water. IFT is the energy expressed per unit area (J/m^2) and is dimensionally equivalent to a force per unit length (N/m or dyne/cm), hence the interfacial tension (σ).

A quantitative relationship describing the effects of surfactant concentrations on surface tension is depicted in Figure 11.5. Surface tension linearly decreases with the increasing logarithmic concentrations ($\log C$) of surfactant. The sharp change in the slope is seen when the concentration of surfactant reaches its CMC value. For example, C_8TMAC has $\log C = -1.1$ at the inflection point, its CMC is then $10^{-1.1} = 0.079$ M.

Mobilization facilitates the removal of residual saturation when surfactant is used to enhance P&T. However, excessive mobilization should be avoided in remediation as well. This is because

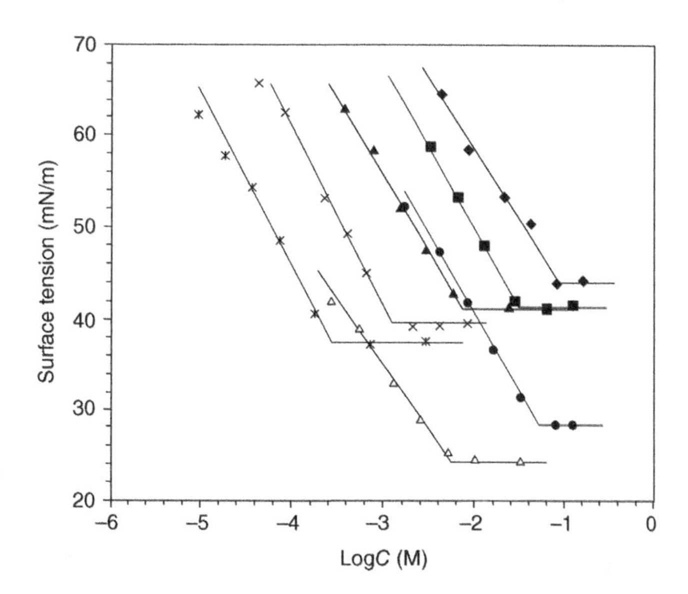

Figure 11.5 Surface tension as a function of log molar concentration ($\log C$) of various cationic surfactants at 25 °C in a simulated river water: ◆, C_8TMAC (octyl-trimethylammonium chloride), $C_8H_{17}N^+(CH_3)_3Cl^-$; ■, C_{10}TMAC, $C_{10}H_{21}N^+(CH_3)_3Cl^-$; ▲, C_{12}TMAC, $C_{12}H_{25}N^+(CH_3)_3Cl^-$); ×, C_{14}TMAC, $C_{14}H_{29}N^+(CH_3)_3Cl^-$; *, C_{16}TMAC, $C_{16}H_{33}N^+(CH_3)_3Cl^-$); ●, C_8OHC (octyl- dimethyl-2-hydroxyethylammonium chloride), $C_8H_{17}N(CH_3)_2C_2H_4OHCl^-$; △, C_{10}OHC, decyl dimethyl-2-hydroxyethylammonium chloride, $C_{10}H_{21}N(CH_3)_3Cl^-C_2H_4OHCl^-$. *Source:* Reprinted with permission from Rosen et al. (2001). Copyright (2001) American Chemical Society.

once the DNAPLs become mobile, they will readily sink. In other words, DNAPLs will tend to move deeper into an aquifer and contaminate regions that were previously clean. Mobilization should not be a concern for LNAPLs.

A quantitative approach describing mobilization is the use of a dimensionless **capillary number** (N_c) that is defined as follows (Dawson and Roberts 1997):

$$N_C = \frac{\mu v}{\gamma n \cos\theta} \qquad \text{Eq. (11.2)}$$

where μ = fluid dynamic viscosity (dyne × s/cm^2), v = pore-water velocity (cm/s), γ = interfacial tension (dyne/cm) between NAPLs and water, n = soil porosity, and θ = contact angle at the solid–water–NAPL interface. The dimensionless capillary number in Eq. (11.2) is, therefore, the ratio of viscous force of the displacing fluid to the capillary force. The displacing fluid (e.g. surfactant-laden solvent) is the wetting fluid, rather than the fluid (residual NAPL) being displaced. The viscous force of the displacing fluid is the driving force to mobilize the residual NAPLs, whereas the capillary force is the resistant force to entrap the residual NAPLs.

The effects of capillary number on residual saturation of contaminants are illustrated in Figure 11.6 for NAPL subject to horizontal water and chemical flooding in sandstone and bead packs. As shown, residual NAPLs is high (immobile) at a low N_c and NAPLs become mobile at a high N_c. Therefore, a high N_c is required to keep residual saturation at minimum. The dimensionless capillary numbers greater than 10^{-5} cannot be achieved solely through water flooding. A significant reduction in residual NAPL can typically be attained by introducing chemicals such as surfactants to lower the interfacial tension between NAPL and water. In a typical pump-and-treat operation, μ (viscosity), v (groundwater flow), or n (porosity) in Eq. (11.2) cannot be reduced significantly. This is when surfactants come into play. The use of surfactants will significantly reduce γ (interfacial tension), sometimes by a factor of 10^4, hence a significant increase in N_c. Consequently, the use of surfactant will significantly decrease the residual saturation and therefore improve the efficiency of pump-and-treat.

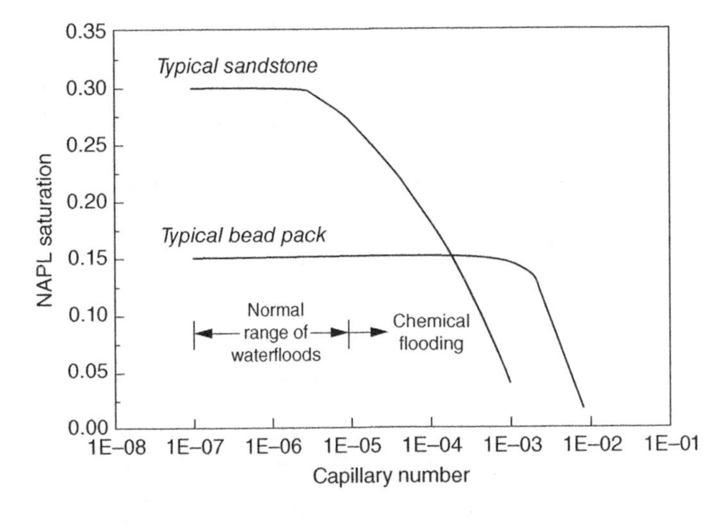

Figure 11.6 The dependence of residual NAPL saturation on capillary number N_c. *Source:* Dawson and Roberts (1997); Reprinted with permission from John Wiley & Sons.

Box 11.1 Dispersant use in the Gulf of Mexico oil spill, USA

The explosion of the BP's Deepwater Horizon Oil Rig in the Gulf of Mexico on 20 April 2010 made the word "dispersant" a household name. The explosion resulted in the release of an estimated 1.47–2.52 million gallons of oil per day into the Gulf of Mexico. This led to the use of chemical dispersants to control the spill, which totaled approximately 2.1 million gallons of dispersants between 15 May and 12 July 2010.

Dispersants are mainly composed of a combination of surface active agents (surfactants) and solvents. The surfactant serves as the active ingredient and reduces the interfacial tension, or surface tension of the water, whereas the solvents and other non-active ingredients are used to help the surfactants reach the oil–water interfaces. They do not change the amount of oil, but they simply interact with the oil and water to change the distribution. Dispersants emulsify, disperse, and dissolve oil. When oil is emulsified, the oil breaks into tiny drops and the drops are suspended in water. Once the surfactant has reached the oil–water interface, the surface tension decreases, allowing the oil to be more easily degraded. Decreasing the surface tension is one of the key reasons for using a dispersant. Dispersants have been available for use in oil spill cleanup since the 1960s. They can be applied by boats, helicopters, or fixed-wing aircrafts.

Corexit EC9500A and Corexit EC9527A, products of Nalco Holding Company, were the dispersants used during the 2010 Deepwater Horizon oil spill. Dispersants must be listed on the National Contingency Plan (NCP) before they can even be considered for use in oil spill relief in the United States. At the beginning of the Gulf spill, the proprietary composition was not public, but Nalco's safety data sheet identified the main components as 2-butoxyethanol and a proprietary organic sulfonate with a small concentration of propylene glycol. In response to public pressure, the EPA and Nalco released the list of the six ingredients in Corexit 9500, revealing constituents including sorbitan, butanedioic acid, and petroleum distillates. Corexit EC9500A is made mainly of hydrotreated light petroleum distillates, propylene glycol, and a proprietary organic sulfonate. Using electrospray (ESI) mass spectrometry in the negative-ion detection mode, the diethylhexylsulfosuccinate surfactant was confirmed. In the positive-ion detection mode, nonionic ethoxylated surfactants were also tentatively identified (Place et al. 2010).

The EPA rated the effectiveness of Corexit EC9500A and Corexit EC9527A in dispersing oils as 55 and 63%, respectively. The controversial debate of the use of dispersants in the Gulf of Mexico had not been about the effectiveness of oil spill control. It is more about the environmental and ecological safety of these dispersants. The media was so overwhelmingly focused on dispersants that the original oil spill problem appears to receive much less attention.

The discussion above regarding the capacity of surfactants to increase solubility and reduce interfacial tension is noteworthy. These characteristics are the basis for surfactants' many applications, including enhanced oil recovery in petroleum industry, enhanced contaminant removal in soil and groundwater remediation, and enhanced oil spill cleanup in water pollution control. Box 11.1 briefly documents the use of surfactant-containing dispersants in the cleanup of the Gulf oil spill.

11.1.4 Cosolvent Effects on Solubility and Mobilization

Cosolvents are organic compounds with a hydrophobic part, usually hydrocarbon chains, and a hydrophilic functional group, such as hydroxyl, carboxylic, and aldehyde groups. Cosolvents most commonly used for *in situ* flushing are alcohols. The amphiphilic nature (i.e. hydrophilic and hydrophobic groups) enables cosolvents to be miscible in both the aqueous phase and

NAPL phase. For the cosolvency effect to be dominant, the volume fraction of cosolvent to groundwater should generally be higher than 10% (Schwarzenbach et al. 2016). At this concentration range, the cosolvent-enhanced solubility is exponentially correlated to the volume fraction of the cosolvent in the mixture by the cosolvency power (Banerjee and Yalkowsky 1988; Schwarzenbach et al. 2016) as follows:

$$S = S_0 \times 10^{\sigma f} \quad \text{Eq. (11.3)}$$

where S is the cosolvent-enhanced solubility (mass or mol/L) in the mixture of cosolvent and contaminant, S_0 is the initial solubility (mass or mol/L) in water, σ is known as the **cosolvency power** (dimensionless), and f is the cosolvent volume fraction.

Similar to surfactants, cosolvents enhance the removal of NAPLs from a porous medium by two mechanisms: solubilization and mobilization. Solubilization of NAPLs by cosolvents is achieved through a reduced polarity by flushing cosolvent solution to the resident groundwater (Jafvert 1996). Mobilization of NAPLs is attributed to one of these three mechanisms: (a) If a significant volume fraction of cosolvent is used, a single phase is produced (e.g. a significant amount of water-miscible methanol is added to NAPLs). (b) The addition of cosolvent causes the interfacial tension to drop. (c) If a low-density cosolvent mixes with DNAPLs, a less dense cosolvent–DNAPL mixture is produced. This causes swelling of DNAPLs.

The presence of cosolvent in a NAPL–water system can alter the physical properties of both water and NAPL by partitioning into both phases. Given a sufficient amount of cosolvent, a single phase (i.e. completely miscibility) is formed, which is one of the mobilization mechanisms mentioned above. This process can be illustrated by a **ternary phase diagram** such as the one shown in Figure 11.7 for isobutanol (IBA)–water–PCE system. The binodal curve (coexistence curve) represents the boundary between the one-phase region and two-phase region. Above this curve, all three components (IBA, water, and PCE) exist in one single phase and interfacial tension equals zero. Below this curve, NAPL and water exist as two phases with each containing some cosolvent. The tie lines under the binodal curve represent a constant phase composition and interfacial tension (i.e. in the range of 3–38 mM/m). The relative proportion of each phase can be read from the intersection of the

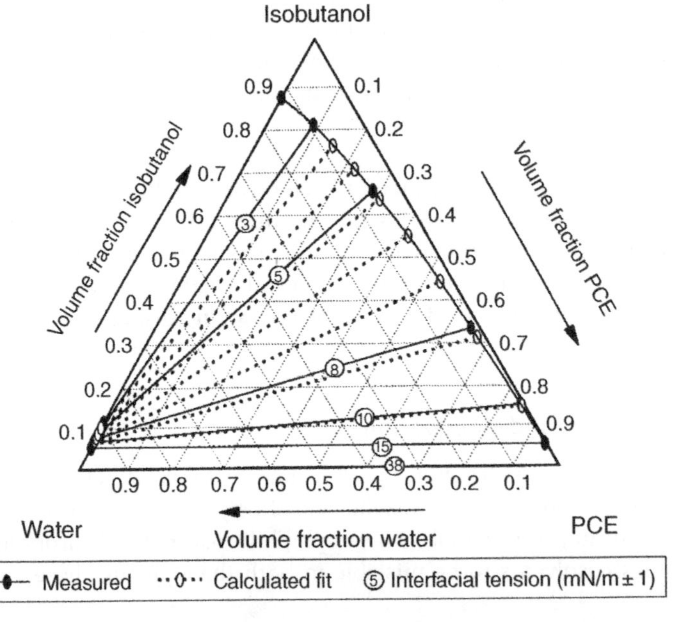

Figure 11.7 Ternary phase diagram for isobutanol-water-PCE system. *Source:* Reprinted with permission from Lunn and Kueper (1999). Copyright (1999) American Chemical Society.

tie lines with the binodal curve. The intersection point on the left hand side defines the aqueous phase composition, and the intersection point on the right hand side of the tie line defines the NAPL (i.e. PCE) phase composition.

The slope of the tie line reflects the equilibrium partitioning of cosolvent into both phases. For shorter-chain alcohols that partition more favorably in the aqueous phase, the tie lines have a negative slope (known as the **type II(−) system**). On the other hand, alcohols with limited water miscibility such as isobutanol or longer-chain or larger-molecule alcohols with more lipophilic nature are partitioned more into NAPL phase. The tie line slopes down toward the water endpoint with a positive slope in the ternary phase diagram. This is known as **type II(+) systems**, such as the one shown in Figure 11.7.

It is obvious that if the cosolvent concentration is above the binodal curve, NAPL is removed mainly through mobilization. If the cosolvent concentration lies below the binodal curve, the dominant NAPL removal mechanism depends on the cosolvent's preference toward water or NAPL. For more water-soluble cosolvents (e.g. methanol, ethanol), the ternary phase diagram has tie lines with negative slopes in a type II(−) system and the primary NAPL removal mechanism under the binodal curve will be dissolution. Some mobilization may occur if interfacial tension is reduced considerably. For cosolvents that partition preferentially into NAPL phase, such as isobutanol, in type II(+) systems, NAPL mobilization is more likely to occur due to a significantly reduced interfacial tension and NAPL swelling. Enhanced NAPL dissolution may also occur but to a less significant extent. Note in Figure 11.7 that the reduced IFT from 38 mN/m for an initial PCE/water to 5 mN/m at a higher isobutanol content will increase the risk of DNAPL downward mobilization. However, this is counteracted by the reduced density of the NAPL–colsolvent mixture (e.g. a 1 : 9 v : v IBA and PCE mixture has a density of 0.985 as a LNAPL, compared to the density of pure IBA of 0.802 and pure PCE of 1.63 as a DNAPL).

The commonly used cosolvents in environmental applications are alcohols (ethanol and pentanol) with densities typically around 0.8 g/cm^3. These alcohols are extremely miscible, and may partition into both water and NAPL phases, and create a mixture that enhances the solubility of contaminants. Like surfactants, they can also be effective in mobilizing contaminants (Bedient et al. 1999). Cosolvents are less expensive than surfactants on a cost per unit mass basis, but they must be used in much greater quantities. For instance, a surfactant concentration of 5–20 wt% would typically be required to solubilize an equal volume of water and oil while 50–80 wt% of a short-chain alcohol would be needed (Miller and Neogi 2008). Therefore, alcohol flushing requires many more pore volumes of flushing solution than a well-designed surfactant flushing scheme to achieve the desired performance objectives.

Furthermore, cosolvents may be restricted to flushing DNAPLs of relatively low molecular weight and viscosity, such as the lighter chlorinated ethanes and ethenes (ITRC 2003). For DNAPLs exhibiting higher molecular weight, viscosity, and compositional complexity, the low-molecular-weight alcohols are less effective at DNAPL dissolution. Higher molecular weight or more complex alcohols are, unfortunately, immiscible and much less soluble in water.

Cosolvent flushing is also affected more by field heterogeneity and hence preferential flow. In the case of surfactant flushing, a more uniform sweep may be induced by mobility control using either polymer, which increases the viscosity of the injected solutions, or surfactant foam addition, which involves the *in situ* generation of foam in high-permeability zones to divert flow into low-permeability zones (ITRC 2003).

11.2 Process Description, Technology Applicability, and Limitations

We describe herein the processes of *ex situ* soil washing, *in situ* soil flushing, and cosolvent flooding regarding their applications and limitations. A calculation example is provided to illustrate how soil washing is used to achieve volume reduction and its dependence on soil texture.

11.2.1 *Ex Situ* Soil Washing

Components of Ex Situ Soil Washing Systems: Regardless of many variations, a typical soil washing system shown previously in Figure 11.1 is composed of the following six distinct process units: (a) pretreatment, (b) separation, (c) coarse-grained treatment, (d) fine-grained treatment, (e) process water treatment, and (f) residual management. Pretreatment is to remove oversized material and prepare a homogeneous feed stream of reasonable size for delivery to the soil washing unit. Unit processes that may be employed include crushing and grinding, mechanical screening, gravity separation, blending and mixing, and magnetic materials removal. Separation systems are designed to separate coarse- and fine-grained solids, usually at the cut point of between 63 and 74 μm (230 and 200 mesh). Hydrocyclones are almost always employed to make this first size separation, although mechanical screens are sometimes used.

After separation, the coarse sands can be returned to the site, and if needed, can go on further treatment using surface washing methods, such as surface attrition, acid or base treatment for solubilization, or specific solvent for dissolving the contaminants. The treatment of fine-grained fractions, which should have greater than 95% of the contaminants by mass, precedes the residuals management step. Because of the small size, gravity settling will be very slow and some will not settle at all owing to their colloidal nature. The contaminated fine particles (silt and clay) and sludge (residual) resulting from the soil washing process may be disposed of in a regulated landfill and/or require further treatment through one or combination of the treatment technologies, such as incineration, low-temperature thermal desorption, chemical extraction/dechlorination, bioremediation, and solidification/stabilization. Finally, the wash water must be recycled back to the process or disposed in a nearby wastewater treatment system if it meets the regulatory discharge limits.

Applicability and Pros and Cons of Ex Situ Soil Washing Systems: There are many factors that should be considered in selecting soil washing as a remedy for contaminated soil. Soil washing technologies can be used independently or in conjunction with other treatment technologies.

The general *applicability* of soil washing can be described as follows:

- Soil washing is considered feasible for the treatment of a wide range of inorganic and organic contaminants including heavy metals, radionuclides, cyanides, polynuclear aromatic compounds, pesticides, and PCBs.
- Soil washing is most appropriate when soils consist of at least 50–70% sands. Soil washing will generally not be cost-effective for soils with fines (silt/clay) content in excess of 30–50%.
- Typically, onsite treatment of soils using soil washing will not be cost-effective unless the site contains at least 5000 tons of contaminated soil.
- Space requirements can be variable based on the design of the soil washing system, system throughput rate, and site logistics. A 20 ton/h unit can be sited on approximately one half acre (2000 m^2), including staging for untreated and treated soils.

The *advantages* of soil washing technology are as follows:

- Soil washing can treat both organics and inorganics in the same treatment system.
- Generally, there are no air or wastewater discharges from the system, making permit processes easier than for many other treatment systems. This attribute should also make the technology attractive to local community stakeholders.
- Soil washing is one of the few permanent treatment alternatives for soils contaminated with metals and radionuclides.
- Most soil washing technologies can treat a broad range of influent contaminant concentrations.
- Depending upon soil matrix characteristics, soil washing can allow for the return of clean coarse fractions of soils to the site at a very low cost.

The *disadvantages* of soil washing technology include the following:

- After treatment, a relatively small volume of contaminated solid media and wash water must be further treated or disposed.
- High humic content in soil, complex mixtures of contaminants, and highly variable influent contaminant concentrations can complicate the treatment process.
- As for any *ex situ* technology, there are space requirements for the treatment system.
- Hydrophobic compounds can be difficult to separate from the soil matrix. Organic compounds with high viscosity, such as No. 6 heating oil, present particular problems.
- For volatile compounds, process components will need to be modified to limit emission of the volatile organics to the air.
- Chelating agents, surfactants, solvents, and other additives are often difficult and expensive to recover from the spent washing fluid.

The calculations detailed in Example 11.1 are helpful for us to understand quantitatively how soil washing can remove contaminant while achieving volume reduction. The subsequent Example 11.2 further demonstrates why soil washing can become less favorable when soil texture changes from a sandy loam to clay loam.

Example 11.1 Soil washing for the volume reduction of treated soil

Soil washing is employed to clean up hydrophobic hydrocarbons from a sandy loam soil consisting of 80% sand, 10% silt, and 10% clay. The soil adsorption coefficients (K_d) were estimated to be 5 L/kg for sand (K_{d1}), 50 L/kg for silt (K_{d2}), and 500 l/kg for clay (K_{d3}). The bulk densities and particle densities of each component are listed in the table. The following assumptions can be made: (i) The soil has an initial contaminant concentration (C_0) of 2000 mg/kg; (ii) A total mass (M_T) of 5 million (5×10^6) kg of soil need to be treated; (iii) A soil to water ratio of 1 : 10 is used, such that the total volume of aqueous surfactant solution is 5×10^7 L (V_T); (iv) After the first wash, the coarse-grained portion of sand can be considered clean and refilled on site after gravity settling, and the fine-grained portions (silt and clay) will undergo further treatment by slurry-phase bioreactors to meet the regulatory requirement. The aqueous wash solution is discharged into a local sewage treatment plant.

a) Demonstrate the percent of volume reduction using soil washing.
b) Calculate the percent of contaminant retained in silt and clay portion, retained in the sand portion, and discharged in the washed water.

Soil components	wt%	K_d (L/kg)	Bulk density (g/cm^3)	Particle density (g/cm^3)
Sand	80	5	1.6	2.6
Silt	10	50	1.3	2.7
Clay	10	500	1.1	2.8

Solution

a) The mass of each soil component can be calculated by multiplying the total soil mass ($M_T = 5 \times 10^6$) by the wt% of each component, and the respective volume can be calculated by dividing the soil mass by the bulk density given in the table above. As shown in the table below, the volume of sand being removed accounts for 74.9%, clearly indicating the efficacy of the volume reduction when soil washing is used to treat sandy soils.

Soil components	Mass (kg)	Volume (m^3)	% Volume
Sand	4.0×10^6	2500	74.9
Silt	5.0×10^5	385	11.5
Clay	5.0×10^5	455	13.6
Total	5.0×10^6	3339	100.0

b) To calculate the percent of contaminant retained in silt and clay, in sand, and in water, one needs to set up the mass balance equation:

$$S_0 M_T = S_1 M_1 + S_2 M_2 + S_3 M_3 + C V_T \quad \text{Eq. (11.4)}$$

where S_0 denotes the initial concentration of contaminant in soil prior to soil washing; S_1, S_2, and S_3 denote sorbed concentration at equilibrium in sand, silt, and clay, respectively; M_1, M_2, and M_3 denote mass of sand, silt, and clay, respectively; C is the contaminant concentration in the liquid fraction at equilibrium; and V_T is the total volume of water used for soil washing. It is important to know that the above mass balance equation has four unknowns, i.e. the aqueous phase contaminant concentration (C) and the contaminant concentration in each solid fraction (S_1, S_2, and S_3). Hence three more equations are needed to solve for these four unknowns. This is where absorption coefficients become important. By assuming a linear sorption isotherm, the concentrations in solid and the concentrations in liquid are related according to the following equations:

$$K_{d1} = \frac{S_1}{C} (\text{for sand}) \quad \text{Eq. (11.5)}$$

$$K_{d2} = \frac{S_2}{C} (\text{for silt}) \quad \text{Eq. (11.6)}$$

$$K_{d3} = \frac{S_3}{C} (\text{for clay}) \quad \text{Eq. (11.7)}$$

Substituting S_1, S_2, and S_3 into the mass balance equation, we have:

$$S_0 M_T = K_{d1} C M_1 + K_{d2} C M_2 + K_{d3} C M_3 + C V_T \quad \text{Eq. (11.8)}$$

By rearranging the above equation, we can solve for the contaminant concentration (C) in the water:

$$C = \frac{S_0 M_T}{K_{d1} M_1 + K_{d2} M_2 + K_{d3} M_3 + V_T}$$

$$= \frac{2000 \frac{\text{mg}}{\text{kg}} \times 5\times10^6 \text{ kg}}{5\frac{\text{L}}{\text{kg}} \times 4\times10^6 \text{kg} + 50\frac{\text{L}}{\text{kg}} \times 5\times10^5 \text{kg} + 500\frac{\text{L}}{\text{kg}} \times 5\times10^5 \text{kg} + 5\times10^7 \text{ L}}$$

$$= 29.0 \frac{\text{mg}}{\text{L}}$$

$$S_1 = C \times K_{d1} = 29.0 \frac{\text{mg}}{\text{L}} \times 5 \frac{\text{L}}{\text{kg}} = 145 \frac{\text{mg}}{\text{kg}} (\text{concentration in sand})$$

$$S_2 = C \times K_{d2} = 29.0\frac{\text{mg}}{\text{L}} \times 50\frac{\text{L}}{\text{kg}} = 1450\frac{\text{mg}}{\text{kg}}(\text{concentration in silt})$$

$$S_3 = C \times K_{d3} = 29.0\frac{\text{mg}}{\text{L}} \times 500\frac{\text{L}}{\text{kg}} = 14\,500\frac{\text{mg}}{\text{kg}}(\text{concentration in clay})$$

With the calculated concentrations (C, S_1, S_2, and S_3), we can further calculate the mass and percent of contaminant present in water (=29.0 mg/L × 5 × 10^7 L = 1.45 × 10^9 mg = 1.45 × 10^6 g) and each soil fraction as given in the table below.

	Concentration	Mass (g)	% of the total contaminant mass
Water phase:	29.0 mg/L	1.45×10^6	14.5
Solid phase:			
Sand	145 mg/kg	5.80×10^5	5.8
Silt	1450 mg/kg	7.25×10^5	7.2
Clay	14 500 mg/kg	7.25×10^6	72.5
Total		1.0×10^7	100.0

The results clearly indicate that soil sand accounts for 74.9% of the total volume, but it is the soil fines (silt and clay) that retain the most (7.2 + 72.5 = 79.7%) of all the contaminant mass for this soil.

Example 11.2 The dependence of soil washing efficiency on soil texture

If the soil texture is changed from 80% sand, 10% silt, and 10% clay to 40% sand, 30% silt, and 30% clay, using other data from Example 11.1, demonstrate how soil texture affects the performance of soil washing.

Solution

If the soil texture is changed to 40% sand, 30% silt, and 30% clay, the same calculation can be done. The results are given below for comparison purpose.

Soil components	Mass (kg)	Volume (m^3)	% Volume
Sand	2.0×10^6	1250	33.2
Silt	1.5×10^6	1154	30.6
Clay	1.5×10^6	1364	36.2
Total	5.0×10^6	3768	100.0

The sand now accounts for only 33.2% of the total volume. It is thus clear that volume reduction by soil washing for this type of soil is not much advantageous.

	Concentration	Mass (g)	% of the total contaminant mass
Water phase:	11.3 mg/L	5.65×10^5	5.6
Solid phase:			
Sand	56 mg/kg	5.80×10^5	1.1
Silt	565 mg/kg	7.25×10^5	8.5
Clay	5650 mg/kg	7.25×10^6	84.8
Total		1.0×10^7	100.0

The results based on the mass balance and linear isotherm indicate that even a higher percentage (8.5% + 84.8% = 93.3%) of contaminants is now retained by soil fines (silt and clay), soil washing does not remove a lot contaminant from the soil and the small volume reduction (33.2%) does not make soil washing an attractive remedial option for soil containing only 40% sand. This example illustrates that soil washing is most appropriate when soils consist of at least 50–70% sand. Soil washing will generally not be cost-effective for soils with fines (silt/clay) content in the range of 30–50%.

11.2.2 *In Situ* Soil Flushing and Cosolvent Flooding

For *in situ* soil flushing, flushing liquids (plain water or chemical additives) can be introduced to soil either through spraying, surface flooding, subsurface leaching, or subsurface injection through wells. The contaminated fluids are often removed and subsequently treated in conjunction with other treatment technologies such as activated carbon, air stripping, biodegradation, or chemical precipitation (see Figure 11.2). The term **cosolvent flooding**, refers to the injection of low-concentration alcohol solutions without the combined use of surfactants.

Like soil washing, soil flushing is generally effective on coarse sand and gravel contaminated with a wide range of organic, inorganic, and reactive contaminants. Soils containing a large amount of clay and silt may not respond well to soil flushing, especially if it is applied as a stand-alone technology.

The removal efficiency by soil flushing depends on site-related factors (type of soil) as well as contaminant-related factors (Table 11.2). For example, soil flushing is most effective in homogeneous and permeable soils (e.g. sands, gravels, and silty sands with permeability $>10^{-4}$ cm/s). Soil flushing efficiency increases with the decrease in soil surface area, carbon contents, and CEC contents, because contaminant sorption generally increases with increasing soil surface area, carbon contents (for organic contaminants), and CEC (for charged metals and organic species). As defined in Section 3.1.2 in Chapter 3, cation exchange capacity (CEC) is the maximum quantity of total cations that a soil is capable of holding at a given pH value, for exchanging with the soil solution. CEC is expressed as milliequivalent of hydrogen per 100 g soil (meq/100 g), or the SI unit centi-mol per kg (cmol/kg).

Typically, soil flushing efficiency increases with the increase in water solubility, and the decrease in sorption coefficient, vapor pressure, and viscosity. Soluble compounds often are easily removed from soil by flushing with water alone. Organic contaminants with $K_{ow} < 10$ are very water soluble, so flushing by water alone should be sufficient. Compounds in this group include lower-molecular-weight alcohols, phenols, and carboxylic acids. Lower-solubility organic compounds can be removed by the use of a compatible surfactant. Examples of such compounds include chlorinated pesticides, polychlorinated biphenyls (PCBs), semivolatiles (chlorinated benzenes and polynuclear aromatic hydrocarbons), petroleum products (gasoline,

Table 11.2 Critical success factors for *in situ* flushing.

Site related factors	Less likely to succeed	Marginal to succeed	More likely to succeed
Dominant contaminant phase*	Vapor	Liquid	Dissolved
Hydraulic conductivity (cm/s)*	Low (<10^{-5})	(10^{-5}–10^{-3})	High (>10^{-3})
Soil surface area (m^2/kg)*	High (>1)	Medium (0.1–1)	Small (<0.1)
Carbon content (% wt.)	High (>10)	Medium (1–10)	Small (<1)
Soil pH and buffering capacity*	NS	NS	NS
CEC and clay content*	High (NS)	Medium (NS)	Low (NS)
Fracture in rock	Present	—	Absent
Contaminant related factors	**Less likely to succeed**	**Marginal to succeed**	**More likely to succeed**
Water solubility (mg/L)*	Low (<100 mg/L)	Medium (100–1 000)	High (>1 000)
Soil sorption coefficient (L/kg)	High (>10 000)	Medium (100–10 000)	Low (<100)
Vapor pressure (mm Hg)	High (>100)	Medium (10–100)	Low (<10)
Liquid viscosity (cPoise)	High (>20)	Medium (2–20)	Low (<2)
Liquid density (g/cm^3)	Low (<1)	Medium (1–2)	High (>2)
K_{ow}	NS	NS	10–1 000

Source: Adapted from Roote (1997), after USEPA (1993).
NS = Indicates no action level specified; * = Indicates higher priority factors.

jet fuel, kerosene, oils, and grease), chlorinated solvents trichloroethene, and aromatic solvents (BTEX). However, the removal of some of these chemicals has not yet been demonstrated.

A wide variety of inorganic and organic contaminants have been successfully treated by soil flushing, including heavy metals (Pb, Cu, and Zn) from battery recycling metal plating facility, halogenated solvents (TCE, TCA) from drying cleaning and electronics assembly, aromatics (benzene, toluene, cresol, and phenol) from wood treating plants, gasoline and fuel oils from petroleum and automobiles, and PCBs and chlorinated phenols from pesticide and electric power facilities. Some inorganic salts such as sulfates and chlorides can be flushed with water alone. Metals in other forms may require acids, chelating agents, or reducing agents for successful soil flushing. In some cases, all three types of chemicals may be used in sequence to improve the removal efficiency of metals. Many inorganic metal salts, such as carbonates of nickel and copper, can be flushed from the soil with dilute acid solutions.

The advantages of *in situ* soil flushing technology are as follows:

- No need to excavate, handle, and transport large quantities of the contaminated soil.
- Enhanced conventional P&T may speed up site remediation and closure.
- Wide applicability to a range of inorganic and organic contaminants in both saturated and unsaturated zones.

The disadvantages of *in situ* soil flushing include the following:

- Lengthy remediation time due to the slow rate of diffusion processes in the liquid phase.
- Potential for spreading contaminants beyond the capture zone, laterally or vertically, if the extraction system is not properly designed or constructed, and hydraulic control is not maintained.
- Limited regulatory acceptance due to the potential for spreading contaminants, and concern with introducing flushing solutions into the subsurface which may remain in residual quantities.

- Flushing solutions adhere to soil, accelerate bacterial growth, or cause precipitation or other reactions with soil or groundwater, thereby reducing effective soil porosity.
- Inability to separate flushing additive from effluent may result in consumption of flushing additive, rendering it cost prohibitive.

11.3 Design and Cost-Effectiveness Considerations

Several design and cost-effectiveness considerations for soil washing and soil flushing are discussed in this section, along with some research progress to address some of these issues. This is followed by the case studies on these two related technologies.

11.3.1 Chemical Additives in Soil Washing and Flushing

A major cost component in soil washing and soil flushing is the cost of chemical additives. A standard protocol would serve this purpose to screen chemical additives for cost-effective use in soil washing and flushing. At present, however, the selection usually requires laboratory batch and column studies, and in some cases, pilot studies before field trials. There are several major groups of chemical additives employed for soil washing and flushing, depending on the type of contaminants to be removed. Generally, acids or chelates are used mainly for the removal of metals, and surfactants or cosolvents are used for hydrophobic organic contaminants (HOCs).

Acids and chelating compounds: Cationic heavy metals (e.g., Cu^{2+}, Pb^{2+}, Cd^{2+}, and Ni^{2+}) in soils can be extracted out with strong inorganic acids (HCl, HNO_3, H_2SO_4, and H_3PO_4) through acid extraction at pH as low as 2. Moreover, H_2SO_4 and H_3PO_4 can dissociate into oxyanions such as PO_4^{3-} and SO_4^{2-} to efficiently remove arsenic by competitive oxyanionic desorption (Ko et al. 2006). Being organic acids and chelate compounds, EDTA (ethylenediaminetetraacetic acid) and DTPA (diethylenetriaminepentaacetate) can remove cationic heavy metals better than inorganic acids. Nitrilotriacetic acid (NTA) can remove metals well, but is not recommended because it is hazardous to human health. The most frequently used chelants are the aminopolycarboxylates, which form very stable and water-soluble chelant–metal complexes leading to the release of the contaminants from soil without precipitation (Ferraro et al. 2016).

Surfactants and cosolvents: The selection of surfactant should consider toxicity, biodegradability, ability to solubilize and mobilize contaminants, foaming, loss in soil, etc. (Table 11.3). Since the capital cost of surfactant use constitutes the single largest cost in soil washing and flushing, surfactants with a lower CMC value (hence the minimal surfactant dosage) is preferred for an effective solubilization and mobilization. As shown in Table 11.3, cationic surfactants are normally not used for remediation, this is because they are mostly toxic to bacteria and not so biodegradable. Cationic surfactants also have strong adsorption because of the negative charge of soil particles. Nonionic surfactants are generally more strongly adsorbed than anionic surfactants, and therefore add significant remediation cost. However, some anionic surfactants may have a significant loss in soil due to precipitation with cations such as Ca^{2+} (Zhang et al. 1998). Although few choices are available, food grade (edible) surfactants are considered favorable for subsurface injection. Edible surfactants are biodegradable, which makes it easier to obtain regulatory acceptance. For the same reason, biosurfactants, if they can be adequately generated on site, can be advantageous than their synthetic counterparts (Box 11.2). Anionic and nonionic surfactants have also been used together. For example, the combined use of 2% Brij 35 (nonionic) and 0.1% sodium dodecylbenzene sulfonate (anionic) was found to remove 70–80% of DDT, whereas individual surfactants had a lower removal rate.

Table 11.3 Surfactant classification based on ion charge and their characteristics.

	Anionic surfactants	Cationic surfactants	Zwitterionic surfactants	Nonionic surfactants
Examples	Sulfonic acid salts, alcohol sulfates, alkylbenzene sulfonates, phosphoric acid esters, and carboxylic acid salts	Polyamines and their salts, quaternary ammonium salts, and amine oxides	β-*N*-alkylaminopripionic acids, *N*-alkyl-β-iminodipropionic acids, *N*-alkylbetains, sulfobetaines, sultaines	Polyoxyethylenated alkylphenols, alcohol ethoxylates, alkylphenol ethoxylates, and alkanotamides
Toxicity	Relatively nontoxic	Toxic	Relatively nontoxic	Relatively nontoxic
Sorption to soil	No sorption	Strong sorption	Can be adsorbed	Not significant sorption
Environmental applications	Good solubilizer, widely used in petroleum recovery and contaminant remediation	Not widely used in environmental application	Can be mixed together as a cosurfactant in petroleum and environmental application	Good solubilizer, can be used as cosurfactant in petroleum and environmental application

Source: AATDF (1997) and Rosen and Kunjappu (2012).

Box 11.2 Green chemicals (surfactants) in soil washing and flushing

Green surfactants are bio-based amphiphilic molecules obtained from nature or synthesized from renewable materials rather than petrochemicals. Certain plants, microbes, and yeasts can produce biosurfactants through biosynthetic processes. Renewable raw materials such as triglycerides/sterols contribute to the hydrophobic part while sugars/amino acids contribute to the hydrophilic part of green surfactants (Rebello et al. 2014). Moreover, agro-industrial wastes such as olive oil mill effluent, soap stock, molasses, starch-rich wastes, and vegetable oils were used for surfactant production. These raw materials can be chemically derived to synthesize biosurfactants by chemical modifications.

Diverse ranges of prokaryotic and eukaryotic microorganisms are capable of producing surfactants. Bacteria capable of producing surfactants include *Psuedomonas aeruginosa* (mono- and di-rhamnolipid biosurfactants), *Corynebacterium, Nocardia, and Rhodococcus* spp. (phospholipids, trehalose dimycolates/dicorynomycolates, glycolipids, etc.), *Bacillus subtilis* (surfactin), *Bacillus licheniformis* (lipopeptide similar to surfactin), and *Arthrobacter paraffineus* (trehalose and sucrose lipids). Fungi involved in surfactant production include the yeasts *Torulopsis* spp. (sophorolipids) and *Candida* spp. (liposan, phospholipids) (Christofi and Ivshina 2002).

Chemically, through various oleochemical transformations such as hydrogenation, hydrolysis, *trans*-esterification as well as certain specific modifications, triglycerides can be used as starting materials to produce various surfactants and surfactant precursors including fatty acid methyl ester, methyl ester sulfonate, fatty alcohols, fatty amines, fatty acid anhydrides, fatty chlorides, fatty acids, fatty acid carboxylates, and alkylpolyglucosides.

Biosurfactants have several inherent advantages over the synthetic surfactants, such as lower toxicity, higher biodegradability, better environmental compatibility, higher foaming ability, higher selectivity and specific activity at extreme temperatures, pH, and salinity. Generally, biosurfactants can be more effective and efficient, because their lower CMCs (about 10–40 times lower than that of commercial surfactants) makes it possible for a lower dose to achieve a maximum decrease in surface tension. The growing demand for biosurfactants has commercialized their productions (Rebello et al. 2014).

This enhancement was attributed to the increased diffusion of surfactant molecules to the soil particles (Ghazali et al. 2010).

In addition to the facilitated contaminant removal by solubilization and mobilization, surfactant foams also improve sweeping efficiency for enhanced recovery of hydrophobic contaminants. Generated through an external foam generator or *in situ* through high-pressure air, foams have a low liquid content and hence a low relative permeability to liquid. The result is the diversion of surfactant solution to zones of lower permeability. Surfactant foams can displace a higher amount of TCE (99 vs. 41% with 25 pore volumes) without lowering the interfacial tension to an ultralow value and causing downward migration (Jeong et al. 2000).

Anionic surfactant formulations used in subsurface injections typically include an electrolyte (e.g. NaCl, $CaCl_2$) and a small amount of cosolvent such as isopropanol alcohol. By varying the electrolyte concentration, surfactant properties can be optimized for solubilization. The addition of cosolvent will stabilize surfactants in solution by preventing surfactant from gel formation (also called liquid crystals) that can be too viscous to pump through an aquifer (ITRC 2003). Nonionic ethoxylated alcohol surfactants can solubilize high amounts of NAPL without the need of salt or alcohol while maintaining low IFTs (Zhou and Rhue 2000).

Surfactants have been recently used together with other chemical additives to further enhance aquifer remediation, including oxidizing agents such as potassium permanganate

($KMnO_4$), sodium persulfate ($Na_2S_2O_8$), and catalyzed hydrogen peroxide (H_2O_2). In this case, surfactants provide optimized contaminant delivery (via desorption and emulsification) to the oxidants for destruction (Dahal et al. 2016). This strategy is termed the surfactant-enhanced ***in situ* chemical oxidation** (S-ISCO). As an example, trichloroethylene (TCE) is oxidized by sodium persulfate as follows (Dugan et al. 2010):

$$C_2HCl_3 + 3Na_2S_2O_8 + 4H_2O \rightarrow 2CO_2 + 9H^+ + 3Cl^- + 6Na^+ + 6SO_4^{2-}$$

When surfactants and oxidants are injected sequentially or simultaneously, S-ISCO addresses the free-phase NAPL and contaminants sorbed to soil. In a batch study, a combination of thermally activated persulfate and oxidant-compatible surfactant C_{12}-MADS (sodium dodecyl diphenyl ether disulfonate) significantly improves not only the oxidation of PAHs contained in coal tar but also the oxidant utilization efficiency. This might show promise for challenging situation in subsurface remediation such as source zone and hot spots with soil-sorbed organic contaminants and NAPLs (Wang et al. 2017).

11.3.2 Recycle of Chemical Additives and Disposal of Flushing Wastes

The extracted washing solution or groundwater from *in situ* flushing can be classified as a RCRA hazardous waste, because of high alcohol (10–50% by volume) or surfactant contents (1–2% by weight) and/or elevated concentrations of NAPL contaminants. Off-site management of a large quantity of hazardous waste may be cost-prohibitive because of the transportation cost. Hence, on-site treatment of recovered groundwater may be needed if reinjection or discharge is not an option. Common treatment processes may be physical separation, such as gravitational settling, decanting, flocculation, sedimentation, membrane filtration, or phase separation based mainly on sorption and volatilization such as carbon adsorption, air stripping, and steam stripping. It should be noted that the presence of surfactant in the waste stream present additional challenges for the treatment of this surfactant-laden waste. For example, foaming in the stripping process severely limits the process operation. Surfactants or cosolvents tend to hold contaminants in solution, which reduces the Henry's law constant and hence the efficacy of air stripping process.

Recycling of surfactants and cosolvents is only economic if higher surfactant/cosolvent concentrations (>3 wt%) and multiple pore volumes (>3) are injected (ITRC 2003). Surfactants can be recovered through ultrafiltration, precipitation, and foam fractionation. For example, anionic surfactant can be precipitated out using $CaCl_2$ to form calcium sulfate derivative of the anionic sodium dodecylsulfate (Venditti et al. 2007). Alternatively, the surfactant–contaminant mixture may be treated by selectively destructing contaminants using direct photolysis if contaminants have appropriate chromophoric moieties such as UV-absorbing PAHs. UV only photolysis and UV/H_2O_2 processes were found to be effective in selectively degrading PAHs without damaging perfluorinated surfactants in the mixture (An 2001).

Case Study 1 Pilot-scale soil washing efficiency for dinitrotoluenes

Two soils were obtained from two former army ammunition plants containing high concentrations of 2,4-dinitrotoluene (2,4-DNT) and 2,6-DNT. One soil from the Badger Army Ammunition Plant (BAAP; Baraboo, WI) in Wisconsin was a sandy soil (90% sand, 8% silt, and 2% clay), and another soil from the Volunteer Army Ammunition Plant (VAAP, Chattanooga, TN) was a silty clay soil (45% sand, 41% silt, and 14% clay).

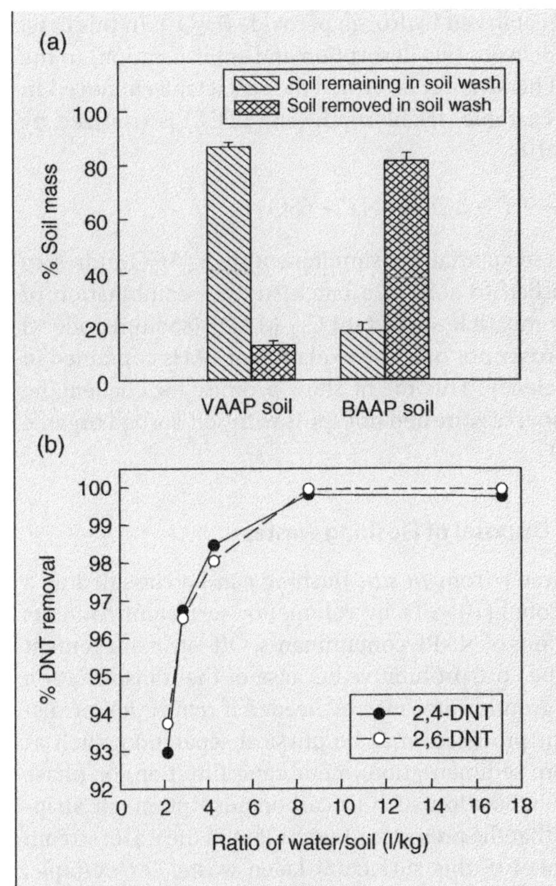

Figure 11.8 Soil washing efficiency as a function of water/soil ratio. *Source:* Zhang et al. (2001); Reprinted with permission from Elsevier. Copyright (2001) Elsevier Science B.V.

Soil washing was performed in a 14-L cylinder with an upward jet flow of warm (60 °C) water to separate DNT-associated fines from the clean sand. The resulting slurry was pumped to a bioslurry reactor. Soil washing efficiency, measured by the % total contaminant removal, increased with the increasing water to soil ratio (Figure 11.8). For the sandy BAAP soil, a water/soil ratio of 10 L/kg or greater resulted in a nearly complete retention of contaminants after removal of sands. For the silty VAAP soil, a water/soil ratio of 17 L/kg was necessary to maintain a 98% or higher reduction in concentration. On soil weight basis, soil washing removed 87% of large particles in BAAP soil, whereas 82% of VAAP soil was fine particles suspended in slurry phase after soil washing (Zhang et al. 2001).

Case Study 2 Soil flushing at the hill air force base, Utah, USA

Two different cells were tested at the Hill Air Force Base (HAFB): a solubilization cell and a mobilization cell. The demonstrations were conducted in cells contained by leakage through the steel sheet piling driven into an underlying impermeable layer. The solubilization cell showed excessive leakage through the sheet piling. Even though Dowfax 8390 (an anionic alkyldiphenyloxide disulfonate surfactant) was not able to flush through the whole cell due to leaking, approximately 50% of the contaminant was removed and 95% of surfactant was recovered after the flushing of 10 pore volumes of 4.3 wt% Dowfax 8390. The mobilization cell was injected with

2.2 wt% Aerosol OT (an anionic surfactant) and 2.1 wt% Tween 80 (a nonionic surfactant) along with 0.43 wt% $CaCl_2$. It was estimated that 85–90% of contaminant was removed in the mobilization cell after 6.6 pore volumes flushing with the surfactant solution. In the control experiment with water alone, less than 1% of the contaminant mass was removed. Overall, the mobilization system was much more efficient than solubilization system; the solubilization system was much easier to design and implement.

Case Study 3 Soil washing at Vineland Chemical Superfund site, Vineland, NJ

This 54-acre contaminated site was a former arsenic-based herbicide manufacturing facility operated from 1950 to 1994. The plant site included a number of manufacturing and storage buildings, a laboratory, several lagoons, and former chicken coops. Prior to 1977, the company stored by-product arsenic salts in open piles and in the chicken coops. As a result of water contacting the exposed piles, arsenic contaminated an adjacent wetland, surface and subsurface soils, groundwater, a nearby river, and a downstream lake. In 1995, the EPA completed the demolition work including the removal and disposal of eight contaminated buildings.

The construction of soil washing facility was completed in the fall of 2003. In early 2004, after a start-up and optimization period, a full-scale operation of the soil washing began. Soil washing treated an estimated amount of 350 000 tons of soils contaminated with arsenic (20–5000 mg/kg). The soil washing system has a treatment capacity of 70 tons per hour (tph) and includes the following unit operations: wet screening, separation (hydrocyclones), soil extraction, arsenic precipitation, leachate regeneration, water clarification, sand dewatering, fines thickening, and filter press dewatering. After treatment, residual arsenic in soil was determined to be below 20 mg/kg, which is classified by USACE (United States Army Corps of Engineers) and the USEPA as a "great success."

Bibliography

AATDF (Advanced Applied Technology Demonstration Facility for Environmental Technology Program) (1997). Technology Practices Manual for Surfactants and Cosolvents.

An, Y.J. (2001). Photochemical treatment of a mixed PAH/surfactant solution for surfactant recovery and reuse. *Environ. Progress* 20 (4): 240–246.

Banerjee, S. and Yalkowsky, S.H. (1988). Cosolvent-induced solubilization of hydrophobic compounds into water. *Anal. Chem.* 60 (19): 2153–2155.

Bedient, P.B., Rifai, H.S., and Newell, C.J. (1999). *Ground Water Contamination: Transport and Remediation*, 2e. Upper Saddle River, NJ: Prentice Hall PTR.

Christofi, N. and Ivshina, I.B. (2002). Microbial surfactants and their use in field studies of soil remediation. *J. Appl. Microbiol.* 93: 915–929.

Dahal, G., Holcomb, J., and Socci, D. (2016). Surfactant-oxidant co-application for soil and groundwater remediation. *Remediation* Spring: 101–108.

Dawson, H.E. and Roberts, P.V. (1997). Influence of viscous, gravitational, and capillary forces on DNAPL saturation. *Ground Water* 35 (2): 261–269.

Dugan, P.J., Siegrist, R.L., and Crimi, M.L. (2010). Coupling surfactants/cosolvents with oxidants for enhanced DNAPL removal: a review. *Remediation* Summer: 27–49.

Ferraro, A., Fabbricino, M., van Hullebusch, E.D. et al. (2016). Effects of soil/contamination characteristics and process operational conditions on aminopolycarboxylates enhanced soil washing for heavy metals removal: a review. *Rev. Environ. Sci. Biotechnol.* 15 (1): 111–145.

Ghazali, M., McBean, E., Shen, H. et al. (2010). Remediation of DDT-contaminated soil using optimized mixtures of surfactants and a mixing system. *Remediation* Autumn: 119–131.
Guha, S., Jaffe, P.R., and Peters, C.A. (1998). Solubilization of PAH mixtures by a nonionic surfactant. *Environ. Sci. Technol.* 32 (7): 930–935.
Hill, A.J. and Ghoshal, S. (2002). Micellar solubilization of naphthalene and phenanthrene from nonaqueous-phase liquids. *Environ. Sci. Technol.* 36 (18): 3901–3907.
Interstate Technology and Regulatory Cooperation Work Group (1997). Technical and Regulatory Guidelines for Soil Washing.
ITRC (Interstate Technology and Regulatory Council) (2003). Technical and Regulatory Guidance for Surfactant/Cosolvent Flushing of NDAPL Source Zones.
Jafvert, C.T. (1996). Technology Evaluation Report: Surfactants/Cosolvents, TE-96-02, Ground-Water Remediation Technologies Analysis Center.
Jeong, S.-W., Corapcioglu, M.Y., and Roosevelt, S.E. (2000). Micromodel study of surfactant foam remediation of residual trichloroethylene. *Environ. Sci. Technol.* 34 (16): 3456–3461.
Ko, L., Lee, C.-H., Lee, K.-P., and Kim, K.-W. (2006). Remediation of soil contaminated with arsenic, zine, and nickel by pilot-scale soil washing. *Environ. Progress* 25 (1): 39–48.
Lee,L. S., Zhai, X., and Lee, J. (2007). INDOT Guidance Document for In-Situ Soil Flushing, Joint Transportation Research Program Technical, Report Series, Purdue University.
Lunn, S.R.D. and Kueper, B.H. (1999). Risk reduction during chemical flooding: preconditioning DNAPL density *in situ* prior to recovery by miscible displacement. *Environ. Sci. Technol.* 33 (10): 1703–1708.
Miller, C.A. and Neogi, P. (2008). *Interfacial Phenomena: Equilibrium and Dynamic Effects.* New York: CRC Press.
Place, B., Anderson, B., Mekebri, A., Furlong, E.T., Gray, J.L., Tjeerdema, R., Field, J. (2010), A role for analytical chemistry in advancing our understanding of the occurrence, fate, and effects of Corexit oil dispersants, *Environ. Sci. Technol.*, 44 (16): 6016–6018.
Rebello, S., Asok, A.K., Mundayoor, S., and Jisha, M.S. (2014). Surfactant: toxicity, remediation and green surfactants. *Environ. Chem. Lett.* 12: 275–287.
Roote, D.S. (1997). *In Situ* Flushing, Technology Overview Report, GWRTAC Series, TO-97-02.
Rosen, M.J. and Kunjappu, J.T. (2012). *Surfactants and Interfacial Phenomena*, 4e. New York, NY: Wiley.
Rosen, M.J., Li, F., Morrall, S.W., and Versteeg, D.J. (2001). The relationship between the interfacial properties of surfactants and their toxicity to aquatic organisms. *Environ. Sci. Technol.* 35 (5): 954–959.
Sabatini, D.A., Knox, R.C., and Harwell, J.H. (eds.) (1995). Surfactant Enhanced Subsurface Remediation: Emerging Technologies. In: *ACS Symposium Series*, vol. 594. Washington, DC: American Chemical Society.
Schwarzenbach, R.P., Gschwend, P.M., and Imboden, D.M. (2016). *Environmental Organic Chemistry*, 3e. New York, NY: Wiley.
Strbak, L. (2000). *In Situ* Flushing with Surfactants and Cosolvents, National Network of Environmental Management Studies Fellow.
USEPA (1993). *Innovative Site Remediation Technology, Soil Washing/Soil Flushing*, vol. 3, EPA 542-B-93-012. USEPA.
USEPA (2001a). *A Citizen's Guide to In Situ Soil Flushing*, EPA 542-F-01-011. USEPA.
USEPA (2001b). *A Citizen's Guide to Soil Washing*, EPA 542-F-01-008. USEPA.
Venditti, F., Angelico, R., Ceglie, A., and Ambrosone, L. (2007). Novel surfactant-based adsorbent material for groundwater remediation. *Environ. Sci. Technol.* 41 (19): 6836–6840.
Wang, L., Peng, L., Xie, L. et al. (2017). Compatibility of surfactants and thermally activated persulfate for enhanced subsurface remediation. *Environ. Sci. Technol.* 51 (12): 7055–7064.

West, C.C. and Harwell, J.H. (1992). Surfactant and subsurface remediation. *Environ. Sci. Technol.* 26 (12): 2324–2330.

Zhang, C., Valsaraj, K.T., Constant, W.D., and Roy, D. (1998). Surfactant screening for soil washing: comparison of foamability and biodegradability of plant-based surfactant with commercial surfactants. *J. Environ. Sci. Health* A33 (7): 1249–1273.

Zhang, C., Daprato, R.C., Nishino, S.F. et al. (2001). Remediation of dinitrotoluene contaminated soils from former ammunition plants: soil washing efficiency and effective process monitoring in bioslurry reactors. *J. Hazardous Mat.* B87: 139–154.

Zhang, C., Zheng, G., and Nichols, C.M. (2006). Micellar partition and its effects on Henry's law constants of chlorinated solvents in anionic and nonionic surfactant solutions. *Environ. Sci. Technol.* 40 (1): 208–214.

Zhou, M. and Rhue, R.D. (2000). Screening commercial surfactants suitable for remediating DNAPL source zone by solubilization. *Environ. Sci. Technol.* 34 (10): 1985–1990.

Questions and Problems

1. Under what soil texture condition, soil washing can achieve the needed volume reduction?

2. Why are contaminants generally associated with the fine soil particles?

3. Why a low silt and clay content is essential for the success of both soil washing and flushing?

4. For Example 11.1 perform the same calculation if the soil has 60% sand, 15% silt, and 25% clay. (a) What would be the volume reduction after soil washing and (b) what would be the percent contaminant retained in each soil fraction as well as the water phase?

5. For Example 11.1, we assumed the soil was washed once. What will be the total percent contaminant retained in the second washing solution and in each soil fraction (sand, silt, and clay). Use the calculated data and justify whether or not the second washing is essential.

6. How is a hydrophobic contaminant's aqueous solubility affected by surfactant concentrations below and above its CMC value?

7. What are the differences/similarities between soil washing and soil flushing? Use a table to make such a comparison between these two related remediation technologies.

8. Researchers indicated that an excessive surfactant-induced mobilization should be avoided especially for DNAPL. Explain why.

9. Both surfactants and cosolvents are employed in soil washing and flushing processes. What are the typical concentration ranges for surfactants and cosolvents, respectively?

10. Define the following terms: HLB, CMC, IFT, and capillary number.

11. Why does the presence of surfactant decrease (rather than increase) the interfacial tension of NAPLs in groundwater?

12. What types of surfactants can/cannot be used to enhance the aquifer remediation?

13 Describe how viscosity, interfacial tension, and soil porosity affect residual saturation of NAPLs.

14 Use the dimensionless capillary number to explain how can surfactants enhance aquifer remediation?

15 What are the major limitations of cosolvent use in remediation as compared to that of surfactant use?

16 How do the following factors affect soil flushing efficiency: hydraulic conductivity, CEC, soil organic matter, K_{ow}, and solubility?

17 From Figure 11.5, (a) Find the critical micelle concentration for surfactant C_{10}TMAC. (b) What % reduction in surface tension of the surfactant solution will occur if surfactant concentration is increased from $10^{-2.4}$ M to its CMC?

18 From Figure 11.5, (a) Find the critical micelle concentration for surfactant C_{14}TMAC. (b) How much surface tension will change if the concentration of surfactant C_{14}TMAC is increased from 10^{-4} to 10^{-3} M?

Chapter 12

Permeable Reactive Barriers

LEARNING OBJECTIVES

1. Understand why passive barrier technologies can become a viable option over the conventional pump-and-treat technologies
2. Delineate the redox (electron transfer) mechanisms responsible for the dechlorination of halogenated solvents and the reduction of other contaminants by zero-valent iron (ZVI)
3. Identify other potentially useful reactive materials suitable for the remediation of selected contaminants
4. Understand the flow regimes in a typical permeable reactive barrier system and the factors affecting groundwater flow
5. Illustrate two common configurations of permeable reactive barriers (PRBs) and the criteria for the selection of a suitable reactive medium
6. Discuss the basic design concepts unique to the reactive barriers and calculate the required thickness of reactive barriers
7. Describe several conventional methods for the excavation of soils and construction of permeable reactive barriers
8. Be aware of some recent research and development progress made on permeable reactive barrier technologies

The energy and labor input required to keep conventional pump-and-treat (P&T) systems operational for many years is a severe economic burden. Passive treatment using permeable reactive barriers (PRBs) is a viable alternative to some long-term conventional P&T systems. A PRB is a wall built below ground to clean up polluted groundwater. The wall is permeable, and allows sufficient flow of groundwater through reactive agents, which is able to trap or destroy contaminants. Clean groundwater flows out the other side of the wall without the potential need of extraction wells. The main advantage of a PRB is that pumping and aboveground treatment are generally eliminated. The purpose of this chapter is to first introduce the reaction mechanisms and hydraulics in the PRB system. Depending on the types of contaminants, various abiotic and biotic reaction mechanisms can be incorporated into a PRB system. This introduction is followed by the discussion regarding the configurations of reactive barriers, types of reactive barrier materials, barrier design concepts, and construction methods.

Soil and Groundwater Remediation: Fundamentals, Practices, and Sustainability, First Edition. Chunlong Zhang.

Companion website: www.wiley.com/go/Zhang/Remediation_1e

12.1 Reaction Mechanisms and Hydraulics in Reactive Barriers

Reactive barriers are so named because "reactions" take place in the barrier materials. The types of reactions depend on the contaminants to be treated and the reactive materials used. Since ZVI is currently the most common reactive barrier, our emphasis is to illustrate the redox mechanisms involved in the dechlorination of halogenated compounds. An overview of PRB using ZVI and additional reactive materials for the treatment of other contaminants will be provided. Finally, the impact of mineral precipitation and biofouling on groundwater flow in PRBs will be described in this section.

12.1.1 Barrier Technologies as a Viable Option for Pump-and-Treat

Subsurface barriers for groundwater remediation include nonreactive barriers and PRBs. The physical barriers described in Section 7.1.1, including slurry walls and sheet piling, belong to the **nonpermeable and nonreactive barriers** used to divert the groundwater flow. As we recall, the keyed-in slurry walls are used for DNAPLs to avoid potential underflow of contaminants, whereas the hanging slurry walls are used for floating LNAPLs (a hanging slurry wall does not extend all the way down to the aquitard). **Permeable reactive barriers (PRBs)**, as shown in Figure 12.1, are constructed by removing soil in a long and narrow trench along the path of the polluted groundwater. The trench is filled with permeable reactive material that can clean up the contaminated groundwater. The reactive materials may be mixed with sand to make it easier for groundwater to flow through the wall (rather than around it). The filled trench or funnel is generally covered with soil, so it usually cannot be seen above ground. The reactive materials can be replenished when the PRB system becomes ineffective after a prolonged operation, or removed when the site is cleaned up.

Two salient features of any PRB system are the exclusion of pumping/extraction wells and the aboveground treatment. PRBs are ideally used for porous sandy soil of shallow aquifers (<50 ft below ground). Like P&T, it is an *in situ* technology most useful for contaminants in dissolved

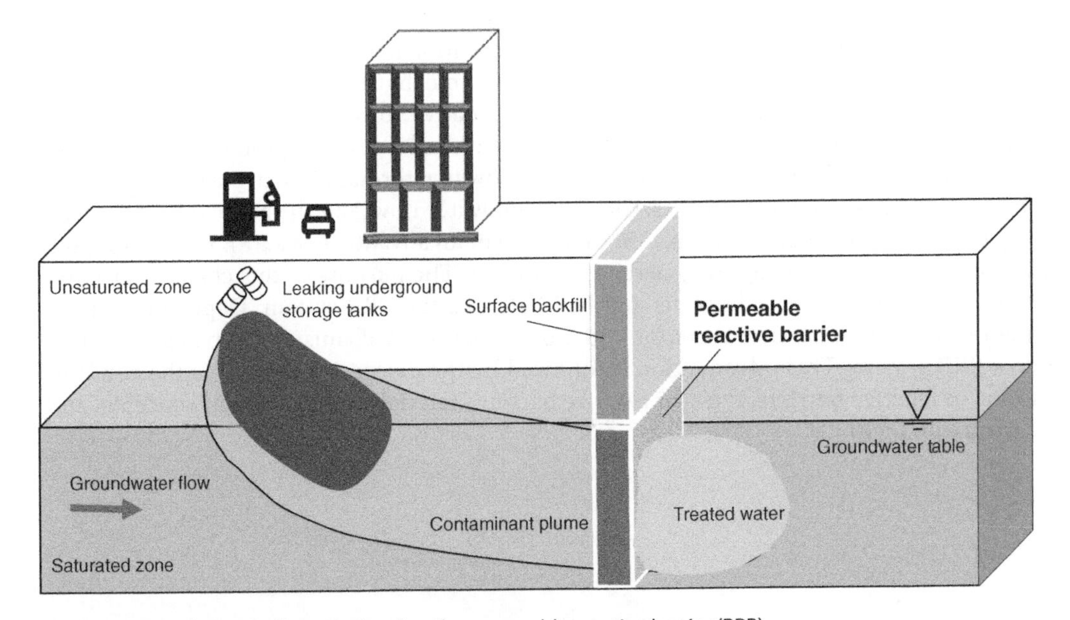

Figure 12.1 A schematic diagram showing the permeable reactive barrier (PRB).

phase. PRB technology comes with a variety of reactive media suitable for different contaminants and possible design configurations as a treatment train. It requires a sufficient flow of groundwater. Therefore, when contaminated groundwater flows through the reactive barrier, it is treated *in situ*. Without the need of active pumping, this passive treatment offers a low operation and maintenance (O&M) costs alternative to the conventional P&T systems that require a long-term monitoring, energy, and labor costs. As a result, a properly designed and operational PRB system should present a significant long-term cost saving. In a hypothetical groundwater remediation system with a 30-year duration, the costs for conventional P&T, PRB using iron and natural attenuation in the cleanup of a halogenated hydrocarbon plume were estimated to be in the order of $9, $3, and $0.5 million dollars, respectively (Blowes 2002).

12.1.2 Dechlorination Mediated through Redox Reactions by Zero-Valent Iron

Since ZVI is by far the most successful reactive medium in the PRB technology, this section highlights the scientific principles of ZVI. Emphasis will be placed on the mechanisms responsible for the dechlorination of chlorinated solvents (RCl, where R denotes the alkyl group) because remediation of this group of compounds has been the most successfully demonstrated and reported to date.

ZVI refers to the iron in its elemental form. Elemental iron has an oxidation state of zero (Fe^0), an iron atom has the tendency to lose two or three electrons at the outmost shell. Losing an electron(s) is an oxidation reaction. In layman's terms, the oxidation of iron metal is also a form of corrosion. If two electrons are lost for each iron atom, ZVI will become ferrous iron (Fe^{2+}). If three electrons are lost, then ferric iron (Fe^{3+}) is produced. As ZVI is oxidized (corroded), the resulting electrons will be transferred to another coupled half-reaction, for example, the transformation (reduction) of chlorinated compounds into potentially nontoxic products. In this reaction, chlorinated compounds are reduced.

The first reported use of the degradation potential of metals for treating chlorinated organic compounds in the environment was by Sweeny and Fischer (1972), who acquired a patent for the degradation of chlorinated pesticides (DDT) by metallic zinc under acidic conditions. The use of ZVI for *in situ* groundwater treatment was conducted by the researchers at the University of Waterloo. Their pioneering work in this area resulted in a patent for the *in situ* application of ZVI (Gillham 1993).

The dechlorination of RCl by ZVI is still an area of many on-going investigations. Although the exact reaction mechanism is still unclear, studies to date have indicated the multiple reactions and pathways potentially involved when ZVI is present in the groundwater. Results also pointed to some predominant reaction mechanisms. Due to its high reactivity, ZVI can react with many groundwater components and contaminants in their oxidizing states. For example, if groundwater has a high DO content (aerobic condition), iron itself can quickly react with oxygen to form Fe^{2+}:

$$2Fe^0(s) + O_2(g) + 2H_2O(l) \rightarrow 2Fe^{2+}(aq) + 4OH^-(aq) \qquad \text{Eq. (12.1)}$$

where (s), (g), (l), and (aq) represent solid, gas, liquid, and aqueous phases, respectively. Hence Fe^0 (s) is the ZVI in its solid (metal) form. For simplicity, we will not specify the state of individual species throughout this chapter unless otherwise noted. Equation (Eq. 12.1) results in the subsequent precipitation of ferrous iron on the iron surface due to the subsequent formation of ferric oxyhydroxide (FeOOH) or ferric hydroxide [$Fe(OH)_3$]. This **passivation** of iron surface will greatly reduce the reactivity of ZVI. It is, therefore, considered that a well oxygenated groundwater is detrimental to the reactivity of ZVI. Fortunately, contaminated groundwater at many sites is not highly oxygenated. After DO is depleted, iron can directly react with water:

$$Fe^0(s) + 2H_2O(l) \rightarrow Fe^{2+}(aq) + 2H_2(g) + 2OH^-(aq) \quad \text{Eq. (12.2)}$$

The above reaction is slower compared to its reaction with other stronger oxidizing agents such as chlorinated solvents. The slow reaction with water (Eq. 12.2) is advantageous to the PRB technology because very little reactive medium (iron) is used up in this side reaction. However, the reaction products including the production of hydrogen gas and the increase in groundwater pH could become a concern. For example, hydrogen formation could lead to the flammability issue, it also will likely reduce the porosity and permeability. Increasing pH is not favorable to the degradation of chlorinated solvents, as it will facilitate the precipitation of iron and reduce the iron reactivity. $Fe(OH)_2$ ($K_{sp} = 8 \times 10^{-16}$) is relatively insoluble, whereas $Fe(OH)_3$ ($K_{sp} = 4 \times 10^{-38}$) is extremely insoluble in water.

Since iron is a strong reducing agent, its reaction with the chlorinated organic compounds such as TCE is very quick via electron transfer (redox reaction). This is because these chlorinated organic solvents (RCl) are in a highly oxidized state. Chlorine in the chlorinated solvents has a high tendency to gain electrons from the carbon (C) atoms it is bonded to. These carbon atoms will tend to gain electrons from the oxidation of iron. It was suggested that the reduction of RCl proceeds primarily by the removal of the halogen atom and its replacement by hydrogen (Eq. 12.3), although other mechanisms probably play a role. This direct reduction of RCl occurs at the metal surface (Figure 12.2a):

$$\text{Oxidation half-reaction}: Fe^0(s) \rightarrow Fe^{2+} + 2e^- \qquad E = +0.440\,V$$

$$\text{Reduction half-reaction}: RX + 2e^- + H^+ \rightarrow RH + X^- \qquad E = +0.5 \text{ to } +1.5\,V$$

$$\text{Overall reaction}: Fe^0(s) + RX + H^+ \rightleftharpoons Fe^{2+} + RH + X^- \qquad E = +0.94 \text{ to } +1.94\,V \quad \text{Eq. (12.3)}$$

In the above two half-reactions, the first half-reaction (as written from left to right) is the oxidation reaction where 2 mol of electrons are lost from ZVI. The second half-reaction is a reduction reaction where each mole of RX gains 2 mol of electrons. The E value with a unit of voltage (V) represents the electrical potential or "electromotive force" for the reaction. Below the two half-reactions is the overall reaction (Eq. 12.3) with electrons balanced between gain and loss, X denotes any halogen atom such as chlorine (Cl) which is the most common halogen in the contaminated environment. A positive E value for the overall reactions is indicative of thermodynamically favorable reaction as written from left to right.

This pathway described in Eq. 12.3 represents a direct electron transfer from metal (Fe^0) to the halocarbon (RCl) sorbed at the metal–water interface, resulting in dechlorination and production of Fe^{2+}. Experimental results suggest that the degradation of chlorinated organics by metals is a surface phenomenon and the rate is governed by the specific surface area of the reactive medium.

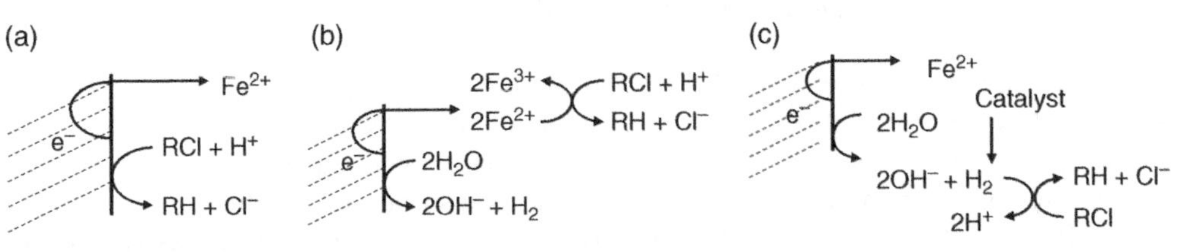

Figure 12.2 Three major electron transfer pathways leading to the reduction of chlorinated solvents (RCl) by zero-valent iron (ZVI): (a) direct reduction by Fe^0, (b) reduction by Fe^{2+}, (c) reduction by H_2.

The second important reaction between ZVI and RCl is through the production of Fe^{2+} in the equation as follows (Eq. 12.4):

$$Fe^0(s) + 2H_2O + 2RCl \rightleftharpoons 2ROH + Fe^{2+} + 2Cl^- + H_2 \qquad \text{Eq. (12.4)}$$

The above equation indicates that the corrosion of elemental iron metal yields ferrous iron and hydrogen, both of which are possible reducing agents relative to contaminants such as chlorinated solvents. The following redox reaction, therefore, is possible between ferrous iron (Fe^{2+}) and RCl:

$$\text{Oxidation half-reaction}: Fe^{2+} \rightarrow Fe^{3+} + e^- \qquad E = -0.77\,V$$

$$\text{Reduction half-reaction}: RCl + 2e^- + H^+ \rightarrow RH + Cl^- \qquad E = +0.5 \text{ to } +1.5\,V$$

$$\text{Overall reaction}: 2Fe^{2+} + RCl + H^+ \rightleftharpoons 2Fe^{3+} + RH + Cl^- \qquad E = -0.27 \text{ to } 0.73\,V \quad \text{Eq. (12.5)}$$

Note that 2 mol of Fe are needed so that the electrons will be balanced (i.e. canceled out in Eq. 12.5). The reduction of RCl by ferrous iron is a slow reaction under anaerobic condition when H_2O is used to serve as an oxidant. This pathway (Figure 12.2b) shows that Fe^{2+} resulting from the corrosion of Fe^0 may also dechlorinate RCl.

The third dechlorination pathway (Figure 12.2c) is the catalytic reduction by H_2, which requires a catalyst:

$$\text{Oxidation half-reaction}: H_2 \rightarrow 2H^+ + 2e^- \qquad E = +0.0\,V$$

$$\text{Reduction half-reaction}: RX + 2e^- + H^+ \rightarrow RH + X^- \qquad E = +0.5 \text{ to } +1.5\,V$$

$$\text{Overall reaction}: H_2 + RX + \text{catalyst} \rightleftharpoons H^+ + RH + X^- \qquad E = +0.5 \text{ to } +1.5\,V \quad \text{Eq. (12.6)}$$

The pathway in Eq. 12.6 shows that H_2O is reduced at the surface, but the product H_2 is subsequently used for the reduction of RCl. H_2 from the anaerobic corrosion of Fe^{2+} might react with RCl if an effective catalyst (typically Pt, Pd, Ni, PtO_2) is present.

The above analysis outlines several reaction pathways potentially involved in the abiotic degradation of RCl by ZVI. Although the exact mechanism and intermediates are unclear, these pathways point to the formation of chloride (Cl^-) in the reaction medium. Experimental evidence also indicates that the reaction at the metal surface is the rate limiting step (Scherer et al. 2000) among these five sequential steps:

a) Transport of reactant to the metal surface
b) Adsorption of reactant to the metal surface
c) Reaction at the metal surface
d) Desorption of product(s) from the metal surface
e) Transport of product(s) away from the metal surface

Moreover, it has become apparent that hydrogenations play a minor role in most systems and that iron surfaces will be covered with precipitates of oxides (or carbonates and sulfides) under most environmental conditions. Therefore, recent research has focused on how the oxide layer mediates transfer of electrons from Fe^0 to the adsorbed RCl. Scherer et al. (2000) formulated a heuristic model to explain the electron transfer at the metal surface. According to this model, an electron transfer (ET) from Fe^0 to RCl takes place through three pathways: (i) the corrosion pit or a similar gap in the oxide film mediates the direct ET from Fe^0 to RCl; (ii) the oxide film serves as a semiconductor to mediate ET from Fe^0 to RCl; and (iii) the oxide film serves as a coordinating surface in which sites of Fe^{2+} reduce RCl. Although the overall redox chemistry involved in ZVI is straightforward (Example 12.1), the reaction mechanism about mass and electron transfer still remains an active area of many on-going studies.

Example 12.1 Redox reactions between ZVI and chlorinated solvents

The dechlorination of trichloroethylene by zero-valent iron was reported to be the result of the following two half-reactions:

$$Fe^0 \rightarrow Fe^{2+} + 2e^-$$

$$C_2HCl_3 + 3H^+ + 6e^- \rightarrow C_2H_4 + 3Cl^-$$

Develop a balanced complete reaction, and describe the oxidation number change and electron transfer for this redox reaction.

Solution

This example is used as a review of some general chemistry about redox reaction. First, it is important to recognize why there are six electrons involved for the second half-reaction for each TCE molecule (C_2HCl_3). To do this, it is helpful to calculate the oxidation number change for each of the two carbon atoms in TCE. In the second half-reaction, C is the only atom having the oxidation number change since the oxidation numbers for both H and Cl atoms remain the same during the redox reaction (H = +1, Cl = −1). In TCE molecule, the C atom attached with 1 H, 1 Cl, and 1 C has a net oxidation number of zero, because Cl is more electronegative than C whereas H is less electronegative than C. For the second C attached to 2 Cl atoms and 1 C, its oxidation number is −2 because Cl is more electronegative than C. The average oxidation number of C in TCE is then (0+2)/2 = +1. Simply put, we can also calculate the average oxidation number directly from its formula C_2HCl_3. That is, $2\times C + 1\times(+1) + 3\times(-1) = 0$, and C = +1.

Since the change of oxidation number is more important, we now determine the oxidation number of two respective C atoms in the product C_2H_4. In C_2H_4, each of its C atom carries the oxidation number of −2 because each C is attached to two less electronegative H atoms. The average oxidation number of C atoms in C_2H_4 is also −2. Hence, for the conversion of C_2HCl_3 to C_2H_4, the first C has an oxidation number change of −2 (from 0 to −2), and the second C has an oxidation number change of −4 (from +2 to −2). The total oxidation number change is −6, corresponding to the gain of 6 mol of electrons per mole of TCE.

To develop a complete reaction, the numbers of electrons in the two half-reactions must be the same. We need to multiply the first half-reaction by 3, and add the two half-reactions to balance the electron loss and gain:

$$\text{Half-reaction (oxidation)}: 3Fe^0 \rightarrow 3Fe^{2+} + 6e^-$$

$$\text{Half-reaction (reduction)}: C_2HCl_3 + 3H^+ + 6e^- \rightarrow C_2H_4 + 3Cl^-$$

$$\text{Overall reaction}: 3Fe^0 + C_2HCl_3 + 3H^+ \rightleftharpoons 3Fe^{2+} + C_2H_4 + 3Cl^- \quad \text{Eq. (12.7)}$$

Equation (Eq. 12.7) is the balanced overall reaction. As shown, there are 6 mol of electrons involved for each mole of TCE. Fe^0 is oxidized, and it serves as a reducing agent. TCE is reduced, and it is an oxidizing agent.

12.1.3 Other Abiotic and Biotic Processes in Reactive Barriers

The ZVI-mediated dechlorination in the preceding section is an abiotic redox process. Of important note, ZVI is used not only for the dechlorination of chlorinated solvents (RCl), but also in the reduction of other contaminants. An example of these reactions involved with ZVI for the removal of toxic chromium is given as follows (Melitas et al. 2001):

$$Fe^0 + CrO_4^{2-} + 4H_2O \rightleftharpoons Cr(OH)_3(s) + Fe(OH)_3 + 2OH^- \quad \text{Eq. (12.8)}$$

$$xFe(OH)_3 + (1-x)Cr(OH)_3 \rightleftharpoons (Fe_xCr_{1-x})(OH)_3 \quad \text{Eq. (12.9)}$$

As shown in Eq. 12.8, the toxic hexavalent Cr(VI), CrO_4^{2-}, is reduced to nontoxic Cr(III). In addition to precipitation of $Cr(OH)_3$ (s) (Eq. 12.8), Cr(III) may also form Cr_2O_3 (s) or solid solutions with Fe(III) according to Eq. 12.9, where x can range from 0 to 1. Similarly, Cr(VI) may also be reduced by atomic hydrogen adsorbed to iron surfaces, by Fe(II) in solution, in the mineral phase, or by dissolved organic compounds, as we have discussed for the reduction of RCl by ZVI.

The reactions indicating the use of ZVI for the reduction of several other contaminants are given in Eqs. 12.9–12.13. Note that in these reactions, all contaminants (U, Pb, and As) are being reduced while Fe^0 is oxidized (Fiedor et al. 1998; Ponder et al. 2000; Su and Puls 2001). The reduced metals (metalloids) are either less toxic or in their precipitated or sorbed species on the iron surface.

- Removal of uranium (U):

$$Fe^0(s) + UO_2(CO_3)_2^{2-} + 2H^+ \rightleftharpoons UO_2(s) + 2HCO_3^- + Fe^{2+} \quad \text{Eq. (12.10)}$$

$$Fe^0 + 1.5UO_2^{2+} + 6H^+ \rightleftharpoons Fe^{3+} + 1.5U^{4+} + 3H_2O \quad \text{Eq. (12.11)}$$

- Removal of lead (Pb):

$$2Fe^0(s) + 3Pb(C_2H_3O_2)_2 + 4H_2O \rightleftharpoons 3Pb^0(s) + 2FeOOH(s) + 6HC_2H_3O_2 \quad \text{Eq. (12.12)}$$

- Removal of arsenate (As):

$$5Fe^0(s) + 2HAsO_4^{2-} + 14H^+ \rightleftharpoons 5Fe^{2+} + 2As^0 + 8H_2O \quad \text{Eq. (12.13)}$$

In principle, however, PRBs are not just limited to redox reactions. PRBs utilizing other physicochemically- and biologically-mediated processes can be employed to retain or remove contaminants, including sorption, precipitation, biodegradation, or the combination of these mechanisms.

Extensive studies have been done on ZVI-based remediation particularly in the lab and bench scale (e.g. Blowes et al. 1997; Butler and Hayes 2001; Agrawal et al. 2002; Alowitz and Scherer 2002). Long-term full-scale PRB technologies have also been demonstrated (Nyer 2001; Naidu and Birke 2015). Table 12.1 is a summary of full-scale PRBs in treating contaminants by various types of reactive materials (ITRC 2011). As discussed previously, the redox-based ZVI has been reported for its use in reducing chlorinated solvents, As (V), Cr(VI), and U(VI). ZVI can also be used together with sorption-based activated carbon for the removal of recalcitrant organics, or together with surfactant-modified clays for metal sorption. Hydroxyapatite and limestone are examples of precipitation-based reactive materials mainly used for metal removal. The biodegradation-based biobarriers can be used to foster bacterial degradation through the addition of bacteria, oxygen, nutrients, or electron acceptor. For example, certain petroleum hydrocarbon plumes from creosote sources are not amenable to ZVI, but can be treated using a containment wall that contains a slow-releasing oxygen compound. The aerobic condition created within the wall allows for biodegradation of the dissolved contaminants as they pass through it.

Table 12.1 Contaminants treated by various types of reactive materials in full-scale PRBs.

Contaminants	ZVI	Biobarriers	Apatite	Zeolite	Slag	ZVI-carbon	Organophilic clay
Chlorinated ethenes, ethanes	●	●				●	
Chlorinated methanes, propanes						●	
BTEX		●					
Energetics		●					
Perchlorate		●	●				
NAPL							●
Creosote							●
Cationic metals (e.g. Cu, Ni, Zn)		●	●			●	
Arsenic	●				●	●	
Chromium(VI)	●					●	
Uranium	●		●				
Stronium-90			●	●			
Nitrate		●	●			●	
Sulfate		●					
MTBE		●					

12.1.4 Hydraulics and Fouling Problems in Reactive Barriers

Groundwater flow in a contaminated aquifer with an installed barrier system can be complicated but it can be simulated through modeling to reveal how various PRB design factors (dimensions and permeability contrasts) will have an impact on it. Figure 12.3 simulates the horizontal (x) and vertical (y) velocity profiles as well as the streamline capture for the reactive barrier (Robertson et al. 2005). The distinctly higher velocities in the permeable barrier indicate that flow is converging into the barrier with a much higher hydraulic conductivity. This simulation also indicates that the average flow velocity in the middle of the permeable barrier is markedly higher (e.g. ~2.7 times) than in the background aquifer undisturbed by the barrier.

The groundwater flow regime depicted in Figure 12.3 will be changed over time because of the geochemically and microbiologically mediated process called **fouling**. For example, in carbonate-rich aquifers, the corrosion of ZVI and mineral precipitation (mostly $CaCO_3$, $FeCO_3$, and FeS) reduce the porosity and hydraulic conductivity over time, thereby causing the preferential flow of groundwater in the barrier (USEPA 1998). The modeling results from Li et al. (2005), however, reveal that significant hydraulic changes should be expected only after approximately 30 years of operation. Under the modeling condition, large regions of PRBs become clogged after 50 years and the barrier becomes less permeable than the aquifer, resulting in the bypass of contaminated groundwater.

Clogging can also be caused microbiologically (**biofouling**). Microbes could enter a reactive cell through groundwater transport and potentially populate in the reactive cell and/or the downgradient aquifer under certain conditions. Microbial growth in the reactive cell can help or hinder the degradation or removal of some types of contaminants. Certain microbes can use H_2 anaerobically generated during iron corrosion as energy for TCE dechlorination. In this

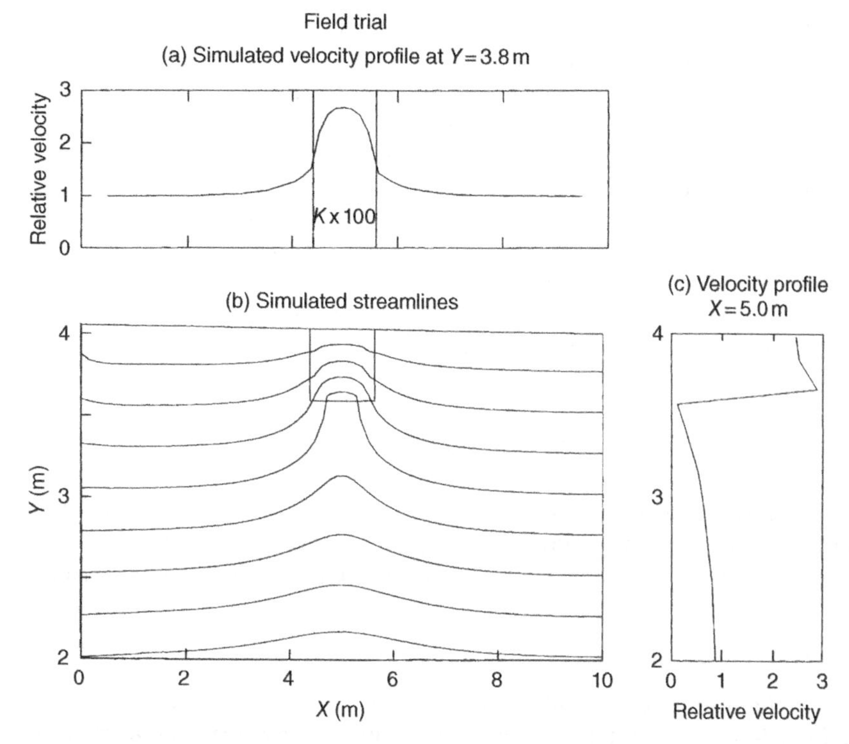

Figure 12.3 Two-dimensional simulation of steady-state groundwater flow through a high-permeability shadow aquifer with a hydraulic conductivity contrast of ×1000 between a permeable reactive barrier and the surrounding aquifer: (a) Horizontal velocity profile through the center of the aquifer layer, parallel to flow, (b) streamlines showing the depth of capture of the high-K layer, and (c) vertical velocity profile through the center of the layer. The modeled domain extends deeper than the regions shown, and curvature of the streamlines at the left model boundary is the result of numerical error. *Source:* Robertson et al. (2005); Reprinted with permission from John Wiley & Sons.

case, microbial activity may be beneficial because it prevents the buildup of H_2 in an iron reactive cell. However, if the growth of microbes is excessive, it could lead to biofouling of the reactive cell in the long term. The biofilm formation and bacterial adhesion on the reactive media could reduce the activity of reactive materials, and limit the groundwater flow rate across the barrier.

Microbes can potentially cause biofouling of iron reactive cells over the long term in three different mechanisms (Tuhela et al. 1997). The first and most common mechanism is the production of Fe(III) by iron bacteria. **Iron bacteria**, such as *Thiobacillus ferrooxidans* and *Leptospirillum ferrooxidans*, can directly use Fe(II) as an energy source. Iron bacteria are acidophiles, and may be present in acidic soils. However, they probably would not be expected to proliferate in alkaline environments produced by ZVI.

The second mechanism relies upon the **stalked and sheathed bacteria** through the oxidization of Fe(II) on sheath surfaces. *Gallionella* and *Leptothrix* spp. are two such bacteria that appear to be involved in Fe(II) oxidation, and sulfide- and thiosulfate-dependent forms also have been reported. Extensive biofouling by stalked and sheathed bacteria has been detected in water wells and sand filters used for iron removal. By this mechanism, growth of stalked and sheathed bacteria potentially can occur in the reactive iron cell or in the downgradient aquifer.

The third mechanism involves **heterotrophic bacteria** that use carbon in organo-ferric complexes. Biodegradation of organo-ferric complexes would liberate Fe(III), resulting in rapid precipitation of ferric hydroxide. However, this third mechanism may not be a primary source of ferric hydroxide precipitation in DNAPL-contaminated groundwater, unless a strong Fe(III) chelant is also present.

A concomitant occurrence is the oxidation of ferrous iron (Fe^{2+}) or Mn(II) by microbially mediated reactions and the subsequent precipitation of ferric iron (Fe^{3+}) or Mn(IV) hydroxides. Iron-related biofouling will clog the walls of barriers and groundwater wells. Ferric hydroxides can precipitate as amorphous $Fe(OH)_3$, or they may develop a crystalline structure such as ferrihydrite ($5\ Fe_2O_3 \bullet 9\ H_2O$). Ferrihydrite has been identified as the solid phase in biofouled groundwater wells (Tuhela et al. 1997). In general, ferric hydroxides have very low solubilities at neutral and alkaline pH; hence, oxidation of Fe(II) is accompanied by nearly complete removal of iron from the aqueous solution by precipitation.

12.2 Process Description of Reactive Barriers

Two aspects of reactive barrier processes are described here. The first is how to configure the reactive barriers in relation to contaminated groundwater plume, and second is the selection of commercially available reactive barrier materials that can be targeted to various contaminants.

12.2.1 Configurations of Reactive Barriers

There are two major configurations commonly used for a PRB system: a continuous reactive barrier system or a funnel-and-gate system (Figure 12.4). In its simplest form, a **continuous PRB** such as granular iron is installed in the path of a dissolved contaminant plume. As the groundwater flows through the reactive zone, contaminants encountered in the entire width of the reactive barrier are degraded to potentially nontoxic compounds. A PRB installed as a **funnel-and-gate system** has an impermeable section (funnel) that directs the captured groundwater flow toward the permeable section (gate). This configuration may sometimes allow a better control over reactive cell placement. PRBs in a funnel-and-gate system can also be installed with multiple gates in series (multimedia) system. In principle, various permeable reactive materials can be installed in series to remove various contaminants or the same contaminant by multiple removal pathways as described in Section 12.1.

Continuous reactive barriers are easier to install and generate less complex flow patterns compared to funnel-and-gate systems. Hence, most recent PRB applications have been continuous reactive barriers. In both barrier configuration systems, the barrier acts passively

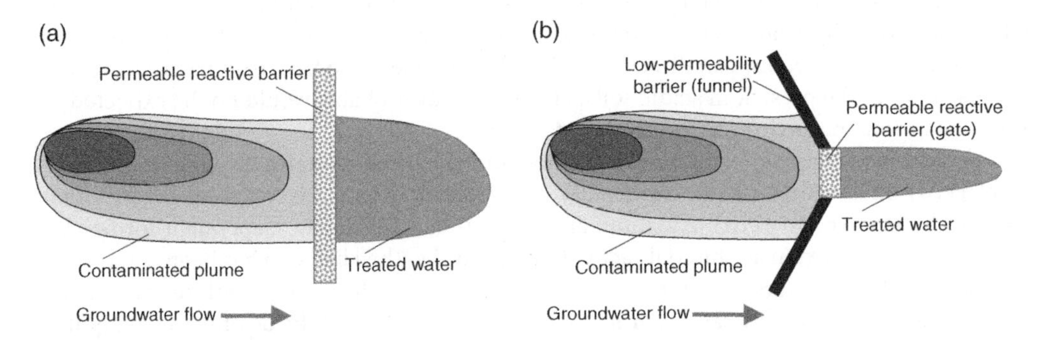

Figure 12.4 Schematic illustration of two major PRB configurations: (a) Continuous PRB (plan view) and (b) Funnel-and-gate system (plan view).

without the need of pumping and structures installed aboveground. Available field data have indicated that the reactive medium is used up very slowly and, therefore, PRBs have the potential to passively treat the plume over several years or decades, which would result in hardly any annual operation and maintenance costs other than site monitoring. Depending on the longevity of the reactive medium, the barrier may have to be rejuvenated or replaced periodically.

Other PRB configurations have also been recently proposed or demonstrated in the field. One such example is the **drain-and-gate reactive barrier** (DG PRBs). The DG PRB system has a collector drain and a distributor drain (or a collector only). The collector drain captures contaminated groundwater, which is then passed through reactive medium and again released into the aquifer by a distributor drain further downgradient (Klammler et al. 2010). Compared to funnel-and-gate PRB, the DG PRB is more appropriate to less conductive aquifer. Another recent development is the horizontal reactive media treatment well (Divine et al. 2018). Conceptually similar to the horizontal well described in Chapter 7, this **horizontal PRB** can access contaminants not accessible by conventional vertical PRB. Since the entry and exit pits for the directionally drilled borings are the only surface disruption, the horizontal PRBs require a minimal footprint on-site and can be implemented without significant modifications to on-site structures.

12.2.2 Available Reactive Media and Selection

In our previous discussion (Table 12.1), several types of reactive materials have been identified with their demonstrated uses in full-scale PRBs. In a survey among various reactive materials packed in a PRB, 29 out of 53 full-scale and pilot-scale PRBs were reported to use ZVI (Fe^0) as reactive medium because of its low cost and high reactivity (USEPA 2002). There are other types of reactive materials with potential use in PRBs. These various types and their selection criteria are discussed as follows.

12.2.2.1 Types of Reactive Media

The most common **granular iron** comes in various forms depending on the sources, such as cast iron fillings, Master Builders, Peerless for field scale, and Fisher Scientific cast iron used for lab scale testing. Iron granules in a large supply are often the by-products of manufacturing processes so their use as a barrier wall material has the added benefit of recycling this material. Ideally, iron in field application should be more than 90% Fe^0 by weight, size range from 8 to 50 mesh, and fines and residual cutting oils and greases are absent. Studies have shown that one determining factor for granular iron is the surface area per unit mass of iron (i.e. the specific area). Generally, commercial irons with a higher specific surface area are preferred. However, higher surface area requirement for reactivity should be balanced with the required hydraulic conductivities of the reactive cells. It is generally considered that the selected particle size should afford to have a reactive cell hydraulic conductivity at least five times (or more) than that of the surrounding aquifer. If needed, granular iron can be mixed with sand and gravel to improve the hydraulic conductivity along the flow path of the PRBs.

Amended granular iron also finds its use in PRBs. One of the cost-effective amendments is the use of **pyrite (FeS_2)**. Pyrite has an added benefit of reducing the pH from the use of Fe^0 because pyrite can be oxidized to produce acid which will offset the acids consumed during the oxidation of Fe^0 (see also Eq. 12.1 and Eq. 12.2):

$$FeS_2 + \frac{7}{2}O_2 + H_2O \rightleftharpoons Fe^{2+} + 2H^+ + 2SO_4{}^{2-} \qquad \text{Eq. (12.14)}$$

Pyrite has an unusual chemistry insofar as the iron is in Fe^{2+} and the sulfur occurs with an oxidation state of −1 forming the $S_2{}^{2-}$ anion. The Fe atoms occur in sixfold coordination (i.e. a coordination number of six that makes up an octahedron with eight faces). Pyrite mixed with Fe^0 has also been found to improve the degradation of carbon tetrachloride. It was also noted that adding pH-controlling amendment could result in the higher levels of dissolved Fe in the downgradient water from the reactive cells.

Zero-valent metals other than pure iron include stainless steel, mild steel, Cu^0, brass (an alloy of copper and zinc), Al^0, and galvanized (Zn^0-coated) metals. Little dehalogenation of RCl occurred with stainless steel, Cu^0, and brass, and no significant advantages of using any of these metals over Fe^0 was found. If the use of other metals is justified economically, batch studies can be conducted to assure the reactivity, complete degradation, and the absence of toxic dissolution products from the selected metal.

Bimetallic media including iron-copper (Fe—Cu), iron-palladium (Fe—Pd), or iron-nickel (Fe—Ni) are capable of reducing RCl at rates significantly higher than ZVI alone. The improved performance for Fe—Cu bimetallic media is the result of galvanic coupling of these two metals (i.e. one metal corrodes preferentially to another when both metals are in an electrical contact), whereas Pd in Fe—Pd system acts as a catalyst. In practical applications, a cost trade-off between the construction of a smaller reactive cell (because of the faster reaction rate) and the higher cost (relative to granular iron) of the new bimetallic media should be considered.

There are reported uses of other innovative reactive media, including Cercona™ iron foam, colloidal iron, nano-sized ZVI (nZVI) (see more details about nZVI in Box 12.1), ferrous iron-containing compounds (FeS, FeS_2), dithionite ($Na_2S_2O_4$), granular activated carbons, hydroapatite, oxygen-releasing compounds (ORC, e.g. magnesium peroxide), etc. **Iron foam** materials provide both porosity and high surface area based on gelation of soluble silicates with soluble aluminates. Other additives to Fe^0 include iron oxides, zeolites, clays, or specialty ceramic materials. Colloidal iron and **nano-ZVI (nZVI)** slurries are designed to be injected into the aquifer, making it possible to install a PRB anywhere a well can be installed. Sodium dithionite is injected into the subsurface to create a permeable treatment zone for ***in situ* redox manipulation** (ISRM). Activated carbons and polymer beads are examples of adsorption-based media in PRB technologies. **Hydroxyapatite** is similar to the materials that are made of animal bone, $Ca_{10}(PO_4)_6Ca(OH)_2$. Its dissolution products can precipitate toxic lead (Pb) in groundwater.

There is also reported use of waste cellulose solids (wood mulch, sawdust, and leaf compost) as the *in situ* reactive barrier for nitrate remediation in groundwater (Robertson et al. 2005). Results showed that such cellulose-containing **biobarriers** could provide NO_3^- treatment for at least a decade or longer without carbon replenishment for heterotrophic denitrifying bacteria.

Zeolites are aluminosilicate minerals, which are framework of silicates consisting of interlocking tetrahedrons of SiO_4 and AlO_4. The ratio of (Si + Al)/O in zeolite must equal 1/2. The aluminosilicate structure is negatively charged and attracts the positive cations that reside within. Unlike most other tectosilicates, zeolites have large vacant spaces or cages in their structures that allow space for large cations (e.g. sodium, potassium, barium, and calcium), and even relatively large molecules and cation groups such as water, ammonia, carbonate ions, and nitrate ions. In the more useful zeolites, the spaces are interconnected and form long wide channels of varying sizes depending on the mineral. These channels allow the easy movement of the resident ions and molecules into and out of the structure. Zeolites are characterized by their ability to lose and absorb water without damage to their crystal structures. The large channels explain the consistent low specific gravity of these minerals. Zeolites and organo-zeolites have been used in the treatment of RCl, Cr(VI), and BTEX.

Box 12.1 Nano-sized zero-valent iron in remediation: Promise and challenges

Nanomaterials are defined as the particles with the size smaller than 100 nm. The nano-ZVI (nZVI) technology has been tested for *in situ* remediation in the United States and Europe. As compared to granular ZVI, the primary advantage of nZVI is the higher reactivity due to a large surface area to volume ratio. For example, the remediation activities for nZVI were 10–1000 times more reactive than granular ZVI. The higher activities mean smaller amounts of iron to treat a contaminated plume at a faster rate, thereby reducing the costs. Unlike granular ZVI which needs a reactive wall in the groundwater flow path, nano-sized iron slurries can be directly injected into the subsurface through existing wells, or through pneumatic or hydraulic fracturing.

However, the higher reactivity of nZVI may not be sufficient to offset other potential limitations. There are several other factors to consider for the practical use of nanomaterials in soil and groundwater remediation. For example, the use of nZVI poses challenges because nZVI quickly agglomerates and oxidizes once released into the environment (Phenrat et al. 2007). The agglomeration prevents nZVI from forming stable dispersions that make their delivery into contaminated plumes very difficult. Sometimes, excessively high reactivities may not be essential in groundwater cleanup scenario because of the restriction of the slow groundwater flow and the reduced longevity of reactive medium. Nanomaterials themselves may also pose adverse impact on indigenous microorganisms, additional environmental and ecological health hazards that we need to evaluate. The toxic effects of nano-particles have been demonstrated in several animal models such as zebrafish and oyster.

Recent studies have been conducted to use different stabilizers, transporters or carriers to overcome the above-mentioned shortcomings of nZVI. For example, certain stabilizing agents can enhance steric or electrostatic repulsions between particles in order to inhibit nZVI aggregation. The surface of nZVI can also be chemically modified by blocking the reactive sites for prolonged use life of nZVI. The addition of a secondary metal such as Pd, Pt, Cu, Ni, or Ag was reported to prevent iron from oxidation and the formation of iron oxide coatings on the particle surface.

Available studies have shown that nZVI has the potential to treat recalcitrant pollutants such as chlorinated organic compounds, nitrates, and hexavalent chromium. It is also clear that much research is still needed to improve the stability and long-term reactivity of nZVI.

12.2.2.2 Reactive Media Selection

In general, suitable reactive media should exhibit the following characteristics (Gavaskar et al. 2000):

- *Sufficiently reactive:* The medium should be sufficiently reactive to allow contaminants degraded before they exit the flow through reactive cell. A medium that affords lower half-lives (faster degradation rates) is preferred.
- *Stable for years and decade:* The reactive medium should retain its reactivity for several years or decades under site-specific geochemical conditions. Since no full- or pilot-scale barrier has been operating for a sufficient length of time to make a direct determination of stability, an understanding of the reaction mechanism can provide some indication of the future behavior of the medium.
- *Highly permeable:* The particle size of the reactive medium should be selected by considering the trade-off between reactivity and hydraulic conductivity. Generally, a higher reactivity requires smaller particle size (higher total surface area), but smaller-sized particles tend to reduce hydraulic conductivity.
- *Environmentally compatible:* The reactive medium should not introduce any harmful by-products into the downgradient environment. Acceptable environmentally compatible reaction products include Fe^{2+}, Fe^{3+}, oxides, oxyhydroxides, and carbonates.

- *Readily available with low cost:* The candidate medium should be readily available in large quantities at a reasonable price. Special site considerations may sometimes justify a higher price. A less expensive medium is preferred, especially if any differences in performance are reported to be slight.
- *Compatible with construction method:* Some innovative construction methods (such as jetting) may require a finer particle size of the reactive medium.

12.3 Design and Construction Considerations

In the brief discussion that follows, the major focus will be the basic design concepts unique to PRBs, and the construction of such barriers. Like any other remediation technologies, site hydrogeological and chemical characterizations (Chapter 5) are the prerequisites for the selection of PRBs suitable for a contaminated site. Bench-, column-, and pilot-scale tests are needed to address other design factors such as the treatability, media selection, reaction rates, and site-specific groundwater chemistry.

12.3.1 Barrier Design Concept

Once data are obtained from site characterization and laboratory testing, they can be used for modeling under different hydrogeologic and geochemical scenarios and determining the location, orientation, configuration, and dimensions of the PRB. To achieve PRB design objectives, several hydrogeologic models are available for the simulation of groundwater flow and contaminant transport in a PRB system, including the widely validated MODFLOW coupled with a particle-tracking model (such as RWLK3D). Hydrogeologic modeling, along with site characterization data, is used for the following design components (Gavaskar et al. 2000):

- *Barrier Locations:* The PRB should be located to optimally capture contaminant plume while considering the site-specific constraints such as property boundary and underground utilities.
- *Barrier Orientation:* The PRB should be oriented such that it will capture the maximum flow with the minimum reactive cell width, considering the seasonal variations in flow directions.
- *Barrier Configuration:* The PRB should be configured either as a continuous reactive barrier, a funnel-and-gate system, or horizontal system suitable for the contaminated plume.
- *Barrier Dimensions:* These include suitable width and thickness of the reactive cell and the width and angle of the funnel for a funnel-and-gate system. Here, reactive cell thickness refers to the length of the groundwater flow path in the reactive medium that provides a sufficient residence (contact) time for the contaminants to be degraded to target cleanup levels.
- *Hydraulic Capture Zone:* The hydraulic capture zone should be estimated for a given PRB design.
- *Design Trade-Offs:* A trade-off between the two interdependent parameters, i.e. a large hydraulic capture zone and a small flow-through thickness of the reactive cell (gate) should be sought with the modeling.
- *Media Selection:* The modeling approach should help in determining the required particle size (hence the hydraulic conductivity of the reactive medium) with respect to the hydraulic conductivity of the aquifer.
- *Longevity Scenarios:* The modeling approach should be used to evaluate future scenarios when clogging due to precipitate formation could potentially reduce porosity and cause flow to bypass the reactive cell. This evaluation gives an indication of the safety factors needed in the design.

- *Monitoring Plan:* Appropriate monitoring well locations and monitoring frequencies should be evaluated with the modeling.

Of critical importance for the design of PRB are the two primary interdependent parameters: the hydraulic capture zone width and residence time. **Capture zone width** refers to the width of the groundwater plume that will pass through the reactive cell or gate (in the case of funnel-and-gate configurations) rather than pass around the ends of the barrier or beneath it (Figure 12.5). Capture zone width can be maximized by maximizing the discharge (groundwater flow volume) through the reactive cell or gate. **Residence time** refers to the amount of time during which contaminated groundwater is in contact with the reactive medium within the gate. Residence times can be maximized either by minimizing the discharge through the reactive cell or by increasing the flow-through thickness of the reactive cell. Thus, the design of PRBs must often balance the need to maximize capture zone width (and discharge) against the desire to increase the residence time. Contamination occurring outside the capture zone will not pass through the reactive cell. Similarly, if the residence time in the reactive cell is too short, contaminant levels may not be reduced sufficiently to meet regulatory requirements.

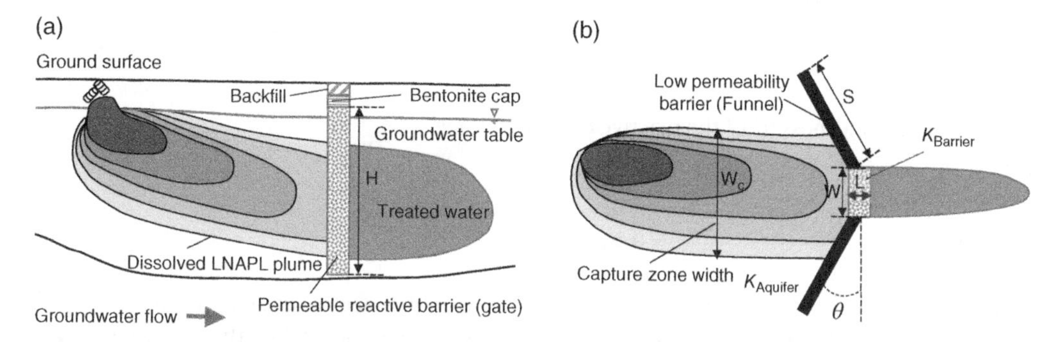

Figure 12.5 Schematics showing the design parameters of a funnel-and-gate system: (a) elevation view and (b) plan view. W = barrier width; L = barrier length (thickness); H = barrier depth; θ = the angle between funnel and gate; $K_{barrier}$ = hydraulic conductivity of the barrier cell; $K_{aquifer}$ = hydraulic conductivity of the surrounding aquifer.

The barrier thickness (L) is related to the residence time (t) and groundwater (pore) velocity (v) as follows:

$$L = \frac{vt}{R} \qquad \text{Eq. (12.15)}$$

where R is the retardation factor (see Eq. 2.18, Chapter 2) of a given contaminant. The resident time in the reactive cell can be estimated from the half-lives of the contaminants (see Example 12.2). The groundwater velocity can be determined through hydrologic modeling of the selected PRB configuration, width, and orientation.

In the detailed design of a reactive barrier system, a groundwater flow model is typically constructed to evaluate the combined effects of barrier configurations, dimensions, and aquifer or barrier hydraulic parameters on the PRB performance. In most cases, numerical models of groundwater flow are used in conjunction with particle tracking codes to construct maps showing travel paths and residence times through the reactive cell. Without going into details, the works by Benner et al. (2001) and Robertson et al. (2005) are used here for illustration.

Example 12.2 Design of a permeable reactive barrier (PRB) system

A laboratory column test has determined the half-life of four hours for the degradation of trichloroethene (TCE) into ethane using commercial zero-valent iron. If the same half-life can be assumed under the field condition, and a 95% removal of TCE is anticipated, what would be the residence time of TCE in the permeable reactive cell? If the groundwater flow velocity is 1 ft/day and the retardation factor for TCE in an iron barrier is 2, what would be the required thickness of the reactive cell?

Solution

Assuming first-order reaction in the reactive cell, we can use the same formula introduced in Chapter 9 (Eqs. 9.8–9.10). The first-order rate constant can be estimated from its half-life as follows:

$$k = \frac{0.693}{t_{1/2}} = \frac{0.693}{4\,\text{h}} = 0.173\,\text{h}^{-1}$$

By rearranging Eq. 9.9 (Chapter 9), we can calculate the residence time in the reactive barrier. Note that a 95% removal is equivalent to C_t/C_0 value of 1 – 0.95 = 0.05 (5%) or $C_t/C_0 = 0.05$.

$$t = \frac{1}{-k}\ln\frac{C_t}{C_0} = \frac{1}{-0.173}\ln(0.05) = 17.3\,\text{h}$$

The required thickness (L) of the reactive cell is as follows:

$$L = \frac{vt}{R} = \frac{\left(\frac{1\text{ft}}{day} \times \frac{1\text{d}}{24\,\text{h}} \times 17.3\,\text{h}\right)}{2} = 0.36\ \text{ft}\ (11\,cm)$$

Note that the term "v" is the groundwater flow velocity in the reactive cell rather than the surrounding aquifer. This velocity term takes porosity into account, rather than the Darcy velocity (see Chapter 3).

Using a 3-D finite element model WATFLOW, Benner et al. (2001) simulated how the geometry of reactive barriers and heterogeneities of hydraulic conductivity in both aquifer and reactive cell impact the flow distribution. Heterogeneities in hydraulic conductivities (K) within the aquifer-barrier system will result in higher groundwater flow velocity and hence the reduced residence times due to the preferential flow through the portions of the barrier. The preferential flow is more significant for heterogeneous K in aquifer for thinner barriers, and homogeneous K in the barrier for thicker barriers. Using a 2-D steady-state finite-element FLONET model, Robertson et al. (2005) demonstrated that the width of reactive cells exhibits more significant effect than the depth on the nitrate removal in a shallow high-permeability sand-and-gravel aquifer. In this case, the PRB does not necessarily have to be installed in full depth of the contaminant plume to be effective.

12.3.2 Construction Methods

The construction of PRB generally requires the excavation of a trench that will house the reactive medium. It is the leading cost component of any PRB system. The conventional methods for excavation include backhoes, clamshell, caissons, continuous trenchers, or augers. These are elaborated as follows.

Backhoes are the most commonly used for conventional trench excavation. The standard backhoe excavation for shallow trenches (<30 ft deep) is the least expensive and fastest method available. This equipment is staged on a crawler-mounted vehicle and consists of a boom, a dipper stick with a mounted bucket, and either cables or hydraulic cylinders to control motion. Bucket widths generally range in sizes up to 5.6 ft. Because the vertical reach of a backhoe is governed by the length of the dipper stick, backhoes can be modified with extended dipper sticks and are capable of reaching depths up to 80 ft.

Deeper excavation can be achieved using a **clamshell**, which can be used for excavation to around 200 ft bgs. Mechanical clamshells are preferred over their hydraulic counterparts because they are more flexible in soils with boulders, can reach greater depths, and involve lower maintenance costs. Clamshell excavation is popular because it is efficient for bulk excavations of almost any type of material except highly consolidated sediment and solid rock. It can also be controlled and operated in small and very confined areas if the boom can reach over the trench. Clamshell excavation, however, has a relatively low production rate compared to a backhoe.

Caissons are load-bearing enclosures that are used to protect an excavation. They are constructed such that the water can be pumped out, keeping the working environment inside the caisson dry. The interior of the caisson is excavated with a large auger to make room for the reactive medium. Caissons as large as 15 ft in diameter have been used in bridge construction; however, smaller diameter caissons are more common for the installation of PRBs. Highly consolidated sediments and cobbles create difficulties in driving in and pulling out the caisson. It also may be difficult to drive a caisson to depths greater than about 45 ft. In the absence of such geotechnical difficulties, caissons have the potential to provide a relatively inexpensive way to install a funnel-and-gate system or a continuous reactive barrier. One significant advantage of using caissons is that they require no internal bracing. Therefore, the caisson can be installed from the ground surface and completed without requiring entry of personnel into the excavation. It also can be installed without significant dewatering in the excavation.

A **continuous trencher** is an option for installing shallow reactive barriers 35–40 ft deep. Although it is not as common as backhoes or clamshells because of the depth restriction, it is capable of simultaneously excavating a narrow, 12- to 24-in-wide trench and immediately refilling it with either a reactive medium and/or a continuous sheet of impermeable, high-density polyethylene (HDPE) liner. The trencher operates by cutting through soil using a chain-saw type apparatus attached to the boom of a crawler-mounted vehicle. The boom is equipped with a trench box, which stabilizes the trench walls as a reactive medium is fed from an attached, overhead hopper into the trailing end of the excavated trench. The hopper contains two compartments, one of which can install media up to gravel size. The other compartment is capable of simultaneously unrolling a continuous sheet of HDPE liner if desired.

A hollow-stem **auger** or a row of hollow-stem augers can also be used to drill holes up to 30 inch in diameter into the ground. When the desired PRB depth is reached, reactive medium is introduced through the stem as the auger is withdrawn. Alternately, the reactive medium can be mixed with a biodegradable slurry and pumped through the hollow stem. By drilling a series of overlapping holes, a continuous PRB can be installed.

The construction methods discussed above all involve the excavation of a trench to house the reactive medium. The economics of excavation methods are strongly correlated with the depth of the PRB installation. The deeper the excavation, the more expensive the effort becomes. **Innovative installation methods** that introduce the reactive medium directly into the ground without first excavating a trench are being tested at some sites. Innovative installation methods, such as jetting, hydraulic fracturing, vibrating beam, deep soil mixing, and the use of mandrels, have been tested at some sites and offer potentially lower-cost alternatives for installing reactive media at greater depths. Jetting or jet grouting involves the injection of grout (sometimes with air and/or water) at high pressures into the ground. The high-velocity jet erodes the soil and replaces some or all of the soil with grout. A mandrel is a hollow steel shaft used to create a vertical void space in the ground for the purpose of placing reactive media. A sacrificial drive shoe is placed over the bottom end of the mandrel prior to being hammered down through the subsurface using a vibratory hammer. Once the void space is created, it can be filled with a reactive medium. Details of these methods can be found in Gavaskar et al. (2000).

Case Study 1 Field-scale PRB at the Denver Federal Center, Denver, Colorado, USA

A 366-m long PRB with ZVI was installed at the Denver Federal Center, Denver, Colorado to intercept groundwater plume containing chlorinated aliphatic hydrocarbons (CAHs), including TCE, TCA, *cis*-1,2-dichloroethane, 1,2-dichloroethene (1,2-DCE), and vinyl chloride. The PRB filled with ZVI consists of four 12.2-m wide gates with thickness between 0.61 and 1.83 m. The funnel was

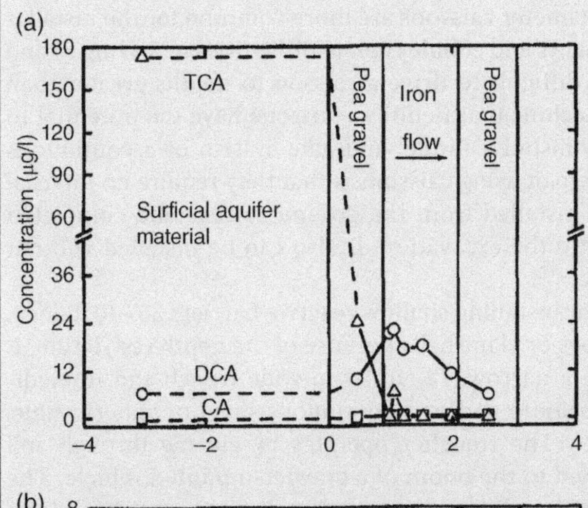

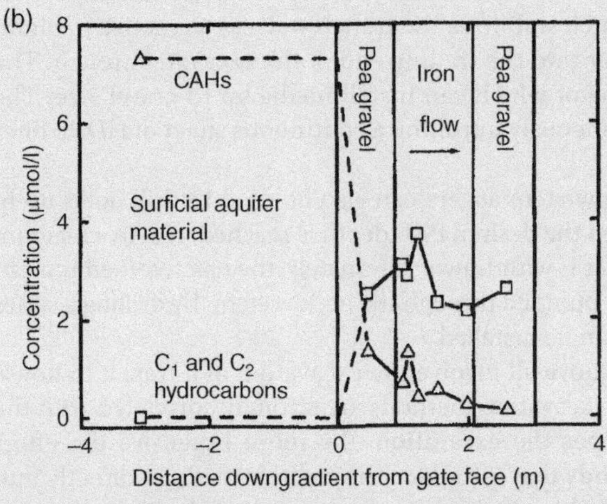

Figure 12.6 Concentrations of (a) trichloroethane (TCA), dichloroethane (DCA), and chloroethane (CA); and (b) organic carbon as chlorinated aliphatic hydrocarbons (CAHs) and as C_1 and C_2 hydrocarbons in groundwater measured in piezometers in gates. Dashed lines infer that the concentration in groundwater entering the upgradient gravel equal the concentration in groundwater 3 m upgradient form the gate face. The decrease in concentration in the upgradient gravel indicates that Fe^0 was not completely excluded from this zone during construction. *Source:* McMahon et al. (1999); Reprinted with permission from John Wiley & Sons.

interlocking metal sheet pile driven into the unweathered bedrock at a depth of 7.5–10 m (McMahon et al. 1999). The funnel-to-gate length ratio was 6.5, and each gate was 3.05 m long in the direction of groundwater flow. The gates contained pea gravel on the upgradient and downgradient ends of the ZVI in the middle. The grain size of the iron was 0.25–2.0 mm with density of 6.98 g/cm^3.

Groundwater mounding on the upgradient side of the PRB resulted in a 10-fold increase in the hydraulic gradient and groundwater velocity through the gates compared to the aquifers unaffected by the PRB. Results show that 75% of the groundwater was moved toward the PRB from the upgradient areas and through the gates. There was some indication of groundwater flow underneath and around certain gates.

Combined groundwater flow and chemical monitoring data over approximately two-year post PRB installation indicated that greater than 99% of the contaminant mass entered the gates, ZVI effectively removed CAHs from groundwater as they moved through the gates. The amount of DCA in the water exiting the gates (Figure 12.6a) represents only 0.7% of the total contaminant mass entering the gates. The presence of DCA is the result of stepwise dehalogenation of TCA and the relative resistance of DCA to degradation by ZVI. The C_1 and C_2 hydrocarbons shown in Figure 10.6b are the end products of the complete dehalogenation of TCA, including acetylene, ethane, ethene, and methane. Monitoring data revealed that 51% of carbon in CAH could be counted for by these dehalogenated end-products. The carbon mass balance unaccounted for may indicate the adsorption onto iron in the gates.

Case Study 2 PRB at the U.S. Coast Guard Support Center in North Carolina, USA

At this site, chromium was released from a hard-chrome plating shop for 30 years until its closure. Soils beneath the shop contained chromium up to 14 500 mg/kg, and dissolved chromate plume with concentration of 10 mg/L extended from beneath the shop to the river only 60 m away. The plume was approximately 35 m wide, extended to a depth of 6.5 m below the ground surface, and extended laterally approximately 60 m to the river. The PRB has a dimension of 46 m long, 7.3 m deep, and 0.6 m wide with a continuous wall configuration of zero-valent iron. This was installed approximately 30 m from the river. The PRB system was operated from 1996 to 2004 for its long-term performance (Wilkin et al. 2005).

After eight years of operation, the PRB remains effective at reducing Cr(VI) from the average concentration of >1500 μg/L in the upgradient of the PRB to the average concentration of <1 μg/L in the groundwater downgradient of the PRB. In multiple well testing over the eight-year period from 1996 to 2004, dissolved Cr concentration downgradient of the PRB never exceeded 5 μg/L in 117 of 119 sampling events. The highest concentration of total Cr observed in the well positioned downgradient of the PRB was 6 μg/L, which is significantly lower than the maximum concentration limit (MCL) of 100 μg/L.

A 2-D concentration contour map from the monitoring data is depicted in Figure 12.7. The long-term trends indicated that Cr continues to be removed from the groundwater plume after eight years of operation. This figure also reveals that the depth of the Cr plume has progressively decreased at a rate of approximately 0.1 m/year, and the concentrations of the Cr entering the PRB appear to be decreasing with time. It was estimated that the PRB removed 4.1 kg Cr per year, or about 33 kg of Cr has been sequestrated into immobile forms in the soil after eight years. Further experimental evidence showed that chromium is predominately in its trivalent Cr(III) sequestrated in the reactive media, and part of the Cr(III) is associated with iron sulfide grains formed as a result of microbially mediated sulfate reduction in and around the PRB.

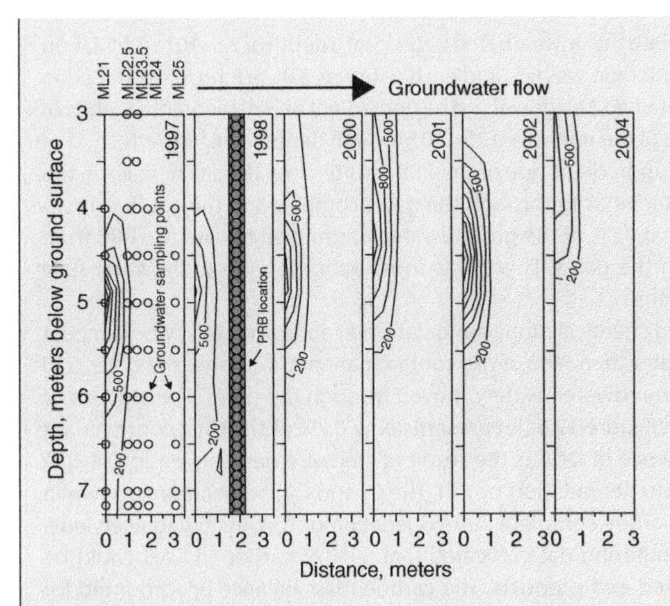

Figure 12.7 Cross-sectional profiles showing the concentration distribution of total Cr in groundwater relative to the position of the PRB. Data collected in 1999 and 2003 are not shown. Contours are based on concentration data from 44 subsurface sampling points. The contour interval is 300 μg Cr/L.
Source: Reprinted with permission from Wilkin et al. (2005). Copyright (2005) American Chemical Society.

The eight-year performance data cited from this case study provided compelling evidence on the longevity of reactive zero-valent iron under the field condition. Life cycle assessment further reveals significant environmental advantages and sustainability of PRB over the labor and energy-intensive pump-and-treat system (Box 12.2, Figure 12.8).

Box 12.2 PRB versus pump-and-treat: Life cycle assessment

Higgins and Olson (2009) performed the comparison between the permeable reactive barrier and the conventional pump-and-treat for the remediation of a chlorinated-solvent-contaminated site. It was surmised that the greater material production requirements to install PRBs may offset the reduced impact during the operational phase. Such trade-offs were quantitatively analyzed using the life cycle assessment (LCA).

The LCA results show that the environmental impacts of P&T technology are driven by the operational energy demand, whereas the ZVI reactive media and the energy usage during the construction of ZVI medium drives the potential environmental impacts of PRB (Figure 12.8). Trade-offs depend on the longevity of the reactive medium. Regardless, a conservative estimate based on a relatively short longevity demonstrates that PRB offers significant environmental advantages in impact categories of human health and ozone depletion. A minimal longevity of 10 years is required to exert benefit in all impact categories of PRB over P&T.

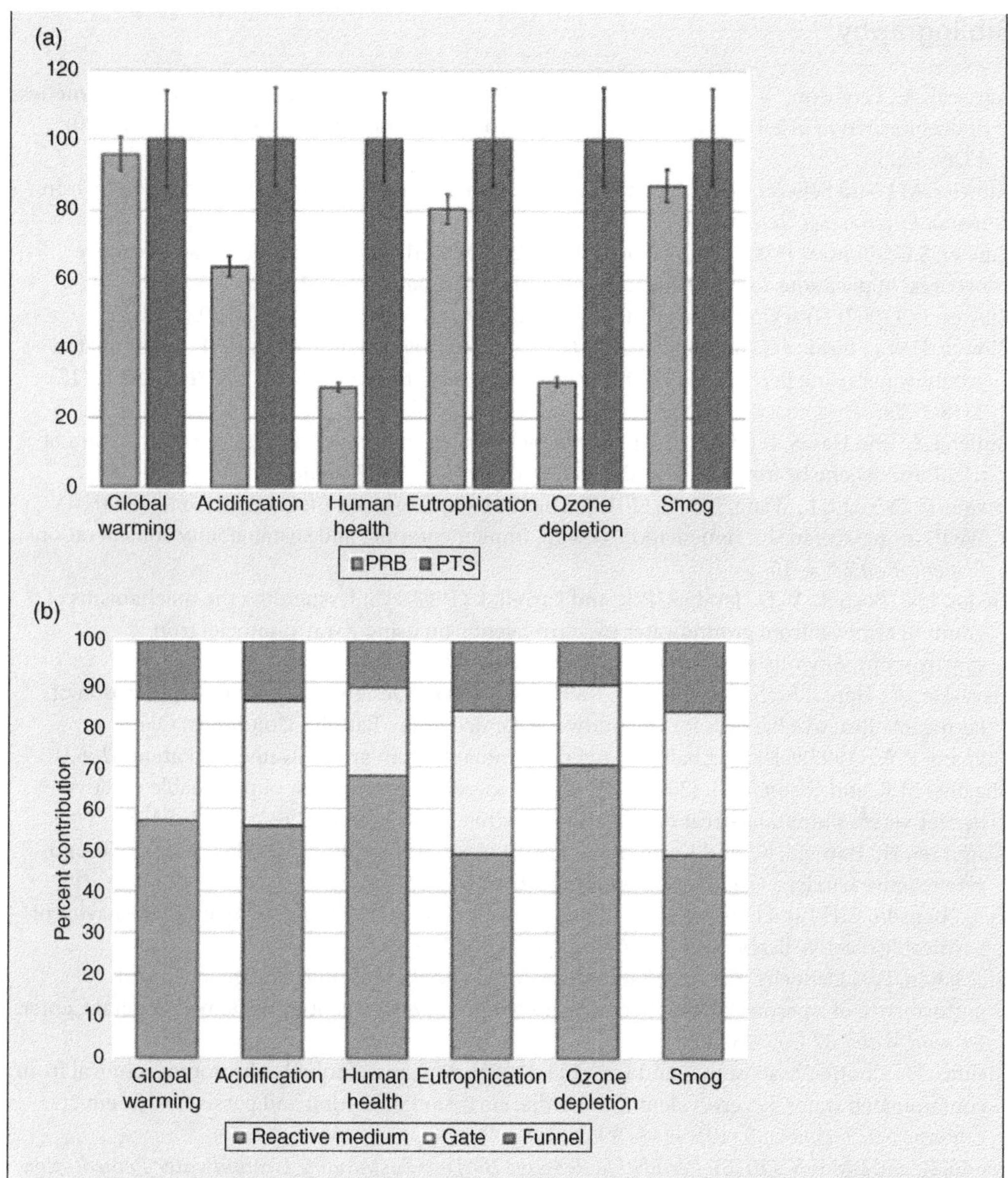

Figure 12.8 (a) Relative impacts of PRB compared with pump-and-treat. Results are normalized by the greatest value in each impact category. Error bars represent 95% confidence intervals as determined by Monte Carlo simulation. (b) Subsystem contributions to impact categories. Replacement of the gate and ZVI are assumed to occur every 10 years. *Source:* Reprinted with permission from Higgins and Olson (2009). Copyright (2009) American Chemical Society.

Bibliography

Agrawal, A., Ferguson, W.J., Gardner, B.O. et al. (2002). Effects of carbonate species on the kinetics of dechlorination of 1,1,2,-trichloroethane by zero-valent iron. *Environ. Sci. Technol.* 36 (20): 4326–4333.

Alowitz, M.J. and Scherer, M.M. (2002). Kinetics of nitrate, nitrite, and Cr (VI) reduction by iron metal. *Environ. Sci. Technol.* 36 (3): 299–306.

Benner, S.G., Blowes, D.W., and Molson, J.W.H. (2001). Modeling preferential flow in reactive barriers: implications for performance and design. *Ground Water* 39 (3): 371–379.

Blowes, D. (2002). Tracking hexavalent Cr in groundwater. *Science* 295 (5562): 2024–2025.

Blowes, D.W., Ptacek, C.J., and Jambor, J.L. (1997). *In-situ* remediation of Cr (VI)-contaminated groundwater using permeable reactive walls: laboratory studies. *Environ. Sci. Technol.* 31 (12): 3348–3356.

Butler, E.C. and Hayes, F. (2001). Factors influencing rates and products in the transformation of trichloroethylene by iron sulfide and iron metal. *Environ. Sci. Technol.* 35 (19): 3884–3891.

Divine, C., Wright, J., Wang, J. et al. (2018). The horizontal reactive media treatment well (HRX Well®) for passive in situ remediation: design, implementation, and sustainability considerations. *Remediation* 28: 5–16.

Fiedor, J.N., Bostick, W.D., Jarabek, R.J., and Farrell, J. (1998). Understanding the mechanisms of uranium removal from groundwater by zero-valent iron using X-ray photoelectron spectroscopy. *Environ. Sci. Technol.* 32 (10): 1466–1473.

Gavaskar, A., Gupta, N., Sass, B., Janosy, R., Hicks, J. (2000).Design Guidance for Application of Permeable Reactive Barriers for Groundwater Remediation, Battelle, Columbus, OH.

Gillham, R.W. (1993).Cleaning halogenated contaminants from groundwater, US Patent 5266213A.

Higgins, M.R. and Olson, T.M. (2009). Life-cycle case study comparison of permeable reactive barrier versus pump-and-treat remediation. *Environ. Sci. Technol.* 43 (24): 9432–9438.

Klammler, H., Hatfield, K., and Kacimov, A. (2010). Analytical solutions for flow near drain-and-gate reactive barriers. *Ground Water* 48 (3): 427–437.

Li, L., Benson, C.H., and Lawson, E.M. (2005). Impact of mineral fouling on hydraulic behavior of permeable reactive barriers. *Ground Water* 43 (4): 582–596.

McMahon, P.B., Dennehy, K.F., and Sandstrom, M.W. (1999). Hydraulic and geochemical performance of a permeable reactive barrier containing zero-valent iron, Denver Federal Center. *Ground Water* 37 (3): 396–404.

Melitas, N., Chuffe-Moscoso, O., and Farrell, J. (2001). Kinetics of soluble chromium removal from contaminated water by zero-valent iron media: corrosion inhibition and passive oxide effects. *Environ. Sci. Technol.* 35 (19): 3948–3953.

Naidu, R. and Birke, V. (2015). *Permeable Reactive Barrier: Sustainable Groundwater Remediation.* Boca Raton, FL: CRC Press.

Nyer, E.K. (2001). Chapter 11: permeable treatment barriers. In: *In Situ Treatment Technology*, 2e (ed. E.K. Nyer, P.L. Palmer, E.P. Carman, et al.). Lewis Publishers.

Phenrat, T., Saleh, N., Sirk, K. et al. (2007). Aggregation and sedimentation of aqueous nanoscale zerovalent iron dispersions. *Environ. Sci. Technol.* 41 (1): 284–290.

Ponder, S.M., Darab, J.G., and Mallouk, T.E. (2000). Remediation of Cr(VI) and Pb(II) aqueous solution using supported, nanoscale zero-valent iron. *Environ. Sci. Technol.* 34 (12): 2564–2569.

Robertson, W.D., Blowes, D.W., Ptacel, C.J., and Cherry, J.A. (2000). Long-term performance of *in situ* reactive barriers for nitrate remediation. *Ground Water* 38 (5): 689–695.

Robertson, W.D., Yeung, N., van Driel, P.W., and Lombardo, P.S. (2005). High-permeability layers for remediation of ground water: go wide, no deep. *Ground Water* 43 (4): 574–581.

Scherer, M.M., Richter, S., Valentine, R.L., and Alvarez, P.J.J. (2000). Chemistry and microbiology of permeable reactive barriers for *in situ* groundwater clean up. *Crit. Rev. Environ. Sci. Technol.* 30 (3): 363–411.

Su, C. and Puls, R.W. (2001). Arsenate and arsenite removal by zerovalent iron: kinetics, redox transformation, and implications for *in situ* groundwater remediation. *Environ. Sci. Technol.* 35 (7): 1487–1492.

SweenyK. H., and FischerJ. R. (1972). Decomposition of halogenated pesticides, U. S. Patent, 3737384.

The Interstate Technology & Regulatory Council (ITRC) (2005). Permeable Reactive Barriers: Lessons Learned/New Directions.

The Interstate Technology & Regulatory Council (ITRC) (2011). Permeable Reactive Barrier: Technology Update.

Tuhela, L., Carlson, L., and Tuovinen, O.H. (1997). Biogeochemical transformations of Fe and Mn in oxic groundwater and well water environments. *J. Environ. Sci. Health. Part A: Environ. Sci. Eng. Toxic Haz Substances Control* A32: 407–426.

USEPA(1998). Permeable Reactive Barrier Technologies for Contaminant Remediation, EPA/600/R-98/125.

USEPA (2002). Field Application of *In Situ* Remediation Technologies: Permeable Reactive Barriers, 68-W-00-084.

Wilkin, R.T., Su, C., Ford, R.G., and Paul, C.J. (2005). Chromium-removal processes during groundwater remediation by a zero valent iron permeable reactive barrier. *Environ. Sci. Technol.* 39 (12): 4599–4605.

Questions and Problems

1 What makes PRBs a viable option over the conventional pump-and-treat systems?

2 Zero-valent iron (ZVI) is reactive to dissolved oxygen (O_2) in an aerobic groundwater condition, and it is reactive to water when O_2 is depleted. Write the potential redox reactions (a) between ZVI and O_2 in a well-oxygenated groundwater, (b) between ZVI and water in an anaerobic groundwater condition. Explain how these reactions may affect groundwater pH?

3 Illustrate dechlorination mechanism of chlorinated solvents (RCl) by zero-valent iron.

4 Describe the electron transfer mechanism between zero-valent iron and chlorinated solvents.

5 The exact mechanism for the dechlorination of carbon tetrachloride (CCl_4) by zero-valent iron is unclear. If the following overall reaction is proposed:

$$CCl_4 + 4Fe + 4H^+ \rightleftharpoons 4Fe^{2+} + CH_4 + 4Cl^-$$

Show the electron transfer and the oxidation number change for C and Fe in the above reaction.

6 Using the knowledge of electron transfer, develop the two-half reactions that lead to the overall reaction (Eq. 12.13) for the reduction of penta-valent $HAsO_4^{2-}$ into zero-valent As^0 using ZVI.

7 How is the reactivity of elemental iron decreased due to the passivation of iron in groundwater condition?

8 Describe the potential use of the following reactive media in groundwater remediation: (a) hydroapatite, (b) Fe-Pd, (c) zeolite, (d) cellulose, (e) dithionite.

9 What is biofouling of PRB, and what are the common mechanisms for biofouling?

10 What is the problem associated with iron bacteria in groundwater remediation?

11 The half-life of carbon tetrachloride (CT) in a ZVI-mediated reactive cell is 5 h for its degradation into methane as the end-product. The initial CT concentration is 1 mg/L. For the target concentration to achieve its drinking water standard of 5 μg/L, what would be the residence time of CT in the permeable reactive cell? If the groundwater flow velocity is 0.5 ft/day, what would be the required thickness of the reactive cell? Assume retardation factor $R = 1$ for a conservative estimate of the thickness of the PRB wall.

12 In a batch study using granular zero-valent iron, the degradation of atrazine was found to be the first-order with a half-life of 8.91 days under neutral pH condition. The granular iron was mixed with gravel and sand in a permeable reactive barrier to remediate atrazine in groundwater in a farm land that receives a great deal of pesticide applications. The average concentration of atrazine in groundwater was approximately 0.03 mg/L. The clean-up goal is to set at the maximum contaminant level (MCL) of 0.003 mg/L. What would be the residence time in the permeable reactive cell? If the groundwater flow velocity is 1 m/day, what would be the required thickness of the reactive cell? Assume retardation factor $R = 10$ for atrazine in iron-gavel-sand mixture in the PRB wall.

13 Is Fe^0 frequently replaced in a well operated reactive cell? What is the primary O&M cost-item for PRB technology?

14 Why the particle size of granular ZVI is critical, and it should not be too big and not too small?

15 How can colloidal iron and nano-ZVI be deployed in groundwater remediation?

16 Describe the important factors in selecting reactive barrier materials.

17 Why are the capture zone width and residence time important in the design of PRB?

18 What are the commonly used installation methods for reactive barriers in groundwater remediation?

19 What are some innovative methods for the construction of reactive barrier materials?

Chapter 13

Modeling of Groundwater Flow and Contaminant Transport

LEARNING OBJECTIVES

1. Apply Darcy's law and the law of conservation of mass (continuity principle) to the development of groundwater flow and contaminant transport equations
2. Develop the governing equations for groundwater flow in saturated zone under steady-state condition and saturated zone under transient condition
3. Develop the governing equations for groundwater flow in unsaturated zone following the needed modifications
4. Develop the governing equations for contaminant transport in saturation zone considering advection and dispersion
5. Be able to incorporate other contaminant transport processes (adsorption and reaction) into the governing equations
6. Understand the conceptual equations governing the contaminant transport in unsaturated zone
7. Understand the concepts and processes relevant to multiphase and multicomponent flow and transport in porous media
8. Understand the framework for developing the governing equations for NAPL transport in saturated and unsaturated zones
9. Be familiar with the analytical solutions basic to the flow and transport study, such as Dupuit equation, 1-D column with slug and continuous injection
10. Be able to depict contaminant plumes derived from the analytical solutions to several flow and transport processes (e.g. 1-D, 2-D slug and continuous sources)
11. Learn the basic framework of solving partial differential equations using numerical methods and specifically the most common finite difference method

Throughout this text, and particularly in Chapters 2 and 3, the hydraulics of groundwater flow and physical, chemical, and biological transport processes of contaminants were discussed. This chapter will apply the law of conservation of mass (continuity principle) to develop the basic mathematical framework by incorporating hydraulic, physical, chemical, and biological processes, such as advection (Darcy's law), diffusion (Fick's law), sorption, and biodegradation. Our primary approach here is to identify the key flow and transport processes under various soil and groundwater conditions (saturated and unsaturated zones), as well as contaminant scenarios (dissolved solutes, nonaqueous phase liquid, and vapor) and then mathematically delineate these equations that govern the flow and transport processes. These equations are often partial differential equations that require a good mathematical comprehension to obtain solutions. Fortunately, from the practical standpoint, understanding these governing equations is more important than solving them mathematically. Free or commercial programs and software imbedding these mathematical models are readily available to solve these equations either

Soil and Groundwater Remediation: Fundamentals, Practices, and Sustainability, First Edition. Chunlong Zhang.

Companion website: www.wiley.com/go/Zhang/Remediation_1e

analytically or numerically. Analytical solutions, available under simplified conditions, will be described for several flow and transport processes. The concept of numerical solutions for more complex partial differential equations will also be introduced briefly. This chapter is for readers who desire to have a quantitative understanding of groundwater flow and contaminant transport, as a step stone for more advanced discussion of flow and transport modeling. Other readers can skip the mathematical details of this chapter.

13.1 Governing Equations for Groundwater Flow

In this section dealing with groundwater flow, we will start with the simple flow condition in a saturated aquifer and steady-state condition. Transient flow, which requires the inclusion of time as an additional independent variable in the equations, will be discussed next. We will proceed our discussion for a more complicated flow in an unsaturated aquifer.

13.1.1 Saturated Groundwater Flow under Steady-State Condition (Laplace Equation)

The law of conservation of mass, or **continuity principle**, states that there is no net change in the mass of a fluid (e.g. water) contained in a given volume of an aquifer. That is, mass flux into a control volume – mass flux out of a control volume = change in mass per unit time. Here **flux** is defined as mass over time.

Let us consider a **control volume** with infinitesimally small distances of dx, dy, and dz. in the x-, y-, and z-axes, and the specific discharges (i.e. Darcy velocity in L/T) are q_x, q_y, and q_z in the x, y, and z directions, respectively (Figure 13.1). In the x direction, the mass flux into the control volume is $\rho q_x \mathrm{d}y\mathrm{d}z$, and the mass flux out of this control volume is $\rho q_x \mathrm{d}y\mathrm{d}z + (\partial/\partial x)\rho q_x \mathrm{d}x\mathrm{d}y\mathrm{d}z$, where ρ is the density of water (M/L^3), dydz is the cross-sectional area (L^2) of the face normal to the x-axis. Therefore, mass flux into a control volume – mass flux out of a control volume along the x-axis is as follows:

$$\rho q_x \mathrm{d}y\mathrm{d}z - \left(\rho q_x \mathrm{d}y\mathrm{d}z + \frac{\partial}{\partial x}\rho q_x \mathrm{d}x\mathrm{d}y\mathrm{d}z\right) = -\frac{\partial}{\partial x}\rho q_x \mathrm{d}x\mathrm{d}y\mathrm{d}z \quad \text{Eq. (13.1)}$$

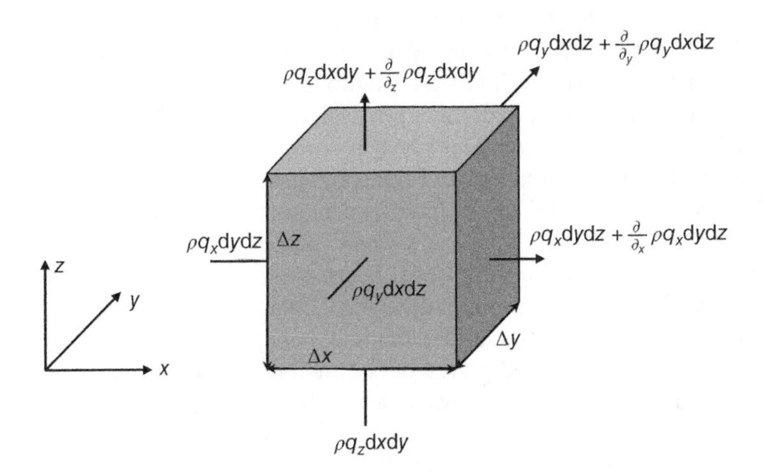

Figure 13.1 Control volume and mass balance for fluid (groundwater) flow in porous media.

The symbol ∂ is cursive type of the letter d, pronounced as dee. It represents a partial derivative, since the flux is a function of more than one independent variable (x, y, and z).

Similarly, we can derive "mass flux into a control volume – mass flux out of a control volume" or the net change of mass flux along the other axes (y and z). Combining these three terms, we can obtain the net total accumulation of mass in the control volume (Fetter 2001):

$$\begin{aligned}\frac{\partial M}{\partial t} &= -\frac{\partial}{\partial x}\rho q_x \mathrm{d}x\mathrm{d}y\mathrm{d}z - \frac{\partial}{\partial y}\rho q_y \mathrm{d}y\mathrm{d}x\mathrm{d}z - \frac{\partial}{\partial z}\rho q_z \mathrm{d}z\mathrm{d}x\mathrm{d}y \\ &= -\left(\frac{\partial}{\partial x}\rho q_x + \frac{\partial}{\partial y}\rho q_y + \frac{\partial}{\partial z}\rho q_z\right)\mathrm{d}x\mathrm{d}y\mathrm{d}z\end{aligned} \quad \text{Eq. (13.2)}$$

If n is used to denote porosity, then the volume of water in the control volume is $n\mathrm{d}x\mathrm{d}y\mathrm{d}z$. The change of mass of water over time is then:

$$\frac{\partial M}{\partial t} = \frac{\partial}{\partial t} n\rho \mathrm{d}x\mathrm{d}y\mathrm{d}z \quad \text{Eq. (13.3)}$$

Now, we return to the mass balance statement by equating Eqs. (13.2) and (13.3):

$$-\left(\frac{\partial}{\partial x}\rho q_x + \frac{\partial}{\partial y}\rho q_y + \frac{\partial}{\partial z}\rho q_z\right)\mathrm{d}x\mathrm{d}y\mathrm{d}z = \frac{\partial}{\partial t} n\rho \mathrm{d}x\mathrm{d}y\mathrm{d}z \quad \text{Eq. (13.4)}$$

Equation (13.4) can be simplified by removing $\mathrm{d}x\mathrm{d}y\mathrm{d}z$ on both sides:

$$-\left(\frac{\partial}{\partial x}\rho q_x + \frac{\partial}{\partial y}\rho q_y + \frac{\partial}{\partial z}\rho q_z\right) = \frac{\partial}{\partial t} n\rho \quad \text{Eq. (13.5)}$$

Equation (13.5) is an important equation based on the mass balance. We will revisit this equation to derive the equations of groundwater flow under various conditions. For now, we consider a special case when the aquifer is at its steady-state. For a steady-state aquifer, the right hand side of the equation "change in mass per unit time"($(\partial/\partial t)n\rho \mathrm{d}x\mathrm{d}y\mathrm{d}z$) is zero, then Eq. (13.5) becomes:

$$\frac{\partial}{\partial x}\rho q_x + \frac{\partial}{\partial y}\rho q_y + \frac{\partial}{\partial z}\rho q_z = 0 \quad \text{Eq. (13.6)}$$

For an incompressible fluid, the density of water (ρ) is constant, Eq. (13.6) becomes:

$$\frac{\partial}{\partial x} q_x + \frac{\partial}{\partial y} q_y + \frac{\partial}{\partial z} q_z = 0 \quad \text{Eq. (13.7)}$$

We now apply Darcy's law by substituting
$q_x = -K_x(\mathrm{d}h/\mathrm{d}x)$, $q_y = -K_y(\mathrm{d}h/\mathrm{d}y)$, $q_z = -K_z(\mathrm{d}h/\mathrm{d}z)$:

$$\frac{\partial}{\partial x}\left(-K_x\frac{\mathrm{d}h}{\mathrm{d}x}\right) + \frac{\partial}{\partial y}\left(-K_y\frac{\mathrm{d}h}{\mathrm{d}y}\right) + \frac{\partial}{\partial z}\left(-K_z\frac{\mathrm{d}h}{\mathrm{d}z}\right) = 0 \quad \text{Eq. (13.8)}$$

If we further assume an **isotropic** ($K_x = K_y = K_z = K$) and **homogeneous** (K = constant) porous medium, then:

$$\frac{\partial^2 h}{\partial x^2} + \frac{\partial^2 h}{\partial y^2} + \frac{\partial^2 h}{\partial z^2} = 0 \quad \text{Eq. (13.9)}$$

Equation (13.9) is **Laplace equation** derived for the flow of incompressible water in an isotropic and homogeneous aquifer. Since it is a steady state, time (t) is no longer an independent variable. The solution to Eq. (13.9) is the hydraulic head (h) at any point in the three-dimensional flow domain, $h = h(x, y, z)$. This allows us to construct a contoured equipotential map of h, and a flow net by adding the flow lines perpendicular to the equipotential lines. Laplace equation is one of the most studied partial differential equation, and the numerical approach to the solution will be discussed in Section 13.4. Laplace equation with source, i.e. the Poisson's equation is given in Box 13.1.

Box 13.1 Laplace equation with source – the Poisson's equation

Equation (13.9) is the general equation used to describe a steady-state flow through an isotropic and homogeneous aquifer without any source of groundwater recharge (e.g. precipitation) or sink of groundwater removal (e.g. extraction well). When sources or sinks are important, the mass balance equation derived need to be modified. This modified Laplace equation is called **Poisson's equation**. In a 2-D x and y system,

$$\frac{\partial^2 h}{\partial x^2} + \frac{\partial^2 h}{\partial y^2} = \frac{-R(x,y)}{T}$$

where $R(x, y)$ is the volume of recharge (water added) per unit time per unit aquifer area (L/t) and T is the transmissivity of the aquifer which is equal to the hydraulic conductivity times the saturated thickness (L^2/t). Note that recharge (source) is defined as negative and discharge (sink) is defined as positive.

13.1.2 Saturated Groundwater Flow under Transient Condition

We now use Eq. (13.5) as a starting equation to derive the transient flow equation by including time (t) as an independent variable. Using the product rule in calculus $((d/dt)(x \cdot y) = x(dy/dt) + y(dx/dt))$, Eq. (13.5) can be rewritten as follows:

$$-\left(\frac{\partial}{\partial x}\rho q_x + \frac{\partial}{\partial y}\rho q_y + \frac{\partial}{\partial z}\rho q_z\right) = \rho\frac{\partial n}{\partial t} + n\frac{\partial \rho}{\partial t} \qquad \text{Eq. (13.10)}$$

The two terms on the right hand side of Eq. (13.10) represent two mechanisms for an aquifer to produce water. The first term is the mass rate of water produced by the compaction of the porous medium due to the change in its porosity (n). The change in porosity reflects the rearrangement of soil particles due to the compaction of an aquifer. The second term is the mass rate of water produced due to the compression of water (hence the change in density ρ). Since the changes in n and ρ are both produced by a change in hydraulic head (h), the volume of water released (per unit aquifer volume) by two mechanisms for a unit decline in head is the specific storage S_s (unit: 1/L). The mass production rate of water (i.e. rate of change of fluid mass storage in the unit of M/[$L^3 \times$ T]) is $\rho S_s \partial h/\partial t$ (Freeze and Cherry 1979). Equation (13.10) now becomes:

$$-\left(\frac{\partial}{\partial x}\rho q_x + \frac{\partial}{\partial y}\rho q_y + \frac{\partial}{\partial z}\rho q_z\right) = \rho S_s \frac{\partial h}{\partial t} \qquad \text{Eq. (13.11)}$$

The specific storage in Eq. (13.11) can be considered as the proportionality constant relating the change in water volume produced per unit decline in head. The specific storage has a dimension of 1/L (e.g. 1/m), which is different from the dimensionless storativity (S) introduced in Section 3.2.3.3 in Chapter 3. The specific storage is related to the **compressibility of aquifer** (α) and the **compressibility of water** (β) by

$$S_s = \rho g(\alpha + n\beta) \qquad \text{Eq. (13.12)}$$

The Addendum at the end of this chapter has the derivation of Eq. (13.12), which should help the readers understand two important mechanisms for the release of groundwater from aquifer storage. Expending the terms on the left hand side and recognizing the terms of the form $\rho\partial q_x/\partial x$ is much greater than terms of the form $q_x\partial\rho/\partial x$ allow us to eliminate ρ from both sides of Eq. (13.11) (Freeze and Cherry 1979). Inserting Darcy's equation, we have:

$$\frac{\partial}{\partial x}\left(K_x\frac{\partial h}{\partial x}\right)+\frac{\partial}{\partial x}\left(K_y\frac{\partial h}{\partial x}\right)+\frac{\partial}{\partial x}\left(K_z\frac{\partial h}{\partial x}\right)=S_s\frac{\partial h}{\partial t} \qquad \text{Eq. (13.13)}$$

Equation (13.13) is applicable for groundwater flow in a saturated anisotropic porous medium. If the aquifer is **isotropic** ($K_x = K_y = K_z = K$) and **homogeneous** (K = constant), Eq. (13.13) becomes:

$$\frac{\partial^2 h}{\partial x^2}+\frac{\partial^2 h}{\partial y^2}+\frac{\partial^2 h}{\partial z^2}=\frac{S_s}{K}\frac{\partial h}{\partial t} \qquad \text{Eq. (13.14)}$$

Different from the Laplace equation (Eq. 13.9) at the steady state, the above equation for transient flow will give a solution of the hydraulic head (h) at any point and time, $h = (x, y, z, t)$. Considering the parameters in Eq. (13.14), we now know that the hydraulic head is affected by two properties of water (ρ, β) and three properties of porous media (n, α, K).

In a confined aquifer with a confined layer thickness (b), storativity (S) relates to the specific storage by $S = S_s \cdot b$ and transmissivity $T = K \cdot b$. Hence, S_s/K in Eq. (13.14) can be replaced by S/T. Therefore, Equation (13.14) becomes:

$$\frac{\partial^2 h}{\partial x^2}+\frac{\partial^2 h}{\partial y^2}+\frac{\partial^2 h}{\partial z^2}=\frac{S}{T}\frac{\partial h}{\partial t} \qquad \text{Eq. (13.15)}$$

Equation (13.15) is the governing equation for groundwater flow in a confined aquifer under transient condition.

13.1.3 Unsaturated Groundwater Flow under Transient Condition (Richards Equation)

In an unsaturated aquifer, the term ρn in Eq. (13.5) can be replaced by $\rho n\theta'$, where θ' is the degree of saturation ($\theta' = \theta/n$). We again use the product rule in calculus, $(d/dt)(x \cdot y \cdot z) = xy(dz/dt) + xz(dy/dt) + yz(dx/dt)$

Equation (13.5) can be rewritten as follows:

$$-\left(\frac{\partial}{\partial x}\rho q_x+\frac{\partial}{\partial y}\rho q_y+\frac{\partial}{\partial z}\rho q_z\right)=\rho n\frac{\partial\theta'}{\partial t}+\rho\theta'\frac{\partial n}{\partial t}+n\theta'\frac{\partial\rho}{\partial t} \qquad \text{Eq. (13.16)}$$

The last two terms on the right hand side are negligible compared to the first term of the moisture change (θ') in the unsaturated zone. Because $nd\theta' = d(n\theta') = d\theta$, Eq. (13.16) becomes:

$$-\left(\frac{\partial}{\partial x}\rho q_x + \frac{\partial}{\partial y}\rho q_y + \frac{\partial}{\partial z}\rho q_z\right) = \rho\frac{\partial\theta}{\partial t} \qquad \text{Eq. (13.17)}$$

After canceling ρ on both sides, and inserting Darcy's equation, we obtain:

$$\frac{\partial}{\partial x}\left[K(\psi)\frac{\partial h}{\partial x}\right] + \frac{\partial}{\partial y}\left[K(\psi)\frac{\partial h}{\partial y}\right] + \frac{\partial}{\partial z}\left[K(\psi)\frac{\partial h}{\partial z}\right] = \frac{\partial\theta}{\partial t} \qquad \text{Eq. (13.18)}$$

Note that the hydraulic conductivity K in an unsaturated aquifer is a function of the pressure head (ψ), and the total head (h) in an unsaturated aquifer is related to ψ by $h = \psi + z$. Thus $dh/dx = d\psi/dx$, $dh/dy = d\psi/dy$, and $dh/dz = d\psi/dz + 1$. The moisture content (θ) is also a function of the pressure head. If we define the slope of θ versus ψ as $C(\psi) = d\theta/d\psi$, we express all terms in Eq. (13.18) in terms of ψ:

$$\frac{\partial}{\partial x}\left[K(\psi)\frac{\partial\psi}{\partial x}\right] + \frac{\partial}{\partial y}\left[K(\psi)\frac{\partial\psi}{\partial y}\right] + \frac{\partial}{\partial z}\left[K(\psi)\frac{\partial\psi}{\partial z} + 1\right] = C(\psi)\frac{\partial\psi}{\partial t} \qquad \text{Eq. (13.19)}$$

Equation (13.19) is the ψ-based equation of transient flow through an unsaturated porous medium. This equation can also be expressed in terms of moisture contents (θ). Equation (13.19) is often called **Richards equation**, which has many other forms. For example, Eq. (13.19) can incorporate water source/sink term, such as transpiration and precipitation, which could be very important particularly in the vadose zone. Regardless, Richards equation is essentially the combined equation of mass balance and the Darcy's law for fluid flow. The solution of this partial differential equation, $\psi(x, y, z, t)$ can be easily converted to hydraulic head solution $h(x, y, z, t)$ using the relation $h = \psi + z$. Richards equation is hardly solved analytically, because the characteristic curves $K(\psi)$, $C(\psi)$ or $\theta(\psi)$ are nonlinear.

13.2 Governing Equations for Contaminant Transport

The equations in the previous section describe the flow of water in an aquifer, or generally fluid in porous media. Our primary intent for these flow equations is to find the rate of flow (velocity q) as a function of hydraulic head, space, and time, as well as other important hydrogeological parameters. In regard to the concern of contaminant transport, we need a set of equations to describe the physical, chemical, and biological parameters in governing transport processes. These transport equations delineate the concentration (C) as a function of property parameters of the chemical and aquifer of concern. In this section, we start with the mass balance approach by first considering advection and dispersion processes. This basic transport equation will be extended to more complicated scenarios, such as unsaturated zone for flow with changing hydraulic conductivity, and the incorporation of adsorption and reactions terms. Finally, the most challenging flow and transport processes for multiphases and multicomponents will be described conceptually followed by some exposure to the basic mathematical framework.

13.2.1 General Mass Balance Equations Considering Advection and Dispersion

In Section 13.1.1, we applied the **law of conservation of mass** of fluid (water) in a control volume with lengths of dx, dy, and dz and specific discharges (L/T) of q_x, q_y, and q_z in the x, y,

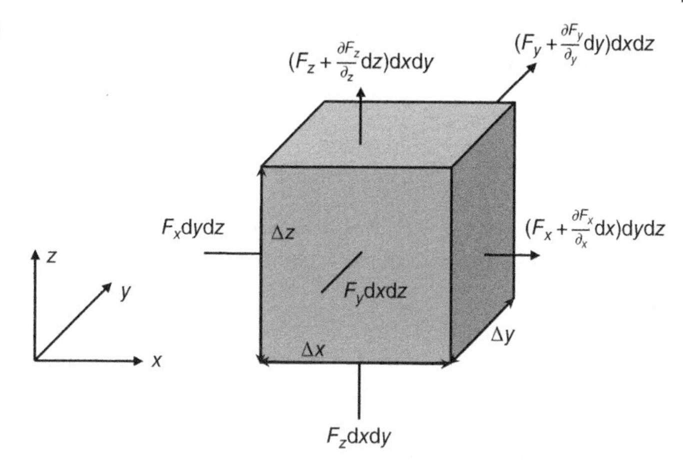

Figure 13.2 Control volume and mass balance for contaminant (chemical) flow via advection and dispersion in porous media.

and z directions, respectively (Figure 13.2). We now consider an inert (conservative) chemical that is dissolved in water and flows into and out of this control volume. For simplicity, we consider only two transport mechanisms as we have defined in Chapter 3, advection and dispersion. In Section 13.1.1, the mass flux of water into the control volume in the x direction, is $\rho q_x dydz$, where ρ is the density of water. For a chemical with concentration C, the mass fluxes by advection and dispersion are both concentration dependent as follows:

$$\text{Advection}(x\text{-direction}) = q_x nC\text{d}y\text{d}z \qquad \text{Eq. (13.20)}$$

$$\text{Dispersion}(x\text{-direction}) = -nD_x \frac{\partial C}{\partial x}\text{d}y\text{d}z \qquad \text{Eq. (13.21)}$$

where dydz is the cross-sectional area normal to the x direction. The negative sign in Eq. (13.21) indicates that the movement is from higher to lower concentrations so that the mass flux (F) is always positive. The total mass of chemical per unit cross-sectional area (dydz = 1) transported in the x direction per unit time, F_x, is the sum of advective and dispersive transport:

$$F_x = q_x nC - nD_x \frac{\partial C}{\partial x} \qquad \text{Eq. (13.22)}$$

Similar equations can be obtained for advection and dispersion in the y and z directions.

$$F_y = q_y nC - nD_y \frac{\partial C}{\partial y} \qquad \text{Eq. (13.23)}$$

$$F_z = q_z nC - nD_z \frac{\partial C}{\partial z} \qquad \text{Eq. (13.24)}$$

The total amount of chemical entering the control volume is as follows:

$$F_i = F_x \text{d}y\text{d}z + F_y \text{d}x\text{d}z + F_z \text{d}x\text{d}y$$

The total amount of chemical leaving the control volume is as follows:

$$F_o = \left(F_x + \frac{\partial F_x}{\partial x}\text{d}x\right)\text{d}y\text{d}z + \left(F_y + \frac{\partial F_y}{\partial y}\text{d}y\right)\text{d}x\text{d}z + \left(F_z + \frac{\partial F_z}{\partial z}\text{d}z\right)\text{d}x\text{d}y$$

The difference between the amount of chemical entering the control volume and the amount of chemical leaving the control volume is as follows:

$$\left(\frac{\partial F_x}{\partial x}+\frac{\partial F_y}{\partial y}+\frac{\partial F_z}{\partial z}\right)\mathrm{d}x\,\mathrm{d}y\,\mathrm{d}z$$

The above term $(F_o - F_i)$ has a dimension of M/T, which is the rate of mass change in the control volume. The rate of mass change (M/T) can also be expressed by:

$$-n\frac{\partial C}{\partial t}\mathrm{d}x\,\mathrm{d}y\,\mathrm{d}z$$

Applying the law of conservation of mass by equating the above two terms and canceling out the control volume $dxdydz$, we obtain:

$$\frac{\partial F_x}{\partial x}+\frac{\partial F_y}{\partial y}+\frac{\partial F_z}{\partial z}=-n\frac{\partial C}{\partial t} \qquad \text{Eq. (13.25)}$$

We now can substitute F_x (Eq. 13.22), F_y (Eq. 13.23), and F_z (Eq. 13.24) into Eq. (13.25). This leads to:

$$\frac{\partial}{\partial x}\left(q_x nC - nD_x\frac{\partial C}{\partial x}\right)+\frac{\partial}{\partial y}\left(q_y nC - nD_y\frac{\partial C}{\partial y}\right)+\frac{\partial}{\partial z}\left(q_z nC - nD_z\frac{\partial C}{\partial z}\right)=-n\frac{\partial C}{\partial t} \qquad \text{Eq. (13.26)}$$

Note that the porosity (n) can be canceled out on both sides. Rearranging the two terms by two transport processes, we have:

$$\left[\frac{\partial}{\partial x}\left[D_x\frac{\partial C}{\partial x}\right]+\frac{\partial}{\partial y}\left[D_y\frac{\partial C}{\partial y}\right]+\frac{\partial}{\partial z}\left[D_z\frac{\partial C}{\partial z}\right]\right]-\left[\frac{\partial}{\partial x}[q_x C]+\frac{\partial}{\partial y}[q_y C]+\frac{\partial}{\partial z}[q_z C]\right]=\frac{\partial C}{\partial t}$$

Eq. (13.27)

Equation (13.27) is a very important governing equation for chemical transport processes by advection and dispersion for three-dimensional flow. Since hydrodynamic dispersion coefficients depend on the flow direction, D_x, D_y, and D_z remain constant in space in a homogeneous porous medium. For two-dimensional flow along x–y plane with the direction of flow parallel to the x-axis ($q_y = q_z = 0$), Eq. (13.27) can be simplified as follows:

$$D_x\frac{\partial^2 C}{\partial x^2}+D_y\frac{\partial^2 C}{\partial y^2}-q_x\frac{\partial C}{\partial x}=\frac{\partial C}{\partial t} \qquad \text{Eq. (13.28)}$$

For one-dimensional flow along x direction, such as in a soil column, the above equation can be further simplified as follows:

$$D_x\frac{\partial^2 C}{\partial x}-q_x\frac{\partial C}{\partial x}=\frac{\partial C}{\partial t} \qquad \text{Eq. (13.29)}$$

Equations (13.28) and (13.29) can be solved analytically in a number of situations. We will discuss their analytical solutions in the subsequent sections.

13.2.2 Governing Equations for Contaminant Transport in Unsaturated Zone

The mass balance equation (Eq. 13.27) is the general equation governing advective and dispersive transport of contaminant in the porous media. For an unsaturated zone, modifications of Eq. (13.27) are needed to delineate the contaminant transport processes unique to an unsaturated zone. The unsaturated zone has several distinct characteristics:

a. Soil moisture content (θ) is less than the porosity ($\theta < n$), so soil pores are occupied by both water and contaminant vapor (θ_a), where θ_a is the soil air content.
b. Assuming the absence of free-phase liquid such as NAPL in the vadose zone, three contaminant phases are possible:

$$C_T = \rho_b S + \theta C + \theta_a C_a \qquad \text{Eq. (13.30)}$$

where C_T = the total concentration of contaminant in all forms (M/L^3), ρ_b = bulk density (M/L^3), θ = volumetric water content (L^3/L^3), θ_a = volumetric air content (L^3/L^3), S = contaminant concentration in the soil (M/M), C = contaminant concentration in the liquid (water) phase (M/L^3), C_a = contaminant concentration in the gaseous phase (M/L^3).
c. Dissolved phase solute and vapor phase solute, corresponding to their respective concentrations C and G, have different mass transport processes, hence, mass balance equations. For example, dispersion predominates molecular diffusion in liquid phase (unless groundwater flow velocity is very low). On the other hand, advective and dispersive transport may be not as important as molecular diffusion in the vapor phase of the unsaturated zone (unless under a forced airflow such as pumping).
d. Unsaturated hydraulic conductivity is not a constant like that in the saturated zone. It is typically a nonlinear function of moisture (θ) or pressure head (ψ), hence a nonlinear function of $K(\theta)$ or $K(\psi)$ should be used (e.g. Eq. 3.16 in Chapter 3).
e. For simplicity, a predominately vertical (z direction) transport of water can be assumed compared to the lateral transport processes in the x–y plane.

Šimůnek and Genuchten (2016) used the following governing equation to describe the dissolved solutes in the vadoze zone:

$$\left[\frac{\partial}{\partial z}\left(\theta D_l \frac{\partial C}{\partial z}\right) + \frac{\partial}{\partial z}\left(\theta_a D_g \frac{\partial C_a}{\partial z}\right)\right] - \left[\frac{\partial}{\partial z}(q_z C)\right] - S = \frac{\partial C_T}{\partial t} \qquad \text{Eq. (13.31)}$$

With the exception of the source/sink term (S, M/L^2/T), the above equation is in essence similar to Eq. (13.27), because the first term on the left hand side relates dispersion in liquid (where D_l is the effective dispersion coefficient), the second term relates diffusion in the gaseous phase (where D_g is the gas phase diffusion coefficient), the third term relates advective transport with the moving water (advective flow with moving air is neglected in Eq. 13.31). Equation (13.31) regards only to the vertical direction (z) of the contaminant transport processes.

If we substitute the C_T term in Eq. (13.30) into (13.31), we have:

$$\left[\frac{\partial}{\partial z}\left(\theta D_l \frac{\partial C}{\partial z}\right) + \frac{\partial}{\partial z}\left(\theta_a D_g \frac{\partial C_a}{\partial z}\right)\right] - \left[\frac{\partial}{\partial z}(q_z C)\right] - S = \frac{\partial(\rho_b S + \theta C + \theta_a C_a)}{\partial t} \qquad \text{Eq. (13.32)}$$

Like Richards equation (Eq. 13.19) as the governing equation for water flow in unsaturated zone, Eq. (13.32) also has many variations for contaminant transport processes reported in the literature. For example, Eq. (13.32) can be simplified for nonvolatile contaminants ($G = 0$) in 1-D transport:

$$\frac{\partial}{\partial z}\left(\theta D_1 \frac{\partial C}{\partial z}\right) - \frac{\partial}{\partial z}(q_z C) - S = \frac{\partial(\rho_b S + \theta C)}{\partial t} \quad \text{Eq. (13.33)}$$

For an inert and nonsorbing contaminant ($S = 0$), Eq. (13.33) can be further simplified as follows:

$$\frac{\partial}{\partial z}\left(D_1 \frac{\partial C}{\partial z}\right) - \frac{\partial}{\partial z}(q_z C) = \frac{\partial C}{\partial t} \quad \text{Eq. (13.34)}$$

where z is the vertical distance along the soil depth, and q_z = pore water velocity in L/T.

13.2.3 Governing Equations Incorporating Adsorption and Reaction

Besides the physical processes (advection and diffusion) we have described so far in this chapter, there are a number of chemical and biological processes that will alter the contaminant concentration in soil and groundwater. These processes include adsorption–desorption, hydrolysis, acid–base reactions, dissolution–precipitation, redox reaction, ion exchange, and biological degradation (see Chapter 2). In the following discussion, we focus on adsorption and first-order reaction. For simplicity, first-order reaction is commonly assumed to model chemical, biological, and radioactive decay. These processes can be incorporated into the existing advection and dispersion models following additional mathematical manipulations. As such, the model can better represent the contaminant transport processes, but becomes mathematically more demanding to solve analytically or numerically.

Adsorption Incorporated into the Advection–Dispersion Model. In a broad sense, sorption includes adsorption, chemisorption, absorption, and ion exchange. These processes can be considered as heterogeneous processes because they involve both the dissolved phase and the solid phase (soil). From the mechanistic standpoint, adsorption is due to the surface retention of contaminant by soil particles. Chemisorption occurs when contaminants react with the components of soil (sediment) particles, and cationic exchange is the electrostatic attraction between the negatively charged soil particles and the positively charged contaminants. Absorption differs from adsorption in that contaminants diffuse into the particle and are retained on the interior surface. Irrespective of these differences, all sorption processes can be regarded as partitioning.

To incorporate sorption into the governing equation, we start with the 1-D advection–dispersion model (Eq. 13.29) by adding an additional term involving the concentration of contaminant sorbed to soil (S):

$$D_x \frac{\partial^2 C}{\partial x} - q_x \frac{\partial C}{\partial x} - \frac{\rho_b}{n}\frac{\partial S}{\partial t} = \frac{\partial C}{\partial t} \quad \text{Eq. (13.35)}$$

where ρ_b is the bulk density of soil (M/L^3), n is dimensionless porosity, and S is the concentration of contaminant sorbed to soil, which is expressed in the unit of mass (contaminant)/mass (soil), such as mg/kg. The negative sign of the third term on the left hand side is to keep the overall term positive. Note that C is the concentration of contaminant in the pore water, which has a unit of mass of contaminant per unit volume of pore water (L^3).

If we examine the dimensions of the sorption term in Eq. (13.35):

$$\frac{\rho_b}{n}\frac{\partial S}{\partial t}=\frac{\frac{\cancel{\text{Mass(soil)}}}{\text{Volume(soil)}}}{n}\times\frac{\frac{\text{Mass(contaminant)}}{\cancel{\text{Mass(soil)}}}}{\text{time}}=\frac{\text{Mass(contaminant)}}{\left[\text{Volume(soil)}\times n\right]\times\text{time}}$$

$$=\frac{\text{Mass(contaminant)}}{\text{Volume of pore water}\times\text{time}}$$

The above dimensional analysis indicates that $(\rho_b/n)(\partial S/\partial t)$represents the change in concentration in pore water caused by adsorption or desorption. The unit of this term (M L^3/T) is then consistent with other terms in Eq. (13.35). Note that porosity (n) is used above for saturated zone. If unsaturated zone is concerned, then volumetric moisture content (θ) should be used in place of n.

Since adsorbed phase concentration S is related to aqueous phase concentration (C), the term $(\rho_b/n)(\partial S/\partial t)$ can be rewritten as follows:

$$\frac{\rho_b}{n}\frac{\partial S}{\partial t}=\frac{\rho_b}{n}\frac{\partial S}{\partial C}\frac{\partial C}{\partial t} \qquad \text{Eq. (13.36)}$$

The term $(\partial S/\partial C)$ on the right hand side of Eq. (13.36) represents the partitioning of the contaminant between the soil and the pore water. If a linear adsorption isotherm is expected, then

$$\frac{dS}{dC}=K_d \qquad \text{Eq. (13.37)}$$

where K_d is known as the distribution coefficient as we have defined in Chapter 3. Thus:

$$\frac{\rho_b}{n}\frac{\partial S}{\partial t}=\frac{\rho_b K_d}{n}\frac{\partial C}{\partial t} \qquad \text{Eq. (13.38)}$$

Substituting Eq. (13.38) into (13.35), and rearranging yields:

$$D_x\frac{\partial^2 C}{\partial x^2}-q_x\frac{\partial C}{\partial x}=\left(1+\frac{\rho_b K_d}{n}\right)\frac{\partial C}{\partial t} \qquad \text{Eq. (13.39)}$$

In Chapter 2, we defined $1+(\rho_b K_d/n)$ as the retardation factor R (Eq. 2.18). We now know this definition only applies to a linear sorption model. The linear sorption model has an obvious limitation because it does not have an upper limit for the amount of contaminant to be sorbed onto the soil. If Freundlich sorption isotherm applies, i.e. $S = KC^N$ (this is equivalent to Eq. 7.13, where $S = X/M$ and $N = 1/n$), then the sorption isotherm will be curvilinear. We substitute S into Eq. (13.35):

$$D_x\frac{\partial^2 C}{\partial x^2}-q_x\frac{\partial C}{\partial x}-\frac{\rho_b}{n}\frac{\partial\left(KC^N\right)}{\partial t}=\frac{\partial C}{\partial t} \qquad \text{Eq. (13.40)}$$

After differentiation and rearrangement, Eq. (13.40) becomes:

$$D_x \frac{\partial^2 C}{\partial x^2} - q_x \frac{\partial C}{\partial x} = \left(1 + \frac{\rho_b K' C^{N-1}}{n}\right) \frac{\partial C}{\partial t} \qquad \text{Eq. (13.41)}$$

It becomes clear that the retardation factor for a Freundlich sorption isotherm is:

$$R = 1 + \frac{\rho_b K C^{N-1}}{n} \qquad \text{Eq. (13.42)}$$

Using the same approach, we can derive the retardation factor if sorption can be described by the Langmuir model. Langmuir sorption assumes a finite number of sorption sites. Sorption will cease when all the sorption sites are saturated with sorbed contaminants.

$$S = \frac{\alpha \beta C}{1 + \alpha C} \qquad \text{Eq. (13.43)}$$

where α is an adsorption constant related to the binding energy (L/mg) and β is the maximum amount of contaminant that can be sorbed by the soil (mg/kg). Substituting S into Eq. (13.35), we obtain:

$$D_x \frac{\partial^2 C}{\partial x^2} - q_x \frac{\partial C}{\partial x} - \frac{\rho_b}{n} \frac{\partial\left(\alpha \beta C / (1 + \alpha C)\right)}{\partial t} = \frac{\partial C}{\partial t} \qquad \text{Eq. (13.44)}$$

After differentiation using the quotient rule and rearrangement, Eq. (13.44) becomes:

$$\left[1 + \frac{\rho_b}{n}\left(\frac{\alpha \beta}{(1 + \alpha C)^2}\right)\right] \frac{\partial C}{\partial t} = D_x \frac{\partial^2 C}{\partial x^2} - q_x \frac{\partial C}{\partial x} \qquad \text{Eq. (13.45)}$$

The retardation factor for the Langmuir sorption isotherm becomes:

$$R = 1 + \frac{\rho_b}{n} \frac{\alpha \beta}{(1 + \alpha C)^2} \qquad \text{Eq. (13.46)}$$

From the above discussion, we can conclude that the one-dimensional advection–dispersion–sorption of contaminant transport can be generally expressed as

$$R \frac{\partial C}{\partial t} = D_x \frac{\partial^2 C}{\partial x^2} - q_x \frac{\partial C}{\partial x} \qquad \text{Eq. (13.47)}$$

where the retardation factor R can be expressed as Eqs. (2.18), (13.42), or (13.46) depending on whether sorption fits linear, Freundlich, or Langmuir model.

Reaction Incorporated into Advection–Dispersion Model. We now can add a reaction term in Eq. (13.47) to conclude our discussion on various governing equations. This generalized equation represents advection, dispersion, sorption, and reaction in one-dimensional flow (Fetter 1993):

$$R \frac{\partial C}{\partial t} = D_x \frac{\partial^2 C}{\partial x^2} - q_x \frac{\partial C}{\partial x} + k_1 C + k_0 \qquad \text{Eq. (13.48)}$$

where k_1 (T^{-1}) and k_0 (M/T) are consolidated rate factors for the first-order ($dC/dt = k_1C$) and zero-order ($dC/dt = k_0$) reactions, respectively. Since k_1 and k_0 have dimensions of T^{-1} and M T^{-1}, the units in Eq. (13.42) are consistent. The first- and zero-order reactions are commonly used, but other reaction orders are also possible. The two reaction terms in Eq. (13.48) can be any biological, chemical, or radioactive decay.

Some reactions are sufficiently fast relative to groundwater flow rate and are reversible. For these reactions, we can assume that locally the contaminant is in chemical equilibrium with the surroundings. Other reactions are insufficiently fast or they are irreversible. For example, acid–base and complexation reactions are typically fast, whereas redox reactions are slow. Dissolution and precipitation reactions can be fast or slow relative to the groundwater flow.

13.2.4 General Concepts and Equations Describing Multiphase Flow and Transport

The multiphase to be discussed here refers to two or more immiscible fluids or phases, such as oil, water and air, in the porous medium of an aquifer. The problems of multiphase flow and transport are often encountered in groundwater hydrology when NAPLs are present. Thus, an unsaturated zone could have three immiscible phases (i.e. air, water, and NAPLs), whereas a saturated zone could have two immiscible phases (i.e. water and NAPLs). Since NAPLs often come with various chemical components, the multiphase problem is complicated by the presence of multicomponents in each corresponding phase. In the discussions that follow, we first define various processes related to multiphase and multiple components. Then, multiphase equations and mass balance equations are described in a more quantitative detail.

13.2.4.1 Processes Relevant to Multiphase and Multiple Components

In delineating the flow and transport of multiphase and multicomponents in an aquifer, one needs to consider all the physical, chemical, and biological processes we have described in the preceeding sections (advection, dispersion, diffusion, adsorption, and chemical reactions). Additionally, the following aspects must be considered: interfacial tension, capillary force, capillary pressure, relative permeability, and reduced volatility and solubility in a NAPL mixture.

a. *Interfacial tension:* Unlike miscible fluids where two fluids dissolve in each other and no interfacial tension is present, a multiphase system has a distinct fluid–fluid interface. Therefore, the interfacial tension between the two fluids is nonzero. Interfacial tension (γ) occurs because molecules near an interface have different molecular interactions than the equivalent molecules within the bulk fluid. The unit of γ is force (capillary force)/unit length (e.g. dyne/cm) parallel to the surface which is exerted perpendicular to any line drawn on the surface.
b. *Capillary force and capillary pressure:* The capillary forces result in the difference in pressure, termed **capillary pressure** (P_c), across the interface between two immiscible fluids. In porous media, P_c is the difference between the pressure in the nonwetting phase (P_{nw}) and the pressure in the wetting phase (P_w): $P_c = P_{nw} - P_w$. In NAPLs (oil)–water systems, water is typically the wetting phase, whereas for gas–NAPL (oil) systems, NAPL is typically the wetting phase. For a drop of water on a surface with a contact angle θ (for $\theta < 90°$ when water is the wetting phase) (Figure 13.3), capillary pressure is related to interfacial tension as follows:

$$P_c = \frac{2\gamma \cos\theta}{R} \qquad \text{Eq. (13.49)}$$

where R is the mean radius of curvature. When NAPL is flowing through a water-filled pore, R is equivalent to the average radius of the pore that NAPL must move through to exit

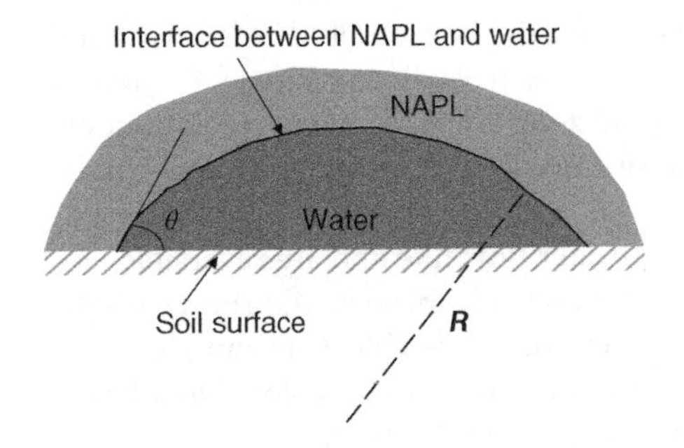

Figure 13.3 Interface and contact angle between NAPLs and water.

or enter pore in an aquifer. Equation (13.49) clearly states that fluid moves at a smaller rate when it has to overcome a larger interfacial tension (γ) and smaller pore (R).

c. *Capillary pressure and Darcy's law:* The capillary pressure can be perceived as the measure of tendency of a porous medium to hold the wetting fluid phase or to repel the nonwetting phase. This pressure head therefore should be considered in applying Darcy's law to the fluid flow along x-axis.

$$q = -K\frac{dh}{dx} = \frac{-K\left(z + \frac{P_c}{g}\right)}{dx} \qquad \text{Eq. (13.50)}$$

If the hydraulic conductivity (K) is replaced by the permeability k (see Eq. 3.5, Chapter 3), the Darcy's law can be written as follows:

$$q = -\frac{k\rho g}{\mu}\frac{d\left(h + \frac{P_c}{\rho g}\right)}{dx} \equiv -\frac{k}{\mu}\left(\rho g\frac{\partial h}{\partial x} + \frac{\partial P_c}{\partial x}\right) \qquad \text{Eq. (13.51)}$$

d. *Relative permeability as a function of residual saturation:* When two fluids coexist in a porous medium, the permeabilities of both fluids (k_0 and k_w for NAPL and water, respectively) are reduced. In other words, when NAPL is mixed with water, the movement of both NAPL and water will be significantly slowed down. A dimensionless measure called relative permeability is used to refer to the ratio of effective permeability to the intrinsic permeability. The **relative permeability** is a function of the saturation of either fluid (water and NAPL) as shown in Figure 13.4.
As shown in the relative permeability curve in Figure 13.4, when NAPL saturation (S_o) is high, NAPL is present as a continuous phase and water becomes discontinuous. The flow of NAPL will dominate the flow of water. If S_o is very high (e.g. the left hand side of Figure 13.4 when S_w is very low, because $S_o + S_w = 1$), water becomes immobile. This critical point of saturation for water is termed the residual situation of water. In a similar way, we can define the residual saturation of the NAPL (right hand side of the figure). This is the noncontinuous phase of NAPLs and as such, the residual NAPLs will be trapped in soil pores. This portion of NAPLs is termed as **residual saturation**.

e. *Reduced volatility and solubility for compounds in a mixture:* NAPL is typically a multicomponent mixture, the fate and transport processes of a chemical in a mixture could be significantly different from that of the same chemical in the mixture. For example, the biodegradation of benzene may be inhibited by toluene and other compounds in the BTEX mixture. Similarly, the sorption of benzene may be reduced by toluene in the mixture through a competitive sorption mechanism. While experimental data are needed to discern such effects on the biodegradation

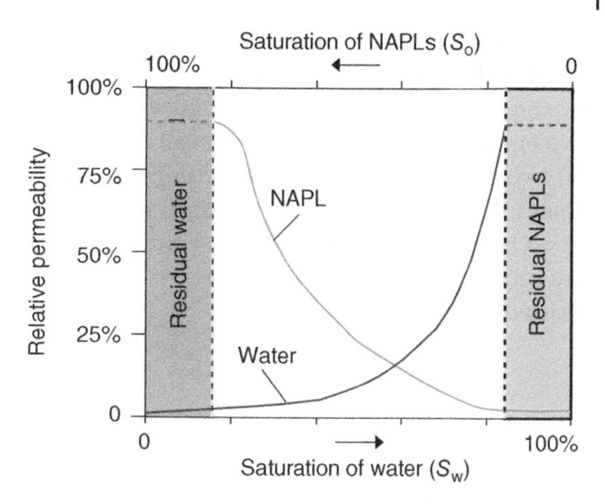

Figure 13.4 Relative permeability as a function of water saturation (S_w) and NAPL saturation (S_o).

and sorption processes, the behavior of compounds in a mixture during the volatilization and dissolution is well known, as described in the following text. For volatilization, **Raoult's law** describes the volatilization of each individual compound in a mixture:

$$P_i = x_i \, P_{i0}^* \qquad \text{Eq. (13.52)}$$

where P_i is the vapor pressure (atm) of the compound i in the NAPL mixture, x_i is the mole fraction of compound i in its mixture (unitless), and P_{i0}^* is the vapor pressure (atm) of pure compound i in equilibrium with its own vapor. The temperature-dependent pure compound vapor pressure P_{i0}^* is readily available from the literature. Raoult's law tells us that, if we know the composition of the mixture (or mole faction of each compound in the mixture), we will be able to estimate the vapor of each compound in the mixture. For a pure compound (i.e. $x_i = 1$), the vapor pressure will remain the same at its own vapor pressure. Otherwise for any compound in a mixture ($x_i < 1$), the partial pressure of each individual compound will be smaller than the vapor pressure of its own at the same temperature.

Raoult's law can also be employed to relate the aqueous solubility of a compound in a mixture to the solubility of its pure phase. The effective solubility (S_i) of an organic compound in groundwater is the product of its mole fraction in the mixture (x_i) and the pure phase solubility (S_{i0}^*):

$$S_i = x_i \, S_{i0}^* \qquad \text{Eq. (13.53)}$$

For example, benzene has an aqueous solubility of 1780 mg/L at 25 °C. If a NAPL mixture contains $x_i = 0.3$ mol fraction of benzene, the effective solubility of benzene is now reduced to 0.3×1780 mg/L = 534 mg/L. This solubility is significantly lower than the solubility of benzene in water when water contains only benzene.

13.2.4.2 Framework of Governing Equations for Multiphase Flow and Transport

Multiphase Flow: When two or more phases are present, the Darcy's law in describing specific discharge (q) in Eq. (13.51) can be applied to the flow of each fluid phase. For either water ($\alpha = 1$) or NAPL ($\alpha = 2$), specific discharge is a function of the intrinsic permeability (k_α), density (ρ_α), viscosity (μ_α), and capillary pressure ($P_{c\alpha}$), where $P_{c\alpha}$ is in turn a function of the saturation (S_α). According to Bear (1972), there are 15 variables for the flow of a two-phase ($\alpha = 1, 2$) and three-component ($i = 1, 2, 3$) system: P_α, P_c, $(q_i)_\alpha$, μ_α, ρ_α, S_α. Hence, 15 equations are needed to derive the complete solution of the flow in time and space.

These include (i) six equations of fluid motion for an anisotropic porous medium (see Eqs. 13.54 and 13.55), (ii) an equation of saturation with respect to two fluids, i.e. $S_1 + S_2 = 1$, (iii) two equations for the state of density which is a function of the fluid pressure, $\rho_\alpha = \rho_\alpha(P_\alpha)$, (iv) two equations for the state of viscosity which is also a function of fluid pressure, $\mu_\alpha = \mu_\alpha(P_\alpha)$, (v) two continuity (mass conservation) equations, one for each phase (see Eq. 13.56), (vi) an equation relating P_c to the pressure difference between two phases, $P_c = P_2 - P_1$ if water ($\alpha = 1$) is the wetting fluid, (vii) an equation relating P_c to saturation, $P_c = P_c(S_1)$, where S_1 is the saturation of the water as the wetting fluid. These six equations of fluid motion can be written from Eq. (13.51).

$$q_{i1} = -\frac{k_{ij1}}{\mu_1}\left(\rho_1 g\frac{\partial h}{\partial x_j} + \frac{\partial P_{c1}}{\partial x_j}\right) \quad i, j = 1, 2, 3 \qquad \text{Eq. (13.54)}$$

$$q_{i2} = -\frac{k_{ij2}}{\mu_2}\left(\rho_2 g\frac{\partial h}{\partial x_j} + \frac{\partial P_{c2}}{\partial x_j}\right) \quad i, j = 1, 2, 3 \qquad \text{Eq. (13.55)}$$

The general mass conservation equation derived previously (Eq. 13.5) can be modified by incorporating the saturation term (S) and phase notation (α) for the two-phase system.

$$-\left(\frac{\partial}{\partial x}\rho q_x + \frac{\partial}{\partial y}\rho q_y + \frac{\partial}{\partial z}\rho q_z\right) = \frac{\partial}{\partial t}n\rho$$

Thus, after rearrangement, the above equation becomes:

$$\frac{\partial}{\partial t}nS_\alpha\rho_\alpha - \left(\frac{\partial}{\partial x}\rho_\alpha q_x + \frac{\partial}{\partial y}\rho_\alpha q_y + \frac{\partial}{\partial z}\rho_\alpha q_z\right) = 0 \qquad \text{Eq. (13.56)}$$

Six dependent variables, $(q_i)_\alpha$, can be further eliminated by inserting Eqs. (13.54) and (13.55) into the mass conservation equation (Eq. 13.56). Thus, the solution for a flow in two homogeneous incompressible fluids (ρ = constant) in an anisotropic medium reduces to the following equations:

$$n\frac{\partial S_1}{\partial t} - \frac{\partial}{\partial x_i}\left(-\frac{k_{ij1}}{\mu_1}\left(\rho_1 g\frac{\partial h}{\partial x_j} + \frac{\partial P_{c1}}{\partial x_j}\right)\right) = 0 \qquad \text{Eq. (13.57)}$$

$$n\frac{\partial S_2}{\partial t} - \frac{\partial}{\partial x_i}\left(-\frac{k_{ij2}}{\mu_2}\left(\rho_2 g\frac{\partial h}{\partial x_j} + \frac{\partial P_{c2}}{\partial x_j}\right)\right) = 0 \qquad \text{Eq. (13.58)}$$

$$S_1 + S_2 = 1 \qquad \text{Eq. (13.59)}$$

$$P_2 - P_1 = P_c(S_1) \qquad \text{Eq. (13.60)}$$

Equations (13.57)–(13.60) can be solved for four unknowns, S_1, S_2, P_1, and P_2 if appropriate boundary and initial conditions for Eqs. (13.57) and (13.60) are supplemented.

Chemical Mass Balance Equations: The mass balance for chemical i is expressed in terms of overall volume of the chemical i per unit pore volume (C_i, unit L^3/L^3). For a three-phase (α = 1, 2, 3) system (USEPA 1999):

$$\nabla\cdot\left[\sum_{\alpha=1}^{3}\rho_i\left(nS_\alpha D_{i\alpha}\nabla C_{i\alpha}\right)\right] - \nabla\cdot\sum_{\alpha=1}^{3}\rho_i C_{i\alpha}q_\alpha \pm R_i = \frac{\partial}{\partial t}\left(n\rho_i C_i\right) \qquad \text{Eq. (13.61)}$$

Note that the above equation is analogous to the mass balance equation we derived earlier (e.g. Eq. 13.27) with additional notations. The first and second terms on the left hand side of the equation relate to dispersion and advection, respectively, whereas R_i reflects the rate of any chemical and biological reactions. Thus, the right hand side of the mass balance equation is the accumulation of mass per unit time (m/T-L^3) for component i. The Fickian dispersive flux in Eq. (13.61) is assumed to be proportional to the first derivative of concentration with respect to the flow direction ($\nabla C_{i\alpha}$) and a linear diffusion coefficient ($D_{i\alpha}$).

Note that the symbol ∇ (**del operator**) is a shorthand notation to simplify long mathematical expressions for gradient or divergent. For example, concentration (C_i) is a **scalar** (a quantity characterized only by its magnitude), and the gradient of concentration for chemical i (C_i):

$$\text{grad}\, C_i \equiv \frac{\partial C_i}{\partial x}\boldsymbol{i} + \frac{\partial C_i}{\partial y}\boldsymbol{j} + \frac{\partial C}{\partial z}\boldsymbol{k} \equiv \nabla C_i \qquad \text{Eq. (13.62)}$$

where $\boldsymbol{i}, \boldsymbol{j}$, and $\boldsymbol{k}$ are the unit vectors in the x, y, and z directions (vector is a quantity characterized by its magnitude and direction). Thus, $\nabla C_{i\alpha}$ in Eq. (13.62) is the gradient of concentration of chemical i in phase α. If the del operator is dotted into a vector such as velocity ($\nabla \cdot q$), it yields a scalar, called the **divergence** (div).

$$\nabla \cdot q \equiv \text{div}\, q \equiv \frac{\partial q_x}{\partial x} + \frac{\partial q_y}{\partial y} + \frac{\partial q_z}{\partial z} \qquad \text{Eq. (13.63)}$$

If the del operator is applied to grad C_i, it results in the second derivative of the concentration:

$$\nabla \cdot (\nabla C_i) \equiv \nabla^2 C_i \equiv \frac{\partial^2 C_i}{\partial x^2} + \frac{\partial^2 C_i}{\partial y^2} + \frac{\partial^2 C_i}{\partial z^2} \qquad \text{Eq. (13.64)}$$

Note also that the overall volume of the chemical i per unit pore volume (C_i, unit L^3/L^3) is the sum of the volume fractions of chemical i in three phases (e.g. $\alpha = 1$ for water, $\alpha = 2$ for NAPL, and $\alpha = 3$ for air) plus the volume fraction of the same chemical in its adsorbed phase ($\hat{C}_i$). For an NAPL containing six compounds of BTEX series, $i = 1, 2, ..., 6$, then (USEPA 1999):

$$C_i = \left(1 - \sum_{i=1}^{6} \hat{C}_i\right) \sum_{\alpha=1}^{3} S_\alpha\, C_{i\alpha} + \hat{C}_i \qquad \text{Eq. (13.65)}$$

In addition to the flow and mass balance equations introduced above, heat (energy) balance equations may be also needed. Furthermore, these generalized equations for multiphase flow and multicomponent transport need to be supplemented with a number of **constitutive equations** to make these governing equations solvable. The constitutive equations express interrelationships and constraints of physical processes, variables, and parameters, and many of these correlations for estimating properties and interrelationships must be determined by experimental studies, including, but not limited to, capillary pressure as a function of saturation ($P_{C\alpha} = P_{C\alpha}(S_\alpha)$), relative permeability, equilibrium partitioning, degradation constant, etc. (Wu and Qin 2009). Without further details, readers should now realize the large number and complexity of equations essential to describing the flow and transport of multiphase and multicomponent in the contaminated soil and groundwater. Three-dimensional flow and transport simulator, for example, UTCHEM, has been developed and validated to predict NAPL flow and transport as well as processes for their remediation.

13.3 Analytical Solutions to Flow and Transport Processes

Analytical solutions (models) are the exact solutions in the form of mathematical expressions. The analytical models are favorable in terms of decreased complexity and input data requirement. However, even some of the simplest analytical models may require complex mathematics. The analytical models described represent some common analytical solutions under simplifying conditions. Unlike the derivation of governing equations in the preceding section, our focus of this section on analytical solutions is not the mathematical derivation but rather the underlying assumptions and the uses of these analytical solutions to depict the particular contaminant transport processes. We will discuss these solutions in the order of their increasing mathematical complexity.

13.3.1 Darcy's Law: 1-D Flow in Unconfined Aquifer (Dupuit Equation)

In a confined aquifer, Darcy's law can be written as $Q = -KA\mathrm{d}h/\mathrm{d}l$, where the cross-sectional area ($A = hw$) is a constant. In an unconfined aquifer, the solution of the equations for groundwater flow is more complicated because the piezometric head at water table and the thickness of flow (h) changes as groundwater is withdrawn (i.e. the removal of groundwater from the aquifer lowers the water table). Thus, the cross-sectional area (A) also changes, as h changes in the flow path. If the flow direction is assumed parallel to the x axis, Darcy's law can be written as follows:

$$Q = -K(hw)\frac{\mathrm{d}h}{\mathrm{d}x} \qquad \text{Eq. (13.66)}$$

Note in Eq. (13.66) that the Darcy velocity at the groundwater table is expressed in terms of $\mathrm{d}h/\mathrm{d}x$ (h over the distance x), rather than $\mathrm{d}h/\mathrm{d}l$ (h over the flow path length). This is a reasonable assumption for small water-table slopes. If we denote q as the flow per unit width ($w = 1$), then:

$$q = -Kh\frac{\mathrm{d}h}{\mathrm{d}x} \qquad \text{Eq. (13.67)}$$

Note that q in the above equation has the unit of flow rate per unit width (L^2/T). It is assumed to be constant over the depth of flow. With the boundary conditions ($x = 0$, $h = h_0$; $x = L$, $h = h_L$, Figure 13.5), we can integrate Eq. (13.67):

$$\int_0^L q\mathrm{d}x = -K\int_{h_0}^{h_L} h\mathrm{d}h \qquad \text{Eq. (13.68)}$$

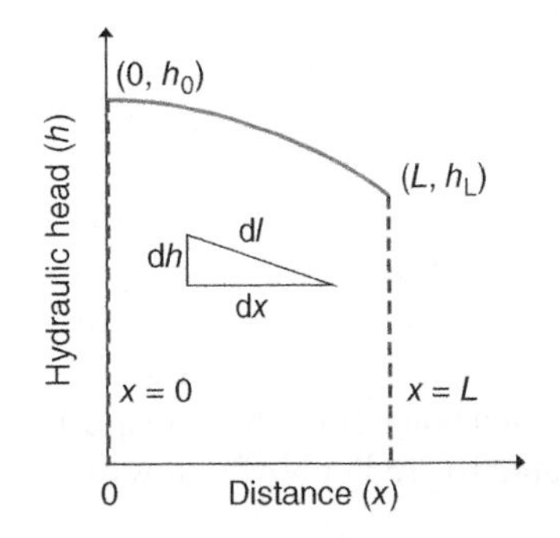

Figure 13.5 Steady-state flow through an unconfined aquifer.

Integration of the above equation yields:

$$qL = -K\left(\frac{h_L^2}{2} - \frac{h_0^2}{2}\right) \quad \text{Eq. (13.69)}$$

Rearrangement of Eq. (13.69) yields the nonlinear **Dupuit equation**:

$$q = \frac{K}{2L}\left(h_0^2 - h_L^2\right) \quad \text{Eq. (13.70)}$$

We are now developing an equation of h as a function of x, as shown in Figure 13.5. Recalling the Laplace equation for 3-D flow (Eq. 13.9), we now apply this to 1-D flow by assuming the flow is only horizontal (i.e. $\partial h/\partial y = 0$, $\partial h/\partial z = 0$):

$$\frac{\partial^2 h}{\partial x^2} = 0 \quad \text{Eq. (13.71)}$$

The solution to Eq. (13.71) is as follows:

$$h^2 = ax + b \quad \text{Eq. (13.72)}$$

Applying boundary conditions ($x = 0$, $h = h_0$; $x = L$, $h = h_L$), we can solve for the constants a and b:

$$b = h_0^2 \quad \text{Eq. (13.73)}$$

$$a = \frac{h_L^2 - h_0^2}{L} \quad \text{Eq. (13.74)}$$

Substituting the values of a and b (Eqs. 13.73 and 13.74) into Eq. (13.72), and rearranging yields:

$$h = \sqrt{h_0^2 - \frac{x}{L}\left(h_0^2 - h_L^2\right)} \quad \text{Eq. (13.75)}$$

The term h^2 in Eq. (13.72) indicates the parabolic form of the water table, hence Eq. (13.75) is called **Dupuit parabola**. The Dupuit equation and Dupuit parabola are the solutions of Darcy's law to flow in the unconfined condition. Example 13.1 illustrates the use of Darcy's law and Dupuit equations in unconfined flow versus confined flow.

Example 13.1 Use of Darcy's law and Dupuit equation: Unconfined versus confined aquifers

Assume: $K = 1$ ft/day, $h_0 = 103$ ft, $h_{100} = 93$ ft, $L = 100$ ft (see Figure 13.5). (a) Calculate q and h at $x = 25$, 50, and 75 ft for a unconfined aquifer. (b) Calculate q and h at $x = 25$, 50, 75, and 100 ft for a confined aquifer.

Solution

a) For an *unconfined* aquifer, use Eq. (13.90) for q and Eq. (13.75) for h.

$$q = \frac{K}{2L}\left(h_0^2 - h_L^2\right) = \frac{1\dfrac{\text{ft}}{\text{day}}}{2 \times 100\,\text{ft}}\left(103^2 - 93^2\right)\text{ft}^2 = 9.8\frac{\text{ft}^2}{\text{day}}$$

$$h(25\,\text{ft})=\sqrt{103^2-\frac{25}{100}\left(103^2-93^2\right)}=100.6\,\text{ft}$$

$$h(50\,\text{ft})=\sqrt{103^2-\frac{50}{100}\left(103^2-93^2\right)}=98.1\,\text{ft}$$

$$h(75\,\text{ft})=\sqrt{103^2-\frac{75}{100}\left(103^2-93^2\right)}=95.6\,\text{ft}$$

b) For a *confined* aquifer, the change in h is linear (Eq. 13.67). The flow per unit width (q) is:

$$q=-Kh\frac{dh}{dl}=-1\frac{\text{ft}}{\text{day}}\left(98\,\text{ft}\right)\times\frac{\left(103-93\right)\text{ft}}{\left(0-100\right)\text{ft}}=9.8\frac{\text{ft}^2}{\text{day}}$$

Note that we use the average thickness for h (103+93)/2 = 98 to be consistent with the Dupuit conditions.

$$h(x)=h_0-\frac{h_0-h_L}{L}\times x$$

$$h(25)=103-\frac{103-93}{100}\times 25=100.5\,\text{ft}$$

$$h(50)=103-\frac{103-93}{100}\times 50=98.0\,\text{ft}$$

$$h(75)=103-\frac{103-93}{100}\times 75=95.5\,\text{ft}$$

By comparing the results of an unconfined and a confined aquifer, we note that the flow rates (q) are the same, and the head (h) values at the starting and the ending locations are the same ($h_0 = 103$ ft, $h_{100} = 93$ ft at $L = 100$ ft). However, the h values at other locations for the unconfined aquifer is always larger than that of the confined aquifer. This is due to the parabolic form of water table surface in the unconfined aquifer relative to the straight line between the fixed heads on both sides of the confined aquifer (see Figure 13.5).

13.3.2 Fick's Second Law: 1-D Diffusion Only Solutions

In this section, we discuss the analytical solution to the simplest 1-D diffusion only process. In the subsequent sections, we will display more complicated contaminant processes such as advection and diffusion (with a continuous source or an instantaneous source) under 1-D, 2-D and 3-D.

In surface water, molecular diffusion is generally not a relevant transport process except at the stagnant boundary layers between water–air interface and in quiescent sediment pore waters. In groundwater, however, flow velocities are much slower, and thus molecular diffusion can be important. By neglecting the advective term in Eq. (13.29), we can obtain the Fick's second law:

$$D_x\frac{\partial^2 C}{\partial x^2}=\frac{\partial C}{\partial t} \qquad \text{Eq. (13.76)}$$

This second-order partial differential equation needs one initial and two boundary conditions (one for each order) for an analytical solution (Van Der Perk 2014). When the concentration field is independent of time (steady-state) and D is independent of C, Fick's second law is reduced to Laplace equation in 1-D.

With an initial condition of $C(x, 0) = 0$ (the porous medium is contaminant free at $t = 0$) and boundary conditions of $C(\infty, 0) = 0$ and $C(0, t) = C_0$, Eq. (13.76) has the following solution (Fetter 1993; Sharma and Reddy 2004):

$$C(x,t) = C_0 \operatorname{erfc}\left(\frac{x}{2\sqrt{D_e t}}\right) \quad \text{Eq. (13.77)}$$

where $C(x, t)$ = concentration at distance x from the source at time t, C_0 = the initial contaminant concentration, and erfc is the complementary error function which is typically tabulated in handbooks. The erfc value can also be easily calculated in Microsoft Excel (e.g. @erfc(1) = 0.157 299; @erfc(−1.5) = 1.966105). The function of erfc is related to erf and is defined as follows:

$$\operatorname{erfc}(x) = 1 - \operatorname{erf}(x) = 1 - \frac{2}{\sqrt{\pi}}\int_x^{\infty} e^{-t^2} dt$$

The effective diffusion coefficient in porous media, D_e, is related to the diffusion coefficient in water (D_x in Eq. 13.76) by a tortuosity factor (ω) according to

$$D_e = \omega D_x \quad \text{Eq. (13.78)}$$

where tortuosity factor (ω) is an intrinsic property of a porous medium usually defined as the ratio of the actual flow path length to the straight distance between the ends of the flow path. Diffusion described in Eq. (13.77) is a slow process, but it can be important because contaminants move through diffusion even if the groundwater is not flowing by advection. Diffusion can be the predominant mechanism in special case, for example, when chemicals in landfill leachate diffuse through a liner or a clay layer. Example 13.2 illustrates the diffusion only process using Eq. (13.77).

Example 13.2 Diffusion through a landfill liner

The diffusion coefficient for benzene through flexible membrane liners (FMLs) used for lining impoundments and landfills was determined to be 5.1×10^{-9} cm^2/s (Prasad et al. 1994). Estimate the concentration ratio, C_t/C_0 at the minimum separation distance of 1 m between liner and groundwater table after 100 years of diffusion.

Solution

Applying Eq. 13.77 at $t = 100\,\text{yr} = 100 \times 365 \times 24 \times 3600\,\text{s} = 3.15 \times 10^9\,\text{s}$

$$\frac{C(1\text{m}, 100\,\text{yr})}{C_0} = \text{efrc}\left(\frac{x}{2\sqrt{D_e t}}\right) = s\left(\frac{1\,\text{m} \times \frac{100\,\text{cm}}{1\,\text{m}}}{2\sqrt{5.1 \times 10^{-9}\frac{\text{cm}^2}{\text{s}} \times 3.15 \times 10^9\,\text{s}}}\right)$$

$$= \operatorname{erfc}(12.47) = 1.18 \times 10^{-69}$$

This example illustrates that the diffusion through landfill liner is a very slow process. This liner is very effective in preventing the contamination to groundwater through diffusion.

13.3.3 Advection and Dispersion: 1-D, 2-D, and 3-D Solutions to Slug Injection

This is an instantaneous injection, also referred to as pulse input or **slug injection**, of contaminant into a porous medium. The analytical solutions are available for certain cases in 1D, 2D, and 3D.

1-D Slug Injection: The solution corresponding to an instantaneous source at $x = 0$ in 1-D (Bedient et al. 1999) is

$$C(x,t)=\frac{M}{2(\pi t)^{1/2}\sqrt{D_x}}\exp\left(-\frac{(x-v_x t)^2}{4D_x t}\right) \quad \text{Eq. (13.79)}$$

where M is the mass of contaminant injected per unit cross-sectional area of the pore space (hence porosity should be used for the correction). If the injection point is $x = x_0$, then x in the above equation can be replaced by $x - x_0$. In 1-D soil column with column length L, if x is replaced by L, then $x\ (L, t)$ represents the concentrations in the column effluent.

The mathematical form of Eq. (13.65) is analogous to the well-known "bell-shaped" normal (or Gaussian) distribution:

$$f(x)=\frac{1}{\sqrt{2\pi\sigma^2}}\exp\left(-\frac{(x-\mu)^2}{2\pi\sigma^2}\right) \quad \text{Eq. (13.80)}$$

where the arithmetic mean equals μ and the standard deviation equals σ. This implies that the center of the contaminant mass at point of $\mu = v_x t$ by advection, whereas the dispersion spreads from this center with a standard deviation of $\sigma = (2D_x t)^{1/2}$. A schematic of concentration profile with change of distance and time after an instantaneous injection of a contaminant solution along the 1-D soil column is given in Figure 13.6. In Chapter 2, we have depicted advection and dispersion in 1-D soil column (Figure 2.12a), but we now have a better quantitative understanding of the mathematical model in describing these processes.

Equation (13.79) can be extended to include reactions. For example, if the contaminant undergoes first-order reaction with a rate constant of k_1, then:

$$C(x,t)=\frac{M}{2(\pi t)^{1/2}\sqrt{D_x}}\exp\left(-\frac{(x-v_x t)^2}{4D_x t}\right)\exp(-k_1 t) \quad \text{Eq. (13.81)}$$

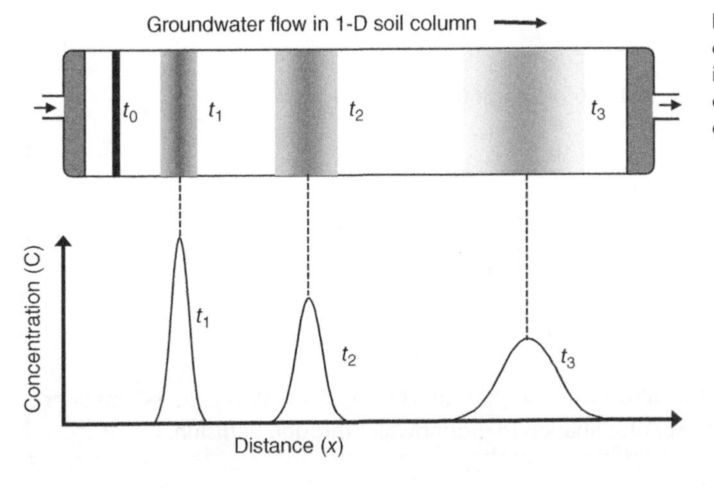

Figure 13.6 Contaminant concentration profile following an instantaneous injection of contaminant solution in 1-D soil column.

2-D Slug Injection: The solution corresponding to an instantaneous source at $x = 0$, $y = 0$ in 2-D (Freeze and Cherry 1979; Fetter 1993) is

$$C(x,y,t) = \frac{C_0 A}{4\pi t\sqrt{D_x D_y}}\left[\exp\left[-\frac{(x - v_x t)^2}{4D_x t} - \frac{y^2}{4D_y t}\right]\right] \quad \text{Eq. (13.82)}$$

where A is the cross-sectional area in the x–y plane, the term prior to exp should have a concentration unit (M/L^3). In Chapter 2, we have depicted the 2-D contaminant plume following an instantaneous injection such as a point source of chemical spills (Figure 2.12c). Figure 13.7 is an elaboration of the graphical representation of Eq. (13.82) in regard to advection and dispersion over two-dimensional space (x and y) and time (t_1, t_2, t_3).

As shown in Figure 13.7, the distances of the plume traveled by advection at t_1, t_2, and t_3 are $v_x t_1$, $v_x t_2$, and $v_x t_3$, respectively, where v_x is the groundwater flow velocity. $3\sigma_x$ and $3\sigma_y$ represent the distance from the center of the plume traveled by dispersion along the longitudinal and transverse direction, respectively, where $\sigma_x = 2\sqrt{D_x t}$ and $\sigma_y = 2\sqrt{D_y t}$. Thus the plume dimensions at the x and y directions can be estimated.

3-D Slug Injection: The solution corresponding to an instantaneous source at $x = 0$, $y = 0$, $z = 0$ in 3-D (Bedient et al. 1999):

$$C(x,y,z,t) = \frac{C_0 V_0}{8(\pi t)^{3/2}\sqrt{D_x D_y D_z}}\left[\exp\left[-\frac{(x - v_x t)^2}{4D_x t} - \frac{y^2}{4D_y t} - \frac{z^2}{4D_z t}\right]\right] \quad \text{Eq. (13.83)}$$

where $C_0 V_0$ is the mass of contaminants involved in a spill. If the spill (injection) is located in a point (x_0, y_0, and z_0), then x, y, and z in the above equations can be replaced by $x - x_0$, $y - y_0$, and $z - z_0$, respectively. At this point, readers should compare Eqs. (13.81), (13.82), and (13.83) for 1-D, 2-D, and 3-D slug injection and deduce the advection and dispersion from the given equation.

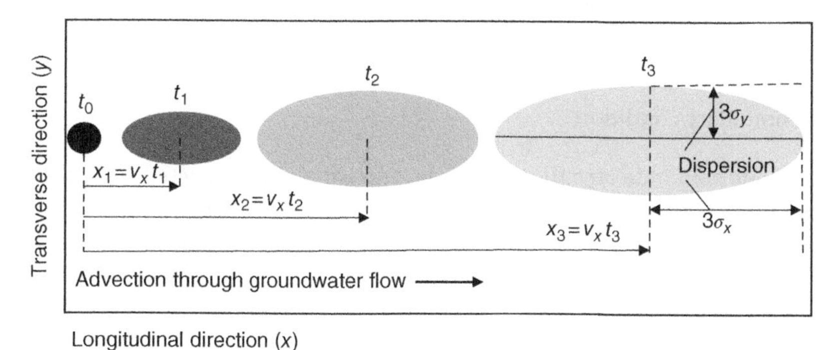

Figure 13.7 Plan view of the contaminant plume from an instantaneous injection in 2-D uniform flow at three different times t_1, t_2, and t_3.

13.3.4 Advection and Dispersion: 1-D Solutions to Continuous Injection

In our previous discussion about one-dimensional advection and dispersion, the governing equation (Eq. 13.29):

$$D_x \frac{\partial^2 C}{\partial x^2} - q_x \frac{\partial C}{\partial x} = \frac{\partial C}{\partial t}$$

Depending on the initial and boundary conditions, the above equation can have several analytical solutions as briefly described in the following text. These one-dimensional advection–dispersion models can be used in soil column studies.

First-Type Boundary Continuous Injection: The solution for 1-D continuous input under the first-type boundary condition is (Bear 1961; USGS 1992):

$$C(x,t)=\frac{C_0}{2}\left[\operatorname{erfc}\left(\frac{x-v_x t}{2\sqrt{D_x t}}\right)+\exp\left(\frac{v_x x}{D_x}\right)\operatorname{erfc}\left(\frac{x+v_x t}{2\sqrt{D_x t}}\right)\right] \quad \text{Eq. (13.84)}$$

The initial and boundary conditions are given as follows:

$C(x,\ 0)=0 \quad x \geq 0$ initial condition

$C(0,\ t)=C_0 \quad t \geq 0$ boundary condition

$C(\infty,\ t)=0 \quad t \geq 0$ boundary condition

The above conditions indicate that the initial concentration in the column is zero and the concentration of the contaminant in the pore water at $x=0$ is C_0. This is a fixed concentration boundary condition, i.e. **the first-type boundary**.

Second-Type Boundary Continuous Injection: The solution for 1-D continuous input under the second-type boundary condition is (Fetter 1993):

$$C(x,t)=\frac{C_0}{2}\left[\mathit{erfc}\left(\frac{x-v_x t}{2\sqrt{D_x t}}\right)-\exp\left(\frac{v_x x}{D_x}\right)\mathit{erfc}\left(\frac{x+v_x t}{2\sqrt{D_x t}}\right)\right] \quad \text{Eq. (13.85)}$$

The initial and boundary conditions for the above solution are as follows:

$C(x,\ 0)=0 \quad -\infty<x<+\infty$ initial condition

$\int_{-\infty}^{+\infty} n_e C(x,t)dx=C_0 n_e v_x t \quad t>0$ boundary condition

$C(\infty,\ t)=0 \quad t\geq 0$ boundary condition

The second boundary condition indicates the initial concentration being injected is C_0 and the injected mass of contaminant over the domain from $-\infty$ to $+\infty$ is proportional to the length of time of injection. The contaminant is free to disperse both upgradient and downgradient. Fetter (1993) indicates that Eq. (13.85) applies to a canal that is discharging contaminated water into an aquifer as a line source. Note that Eqs. (13.84) and (13.85) differ in the sign of the second term.

Third-Type Boundary Continuous Injection: A third solution for Eq. (13.29) under the third boundary condition is (van Genuchten and Alves, 1982; Fetter 1993):

$$C(x,t)=\frac{C_0}{2}\left[\operatorname{erfc}\left[\frac{x-v_x t}{2\sqrt{D_x t}}\right]+\left[\frac{v_x^2 t}{\pi D_x}\right]^{1/2}\exp\left[-\frac{(x-v_x t)^2}{4D_x t}\right]\operatorname{erfc}\left[\frac{x+v_x t}{2\sqrt{D_x t}}\right]-\frac{1}{2}\left[1+\frac{v_x x}{D_x}+\frac{v_x^2 t}{D_x}\right]\exp\left[\frac{v_x x}{D_x}\right]\operatorname{erfc}\left[\frac{x-v_x t}{2\sqrt{D_x t}}\right]\right]$$

Eq. (13.86)

with the following initial and boundary conditions:

$C(x,0)=0$ initial condition

$$\left. \left(-D\frac{\partial C}{\partial x} + v_x C \right) \right|_{x=0} = v_x C_0 \text{ boundary condition}$$

$$\left. \frac{\partial C}{\partial x} \right|_{x \to \infty} = (\text{finite}) \text{ boundary condition}$$

The third boundary condition specifies that as x approaches infinity, the concentration gradient will still be finite.

Note that Eqs. (13.84)–(13.86) all reduce to the approximate solution as the flow path increases:

$$C(x,t) = \frac{C_0}{2}\left[\text{erfc}\left[\frac{x - v_x t}{2\sqrt{D_x t}} \right] \right] \quad \text{Eq. (13.87)}$$

The graphic representation of Eq. (13.87) corresponding to a tracer experiment in a sand column is shown in Figure 13.8 (Freeze and Cherry 1979; Hemond and Fechner-Levy 2014).

Note that Eqs. (13.84) through (13.86) consider only advection–dispersion processes occurred in the column. If sorption is to be included, then substitute x by Rx and D_x by RD_x, where R is the retardation factor for the contaminant of interest. For example, the 1-D solution to a continuous injection in Eq. (13.87) can be modified as follows:

$$C(x,t) = \frac{C_0}{2}\left[\text{erfc}\left[\frac{Rx - v_x t}{2\sqrt{RD_x t}} \right] \right] \quad \text{Eq. (13.88)}$$

13.3.5 Advection and Dispersion: 2-D and 3-D Solutions to Continuous Injection

2-D Continuous Injection: In our previous discussion (Eq. 13.28), the governing equation for two-dimensional advection and dispersion is:

$$D_x \frac{\partial^2 C}{\partial x^2} + D_y \frac{\partial^2 C}{\partial y^2} - q_x \frac{\partial C}{\partial x} = \frac{\partial C}{\partial t}$$

The analytical solution for 2-D continuous input at steady-state (Fetter 1993; USGS 1992; modified from Bear 1979):

$$C(x,y) = \frac{C_0 Q}{2\pi\sqrt{D_x D_y}} \exp\left(\frac{v_x x}{2D_x} \right) K_0 \left[\left[\frac{v_x^{\,2}}{4D_x} \right] \left[\frac{x^2}{D_x} + \frac{y^2}{D_y} \right] \right]^{1/2} \quad \text{Eq. (13.89)}$$

where K_0 is the modified Bessel function of the second kind and zero order. Its value can be obtained in Excel using @BESSELK(k, 0). For example, at $k = 0.05$, @BESSELK(0.05, 0) = 3.114. Q is the rate at which a tracer of contaminant at concentration C_0 is injected.

The 2-D plume can be tracked at any given time t from a continuous source. Incorporating t into the governing equation (Eq. 13.28), we obtain its solution (Fetter 1993, p. 63):

$$C(x,y,t) = \frac{C_0 Q}{4\pi\sqrt{D_x D_y}} \exp\left(\frac{v_x x}{2D_x} \right) \left[W[0,B] - W[t_D, B] \right] \quad \text{Eq. (13.90)}$$

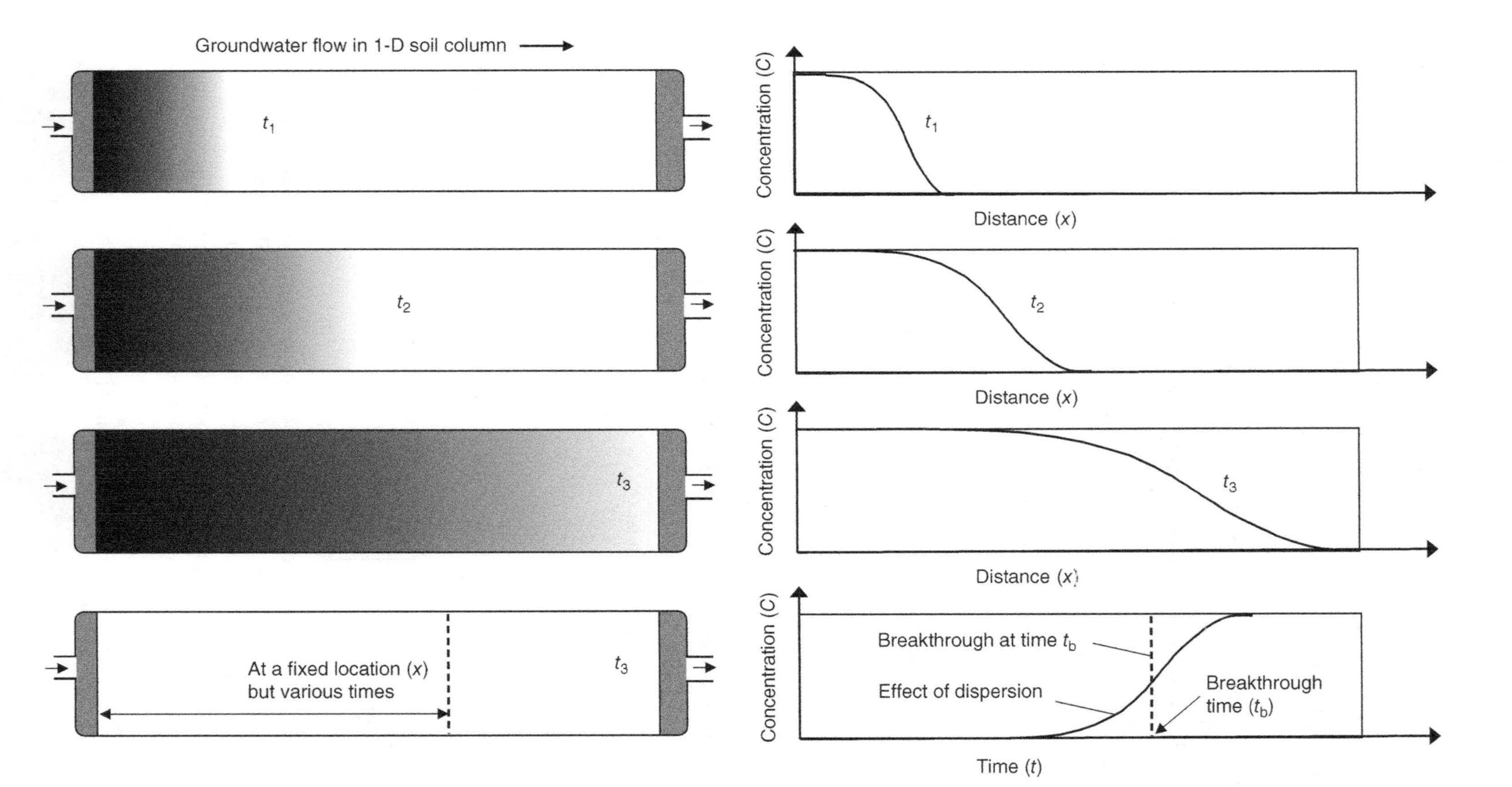

Figure 13.8 Contaminant plume in 1-D soil column following a continuous injection. The first three rows from the top to the bottom depict the contaminant concentrations as a function of the distance from the column inlet at three different times (t_1, t_2, and t_3). The last row depicts the contaminant concentrations at a fixed cross-section of the column at distance x from the column inlet as a function of time.

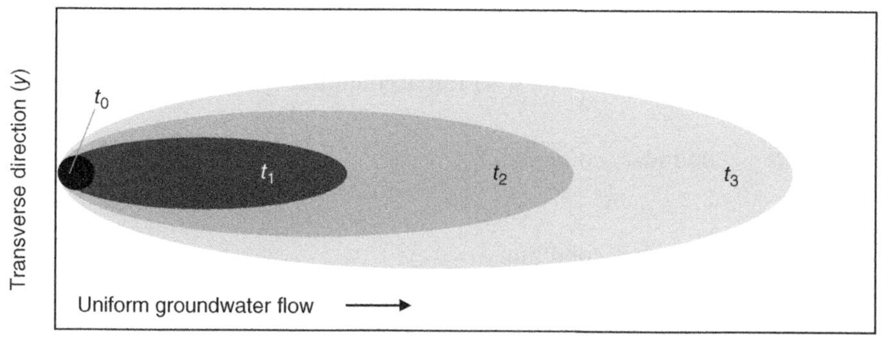

Figure 13.9 Development of contaminant plume in 2-D uniform flow from a continuous source.

where t_D is the dimensionless form of time, W is known as the well function given by Hantush (1956) and Fetter (1999).

$$t_D = \frac{v_x^2 t}{D_x} \qquad \text{Eq. (13.91)a}$$

$$B = \left[\frac{(v_x x)^2}{4D_x^2} + \frac{(v_x y)^2}{4D_x D_y}\right]^{1/2} \qquad \text{Eq. (13.91)b}$$

Equation (13.90) is depicted in Figure 13.9 for a uniform groundwater flow in a 2-D system with a continuous source (Freeze and Cherry 1979). Readers should compare this figure with Figure 13.7 to visualize the striking difference in the 2-D contaminant plumes between an instantaneous source and a continuous source.

3-D Continuous Injection: By neglecting advective flow in the y and z flow directions (q_y and q_z), the 3-D solute transport equation (Eq. 13.27) can be simplified to:

$$\frac{\partial}{\partial x}\left(D_x \frac{\partial C}{\partial x}\right) + \frac{\partial}{\partial y}\left(D_y \frac{\partial C}{\partial y}\right) + \frac{\partial}{\partial z}\left(D_z \frac{\partial C}{\partial z}\right) - \frac{\partial}{\partial x}(q_x C) = \frac{\partial C}{\partial t} \qquad \text{Eq. (13.92)}$$

Hunt (1978) presents an analytical solution for a point source with a conservative solute (Hunt 1978; USGS 1992):

$$C(x,y,z,t) = \frac{C_0 Q}{4n\pi r\sqrt{D_y D_z}} \exp\left(\frac{v_x(x-x_0)}{2D_x}\right)\left\{\exp\left[\frac{r v_x}{2D_x}\right] erfc\left[\frac{r+v_x t}{2\sqrt{D_x t}}\right] + \exp\left[\frac{-r v_x}{2D_x}\right] \text{erfc}\left[\frac{r-v_x t}{2\sqrt{D_x t}}\right]\right\} \qquad \text{Eq. (13.93)}$$

where

$$r = \left[(x-x_0)^2 + \frac{D_x(y-y_0)^2}{D_y} + \frac{D_x(z-z_0)^2}{D_z}\right]^{1/2} \qquad \text{Eq. (13.94)}$$

x_0, y_0, z_0 = coordinates of the point source

$$C = 0\left(-\infty < x < \infty,\quad -\infty < y < \infty,\quad -\infty < z < \infty\right) \text{ Initial condition}$$

$$\frac{\partial C}{\partial x} = 0 \qquad x = \pm\infty \qquad \text{Boundary condition}$$

$$\frac{\partial C}{\partial y} = 0 \qquad y = \pm\infty \qquad \text{Boundary condition}$$

$$\frac{\partial C}{\partial z} = 0 \qquad z = \pm\infty \qquad \text{Boundary condition}$$

In the preceding three sections (Sections 13.3.3–13.3.5), we have presented the mathematical formulas for 1-D, 2-D, and 3-D contaminant plumes under both slug injection and continuous injection. While the 1-D and 2-D graphical comparisons given in Chapter 2 (Figure 2.12) are self-evident, the fundamentals of mass balance in deriving these equations are what we supposed to learn here.

13.4 Numerical Solutions to Flow and Transport Processes

Numerical solutions are the numerical approximates to solutions of ordinary or partial differential equations. Numerical solutions are essential in groundwater modeling, because many differential equations cannot be solved exactly (analytic or closed form solution) and we must rely on numerical techniques to solve them. Since they are not the exact mathematical expressions, errors are inherent. Numerical approaches typically generate matrices that are so large that they require the use of computers to perform multiple iterations to converge a solution.

13.4.1 Partial Differential Equations and Numerical Methods

The governing equations we have seen so far in this chapter for the flow and transport processes are all **partial differential equations** (PDEs). PDEs involve two or more independent variables such as space (x, y, z) and/or time (t), an unknown function such as head (h) or concentration (C) as dependent variable, and partial derivatives of the unknown function with respect to the independent variables ($\partial h/\partial x$, $\partial C/\partial t$). Recall that ordinary differential equations (ODEs) are expressed using d (e.g. dC/dt), while partial derivatives are expressed using the curly ∂ (e.g. $\partial C/\partial t$). In flow equations, the unknown function is the hydraulic head (h) as a function of the space and/or time, such as the Laplace equation (Eq. 13.14) for flow in the saturated zone and Richards equation (Eq. 13.19) for flow in the unsaturated zone. In the transport equations, the unknown function is the contaminant concentration (C) as a function of the space and/or time as well as other parameters, such as the equations involving advection–dispersion (Eq. 13.27) with added adsorption and reaction terms (Eq. 13.48).

Three most popular numerical techniques for solving PDEs in groundwater modeling studies are finite difference method (FDM), finite element method (FEM), and the method of characteristics (MOC). Both FDM and FEM convert an PDE problem into its matrix form, and the PDE describing the groundwater flow and contaminant transport is then solved using matrix algebra. In approaching the numerical approximation, FDM uses derivative (detailed below), whereas FEM uses integration (specifically integration by parts) to convert PDE into a more manageable form. **Finite element method** divides groundwater basin into mesh of triangular or quadrilateral shapes (element). The triangular element is particularly advantageous to simulate irregular geographic and geologic features such as recharging rivers, faults, and aquifer boundaries. FEM thus requires more computation power and run time.

The **method of characteristics** (MOC) is used for contaminant transport modeling especially when convective transport dominates. It solves the PDE by converting PDE into ODE by rewriting the transport equation using fluid particles as the reference. In 2-D domain, there will be three equations including x-velocity, y-velocity, and concentration. The solutions of these equations are called the characteristic curves, hence the name.

Further discussion of finite element method and method of characteristics will be beyond the scope of this book. In the following discussions, our focus will be on finite difference method to illustrate the basics of numerical approaches as a step stone for numerical computation. Compared to FEM and MOC, FDM requires less computation power and run time. It is also the most intuitive and can be considered the most commonly used approach in groundwater modeling,

Finite difference method divides groundwater basin into a mesh of polygons, usually square/rectangular. Within each grid, hydraulic parameters can be considered to be constant. **Nodes** are then defined as either in the center of each square/rectangular mesh or at the intersections of the grid lines. Partial derivatives are substituted by the finite difference. For example, the **first derivative** of hydraulic head (h) in PDE, e.g., $\partial h/\partial x$ is approximated by the difference in hydraulic head between the two consecutive nodal points ($\Delta h / \Delta x = (h_{i+1}^n - h_i^n) / \Delta x$). Similarly, a **second derivative** is obtained by taking the difference between the first derivatives at the two consecutive nodal points (Table 13.1).

To understand the approximation using finite difference method, attention should be paid carefully about the notations in Table 13.1.

- Independent variable: h = hydraulic head; C = concentration;
- Time domain: n and $n+1$ = two consecutive grids in time; Δt = grid size in time;

Table 13.1 Examples of finite difference approximations for partial derivatives.

Partial derivative of hydraulic head (*h*)	Finite difference approximation	Partial derivative of concentration (*C*)	Finite difference approximation
$\frac{\partial h}{\partial x}$	$\frac{h_{i+1}^n - h_i^n}{\Delta x}$	$\frac{\partial C}{\partial x}$	$\frac{C_{i+1}^n - C_i^n}{\Delta x}$
$\frac{\partial h}{\partial y}$	$\frac{h_{j+1}^n - h_j^n}{\Delta y}$	$\frac{\partial C}{\partial y}$	$\frac{C_{j+1}^n - C_j^n}{\Delta y}$
$\frac{\partial h}{\partial z}$	$\frac{h_{k+1}^n - h_k^n}{\Delta z}$	$\frac{\partial C}{\partial z}$	$\frac{C_{k+1}^n - C_k^n}{\Delta z}$
$\frac{\partial h}{\partial t}$	$\frac{h_i^{n+1} - h_i^n}{\Delta t}$	$\frac{\partial C}{\partial t}$	$\frac{C_i^{n+1} - C_i^n}{\Delta t}$
$\frac{\partial^2 h}{\partial x^2}$	$\frac{h_{i+1}^n - 2h_i^n + h_{i-1}^n}{\Delta x^2}$	$\frac{\partial^2 C}{\partial x^2}$	$\frac{C_{i+1}^n - 2C_i^n + C_{i-1}^n}{\Delta x^2}$
$\frac{\partial^2 h}{\partial y^2}$	$\frac{h_{j+1}^n - 2h_j^n + h_{j-1}^n}{\Delta y^2}$	$\frac{\partial^2 C}{\partial y^2}$	$\frac{C_{j+1}^n - 2C_j^n + C_{j-1}^n}{\Delta y^2}$
$\frac{\partial^2 h}{\partial z^2}$	$\frac{h_{k+1}^n - 2h_k^n + h_{k-1}^n}{\Delta z^2}$	$\frac{\partial^2 C}{\partial z^2}$	$\frac{C_{k+1}^n - 2C_k^n + C_{k-1}^n}{\Delta z^2}$
$\frac{\partial^2 h}{\partial t^2}$	$\frac{h_i^{n+1} - 2h_i^n + h_i^{n-1}}{\Delta t^2}$	$\frac{\partial^2 C}{\partial t^2}$	$\frac{C_i^{n+1} - 2C_i^n + C_i^{n-1}}{\Delta t^2}$

- Space domain: i, j, k = three-dimensional spatial grids at a node where $x = i$, $y = j$, and $z = k$; and Δx, Δy, and Δz = grid sizes in their respective x, y, and z directions. For simplicity, we assume $\Delta x = \Delta y = \Delta z = a$.

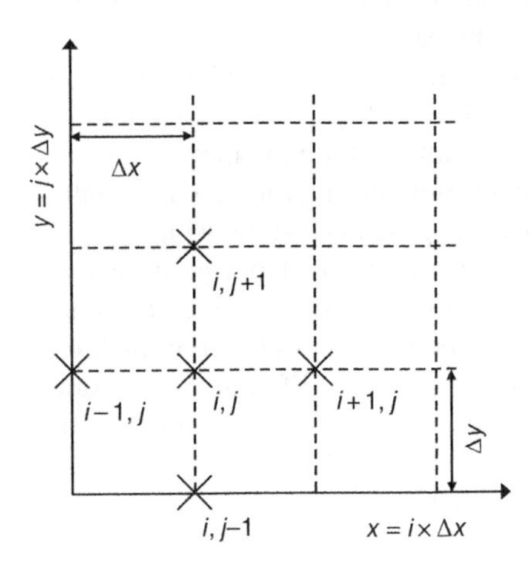

Figure 13.10 Finite difference with mesh centered nodes in two dimensions (*x* and *y*).

The approximations for the first derivatives in Table 13.1 are self-explanatory. The approximations for the second derivatives are less evident, but can be obtained by taking a difference between the first derivatives at two consecutive nodal points in space or time. For example, the second derivative of concentration in 2-D x–y plane (Figure 13.10) with respect to x direction is:

$$\frac{\partial^2 C}{\partial x^2} = \frac{\dfrac{C_{i+1,j} - C_{i,j}}{\Delta x} - \dfrac{C_{i,j} - C_{i-1,j}}{\Delta x}}{\Delta x} = \frac{C_{i+1,j} - 2C_{i,j} + C_{i-1,j}}{\Delta x^2} \quad \text{Eq. (13.95)}$$

Similarly in the y direction:

$$\frac{\partial^2 C}{\partial x^2} = \frac{\dfrac{C_{i,j+1} - C_{i,j}}{\Delta y} - \dfrac{C_{i,j} - C_{i,j-1}}{\Delta y}}{\Delta y} = \frac{C_{i,j+1} - 2C_{i,j} + C_{i,j-1}}{\Delta y^2} \quad \text{Eq. (13.96)}$$

Example 13.3 illustrates the finite difference method involving the first- and second-derivatives.

Example 13.3 Illustration of finite difference method (FDM)

Write the finite difference approximation for the two partial deferential equations we introduced in this chapter:

a) Eq. (13.76): $D_x \dfrac{\partial^2 C}{\partial x^2} = \dfrac{\partial C}{\partial t}$

b) Eq. (13.15): $\dfrac{\partial^2 h}{\partial x^2} + \dfrac{\partial^2 h}{\partial y^2} + \dfrac{\partial^2 h}{\partial z^2} = \dfrac{S}{T}\dfrac{\partial h}{\partial t}$

Solution

a) Using Table 13.1, we can write the finite difference approximation as follows:

$$D_x \frac{C_{i+1}^n - 2C_i^n + C_{i-1}^n}{\Delta x^2} = \frac{C_i^{n+1} - C_i^n}{\Delta t}$$

b) The second derivatives on the left hand side of Eq. (13.15) are as follows:

$$\frac{\partial^2 h}{\partial x^2} = \frac{h_{i+1,j,k} - 2h_{i,j,k} + h_{i-1,j,k}}{\Delta x^2} \quad \text{Eq. (13.97)}$$

$$\frac{\partial^2 h}{\partial y^2} = \frac{h_{i,j+1,k} - 2h_{i,j,k} + h_{i,j-1,k}}{\Delta y^2} \quad \text{Eq. (13.98)}$$

$$\frac{\partial^2 h}{\partial z^2} = \frac{h_{i,j,k+1} - 2h_{i,j,k} + h_{i,j,k-1}}{\Delta z^2} \qquad \text{Eq. (13.99)}$$

Note that the superscript n for the time domain is not included in the above three equations (Eqs. 13.97–13.99) for simplicity. On the right hand side of Eq. (13.15), the first derivative is:

$$\frac{\partial h}{\partial t} = \frac{h_i^{n+1} - h_i^n}{\Delta t}$$

If we further assume $\Delta x = \Delta y = \Delta z = a$, and substitute all the above equations (Eqs. 13.97–13.99) into Eq. (13.15), we obtain:

$$h_{i-1,j,k} - 6h_{i,j,k} + h_{i+1,j,\,k} + h_{i,j-1,k} + h_{i+1,j+1,k} + h_{i,j,k-1} + h_{i,j,k+1}$$
$$\frac{a^2 S}{T} \frac{h_i^{n+1} - h_i^n}{\Delta t}$$

The above equation will generate a set of matrix equations, which can be easily solved.

13.4.2 2-D Laplace Equation Using Finite Difference Method

The intent of this section is to illustrate an example for the use of numerical approach in solving 2-D Laplace equation. The solution to 2-D Laplace equation can be easily obtained using Excel spreadsheet without prior knowledge in numerical solution. We will derive the general formula and then use actual data to obtain the numerical approximations of the solution to 2-D Laplace equation.

We substitute Eqs. (13.97) and (13.98) into Eq. (13.9) (all in their 2-D format in the x–y plane):

$$\frac{\partial^2 h}{\partial x^2} + \frac{\partial^2 h}{\partial y^2} = 0$$

$$\frac{h_{i+1,j} - 2h_{i,j} + h_{i-1,j}}{\Delta x^2} + \frac{h_{i,j+1} - 2h_{i,j} + h_{i,j-1}}{\Delta y^2} = 0 \qquad \text{Eq. (13.100)}$$

For convenience and simplicity, by assuming $\Delta x = \Delta y$, we can solve for $h_{i,\,j}$:

$$h_{i,j} = \frac{h_{i+1,j} + h_{i+1,j} + h_{i,j-1} + h_{i,j+1}}{4} \qquad \text{Eq. (13.101)}$$

The general formula given by Eq. (13.101) states that the value at any grid point (i, j) is equal to the average of the values at the four adjacent grid points, i.e. $i-1$, j (left), $i+1$, j (right), i, $j-1$ (top), and i, $j+1$ (bottom). To start the numerical solution, the values of hydraulic head at the boundary points must be given. For all the grid points within the boundary, which have the four adjacent points (left, right, top, and bottom), the values can be calculated using Eq. (13.101). For the corner and edge grid points, which do not have all four adjacent points, an adjustment can be made to calculate the average. For example, if the bottom point is missed, the value at the top mesh point can be substituted. In Excel spreadsheet, formula of Eq. (13.101) can be easily inserted into any cell and then copied to all the cells within the boundary. Following the setup of these formulas, an iterative calculation should be enabled under the menu: Excel > Options > Formulas. Under Formulas, Enable Iterative Calculation must be turned on (checked), and Maximum Iterations and Maximum Changes must be specified. The higher the number of iterations, the more accurate the result. Sometimes the "Calculate" button on the left bottom of the spreadsheet must be pressed multiple times to get the result. A chart can be generated using Excel to show the flow net and equipotential lines. The example can be found through the links provided on the book's website.

Box 13.2 below provides a brief synopsis that bridges what we have discussed in this chapter with what we can do with a variety of models dealing with groundwater flow and contaminant transport.

Box 13.2 Selection of computer models for flow and transport

Groundwater modeling uses a computer-based methodology for mathematical analysis. It is an essential tool for investigating and managing the mechanisms and controls of groundwater systems. It is also a valuable tool in the screening of alternative remediation technologies and strategies in cleaning up contaminated groundwater. There are hundreds of groundwater models, and these models can be categorized in various ways, depending on the purpose of the model, the nature of the groundwater system, and the mathematical methods employed. For practical purpose, groundwater models can be grouped into flow models, transport models, chemical reaction models, stochastic models, models for fractured rocks, and groundwater management models (USEPA 1993).

We have briefly mentioned the use of some of the models in previous chapters. Our emphases in this chapter are the mathematical frameworks of flow models for porous media (saturated flow, unsaturated flow, and multiphase flow), and transport models for porous media (solute transport and vapor transport which are discussed in Chapter 8). Readers interested in many other models such as heat-related models, hydrogeochemical models, stochastic models, and models for fractured rocks should consult the USEPA's Center for Subsurface Modeling Support (http://www.epa.gov/water-research), the USGS's groundwater software information center at http://water.usgs.gov/software. Some of these software packages are in the public domain for free download (http://www.ehsfreeware.com). A review of this rapidly growing amount of information regarding each model's characteristics and uses is beyond the scope of this chapter.

One of the most widely used **groundwater flow models** MODFLOW is worth noting. It is so named because MODFLOW is a modular 3-D finite difference flow model, and each module deals with a specific feature of hydrological system to be simulated, such as wells, recharges, and rivers. MODFLOW works for various confined, unconfined, or mixed confined/unconfined aquifers. The primary output from the MODFLOW is the distribution of hydraulic head.

The flow results from MODFLOW can be incorporated into other models for contaminant transport calculations, such as MODPATH, MT3D, and RT3D. These **chemical transport models** may be the conservative models, which consider only diffusion, advective, and/or dispersive processes. Nonconservative contaminant transport models consider reactions, including sorption, redox, and biological degradation.

The governing equations of the flow and transport processes discussed in this chapter will help readers better understand the various models to be explored.

Addendum: Specific storage

Specific storage (S_s) is defined as the volume of water released from a unit volume of aquifer under a unit decline in hydraulic head (Freeze and Cherry 1979). Water is released from the aquifer storage in two mechanisms: (i) the compaction of aquifer by an increased stress and (ii) the expansion of water by a decreased pressure. The first mechanism is related to aquifer compressibility (α) and the second by the water compressibility (β). Aquifer compressibility is defined as follows:

$$\alpha = \frac{\frac{-\mathrm{d}V_T}{V_T}}{\mathrm{d}\sigma_e} \quad \text{Eq. (A13.1)}$$

where the volume reduction ($-\mathrm{d}V_T$) of aquifer by compaction is equal to the water expelled from the aquifer ($\mathrm{d}V_W = -\mathrm{d}V_T$). Note that the volume reduction $\mathrm{d}V_T$ is negative, but the amount of water produced $\mathrm{d}V_W$ is positive. From Eq. (A13.1), we therefore have:

$$\mathrm{d}V_w = -\mathrm{d}V_T = \alpha V_T \mathrm{d}\sigma_e \quad \text{Eq. (A13.2)}$$

The total stress (σ_T) overlying an arbitrary plane of a saturated porous aquifer is the weight of rock and water. This total stress is borne by the upward fluid pressure (p) and the effective stress (σ_e) which is not borne by the fluid, i.e. $\sigma_T = p + \sigma_e$, or $\mathrm{d}\sigma_T = \mathrm{d}p + \mathrm{d}\sigma_e$. The effective stress is due to the rearrangement of the soil grains which results in the compaction of the granular skeleton. Since total stress (σ_T) can be essentially considered constant over time ($\mathrm{d}\sigma_T = 0$), we have $\mathrm{d}\sigma_e = -\mathrm{d}p$. Substituting $\mathrm{d}p = -\rho gh$, we obtain: $\mathrm{d}\sigma_e = \rho g\mathrm{d}h$. Substituting $\mathrm{d}\sigma_e$ into Eq. (A13.2):

$$\mathrm{d}V_w = \alpha V_T \rho g \mathrm{d}h \quad \text{Eq. (A13.3)}$$

For a unit decline of hydraulic head ($\mathrm{d}h = 1$) and unit aquifer volume ($V_T = 1$), we can deduce the water released due to the compaction of aquifer as follows:

$$\mathrm{d}V_w = \alpha \rho g \quad \text{Eq. (A13.4)}$$

We now examine how the amount of released water (V_W) responds to the reduced aquifer pressure. Water compressibility (β) is a property of water itself under pressure, and it is defined as follows:

$$\beta = \frac{\frac{-\mathrm{d}V_W}{V_W}}{\mathrm{d}p} \quad \text{Eq. (A13.5)}$$

By replacing $V_W = nV_T$ and $\mathrm{d}p = -\rho g\mathrm{d}h$, and rearranging Eq. (A13.5), we have:

$$dV_W = -\beta(nV_T)(-\rho g\mathrm{d}h) \quad \text{Eq. (A13.6)}$$

For a unit decline of hydraulic head ($\mathrm{d}h = 1$) and unit aquifer volume ($V_T = 1$), Eq. (A13.6) becomes:

$$\mathrm{d}V_w = \beta n \rho g \quad \text{Eq. (A13.7)}$$

By combining Eqs. (A13.4) and (A13.7), we can obtain the change of water volume per unit decline of hydraulic head ($\mathrm{d}h$) and unit aquifer volume through two mechanisms previously described. This volume is what we defined early as specific storativity (S_s).

$$S_S = \alpha\rho g + \beta n\rho g = \rho g(\alpha + n\beta) \quad \text{Eq. (A13.8)}$$

The above formula should be dimensionally consistent. The dimension of α and β is the inverse of the pressure or stress (such as m^2/N, or $L\,T^2\,M^{-1}$), so the terms on right hand side result in a unit of $(M/L^3)\,(L/T^2)\,(L\,T^2/M) = L^{-1}$. This is consistent to the unit of S_s on the left hand side, which is defined as the volume change per unit volume and head $(L^3/L^3\text{–}L = L^{-1})$.

See also Fetter (2001) for a slightly different way of the derivation.

Bibliography

Anderson, M., Woessner, W., and Hunt, R. (2015). *Applied Groundwater Modeling*, 2e, 630 pp. Elsevier.

Bear, J. (1961). *Dynamics of Fluids in Porous Media*. New York: Elsevier Publishing Company, Inc.

Bear, J. (1972). *Dynamics of Fluids in Porous Media*. New York: Dover Publications, Inc.

Bear, J. (1979). *Hydraulics of Groundwater*. New York: McGraw-Hill.

Bear, J. and Cheng, A.H.-D. (2010). *Modeling Groundwater Flow and Contaminant Transport*, 834 pp. Springer.

Bedient, P.B., Rifai, H.S., and Newell, C.J. (1999). *Ground Water Contamination: Transport and Remediation*, 2e. Upper Saddle River, NJ: Prentice Hall PTR.

Cushman, J.H. and Tartakovsky, D.M. (2017). *The Handbook of Groundwater Engineering*, 3e. CRC Press.

Fetter, C.W. (1993). *Contaminant Hydrogeology*. New York: Macmillan Publishing Company.

Fetter, C.W. (1999). *Contaminant Hydrogeology*. Upper Saddle River, NJ: Prentice Hall.

Fetter, C.W. (2001). *Applied Hydrogeology*, 4e. Upper Saddle River, NJ: Prentice Hall.

Freeze, R.A. and Cherry, J.A. (1979). *Ground Water*. Englewood Cliffs, NJ: Prentice-Hall.

Grasso, D. (1993). *Hazardous Waste Site Remediation – Source Control*. Lewis Publishers.

Hantush, M.S. (1956). Analysis of data from pumping tests in leaky aquifers. *Eos, Transactions American Geophysical Union* 37 (6): 702–714.

Helmig, R. (1997). *Multiphase Flow and Transport Processes in the Subsurface – a Contribution to the Modeling of Hydrosystems*. Berlin Heidelberg: Springer-Verlag.

Hemond, H.F. and Fechner-Levy, E.J. (2014). *Chemical Fate and Transport in the Environment*, 3e. Academic Press Pages.

Hunt, B. (1978). Dispersive sources in uniform ground-water flow: American Society of Civil Engineers. *J. Hydraulic Division* 104 (HY1): 75–85.

Logan, B.E. (2012). *Environmental Transport Processes*, 2e. John Wiley & Sons Inc.

Prasad, T., Brown, K., and Thomas, J. (1994). Diffusion coefficients of organics in high density polyethylene (HDPE). *Waste Manage. Res.* 12: 61–71.

Schnoor, J.L. (1996). *Environmental Modeling: Fate and Transport of Pollutants in Water, Air, and Soil*. New York: John-Wiley & Sons.

Šimůnek, J. and van Genuchten, M.T. (2016). Chapter 22: contaminant transport in the unsaturated zone: theory and modeling. In: *The Handbook of Groundwater Engineering*, 3e (ed. J.H. Cushman and D.M. Tartakovsky). Boca Raton: CRC Press.

Sharma, H.D. and Reddy, K.R. (2004). *Geoenvironmental Engineering: Site Remediation, Waste Containment, and Emerging Waste Management Technologies*. John Wiley & Sons, Inc.

USEPA (1993). *Compilation of Ground-Water Models*, EPA/600/R-93/118. USEPA.

USEPA (1999). *Three-Dimensional NAPL Fate and Transport Model*, EPA/600/R-99/011. USEPA.

USGS (1992). *Techniques of Water-Resources Investigations of the United States Geological Survey, Chapter B7: Analytical Solutions for One-, Two-, and Three-Dimensional Solute Transport in Ground-Water Systems with Uniform Flow*. United States Government Printing Office.

Van Der Perk, M. (2014). *Soil and Water Contamination*, 2e. Boca Raton: CRC Press.
Van Genuchten, M.T. (1981). Analytical solutions for chemical transport with simultaneous adsorption zero-order production, and first-order decay. *J. Hydrol.* 49: 213–233.
Van Genuchten, M.T. and Alves, W.J. (1982). Analytical Solutions of the One-Dimensional Convective-Dispersive Solute Transport Equation: U.S. Department of Agriculture Technical Bulletin 1661, 151 pp.
Wu, Y.-S. and Qin, G. (2009). A generalized numerical approach for modeling multiphase flow and transport in fractured porous media. *Commun. Comput. Phys.* 6 (1): 85–108.

Questions and Problems

1 Check the consistency of units of each term in the flow equation (Eq. 13.17)

$$-\left(\frac{\partial}{\partial x}\rho q_x + \frac{\partial}{\partial y}\rho q_y + \frac{\partial}{\partial z}\rho q_z\right) = \rho\frac{\partial \theta}{\partial t} \qquad \text{Eq. (13.17)}$$

2 Check the consistency of units of each term in the transport equation (Eq. 13.48)

$$R\frac{\partial C}{\partial t} = D_x\frac{\partial^2 C}{\partial x^2} - q_x\frac{\partial C}{\partial x} + k_1 C + k_0 \qquad \text{Eq. (13.48)}$$

3 Given the values of the following parameters for Dupuit equation: $K = 0.5\,\text{m/day}$, $h_0 = 85\,\text{m}$, $h_{150} = 72\,\text{m}$, $L = 150\,\text{m}$ (refer to Figure 13.5), determine the specific discharge (q) and hydraulic head at $x = 50$, 100, and 125 m in an unconfined aquifer.

4 Using the same data from the question above, determine the value of q and hydraulic head at $x = 50$, 100, and 125 m. if the aquifer is confined with a linear change of hydraulic head.

5 The diffusion coefficient for benzene through flexible membrane liners (FMLs) used for lining impoundments and landfills was determined to be $5.1 \times 10^{-9}\,\text{cm}^2/\text{s}$ (Prasad et al. 1994). Estimate the concentration ratio C_t/C_0 at the minimum separation distance of 0.75 m (75 cm) between liner and groundwater table after 50 years of diffusion.

6 In a laboratory study on diffusion of chromium in one-dimensional transport, chromium concentration was changed from 55 to 2 mg/L with a diffusion distance of 5 cm in six months. Find the diffusion coefficient of chromium in this lab study. Assume this is a diffusion only process without other transport processes such as advection and adsorption.

7 A cylindrical column packed with sand has a length of 1.2 m and a diameter of 8 cm (0.04 m). The packed column has a porosity of 0.35, and a seepage velocity of 0.5 m/h. A total of 10 mg of the non-adsorbing and nonreactive chemical is injected into the column as a slug injection. The dispersion coefficient $D = 4.50 \times 10^{-4}\,\text{m}^2/\text{h}$.

a. Calculate the time when the bulk of chemical exits the column via advection through the flow of water

b. Determine the concentration of chemical at the exit of the column at the time calculated from (a)

c. Determine the exit concentration of the chemical at the time of ±0.3 h of the time calculated from (a)

8 Using the parameters in Problem 7, construct a concentration profile at the column exit for various times (i.e. C versus t at $x = 1.2\,\text{m}$).

9 Using the parameters in Problem 7, construct a concentration profile at $t = 1.2\,\text{h}$ along the flow direction of the column (i.e. C versus x at $t = 1.2\,\text{h}$).

10 Write the finite difference approximation for the partial differential equation (see Box 13.1) introduced in this chapter.

11 Write the finite difference approximation for the following partial differential equation (See box 13.1) introduced in this chapter:

$$\frac{\partial^2 h}{\partial x^2} + \frac{\partial^2 h}{\partial y^2} = \frac{-R(x,y)}{T}$$

12 A small underground container suddenly ruptured and released 10 kg of ethylbenzene over a cross-sectional area of $20\,\text{m}^2$ of the groundwater flow path. Ethylbenzene has an aerobic biodegradation half-life of 25 days and longitudinal dispersion coefficient of $0.075\,\text{m}^2/\text{day}$. The aquifer has a seepage (pore water) velocity of 0.5 m/day and a porosity of 0.40. Estimate the downgradient concentration (mg/L) of ethylbenzene in the groundwater after one month of leakage at 25 m away from the rupture location.

13 Use the same base values of the parameters in Problem 12, but do the following with Excel for time-saving of problem solving to see how the following parameters change the concentration in the flow path.

a. Change the value of x but keep all other parameters the same. At what distance x, the concentration is at its peak after 30 days of the spill?

b. Change the value of D but keep all other parameters the same. How does D impact the downgradient concentration?

14 Using the solubility data in Appendix C (ignore the temperature variation), estimate the solubility of each BTEX component in a dilute aqueous solution containing the mixture of BTEX with mole fractions of 0.1, 0.2, 0.3, and 0.2 for benzene, toluene, ethylbenzene, and *m*-xylene, respectively.

15 Using the vapor pressure data in Appendix C (ignore the temperature variation), estimate the equilibrium vapor pressure of each BTEX component in saturated air immediately adjacent to the pool of BTEX mixture (such as the case of chemical spill). The mixture has the mole fractions of 0.3, 0.2, 0.3, and 0.2 for benzene, toluene, ethylbenzene, and *m*-xylene, respectively.

Appendix A

Common Abbreviations and Acronyms

1-D, 2-D, 3-D	One-, Two-, Three-dimensional
ABS	Absorption Rate (Dust)
acfm	Actual Cubic Foot per Minute
ACGIH	American Conference of Governmental Industrial Hygienists
ACS	American Chemical Society
AFCEE	Air Force Center for Environmental Excellence
AOPs	Advanced Oxidation Processes
APHA	American Public Health Association
ASTM	American Society for Testing and Materials
AT	Averaging Time
ATSDR	Agency for Toxic Substance and Disease Registry
AWQC	Ambient Water Quality Criteria
AWWA	American Water Works Association
BCF	Bio-Concentration Factor
BF	Brownfields
Bgal/d	Billion Gallons per Day
BGS	Below Ground Surface
BHC	Hexachlorocyclohexane (HCH)
BMP	Best Management Practice
BR	Breathing Rate (Dust)
BTEX	Benzene, Toluene, Ethylbenzene, Xylene
BW	Body Weight
CAA	Clean Air Act
CAH	Chlorinated Aliphatic Hydrocarbon
CCL	Contaminant Candidate List
CDI	Chronic Daily Intake
CE	Combustion Efficiency
CEC	Cation Exchange Capacity
CERCLA	Comprehensive Environmental Response, Compensation, and Liability Act
CFR	Code of Federal Regulations
CFU	Colony Forming Units
CMC	Critical Micelle Concentration
CoA	Coenzyme A
CPT	Cone Penetrometer Test

Soil and Groundwater Remediation: Fundamentals, Practices, and Sustainability, First Edition. Chunlong Zhang.

Companion website: www.wiley.com/go/Zhang/Remediation_1e

CSF	Cancer Slope Factor
CWA	Clean Water Act
DCE	Dichloroethlene
DDE	Dichlorodiphenyldichloroethylene (1,1-bis-(4-Chlorophenyl)-2, 2-Dichloroethene)
DDT	1,1,1-Trichloro-2,2-bis(*p*-chlorophenyl)ethane and Dichloro-Diphenyl-Trichloroethane
DENIX	Defense Environmental Network & Information Exchange
DNAPL	Dense NonAqueous Phase Liquid
DNT	Dinitrotoluene
DO	Dissolved Oxygen
DOC	Dissolved Organic Carbon
DoD	Department of Defense
DoE	Department of Energy
DQOs	Data Quality Objectives
DRE	Destruction and Removal Efficiency
DTW	Depth to Water
EBCT	Empty Bed Contact Time
EC	Emerging Contaminants/Exposure Concentration
ED	Exposure Duration
EF	Exposure Frequency
EM	Environmental Management/Electromagnetic Conductivity
ERH	Electrical Resistance Heating
ERT	Electrical Resistance Tomography
ESA	Environmental Site Assessment
EU	European Union
FDM	Finite Difference Method
FEM	Finite Element Method
FIFRA	The Federal Insecticide, Fungicide, and Rodenticide Act
FRTR	Federal Remediation Technologies Roundtable
GAC	Granular Activated Carbon
GC	Gas Chromatography
GC–MS	Gas Chromatography–Mass Spectrometry
gpm	Gallons per Minute
GPR	Ground Penetrating Radar
HAZWOPER	Hazardous Waste Operations and Emergency Response
HBSL	Health-Based Screening Level
HDD	Horizontal Directional Drilling
HI	Hazard Index
HLB	Hydrophile–Lipophile Balance
HLWs	High-Level Radioactive Wastes
HMX	High Melting eXplosive (Octahydro-1,3,5,7-Tetranitro-1,3,5,7-Tetrazocine)
HQ	Hazard Quotient
HRS	Hazard Ranking System
HSWA	Hazardous and Solid Waste Amendment
HV	Heating Value
IAS	*In Situ* Air Sparging
IBA	Isobutanol

ICs	Institutional Controls
IFT	Interfacial Tension
IR	Ingestion Rate
ISCO	*In Situ* Chemical Oxidation
ITRC	Interstate Technology and Regulatory Council
LABS	Linear Alkyl Benzene Sulfonate
LCA	Life Cycle Assessment
LNAPL	Light Nonaqueous Phase Liquid
LOAEL	Lowest-Observed-Adverse-Effect Level
LUST	Leaking Underground Storage Tank
MCL	Maximum Contaminant Level
MCLG	Maximum Contaminant Level Goals
MDL	Method Detection Limit
MGD	Million Gallons per Day
MNA	Monitored Natural Attenuation
MOC	Method of Characteristics
MODFLOW	Modular Finite-Difference Flow Model
MSL	Mean Sea Level
MTBE	Methyl Tertiary Butyl Ether
MW	Molecular Weight
NAC	Nitroaromatic Compound
NAPL	Nonaqueous Phase Liquid
NCP	National Contingency Plan
NIMBY	Not In My Backyard
NIOSH	National Institute for Occupational Safety and Health
NMs	Nanomaterials
NOAA	National Oceanic and Atmospheric Administration
NOAEL	No-Observed-Adverse-Effect Level
NPDWR	National Primary Drinking Water Regulations
NPL	National Priority List
NSDWR	National Secondary Drinking Water Regulations
NSTP	Normal Standard Temperature and Pressure (1 atm 20 °C)
NTE	Not to Exceed
NTU	Nephelometric Turbidity Units
nZVI	Nano-Zero-Valent Iron
O&M	Operation and Maintenance
ODE	Ordinary Differential Equation
ORP	Oxidation–Reduction Potential
OSHA	Occupational Safety and Health Administration
OSWER	Office of Solid Waste and Emergency Response
P&T	Pump-and-Treat
PAC	Pulverized Activated Carbon
PAH	Polynuclear Aromatic Hydrocarbon
PBDE	Polybrominated Diphenyl Ether
PCB	Polychlorinated Biphenyl
PCDD	Polychlorinated Dibenzo-*p*-Dioxin
PCDF	Polychlorinated Dibenzofurans
PCE	Tetrachloroethylene (Perchloroethylene)

PCP	Pentachlorophenol
PDE	Partial Differential Equation
PF	Potency Factor
PFOA	Perflurooctanoic Acid
PFOS	Perflurooctane Sulfonate
PICs	Products of Incomplete Combustion
POHC	Principal Organic Hazardous Constituent
POP	Persistent Organic Pollutants
ppb	Parts per Billion
PPCPs	Pharmaceutical and Personal Care Products
PPE	Personal Protection Equipment
ppm	Parts per Million
ppt	Parts per Trillion/Parts per Thousands
PRB	Permeable Reactive Barrier/Potentially Responsible Party
PV	Present Value
QA/QC	Quality Assurance/Quality Control
QAPP	Quality Assurance Project Plan
R&D	Research and Development
RBCA	Risk-Based Corrective Action
RCRA	Resource Conservation and Recovery Act
RDX	Research Department Explosive (1,3,5-Trinitro-1,3,5-Triazacyclohexane)
REC	Recognized Environmental Condition
Redox	Oxidation–Reduction
RF	Radio Frequency
RfC	Reference Concentration
RfD	Reference Dose
RFH	Radio Frequency Heating
RI/FS	Remedial Investigation/Feasibility Study
RI/RF	Remedial Investigation/Remedial Feasibility
ROD	Record of Decision
ROI	Radius of Influence
RPM	Remedial Project Manager
S/S	Stabilization/Solidification
SARA	Superfund Amendment and Reauthorization Act
SBLR&BRA	Small Business Liability and Brownfields Revitalization Act
scfm	Standard Cubic Foot per Minute
SDWA	Safe Drinking Water Act
SEAR	Surfactant-Enhanced Aquifer Remediation
SEE	Steam-Enhanced Extraction
SERDP	Strategic Environmental Research & Development Program
SMCLs	Secondary Maximum Contaminant Levels
SPT	Standard Penetration Test
SSLs	Soil Screening Levels
SSTLs	Site-Specific Target Levels
SVE	Soil Vapor Extraction
SVOC	Semivolatile Organic Compound
SW-846	Test Methods for Evaluating Solid Waste, Physical/Chemical Methods
SWMU	Solid Waste Management Unit (smoos)

TBA	*t*-Butyl Alcohol
TCA	Trichloroethane
TCE	Trichloroethylene
TCH	Thermal Conductivity Heating
TEA	Terminal Electron Acceptors
TLV	Threshold Limit Value
TNT	2,4,6-Trinitrotoluene
TPHs	Total Petroleum Hydrocarbons
TSCA	Toxic Substances Control Act
TSCF	Transpiration Stream Concentration Factor
TSD	Treatment, Storage, and Disposal
TWA	Time Weighted Average
UF	Uncertainty Factor
UIC	Undergroundwater Injection Control
UR	Unit Risk
USACE	US Army Corps of Engineers
USCS	Unified Soil Classification System
USDA	United State Department of Agriculture
USEPA	U.S. Environmental Protection Agency
USGS	United States Geological Survey
UST	Underground Storage Tank
UV	Ultraviolet
VC	Vinyl Chloride
VOC	Volatile Organic Compound
WFD	Water Framework Directive
WOE	Weight of Evidence
ZOC	Zone of Capture
ZOI	Zone of Influence
ZVI	Zero-Valent Iron

Appendix B

Definition of Soil and Groundwater Remediation Technologies

Technology	Description
Soil, sediment, and sludge	
In situ biological treatment	
Biodegradation	The activity of naturally occurring microbes is stimulated by circulating water-based solutions through contaminated soils to enhance *in situ* biological degradation of organic contaminants. Nutrients, oxygen, or other amendments may be used to enhance biodegradation and contaminant desorption from subsurface materials.
Bioventing	Oxygen is delivered to contaminated unsaturated soils by forced air movement (either extraction or injection of air) to increase oxygen concentrations and stimulate biodegradation.
White rot fungus	White rot fungus has been reported to degrade a wide variety of organopollutants by using their lignin-degrading or wood-rotting enzyme system. Two different treatment configurations have been tested for white rot fungus, *in situ* and bioreactor.
In situ physical/chemical treatment	
Pneumatic fracturing	Pressurized air is injected beneath the surface to develop cracks in low-permeability and over-consolidated sediments, opening new passage ways that increase the effectiveness of many *in situ* processes and enhance extraction efficiencies.
Soil flushing	Water, or water containing an additive to enhance contaminant solubility, is applied to the soil or injected into the groundwater to raise the water table into the contaminated soil zone. Contaminants are leached into the groundwater, which is then extracted and treated.
Soil vapor extraction	Vacuum is applied through extraction wells to create a pressure/concentration gradient that induces gas-phase volatiles to diffuse through soil to extraction wells. The process includes a system for handling off-gases. This technology is also known as *in situ* soil venting, in situ volatilization, enhanced volatilization, or soil vacuum extraction.
Solidification/ stabilization	Contaminants are physically bound or enclosed within a stabilized mass (solidification), or chemical reactions are induced between the stabilizing agent and contaminants to reduce their mobility (stabilization).
In situ thermal treatment	
Thermally enhanced soil vapor extraction	Steam/hot air injection or electric/radio frequency heating is used to increase the mobility of volatiles and facilitate extraction. The process includes a system for handling off-gases.

(*Continued*)

Soil and Groundwater Remediation: Fundamentals, Practices, and Sustainability, First Edition. Chunlong Zhang.

Companion website: www.wiley.com/go/Zhang/Remediation_1e

(Continued)

Technology	Description
Vitrification	Electrodes for applying electricity are used to melt contaminated soils and sludges, producing a glass and crystalline structure with very low leaching characteristics.
Ex situ biological treatment (assuming excavation)	
Composting	Contaminated soil is excavated and mixed with bulking agents and organic amendments such as wood chips, animal and vegetative wastes, which are added to enhance the porosity and organic content of the mixture to be decomposed.
Controlled solid phase biological treatment	Excavated soils are mixed with soil amendments and placed in aboveground enclosures. Processes include prepared treatment beds, biotreatment cells, soil piles, and composting.
Landfarming	Contaminated soils are applied onto the soil surface and periodically turned over or tilled into the soil to aerate the waste.
Slurry phase biological treatment	An aqueous slurry is created by combining soil or sludge with water and other additives. The slurry is mixed to keep solids suspended and microorganisms in contact with the soil contaminants. Upon completion of the process, the slurry is dewatered and the treated soil is disposed of.
Ex situ physical/chemical treatment (assuming excavation)	
Chemical reduction/ oxidation	Reduction/oxidation chemically converts hazardous contaminants to nonhazardous or less toxic compounds that are more stable, less mobile, and/or inert. The oxidizing agents most commonly used are ozone, hydrogen peroxide, hypochlorites, chlorine, and chlorine dioxide.
Base catalyzed decomposition dehalogenation	Contaminated soil is screened, processed with a crusher and pug mill, and mixed with NaOH and catalysts. The mixture is heated in a rotary reactor to dehalogenate and partially volatilize the contaminants.
Glycolate dehalogenaiton	An alkaline polyethylene glycol (APEG) reagent is used to dehalogenate halogenated aromatic compounds in a batch reactor. Potassium polyethylene glycol (KPEG) is the most common APEG reagent. Contaminated soils and the reagent are mixed and heated in a treatment vessel. In the APEG process, the reaction causes the polyethylene glycol to replace halogen molecules and render the compound nonhazardous. For example, the reaction between chlorinated organics and KPEG causes replacement of a chlorine molecule and results in a reduction in toxicity.
Soil washing	Contaminants sorbed onto fine soil particles are separated from bulk soil in an aqueous-based system on the basis of particle size. The wash water may be augmented with a basic leaching agent, surfactant, pH adjustment, or chelating agent to help remove organics and heavy metals.
Soil vapor extraction	A vacuum is applied to a network of aboveground piping to encourage volatilization of organics from the excavated media. The process includes a system for handling off-gases.
Solidification/ stabilization	Contaminants are physically bound or enclosed within a stabilized mass (solidification), or chemical reactions are induced between the stabilizing agent and contaminants to reduce their mobility (stabilization).
Solvent extraction	Waste and solvent are mixed in an extractor, dissolving the organic contaminant into the solvent. The extracted organics and solvent are then placed in a separator, where the contaminants and solvent are separated for treatment and further use.
Ex situ thermal treatment (assuming excavation)	
High-temperature thermal desorption	Wastes are heated to 315–538 °C (600–1000 °F) to volatilize water and organic contaminants. A carrier gas or vacuum system transports volatilized water and organics to the gas treatment system.

Technology	Description
Hot gas decontamination	The process involves raising the temperature of the contaminated equipment or material for a specified period of time. The gas effluent from the material is treated in an afterburner system to destroy all volatilized contaminants.
Incineration	High temperatures, 871–1204 °C (1600–2200 °F), are used to combust (in the presence of oxygen) organic constituents in hazardous wastes.
Low-temperature thermal desorption	Wastes are heated to 93–315 °C (200–600 °F) to volatilize water and organic contaminants. A carrier gas or vacuum system transports volatilized water and organics to the gas treatment system.
Open burn/open detonation (OB/OD)	In OB operations, explosives or munitions are destroyed by self-sustained combustion, which is ignited by an external source, such as flame, heat, or a detonatable wave (that does not result in a detonation). In OD operations, detonatable explosives and munitions are destroyed by a detonation, which is initiated by the detonation of a disposal charge.
Pyrolysis	Chemical decomposition is induced in organic materials by heat in the absence of oxygen. Organic materials are transformed into gaseous components and a solid residue (coke) containing fixed carbon and ash.
Vitrification	Contaminated soils and sludges are melted at high temperature to form a glass and crystalline structure with very low leaching characteristics.
Other treatment	
Excavation and off-site disposal	Contaminated material is removed and transported to permitted off-site treatment and disposal facilities. Pretreatment may be required.
Natural attenuation	Natural subsurface processes – such as dilution, volatilization, biodegradation, adsorption, and chemical reactions with subsurface materials – are allowed to reduce contaminant concentrations to acceptable levels.
Groundwater, surface water, and leachate	
In situ biological treatment	
Cometabolic processes	An emerging application involves the injection of water containing dissolved methane and oxygen into groundwater to enhance methanotrophic biological degradation.
Nitrate enhancement	Nitrate is circulated throughout groundwater contamination zones as an alternative electron acceptor for biological oxidation of organic contaminants by microbes.
Oxygen enhancement with air sparging	Air is injected under pressure below the water table to increase groundwater oxygen concentrations and enhance the rate of biological degradation of organic contaminants by naturally occurring microbes.
Oxygen enhancement with hydrogen peroxide	A dilute solution of hydrogen peroxide is circulated throughout a contaminated groundwater zone to increase the oxygen content of groundwater and enhance the rate of aerobic biodegradation of organic contaminants by microbes.
In situ physical/chemical treatment	
Air sparging	Air is injected into saturated matrices to remove contaminants through volatilization.
Directional wells (enhancement)	Drilling techniques are used to position wells horizontally, or at an angle, in order to reach contaminants not accessible via direct vertical drilling.
Dual-phase extraction	A high-vacuum system is applied to simultaneously remove liquid and gas from low-permeability or heterogeneous formations.
Free product recovery	Undissolved liquid phase organics are removed from subsurface formations, either by active methods (e.g. pumping) or a passive collection system.
Free water or steam flushing/ stripping	Steam is forced into an aquifer through injection wells to vaporize volatile and semivolatile contaminants. Vaporized components rise to the unsaturated zone where they are removed by vacuum extraction and then treated.

(*Continued*)

(Continued)

Technology	Description
Hydrofracturing (enhancement)	Injection of pressurized water through wells cracks low-permeability and overconsolidated sediments. Cracks are filled with porous media that serve as avenues for bioremediation or to improve pumping efficiency.
Passive treatment walls	These barriers allow the passage of water while prohibiting the movement of contaminants by employing such agents as chelators (ligands selected for their specificity for a given metal), sorbents, microbes, and others.
Slurry walls	These subsurface barriers consist of vertically excavated trenches filled with slurry. The slurry, usually a mixture of bentonite and water, hydraulically shores the trench to prevent collapse and retards groundwater flow.
Vacuum vapor extraction	Air is injected into a well, lifting contaminated groundwater in the well and allowing additional groundwater flow into the well. Once inside the well, some of the VOCs in the contaminated groundwater are transferred from the water to air bubbles, which rise and are collected at the top of the well by vapor extraction.
Ex situ biological treatment (assuming pumping)	
Bioreactors	Contaminants in extracted groundwater are put into contact with microorganisms in attached or suspended growth biological reactors. In suspended systems, such as activated sludge, contaminated groundwater is circulated in an aeration basin. In attached systems, such as rotating biological contractors and trickling filters, microorganisms are established on an inert support matrix.
Ex situ physical/chemical treatment (assuming pumping)	
Air stripping	Volatile organics are partitioned from groundwater by increasing the surface area of the contaminated water exposed to air. Aeration methods include packed towers, diffused aeration, tray aeration, and spray aeration.
Filtration	Filtration isolates solid particles by running a fluid stream through a porous medium. The driving force is either gravity or pressure differential across the filtration medium.
Ion exchange	Ion exchange removes ions from the aqueous phase by exchange with innocuous ions on the exchange medium.
Liquid phase carbon adsorption	Groundwater is pumped through a series of canisters or columns containing activated carbon to which dissolved organic contaminants adsorb. Periodic replacement or regeneration of saturated carbon is required.
Precipitation	This process transforms dissolved contaminants into an insoluble solid, facilitating the contaminant's subsequent removal from the liquid phase by sedimentation or filtration. The process usually uses pH adjustment, addition of a chemical precipitant, and flocculation.
UV oxidation	Ultraviolet (UV) radiation, ozone, and/or hydrogen peroxide are used to destroy organic contaminants as water flows into a treatment tank. An ozone destruction unit is used to treat off-gases from the treatment tank.
Other treatment	
Natural attenuation	Natural subsurface processes – such as dilution, volatilization, biodegradation, adsorption, and chemical reactions with subsurface materials – are allowed to reduce contaminant concentrations to acceptable levels.
Air emission/off-gas treatment	
Biofiltration	Vapor phase organic contaminants are pumped through a soil bed and sorbed to the soil surface where they are degraded by microorganisms in the soil.
High-energy corona	The HEC process uses high-voltage electricity to destroy VOCs at room temperature.

Technology	Description
Membrane separation	This organic vapor/air separation technology involves the preferential transport of organic vapors through a nonporous gas separation membrane (a diffusion process analogous to putting hot oil on a piece of waxed paper).
Oxidation	Organic contaminants are destroyed in a high-temperature 1000 °C (1832 °F) combustor. Trace organics in contaminated air streams are destroyed at lower temperatures, 450 °C (842 °F), than conventional combustion by passing the mixture through a catalyst.
Vapor phase carbon adsorption	Off-gases are pumped through a series of canisters or columns containing activated carbon to which organic contaminants adsorb. Periodic replacement or regeneration of saturated carbon is required.

Source: Adapted from Federal Remediation Technologies Roundtable (FRTR), Remediation Technologies Screening Matrix and Reference Guide, Version 4 (http://www.frtr.gov/matrix2/top_page.html).

Appendix C

Structures and Properties of Important Organic Pollutants in Soil and Groundwater

Soil and Groundwater Remediation: Fundamentals, Practices, and Sustainability, First Edition. Chunlong Zhang.

Companion website: www.wiley.com/go/Zhang/Remediation_1e

Chemical name CAS registry number	Structure	MW	BP(°C)	Density (g/cm^3)	Solubility (mg/L)	Vapor pressure (atm)	Henry's law constant (unitless)	Log K_{ow}
BTEX compounds (6)								
Benzene (71-43-2)		78.11	80.1	0.88	1780 (20 °C)	0.079 (15 °C)	0.22 (25 °C)	2.13 (25 °C)
Toluene (108-88-3)	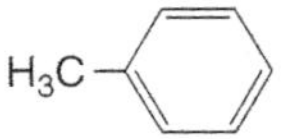 H_3C	92.1	110.8	0.87	515 (20 °C)	0.029 (20 °C)	0.28 (20 °C)[a]	2.69 (20 °C)
Ethylbenzene (100-41-4)	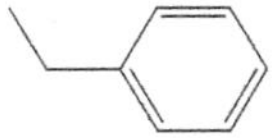	106.17	136.2	0.87	206 (15 °C)	0.0092 (15 °C)	0.35	3.15
o-Xylene (95-47-6)	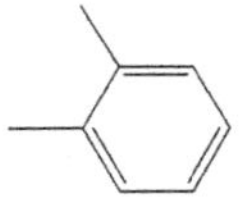	106.17	144	0.88	175 (20 °C)	0.0066 (20 °C)	0.22 (20 °C)[a]	2.77
m-Xylene (108-38-3)	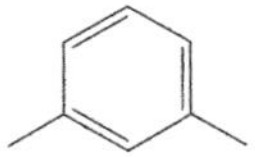	106.17	139	0.86	161[d]	0.0079 (20 °C)	0.294[d]	3.20
p-Xylene (106-42-3)	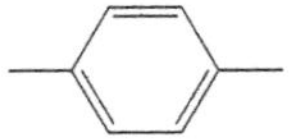	106.17	138.4	0.86	198	0.0086 (20 °C)	0.282[d]	3.15
PAHs (7)								
Anthracene (120-12-7)	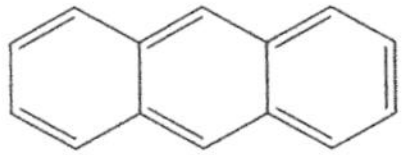	178.23	340	1.25	1.29	7.70E-07[b]	9.26E-04[b]	4.54
Benz(a)anthracene (56-55-3)	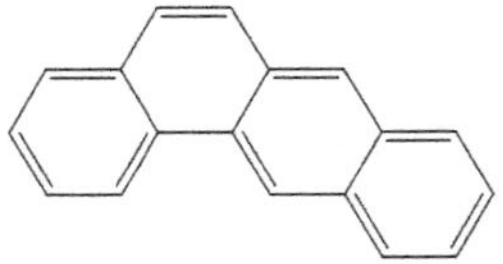	228	437	1.274[c]	0.014	6.30E-09[a]	2.40E-04[a]	5.91[a]
Benzo(a)pyrene (50-32-8)	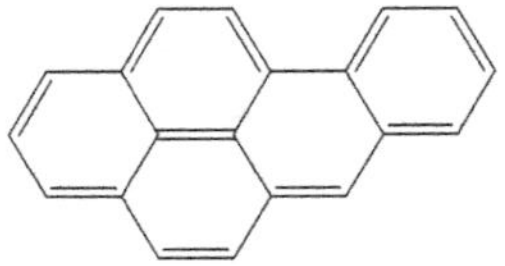	253.2	495	1.04(20 °C)[c]	0.003	2.30E-10[a]	4.90E-05[a]	6.50[a]
Naphthalene (91-20-3)	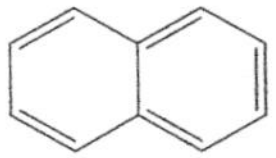	128.16	217.9	1.03[a]	30	3.0E-04[a]	4.90E-02[a]	3.36[a]

Phenanthrene (85-01-08)	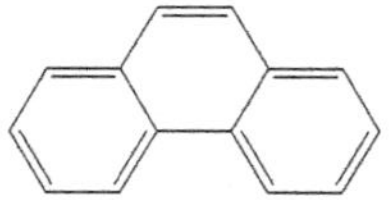	178.22	340	1.179[c]	0.82 (21 °C)	8.88E-07[b]	1.43E-03[b]	4.57[b]
Pyrene (129-00-0)	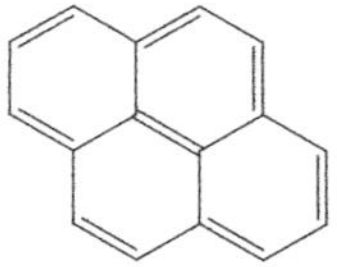	202.26	360	1.27[c]	0.16 (26 °C)	3.95E-08[b]	3.68E-04[b]	5.13[b]
Styrene (100-42-5)	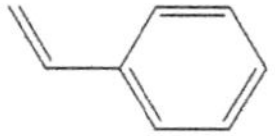	104.15	145.2	0.91[a]	250b	6.22E-03[b]	0.106[b]	3.05[b]
Halogenated alipahtic hydrocarbons (9)								
Carbon tetrachloride (56-23-5)	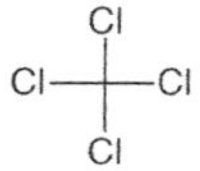	153.82	76.7	1.59[a]	1160	0.12[a]	0.97[a]	2.83 (20 °C)[a]
Chloroform (67-66-3)	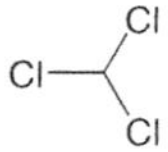	119.38	62	1.48[a]	9300	0.32a	0.2[a]	1.97 (20 °C)[a]
Dichlorodifluromethane (75-71-8)	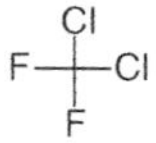	121	8.9[c]	1.48[c]	1.94E-03[b]	148[b]	16.1[b]	2.53[b]
1,1-Dichloroethane (75-34-3)	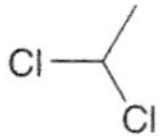	98.96	57.3	1.18[a]	5500 (20 °C)	0.3[a]	0.24[a]	1.79[a]
1,2-Dichloroethane (107-06-2)		98.96	83.5	1.24[a]	8690 (20 °C)	0.091[a]	0.041[a]	1.47[a]
Tetrachloroethylene (127-18-4)	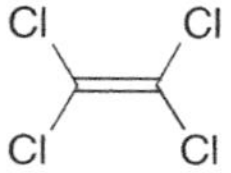	165.83	121	1.62[a]	400[a]	0.02[a]	0.34[a]	2.68[a]
Trichloroethene (79-01-6)	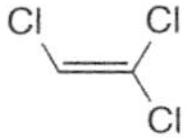	131.39	87[c]	1.46[a]	1000[a]	0.08[a]	0.42[a]	2.42[a]
1,1,1-Trichloroethane (71-55-6)	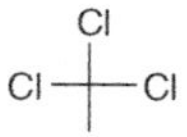	133.41	71–81	1.34[a]	4400 (20 °C)	0.13[a]	0.77[a]	2.48[a]
Vinyl chloride (75-1-4)		62.5	−14	0.91[a]	1100	3.4[a]	99[a]	0.60[a]

(*Continued*)

(Continued)

Chemical name CAS registry number	Structure	MW	BP(°C)	Density (g/cm^3)	Solubility (mg/L)	Vapor pressure (atm)	Henry's law constant (unitless)	Log K_{ow}
Halogenated aromatic hydrocarbons (3)								
Chlorobenzene (108-90-7)	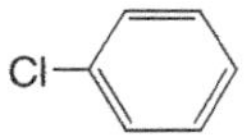	112.56	132	1.11[a]	500 (20 °C)	0.016[a]	0.165[a]	2.92[a]
1,2-Dichlorobenzene (95-50-1)	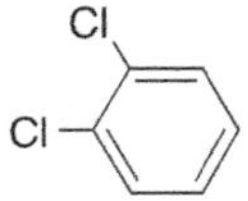	147.01	179	1.3	145	1.88E-03[b]	7.70E-02[b]	3.38[b]
Hexachlorobenzene (118-74-1)	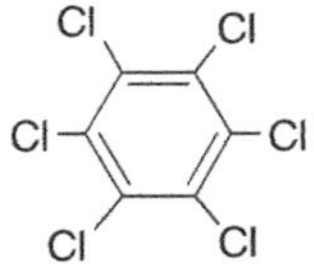	284.80	326/322	1.21[c]	0.004-0.006	3.45E-06[b]	6.11E-02[b]	5.50[b]
DDTs and other chlorinated pesticides (4)								
Aldrin (309-00-2)	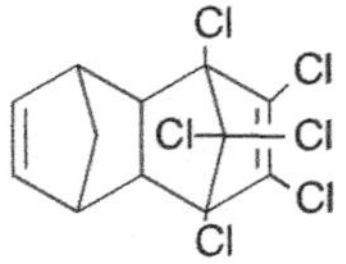	364.93	145[c]	1.6[c]	0.01	7.90E-09[b]	5.98E-04[b]	6.50[b]
4,4′-DDT (50-29-3)	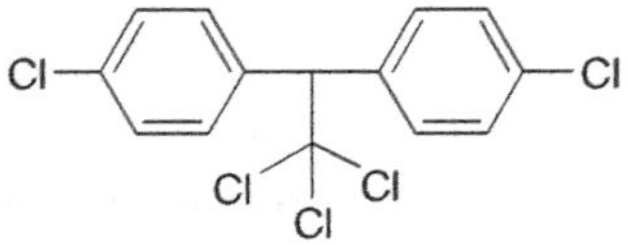	354.5	260[c]	1.6[c]	0.0031–0.0034	9.38E-10[b]	4.23E-04[b]	6.37[b]
Endrin (72-20-8)	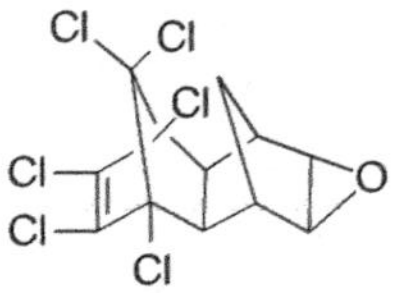	380.90	245(deco)[c]	1.7[c]	2.50E-04[c]	3.95E-09[b]	3.06E-04[b]	4.56[b]
α-Lindane (α-HCH) (319–84-6)	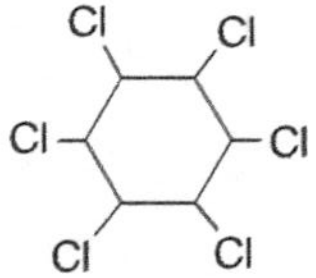	290.82	311[c]	1.9[c]	7.3[a]	1.30E-08[a]	2.20E-05[a]	3.78[b]

Polychlorinated biphenyl (PCBs) (2)								
Aroclor 1254 (11097-69-1)	Cl_4–Cl_6	325.06	290–325[c]	1.50[a]	0.012[a]	1.0E-07[a]	0.12[a]	6.5[a]
Aroclor 1260 (11096-82-5)	Cl_5–Cl_8	371.22	386[c]	1.57[a]	2.70E-03[a]	5.30E-03[a]	0.30[a]	6.7[a]
Dioxins and furans (2)								
2,3,7,8-Tetrachlorodibenzo-*p*-dioxin (TCDD) (1746-1701-6)	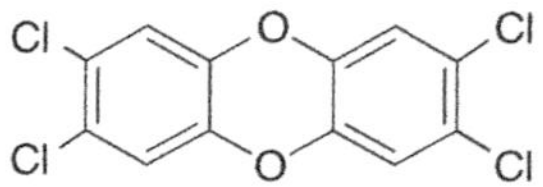	321.96	NA	1.8[c]	2.0E-04[c]	1.65E-09[b]	2.02E-03[b]	6.64[b]
2,3,7,8-Tetrachlorodibenzofuran (89059-46-1)	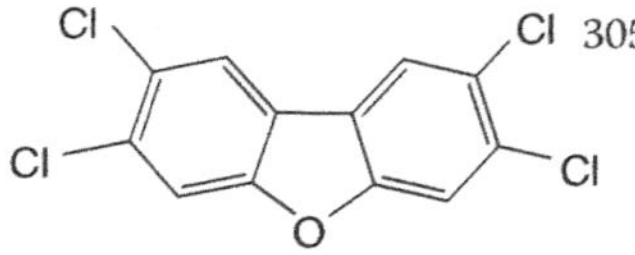	305.96	377	1.74	1.93E-03[d]	2.91E-09[d]	1.44E-04[d]	6.92[d]
Nitrogen-containing compounds (8)								
Atrazine (1912-1924-9)	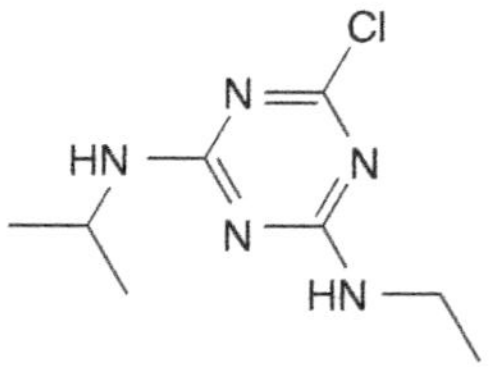	215.7	313[b]	1.2	70	3.95E-10 (20 °C)	1.0E-07[a]	2.51
2,6-Dinitrotoluene (606-20-2)	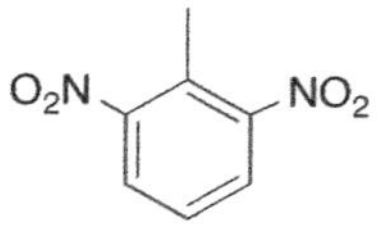	182.14	285	1.54 (15 °C)[d]	208	1.06E-06[d]	3.39E-05	2.18[d]
2,4-Dinitrotoluene (121-14-2)	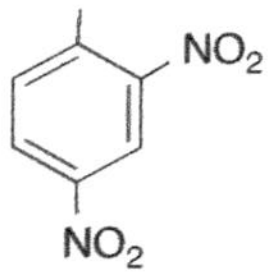	182.13	300	1.52 (15 °C)[d]	270 (22 °C)	1.93E-07[d]	3.79E-06[d]	2.18[d]
Nitrobenzene (98-95-3)	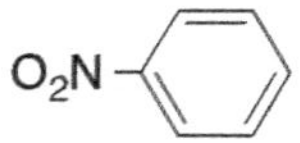	123.1	211	1.20	1900 (20 °C)	1.45E-04	2.95E-06[d]	1.85

(*Continued*)

(Continued)

Chemical name CAS registry number	Structure	MW	BP(°C)	Density (g/cm^3)	Solubility (mg/L)	Vapor pressure (atm)	Henry's law constant (unitless)	Log K_{ow}
2-Nitrophenol (88-75-5)		139.11	214/217	1.5[c]	2100 (20 °C)	1.78E-04[b]	5.71E-04[b]	1.89[b]
2,4,6-Trinitrotoluene (118-96-7)		227.13	240	1.65	140	2.63E-07[c]	1.10E-06	1.84
RDX (121-82-4)		222.1	276–280[c]	1.82[c]	59.7[c]	1.76E-09[d]	2.59E-06[d]	0.87[d]
HMX (2691-41-0)		296.2	436[c]	1.9[c]	5[c]	3.17E-11[d]	3.55E-08[d]	0.1d[d]
Oxygenated organic compounds (7)								
2,4-Dichlorophenol (120-83-2)		163.01	210	1.38	4500	1.18E-04[c]	1.26E-05[d]	2.75[d]
Diethyl phthalate (84-66-2)		222.2	298	1.12	210	8.19E-06[b]	8.06E-05[b]	2.35[b]

Methyl *tert* butyl ether (MTBE) (1634-04-4)	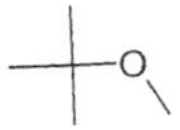	88.15	47.04d	0.7353c	1.98E04	0.332[d]	8.24E-02[d]	0.94[d]
Nonylphenol (104-40-5)	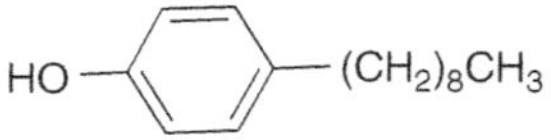 HO–C6H4–(CH2)8CH3	220.34	290–297	0.95	11 (20 °C)	1.08E-04[d]	2.44E-04[d]	5.76[d]
Pentachlorophenol (87-86-5)	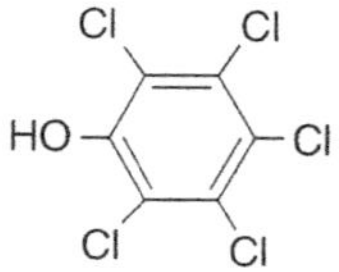	266.35	310	1.98	14 (20 °C)	1.80E-07[a]	1.50E-04[a]	5.04[b]
Phenol (108-95-2)	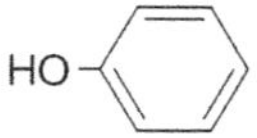	94.11	182	1.07	82 000 (15 °C)	6.81E-04[b]	1.85E-05[b]	1.48[b]
2,4,6-Trichlorophenol (88-06-2)	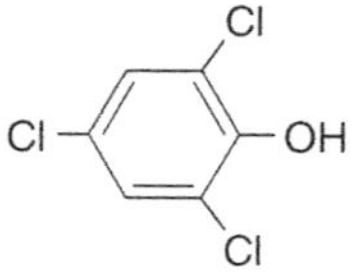	197.46	244.5	1.7	800	1.09E-05[b]	2.02E-06[b]	3.38[b]
Sulfur- and phosphorous-containing organic compounds (2)								
Parathion (56-38-2)	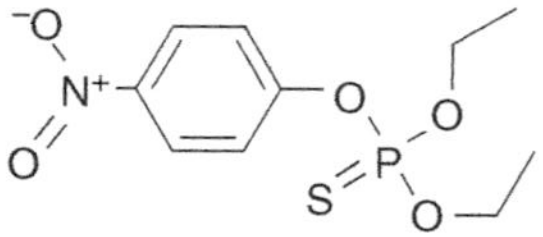	291.3	157–162	1.26	24	3.95E-06[d]	1.21E-05[d]	3.81 (20 °C)
Glyphosate (1071-83-6)	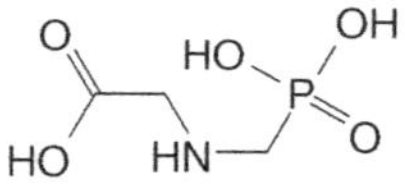	169.07	417	1.7	12 000	1.29E-08[c]	1.67E-17[d]	−3.4[d]

Note: (1) The selected chemicals are the frequent organic pollutants in soil and groundwater. The list is not exhaustive. They are arranged in an alphabetical order in each major contaminant group discussed in Chapter 2. Inorganic pollutants are not included in the table.
(2) Molecular weight (MW), boiling point, solubility, vapor pressure, Henry's law, and log K_{ow} data are compiled from Verschueren (2001). Solubility values are at 25 °C unless otherwise indicated. Other sources of property data are as follows:
[a] Hemond and Fechner-Levy (2014).
[b] Valsaraj (2009). Vapor pressures in kPa are converted into atm by multiplying 1/101.325. Henry's law constants in kPa-dm^3/mol (i.e. Pa-m^3/mol) are converted into dimensionless Henry's law constant by multiplying 4.04×10^{-4}.
[c] PubChem, http://pubchem.ncbi.nlm.nih.gov, U.S. National Library of Medicine.
[d] USEPA (2018); Henry's law constants in atm × m^3/mol was converted into dimensionless Henry's law constant using EPA Online-Tools for Site Assessment Calculation.

References

Hemond, H.F. and Fechner-Levy, E.J. (2014). *Chemical Fate and Transport, 3e*. London, UK: Academic Press.

US EPA (2018). *The Estimation Programs Interface (EPI) Suite (v4.11)*.

Valsaraj, K.T. (2009). *Elements of Environmental Engineering: Thermodynamics and Kinetics, 3e*. CRC Press.

Verschueren, K. (2001). *Handbook of Environmental Data on Organic Chemicals, 4e, vol. 1–2*. John Wiley & Sons.

Appendix D

Unit Conversion Factors

To convert from:	To:	Multiply by:
Length		
foot (ft)	meter (m)	0.305
inch (in)	centimeter (cm)	2.54
kilometer (km)	mile (mi)	0.6214
meter (m)	micrometer (μm)	1×10^6
meter (m)	nanometer (nm)	1×10^9
yard (yd)	meter (m)	0.9144
Area		
acre (ac)	square foot (ft^2)	43 560
acre (ac)	square meter (m^2)	4 047
hectare (ha)	square meter (m^2)	10 000
hectare (ha)	acre (ac)	2.471
square foot (ft^2)	square meter (m^2)	0.0929
Volume		
cubic foot (ft^3)	cubic meter (m^3)	0.0283
cubic foot (ft^3)	gallon (gal)	7.48
cubic meter (m^3)	litter (L)	1 000
cubic yard (yd^3)	gallon (gal)	202
gallon (gal)	litter (L)	3.785
gallon (gal)	cubic meters (m^3)	0.003 785
litter (L)	cubic decimeter (dm^3)	1
litter (L)	cubic centimeter (cm^3)	1 000
Mass		
kilogram (kg)	pound (lb)	2.205
pound (lb)	gram (g)	453.59
ton (metric)	kilogram	1 000
ton (metric)	pound (lb)	2 204
Speed, velocity, hydraulic conductivity		
foot per day (ft/d)	meter per day (m/d)	0.3048
foot per day (ft/d)	kilometer per year (km/yr)	1.2
Flow rate		
cubic foot per second (cfs)	gallon per minute (gpm)	448.8
cubic meter per second (m^3/s)	acre feet per day (ac-ft/d)	70.07
gallon per minute (gpm)	cubic meter per second (m^3/s)	6.31×10^{-5}
million acre feet per year	cubic meter per second (m^3/s)	39.107
million gallons per day (MGD)	cubic meter per second (m^3/s)	0.0438

(*Continued*)

Soil and Groundwater Remediation: Fundamentals, Practices, and Sustainability, First Edition. Chunlong Zhang.

Companion website: www.wiley.com/go/Zhang/Remediation_1e

(Continued)

To convert from:	To:	Multiply by:
Density		
gram per milliliter (g/mL)	pound per cubic foot (lb/ft^3)	62.5
kilogram per cubic meter (kg/m^3)	pound per cubic foot (lb/ft^3)	0.0624
pound per cubic foot (lb/ft^3)	kilogram per cubic meter (kg/m^3)	16.018
Force		
Newton ($1\,N = 1\,kg{\cdot}m/s^2$)	Dyne ($1\,dyn = 1\,g{\cdot}cm/s^2$)	1×10^5
Pound-force (lbf)	Newton (N)	4.448
Pressure or stress		
atmosphere (atm)	mm of Hg	760
atmosphere (atm)	kilopascal (kPa)	101.325
atmosphere (atm)	pound per square inch (psi)	14.7
bar	atmosphere (atm)	0.987
inch of water	mm of Hg	1.86
Pascal (Pa)	Newton per square meter (N/m^2)	1
torr	mm of Hg	1
Work or energy		
British thermal unit (BTU)	calories (cal)	252
calories (cal)	Joules (J)	4.18
kilocalories (kcal)	British thermal unit (BTU)	3.97
Newton-meter (N-m)	Watt-hours (Wh)	2.78×10^{-4}
calories (cal)	Newton-meter (N-m)	1
Power		
kilowatt (kw)	Joules per second (J/s)	1 000
kilowatt (kw)	British thermal unit per hour (BTU/h)	3412
horsepower (hp)	Watt (W)	746
Concentration		
parts per million (ppm)	parts per billion (ppb)	1×10^3
parts per million (ppm)	parts per trillion (ppt)	1×10^6
Molarity (M)	moles per litter (mol/L)	1
parts per million (ppm) (in water)	milligram per litter (mg/L)	1
parts per million (ppm) (in soil)	milligram per kilogram (mg/kg)	1
percent (%)	parts per millions (ppm)	10000
Viscosity		
centipoise (cP, dynamic viscosity)	Pascal·second (Pa·S) (N·s/m2) (kg/m·s)	1×10^{-3}
centistoke (cSt, kinematic viscosity)	Square millimeter per second (mm^2/s)	1

Formulas for Temperature Conversions

$°C\left(\text{degree Celsius}\right) = \frac{5}{9}\left(°F - 32\right)$	$°F\left(\text{degree Fahrenheit}\right) = \frac{9}{5}°C + 32$
$°K\ (\text{degree Kelvin}) = °C + 273.15$	$°K\left(\text{degree Kelvin}\right) = \frac{5}{9}\left(°F - 32\right) + 273.15$
$°R\left(\text{degree Rankine}\right) = \frac{9}{5}\left(°C + 273.15\right)$	$°R\ (\text{degree Rankine}) = °F + 459.67$

References

Spellman, F.R. and Whiting, N.E. (2005). *Environmental Engineer's Mathematics Handbook*. CRC Press.

Masters, G.M. and Ela, W.P. (2007). *Introduction to Environmental Engineering and Science*, 3e, Prentice Hall, Upper Saddle River, NJ.

Appendix E

Answers to Selected Problems

Chapter 2

15. $CaCO_3$: 7.746×10^{-5} mol/L; $CaSO_4$: 7.021×10^{-3} mol/L
16. (a) CdS: 4.57×10^{-9} mg/L; (b) $Ca_3(PO_4)_2$: 0.035 mg/L
17. $\lg S = \frac{1}{3}\lg\left[Cd^{2+}\right] - \frac{2}{3}\text{pH} + 9.13$
20. (a) f_{w1} = 10.6%, f_{w2} = 2.2%; (b) K_{oc1} = 80 mL/g, K_{oc2} = 415 mL/g; (c) R_1 = 12.2, R_2 = 59.1 (subscript 1 = naphthalene, 2 = pyrene).
22. (a) K_d = 32.5 L/kg (Lake Michigan sediment), 8 L/kg (Marlette soil), 4 L/kg (Woodburn soil), 2 L/kg (Mississippi River sediment); (b) Lake Michigan sediment > Marlette soil > Woodburn soil > Mississippi River sediment, (c) Corresponding K_{oc} values are: 808, 444, 317, and 500 L/kg.
23. (a) 4.36; (b) 1.67; (c) 1801

Chapter 3

2. (b) silty clay; (d) silty clay loam
3. (a) sandy clay; (c) sandy loam
5. 2.07%
15. 43%
16. 22%
17. 43 ft.
18. 1.4×10^{8} gal
19. (a) 20 m^3/day; (b) 1 m^2/day
20. (a) 773 gal/min; (b) 2576 gal/min
23. (a) 1.27 ft/ft; (b) 0.415 gal/min
24. 2.31×10^{-4} ft/s
25. (a) 0.0217 m^2/min, (b) 0.00062 m/min
27. $Z = 0.5\,\text{m}$, $\frac{p}{\rho g} = 0.097\,\text{m}$, $\frac{v^2}{2g} = 3.19 \times 10^{-13}\,\text{m}$, total head $m \approx 0.5\,\text{m}$

Soil and Groundwater Remediation: Fundamentals, Practices, and Sustainability, First Edition. Chunlong Zhang.

Companion website: www.wiley.com/go/Zhang/Remediation_1e

Chapter 4

7. (a) \$10 167
8. (a) \$12 879; (c) 9 704
9. \$292 645
11. Option 2 has a net saving of \$634 806 – \$483 575 = \$151 231.
13. (a) 17 in 1 million; (b) 9 cancers/yr.; (c) 965 cancers/yr
14. (a) 6.35×10^{-5}; (b) 8.88×10^{-5}
15. (a) 0.000 06; (b) 0.0009
16. 0.001 076 (11 out of 10 000 chances)
17. HQ of $0.172 < 0.2$, no adverse human health effects (noncancer) are expected.
18. 0.140 mg/L
21. 0.001 672 or 1672 per 1 000 000 population
22. For PCE, the risk = 9.15×10^{-6}
23. For PCE, the risk = 3.05×10^{-8}

Chapter 5

15. 5.10 gal
19. 0.00283
20. 0.0147 cm/s
21. (a) $W(u) = 8.2278$ when $u = 1.5 \times 10^{-4}$; (b) $W(u) = 4.9483$ when $u = 4 \times 10^{-3}$
22. $u = 0.001 07$; (b) $W(u) = 6.26$. Drawdown = 3.06 m

Chapter 6

18. PCBs are very low in volatility, so air striping does not work. *In situ* anaerobic bioremediation works well for PCE, but not for PCB in general.
19. Thermal desorption will work only if the chemical is sufficiently volatile.

Chapter 7

9. $Q/\pi bv$ (two wells), and $\sqrt[2]{2}Q / \pi bv$ (three wells).
11. (a) 1.31×10^5 kg; (b) 8.65×10^4 gal, (c) 7.14%
12. (a) 0.0044 kg (total), 0.015%; (b) 14.6 years
13. (a) 3000 ft^3; (b) 382.5 ft^3; (c) 105 ft^3
14. 7.48 m
15. Yes, based on $2Y^0_{max} = 47.6$ m at $x = 0$ and $2Y_{max} = 95.2$ m at $x = \infty$
16. (a) 30.3 m; (b) 0.00236 m^3/s; (c) 8.46 years
17. 52.1%
18. 6.14 m
19. 3209 lb
20. 11.8%

Chapter 8

10. 1.8×10^{-11} cm^2 or 1.8×10^{-3} darcy.
18. 23.7 mmHg.
19. 36.5 Torr cyclohexane, 28.5 Torr ethanol
20. 0.039077 mmHg (benzene), 23.79021 mmHg (water)
21. 229580 mg/m^3 (PCE), 24166 mg/m^3 (TCE)
25. (a) 3.17 mg/L; (b) not sufficient
26. 141895 mg/m^3
27. (a) 2.74×10^{-5} mol/L; (b) 0.438 mg/L as CH_4
28. (a) 28.8 mg/L; (b) 2.88×10^{-6} mg; (c) 2.2×10^{-6}%, (d) very low percent due to low H
30. 1.02 kg/d
31. 10.19 kg/d
32. (a) 0.409 m^3/m-min; (b) 4330 cfm
33. 176.5 ft^3/min

Chapter 9

10. 0.396 lb O_2, 0.0615 lb N (from 2,4-DNT rather than external N source), 0.010 lb P
11. (a) and (b) both <1% error; (d) 0.223 g cell/g DNT
12. $\Delta pH = 2.56$
13. (a) 94.98 mg O_2/L ($C_{10}H_8$), 3.82 mg O_2/L ($C_{14}H_{10}$); (b) air sparging is needed for $C_{10}H_8$. (c) 7.49 kg O_2.
14. (a) 0.0495 day^{-1}; (b) 60.5 days; (c) 70.7, 50.0, 35.4, and 25.0% after first, second, third and fourth weeks, respectively
15. (a) 158 days; (b) 70.7%; (c) 1046 lb/day (first two weeks), 740 lb/day (second two weeks)
16. (a) $C_{15}H_{32}O + 25/2\ O_2 + 2\ NH_3 \rightleftharpoons 5\ CO_2 + 2\ C_5H_7O_2N + 12H_2O$; (b) 47250 lb O_2, 1658 lb N, and 275 lb P
17. 2500 kg N and 250 kg P.
18. (a) 4.2 years; (b), 0.035 day^{-1}, 131.58 days; (c) 10.32 years; (d) not feasible for hexachlorobenzene.
19. 88163 lb O_2, 5496 lb N, 916 lb P
21. 18180 gal
22. 1.79 kg
59. 3.61 years
60. 106 years

Chapter 10

2. $C_3H_5OCl + 7/2\ O_2 + 79/6\ N_2 \rightarrow 3CO_2 + 2\ H_2O + HCl + 79/6\ N_2$
6. 0.02 kg/h (four nines rule), and 2 mg/h (six nines rule)
7. 99.9808% (by volume), 99.9878% (by mass)
8. 819 ft^3
9. 15064 acfm

12. Benzene: 390.57 lb, Octane: 571.15 lb, Ethanol: 230.35 lb, Butyl acetate: 580.8 lb.
13. 4770 L/kg
16. (a) 87.5%; (b) 91.67%
17. (a) 15.6 lb/lb C_3H_8; (b) 15.6 lb/lb C_3H_8; (c) 31.3 lb/lb C_3H_8
18. 22.039 Btu/lb
19. 587 lb O_2/h, 1950 lb N_2/h, 2537 lb air/h
20. 881 lb O_2/h, 2925 lb N_2/h, 3806 lb air/h
21. (a) $C_6H_{10}O_5 + 7.2\ O_2 + 27.144 N_2 \rightarrow 6CO_2 + 5H_2O + 1.2\ O_2 + 27.144 N_2$
    (b) $C_6H_{10}O_5 + 12\ O_2 + 45.24 N_2 \rightarrow 6CO_2 + 5H_2O + 6\ O_2 + 45.24 N_2$
22. 17 385 acfm

Chapter 11

4. (a) 52.1% volume reduction; (b) 6.9% in water, 2.1% in sand, 5.1% in silt, and 85.9% in clay
5. 6.9% in water, 2.1% in sand, 5.1% in silt, and 85.9% in clay. The change is minimal in second wash comparing to Problem 4
17. (a) 0.025 M; (b) 25% reduction in surface tension
18. (a) 0.0015 M, 31% reduction in surface tension

Chapter 12

11. $k = 0.1386\ h^{-1}$, $t = 38.2$ h, $b = 0.80$ ft
12. $k = 0.0778\ day^{-1}$, $t = 29.6$ days, $b = 2.96$ m

Chapter 13

3. $q = 3.4\ m^2/d$; $h = 80.0$, 76.6, and 74.3 m
4. $q = 3.4\ m^2/d$; $h = 80.7$, 76.3, and 74.2 m
5. 2.08×10^{-79}
6. $2.94 \times 10^{-6}\ cm^2/s$
7. (a) 2.4 h; (b) 12.2 mg/L; (c) 0.0339 mg/L at 2.1 hours and 0.112 mg/L at 2.7 hours
10. $D_x \frac{C_{i+1}^n - 2C_i^n + C_{i-1}^n}{\Delta x^2} + q_x \frac{C_{i+1}^n - C_i^n}{\Delta x} = \frac{C_i^{n+1} - C_i^n}{\Delta t}$
12. 0.24 mg/L
14. 178, 103, 61.8, and 32.2 mg/L for benzene, toluene, ethylbenzene, and *m*-xylene
15. 0.0237, 0.0058, 0.0276, and 0.00158 atm for benzene, toluene, ethylbenzene, and *m*-xylene

Index

Soil and Groundwater Remediation: Fundamentals, Practices, and Sustainability, First Edition. Chunlong Zhang.

Companion website: www.wiley.com/go/Zhang/Remediation_1e

b

c

f

g

h

m

n

o

p

q

r

s

t

u

v

IUPAC Periodic Table of the Elements

IUPAC Periodic Table of the Elements

Key:

atomic number **Symbol** name conventional atomic weight standard atomic weight

1	2	3	4	5	6	7	8	9	10	11	12	13	14	15	16	17	18
1 **H** hydrogen 1.008 [1.0078, 1.0082]																	2 **He** helium 4.0026
3 **Li** lithium 6.94 [6.938, 6.997]	4 **Be** beryllium 9.0122											5 **B** boron 10.81 [10.806, 10.821]	6 **C** carbon 12.011 [12.009, 12.012]	7 **N** nitrogen 14.007 [14.006, 14.008]	8 **O** oxygen 15.999 [15.999, 16.000]	9 **F** fluorine 18.998	10 **Ne** neon 20.180
11 **Na** sodium 22.990	12 **Mg** magnesium 24.305 [24.304, 24.307]											13 **Al** aluminium 26.982	14 **Si** silicon 28.085 [28.084, 28.086]	15 **P** phosphorus 30.974	16 **S** sulfur 32.06 [32.059, 32.076]	17 **Cl** chlorine 35.45 [35.446, 35.457]	18 **Ar** argon 39.95 [39.792, 39.963]
19 **K** potassium 39.098	20 **Ca** calcium 40.078(4)	21 **Sc** scandium 44.956	22 **Ti** titanium 47.867	23 **V** vanadium 50.942	24 **Cr** chromium 51.996	25 **Mn** manganese 54.938	26 **Fe** iron 55.845(2)	27 **Co** cobalt 58.933	28 **Ni** nickel 58.693	29 **Cu** copper 63.546(3)	30 **Zn** zinc 65.38(2)	31 **Ga** gallium 69.723	32 **Ge** germanium 72.630(8)	33 **As** arsenic 74.922	34 **Se** selenium 78.971(8)	35 **Br** bromine 79.904 [79.901, 79.907]	36 **Kr** krypton 83.798(2)
37 **Rb** rubidium 85.468	38 **Sr** strontium 87.62	39 **Y** yttrium 88.906	40 **Zr** zirconium 91.224(2)	41 **Nb** niobium 92.906	42 **Mo** molybdenum 95.95	43 **Tc** technetium	44 **Ru** ruthenium 101.07(2)	45 **Rh** rhodium 102.91	46 **Pd** palladium 106.42	47 **Ag** silver 107.87	48 **Cd** cadmium 112.41	49 **In** indium 114.82	50 **Sn** tin 118.71	51 **Sb** antimony 121.76	52 **Te** tellurium 127.60(3)	53 **I** iodine 126.90	54 **Xe** xenon 131.29
55 **Cs** caesium 132.91	56 **Ba** barium 137.33	57-71 lanthanoids	72 **Hf** hafnium 178.49(2)	73 **Ta** tantalum 180.95	74 **W** tungsten 183.84	75 **Re** rhenium 186.21	76 **Os** osmium 190.23(3)	77 **Ir** iridium 192.22	78 **Pt** platinum 195.08	79 **Au** gold 196.97	80 **Hg** mercury 200.59	81 **Tl** thallium 204.38 [204.38, 204.39]	82 **Pb** lead 207.2	83 **Bi** bismuth 208.98	84 **Po** polonium	85 **At** astatine	86 **Rn** radon
87 **Fr** francium	88 **Ra** radium	89-103 actinoids	104 **Rf** rutherfordium	105 **Db** dubnium	106 **Sg** seaborgium	107 **Bh** bohrium	108 **Hs** hassium	109 **Mt** meitnerium	110 **Ds** darmstadtium	111 **Rg** roentgenium	112 **Cn** copernicium	113 **Nh** nihonium	114 **Fl** flerovium	115 **Mc** moscovium	116 **Lv** livermorium	117 **Ts** tennessine	118 **Og** oganesson

57 **La** lanthanum 138.91	58 **Ce** cerium 140.12	59 **Pr** praseodymium 140.91	60 **Nd** neodymium 144.24	61 **Pm** promethium	62 **Sm** samarium 150.36(2)	63 **Eu** europium 151.96	64 **Gd** gadolinium 157.25(3)	65 **Tb** terbium 158.93	66 **Dy** dysprosium 162.50	67 **Ho** holmium 164.93	68 **Er** erbium 167.26	69 **Tm** thulium 168.93	70 **Yb** ytterbium 173.05	71 **Lu** lutetium 174.97
89 **Ac** actinium	90 **Th** thorium 232.04	91 **Pa** protactinium 231.04	92 **U** uranium 238.03	93 **Np** neptunium	94 **Pu** plutonium	95 **Am** americium	96 **Cm** curium	97 **Bk** berkelium	98 **Cf** californium	99 **Es** einsteinium	100 **Fm** fermium	101 **Md** mendelevium	102 **No** nobelium	103 **Lr** lawrencium

Source: IUPAC 2018, Reprinted with permission from the International Union of Pure and Applied Chemistry.

Printed and bound by CPI Group (UK) Ltd, Croydon, CR0 4YY

01/03/2023

03196469-0001